Chemical Ecology

Chemical Ecology

Chemical Ecology
The Ecological Impacts of Marine Natural Products

Edited by
Melany P. Puglisi
Mikel A. Becerro

CRC Press
Taylor & Francis Group
Boca Raton London New York

CRC Press is an imprint of the
Taylor & Francis Group, an **informa** business

CRC Press
Taylor & Francis Group
6000 Broken Sound Parkway NW, Suite 300
Boca Raton, FL 33487-2742

First issued in paperback 2020

© 2019 by Taylor & Francis Group, LLC
CRC Press is an imprint of Taylor & Francis Group, an Informa business

No claim to original U.S. Government works

ISBN-13: 978-1-4822-4880-7 (hbk)
ISBN-13: 978-0-367-73330-8 (pbk)

Visit the Taylor & Francis Web site at
http://www.taylorandfrancis.com

and the CRC Press Web site at
http://www.crcpress.com

Dedication

"The delicate balance of mentoring
someone is not to create them in
your own image, but giving them the
opportunity to create themselves."

Steven Spielberg

This book is dedicated to Dr. Valerie Paul, a lifelong mentor and friend who has encouraged us and provided us with so many opportunities to learn and grow into our own careers. Her scientific contributions make her one of the most recognized experts in the field of marine chemical ecology, however her kind personality, life attitude, and friendship touches the lives of all who have had the opportunity to get to know her. Dr. Paul is a former Professor of Biology at the University of Guam Marine Laboratory, Mangilao, GU and currently the Director of the Smithsonian Marine Station at Fort Piece, FL.

Contents

Foreword

"The Dawn and Evolution of *Marine* Chemical Ecology"

On land, perhaps beginning with the acceptance of herbal medicines, it was clear that plants, insects, and animals had evolved the production of bioactive metabolites. Although not well articulated in the early years, it became clear much later that these compounds formed the foundation of defense and communication. Decades later, Whittaker and Feeny's classic 1971 paper, "Allelochemics: Chemical Interactions between Species," provided an organizational foundation for describing the various functions of these adaptive chemical compounds. During the same period, and in stark contrast, knowledge of the chemistry of marine life was fundamentally non-existent. Early chemical studies by Richard Moore and Paul Scheuer in Hawaii focused on the fascinating pigments of the brightly colored echinoderms, while studies of toxic edible marine species formed a major area of research in Japan. However, the natural functions of these elaborate metabolites remained unexplored for at least another decade.

One of the first areas to be explored was the role of marine algal metabolites in structuring predator prey relationships. The unique chemical components of the seaweeds, many of which were produced by a revolutionary halogen-based biosynthesis, provided ample rationale to consider these compounds as providing an adaptive role. Chemists, whose background was limited in ecology, began to agitate their ecologist colleagues to accept this reasonable hypothesis. As time passed, chemists openly interacted with young marine ecologists to emphasize how important this work could be in shaping marine ecology.

A story comes to mind that demonstrates the level of understanding in the early 1970s. For several years, biological oceanographers had mapped the world's oceans, measuring chlorophyll and relating their measurements to overall oceanic productivity assuming the food chain transfer of phytoplankton nutrients into the higher trophic levels. They quantified their chlorophyll data to high resolution, never questioning the validity of their assumptions. In the late 1970s, copepod feeding studies clearly showed that these herbivores carefully select phytoplankton, avoiding a large abundance of chemically rich and un-preferred prey. Alas, the very foundation of their work was now considered questionable if not completely useless. The roles of chemical feeding inhibitors had simply not been considered in assessing how the oceans function.

In the mid-to-late 1970s, young marine ecologists took note of the potential of this new, developing field. The opisthobranchs were found to be specialists preying on chemically defended plants and animals in order to use their defenses for their own. The hypothesis that the shell-less opisthobranch mollusks had, over evolutionary time, lost their shell provided great insight into the complexity of chemical mediators in marine evolution.

At first, bioassays were confined to simple aquaria measurements, but even then, these were clearly recognized as ecologically unrealistic. This was followed by direct measurements in the sea using highly controlled and validated statistical methods. What surfaced some 40 years later was a sophisticated understanding of the roles of secondary metabolites (allelochemicals) in the adaptations of marine plants and animals. Particularly on coral reefs, it is now relatively easy to see those soft-bodied organisms that flourish despite an abundance of predators, and to predict their reliance on chemical defenses.

As this book is in construction, we are experiencing yet another major challenge about how marine systems function. In a fashion analogous to the significance of the human microbiome, it is now clear that the vast microbial diversity of the marine environment and the specific relationships that exist among plants, animals, and microbes have a major influence on how the oceans function. We understand how some sponge metabolites are produced by symbiotic bacteria and can further predict that microorganisms are likely to be involved in the adaptations of marine organisms in a much broader sense. This is an amazing area for growth in the future, but since microbial studies have been confined largely to the laboratory, major advancements in microbial ecological methods will need to be developed. The vast new capacity we now have in metagenomics and DNA sequencing will clearly be important in this new endeavor.

This timely treatise, *Marine Chemical Ecology,* is an up-to-date compilation of chapters from experts in the field of marine chemical ecology. The chapters encompass plant–herbivore chemical evolution, animal chemical defenses, larval settlement inducers, seagrass chemical ecology, predator–prey interactions in marine plankton, and microbial interactions. The chapters focus their comments on diverse marine ecosystems from the tropics to the polar regions and provide effective overviews of the current status of research. While no single text can be considered comprehensive, this is an excellent reference that interested students should consider owning.

William Fenical
Scripps Institution of Oceanography
University of California at San Diego

Acknowledgment

The editors with to thank the many anonymous reviewers who took part in the preparation of this book. Their contribution greatly improved the manuscripts.

Editors

Melany P. Puglisi was born in the United States and obtained her undergraduate training at Southampton College and graduate training in biology and pharmacognosy at the University of Guam and the University of Mississippi, respectively. After two years of postdoctoral studies at Scripps Institution of Oceanography, she joined the new pharmacy program at Lake Erie College of Osteopathic Medicine as a faculty member in the Department of Pharmaceutical Sciences. After a short time, she left Erie to pursue research efforts at the Smithsonian Marine Station in Fort Pierce, Florida followed by Shannon Point Marine Laboratory in Anacortes, Washington, where she established a research program in marine microbial chemical ecology. In 2008, she joined Chicago State University as a founding faculty member in the College of Pharmacy. Chicago State University is a small teaching institution that provides opportunities to underserved populations. She has received recognition from the student pharmacists as Teacher of the Year and she involves student pharmacists in Capstone research projects in marine natural products and chemical ecology. In 2016, she joined the Editorial Board of the *Journal of Natural Products* as the Book Review Editor. She is a member of the Executive Committee of the American Society of Pharmacognosy and was recently appointed as Secretary to the ASP Foundation. Her research interests involve the chemical interactions between benthic marine organisms and microorganisms. She has numerous publications in algal, invertebrate, and microbial chemical ecology.

Mikel A. Becerro is a tenured scientist at the Spanish Research Council (CSIC). Born and raised in the Basque Country, he moved to the Canary Islands to study Marine Biology and then to Catalonia to earn, in 1994, his PhD in Marine Sciences from the University of Barcelona. Dr. Becerro is based at the Center for Advanced Studies of Blanes (CEAB-CSIC), where he is the current deputy director. Dr. Becerro is the leader of the *BITES* lab, a small research group focused on *Biodiversity, Interactions, Threats, and Ecosystem Shifts.* The name of his research lab clearly reflects Dr. Becerro's research activities and interests. As an ecologist, Dr. Becerro has expanded his approaches from understanding specific species interactions and their role in community ecology to broader social-ecological systems and the current biodiversity crisis. Over the years, he has received numerous awards from the Basque, Catalan, and Spanish governments that have opened doors for him to collaborate with world leaders in marine chemical ecology including Dr. Paul Scheuer at the University of Hawai'i Chemistry Department, Dr. Valerie Paul at the University of Guam Marine Laboratory, and Dr. Guido Cimino at the Italian Research Council Institute of Biomolecular Chemistry. In 2002, he was awarded a prestigious Marie Curie Postdoctoral fellowship to conduct research with Dr. Bernard Banaigs at the University of Perpignan Centre of Phytopharmacy. Dr. Becerro was a visiting scientist at the Smithsonian Marine Station at Fort Pierce. His research activities include numerous publications in international journals and communications in national and international conferences, which has led Dr. Becerro to be included in the list of the best Spanish researchers in Marine Biology. Dr. Becerro participates in several master and doctorate programs, including the Erasmus Mundus MSc in Marine Environment and Resources led by the University of the Basque Country, and has supervised many master and PhD students. An active science communicator, Dr. Becerro has participated in numerous forums such as "Tuesdays at the Oceanographic" at the Oceanographic Valencia, or the "Ciudad Ciencia" program promoted by the La Caixa Social Work, and the CSIC, among others. Mikel also has been invited to be an editor of prestigious journals of marine biology like *Advances in Marine Biology* and *Hydrobiologia*.

Contributors

Charles D. Amsler
Department of Biology
University of Alabama at Birmingham
Birmingham, Alabama

Conxita Avila
Department of Animal Biology
    (Invertebrates) and Biodiversity
    Research Institute (IrBIO)
University of Barcelona
Barcelona, Catalonia

Bill J. Baker
Department of Chemistry
University of South Florida
Tampa, Florida

Gordon M. Cragg
Retired Chief, Natural Products Branch
National Institutes of Health/National
    Cancer Institute
Bethesda, Maryland

William Fenical
Scripps Institution of Oceanography
University of California at San Diego
San Diego, California

Jason R. Graff
Department of Botany and Plant Biology
Oregon State University
Corvallis, Oregon

Tilmann Harder
Bremen Marine Ecology—Center for
    Research and Education
University of Bremen
Bremen, Germany

Veijo Jormalainen
Division of Ecology
University of Turku
Turku, Finland

Rian Kabir
Biology Department
Randolph-Macon College
Ashland, Virginia

Grace Lim-Fong
Biology Department
Randolph-Macon College
Ashland, Virginia

Amy L. Lane
Chemistry Department
University of North Florida
Jacksonville, Florida

Taylor A. Lundy
Biological Sciences Department
University of North Florida
Jacksonville, Florida

James B. McClintock
Department of Biology
University of Alabama at Birmingham
Birmingham, Alabama

Juan Moles
Museum of Comparative Zoology &
    Department of Organismic and
    Evolutionary Biology Harvard
    University
Cambridge, Massachusetts

Mareen Möller
Environmental Biochemistry, Institute
    of Chemistry and Biology of the
    Marine Environment
University of Oldenburg
Wilhelmshaven, Germany

David J. Newman
Retired Chief, Natural Products Branch
National Institutes of Health/National
   Cancer Institute
Bethesda, Maryland

Laura Núñez-Pons
Hawai'i Institute of Marine Biology
   (HIMB)
University of Hawai'i at Manöa
Kane'ohe, Hawai'i

Dianna K. Padilla
Department of Ecology and Evolution
Stony Brook University
Stony Brook, New York

Alistair G. B. Poore
Evolution & Ecology Research Centre
School of Biological, Earth and
   Environmental Sciences
University of New South Wales
Sydney, Australia

Kelsey L. Poulson-Ellestad
Department of Biological, Chemical
   and Physical Sciences
Roosevelt University
Chicago, Illinois

Emily K. Prince
Department of Biology
Lander University
Greenwood, South Carolina

Melany P. Puglisi
Department of Pharmaceutical Sciences
Chicago State University
Chicago, Illinois

Sven Rohde
Environmental Biochemistry, Institute
   of Chemistry and Biology of the
   Marine Environment
University of Oldenburg
Wilhelmshaven, Germany

David C. Rowley
Biomedical and Pharmaceutical
   Sciences
The University of Rhode Island
Kingston, Rhode Island

Kathryn M. Schoenrock
Department of Biology
University of Alabama at Birmingham
Birmingham, Alabama

Peter J. Schupp
Environmental Biochemistry, Institute
   of Chemistry and Biology of the
   Marine Environment
University of Oldenburg
Wilhelmshaven, Germany

Jennifer M. Sneed
Smithsonian Museum of Natural
   History
Smithsonian Marine Station at Fort
   Pierce
Fort Pierce, Florida

Erik E. Sotka
Department of Biology
College of Charleston
Charleston, South Carolina

Jan Tebben
Helmholtz Centre for Polar and Marine
   Research
Alfred Wegener Institute
Bremerhaven, Germany

Kathryn L. Van Alstyne
Shannon Point Marine Center
Western Washington University
Anacortes, Washington

Jacqueline L. von Salm
Department of Chemistry
University of South Florida
Tampa, Florida

1 Marine Natural Products with Pharmacological Properties

David J. Newman and Gordon M. Cragg

CONTENTS

1.1   INTRODUCTION

In contrast to the many reviews on the subject of marine-derived natural products that we have published over the last 25-plus years, this chapter will focus more on pharmacological areas that are not primarily related to cancer. However, we will mention some anti-tumor active agents from invertebrate sources but from a target-perspective assay rather than the more usual initial cytotoxicity assays. In addition, we will use data where the compound(s) in question were isolated and purified from an invertebrate source, even though in some cases, the compound(s) may have been produced by either an endophytic microbe (culturable or not yet cultured) or from ingestion of the molecules from a food source (predation on other organisms in most cases). We will not, however, comment, other than *en passant*, on compounds now known to be produced by such microbes, though compounds with the same basic skeleton may well be the products of such interactions between microbe and host.

We will cover toxins, generally isolated from mollusca, and the newer data that demonstrates that some modifications may aid in developing orally active agents derived from these cystine-bridged peptides. There will also be some comments on other toxins from "non-mollusks." There will be a significant section on brominated compounds, in particular pyrroles where different biological activities have recently been reported for both the natural product and (semi-)synthetic modifications and on compounds such as hymenialdisine (**1**) that now are demonstrating some interesting biological activities. In addition, there will be commentary on antiviral compounds that will include agents isolated from sponges, although it has recently been demonstrated by the Gerwick group (Bertin et al. 2015) that spongosine (**2**) has now been isolated from a commensal *Vibrio harveyi* strain. This microbe was isolated from nominally the same sponge used by the Bergmann group in their initial reports on arabinose nucleosides in the early 1950s (Bergmann and Feeney 1950, 1951, Bergmann and Burke 1955), which can be considered as leading to the large number of semisynthetic agents based on nucleosides that have entered the armamentarium of antiviral agents since those reports. Another section will cover antibacterial and antifungal agents, including an interesting comment on biofilm "busters" where there are isomeric differences in activities.

1.2   MOLLUSCAN-DERIVED PEPTIDE TOXINS

1.2.1   ZICONOTIDE

Although the knowledge that cone snails in particular produced a number of toxins was well known, the complexity of these agents was not realized until the pioneering work of Olivera and colleagues over 30 years ago, when they described a peptide toxin from the fish-hunting cone snail *Conus geographus* collected originally in the Philippines (Cruz et al. 1978, Olivera et al. 1985). Little was it realized at that time that this initial discovery and report would lead to the first "direct from the sea" drug, ziconotide (Prialt(R); **3**) approved by the US FDA in late December 2004 for intractable pain. The original compound that ultimately became known as ziconotide was not from *C. geographus* but was originally isolated from a different species *Conus magus* under the name ω-conopeptide MVIIA. Over 200 derivatives of MV111A were synthesized by the small company originally set up by Olivera and its successor but in the end, the natural product structure was the one chosen for clinical trials. Due to its chemical nature, the compound had to be delivered via intrathecal injection, but in spite of this vicissitude, it is still used. The efforts around the administration and methods to alleviate the problems were discussed at length by Alicino et al. (2012).

1.2.2 Toxin Targets (Cabals, Constellation Pharmacology, and Others)

The target of these agents are the ion channels in neural cells and some other targets in similar and different parts of the body, but usually linked to nerve function. Such an example would be the effects on acetylcholine (ACh) signaling. An excellent review paper was published by Fedosov et al. (2012) that also gives an excellent coverage of these toxins and the biological activities that have been or should be investigated, which should be read by interested parties.

A little earlier than the Fedosov et al. review, in an excellent paper in 2010, Teichert and Olivera (2010) discussed the concepts of a "motor cabal" that included a nicotinic receptor with an acetylcholinesterase site inhibitor, another nicotinic receptor with an ion channel blocker, and then blockade of particular sodium and calcium transport channels (ziconotide blocks a Ca^{2+} channel) by some of the components in the overall toxin mix. Then they also demonstrated another "cabal," the so-called "lightning strike cabal," also as a result of the toxin mixtures. In this there are targets such as sodium channels where there is an inactivation inhibitor, potassium channel blockers, and a desensitization inhibitor of the glutamate receptor. The cabal names were derived from the observed effect on fish when "hit" by the mix of cone snail toxins. More thorough descriptions and relevant references are given in the 2010 paper referred to previously and the concept was further expanded into "Constellation Pharmacology" whereby agents come from a variety of snail sources, not just *Conus* but also crassispirine snails that differ from the effects shown by cone snails (Imperial et al. 2014). A much fuller description of the "Constellation Pharmacology" concept with a large number of examples was published by Teichert et al. (2015) and should be consulted by those interested, as one could claim that the cone snails used a mix of different pharmacological targets/toxins to catch their prey.

In addition to the reviews and papers noted in the previous paragraphs, there are also some recent (2014/5) published articles that cover some interesting aspects of conotoxin interactions with targets. Thus, Azam et al. (2015) covered the molecular interactions of the α-conotoxin RgIA with its target and discovered that the interaction of a-CTx RgIA was at the α10/α9 rather than the α9/α10 nAChR subunit interface. Thus, this discovery may facilitate the development of selective ligands with therapeutic potential. By contrast, the αS-conotoxin GVIIIB isolated from *Conus geographus* was found to be over 100 times more selective for the α9/α10 nAChR than other subtypes of this receptor (Christensen et al. 2015). Since voltage-gated sodium channels are also targets of conotoxins, the recent review (December 2015) by Munasinghe and Christie (2015) covering this topic should be read in conjunction with that from the Fedosov group in 2012 (Fedosov et al. 2012) and compared with the more general review that covers all sources of toxins, not just from cone snails by the Stonik group from Vladivostok that interact with nAChR targets (Kudryavtsev et al. 2015).

1.2.3   Australian Research in Cone Snail Toxins

In addition to the group led by Olivera at the University of Utah in the USA, there are two other major academic groups working with cone snail toxins. They are both at the University of Queensland in Brisbane, Australia, with one led by David Craik and the other by Paul Alewood. They are effectively in the same building but run independently, though they do collaborate and publish together at times.

In 2014 they collaborated on an excellent article in *Chemical Reviews* (Akondi et al. 2014) that gave a thorough rendering of the conotoxins known to that date, but they also included an excellent discussion of the classification of the conotoxins using a cyclic format that was reminiscent of the genetic "minute" systems used in the identification of microbial genes. This diagram also included an outer ring showing the target(s) of each classification. In addition, there is an excellent table (Table 4 in Akondi et al. 2014) that lists the known folds and subfolds for both natural and synthetic conotoxins, together with a notation as to the number of amino acids for each toxin and in particular, the number of cysteines together with a spatial rendition of the individual class(es). Later in the review, they cover the synthetic strategies that have been utilized to make known and semi/totally synthetic analogues. This section alone is an excellent one for peptide chemists to investigate. They also have a discussion of conotoxin mimetics that is worth reading particularly when linked to the 3-D solid representations of selected agents.

1.2.4   "Cyclization" of Peptide Toxins

In addition to the *Chemical Reviews* paper, we will comment on others that have come from the Craik and Alewood groups in the last few years as these reports are in general complementary rather than competitive. Schroeder and Craik (2012) briefly described the therapeutic potential of conotoxins and then followed up with a review in the *European Journal of Medicinal Chemistry* on the template possibilities of disulfide-rich macrocyclic peptides in drug design (Northfield et al. 2014). In this paper, in addition to describing the peptide drugs approved by the FDA from 2011–2013 that contained disulfide linkages, they also referred to the synthesis of a cyclized version of the α-conotoxin Vc 1.1 that demonstrated oral activity (Clark et al. 2010). Interestingly, in 2015, they reported further work on the cyclized variant whereby two disulfide bridges in the structure were reduced to one, thus eliminating disulfide shuttling and delivering an orally active molecule with only a slight loss in activity. Removal of all disulfides gave a 30-fold less active moiety (Yu et al. 2015), demonstrating that although only one conotoxin had been approved as a drug, these agents are still being worked on, Durek and Craik (2015) recently published data on the number of US patents on the subject of conotoxins. demonstrating that there were still 38 valid US-issued patents as of the date of the article in June 2015, and this number did not include initial applications.

Further work, though not with conotoxins, described how the plant cyclotides such as kalata A that also are "knotted cyclic peptides" can be taken up by cells, and the linkage of this information with the information on conotoxins is under way in order to devise delivery systems for these agents (Henriques et al. 2015, Qu et al. 2017, de Veer et al. 2017). In late 2014, the Craik group described methods for generating and assaying N-methylated cyclic peptides where the amide –NH– group is converted to –NCH$_3$– demonstrating up to 35% transport into cells in selected cases (Wang et al. 2014).

1.2.5 "Genomic Mining" of Conotoxin Genes

There have been reports demonstrating the use of what might be considered "genetic mining" in order to investigate the potential for the production/isolation of conotoxins from snails by use of genomic techniques. An excellent example would be the report by a Chinese group in 2011 reporting the isolation, using cDNA techniques from a cDNA library derived from *Conus litteratus* of the small conotoxin, It14a, that contains 13 amino acids and has a C–C–C–C framework with an amidated C-terminus (Sun et al. 2011). The group produced this particular peptide together with others resembling it by solid state peptide syntheses, rather than using genetic-based production systems. This particular peptide exhibited analgesic effects in mice and, interestingly, replacing the lysine in position 7 with an alanine gave a more active version. The mechanism of action appears to be inhibition of the nicotinic acetylcholine receptor and the increased efficacy may be due to better hydrophobic binding following removal of the positively charged lysine moiety.

In the middle of 2015 Lavergne et al. published a very interesting report in the *Proceedings of the National Academy of Sciences* (Lavergne et al. 2015) demonstrating the numbers of toxins that could be "identified" by genomic techniques from a single *Conus* specimen. They demonstrated 3,305 novel precursor toxin sequences from one specimen of *Conus episcopatus* by use of a high resolution "interrogation" of the transcriptomes and proteomes in the diverse venom compartments of this gastropod. These precursor sequences separated into 9 known superfamilies and they postulated the presence of 16 previously unknown superfamilies. The paper should definitely be read by investigators interested in these agents, as they identified 168 toxins with the 6-cysteine motif where there is the so-called "cystine knot" known in the pharmacologically active conotoxins so far studied. There were also 208 identified "toxins" with uneven numbers of cysteines that were derived from known motifs. Thus, the potential numbers of conotoxins may have significantly increased from the approximately 70,000 so far assumed to exist.

1.2.6 Synthetic "Conotoxins"

Chemists have also been active in synthesizing potential conotoxins via combinatorial techniques built around peptide syntheses. A recent paper from Chang et al.

(2014) demonstrated how a selective α3β4 nicotinic acetylcholine receptor antagonist was synthesized based upon the structure of the unusual BuIA α-conotoxin that has an unusual 4/4 loop framework. 132 peptides synthesized based on these prototypes were screened followed by synthesis of a second-generation library composed of 62 analogues. Further screening led to 11 peptides, four of which were then screened in a competitive binding assay against ^{3}H epibatidine binding to α3β4 nicotinic acetylcholine receptors expressed in HEK293 membranes. This was an interesting juxtaposition of ant-amphibian sourced natural product versus a marine-gastropod derived agent and led to the identification of novel analogues that are currently undergoing further biological evaluation.

1.2.7 CLINICAL STATUS OF OTHER CONOTOXINS

One may ask what is the status of conotoxins as approved pharmacological agents? To date, only one agent has been approved (Ziconotide [3]), though a number had entered clinical trials over the last 10-plus years. At the original time of writing (December 2015) there were two agents "listed" as being under development in various databases. The first is XEN-2174 (4; Phase II?), a compound that is a very slight modification of the naturally occurring χ-conotoxin MrIA. That compound was originally isolated from *C. marmoreus* and then optimized by medicinal chemistry (Brust et al. 2009). Unlike other conotoxins, either approved or in various levels of testing, this particular agent is a 13-residue peptide and is a noncompetitive inhibitor of the neuronal norepinephrine transporter (NET) (Sharpe et al. 2003). Although the source company, Xenome, is still listed on various sites as having this compound in Phase II trials against cancer pain, and in an abrogated trial at Phase II countering pain from bunion surgery in Bulgaria under the EudraCT protocol 010-019109-40-BG that was terminated by the FDA, no data from this latter trial has yet been published. However, a search of the internet undertaken in August 2015, showed that the web site "www.xenome.com" was up for sale, so the current status of this compound is unknown, but fuller details of the development of this compound were published (February 2015) in a book chapter that can be consulted for further information (Lewis 2015). What is of interest however, is that very recently, a clinical trial report was published from Holland demonstrating the pharmacodynamics and pharmacokinetics in man. However, the date of the trial was not given (Okkerse et al. 2017), so the current status is still unconfirmed.

There is a similar story with the second compound, Leconotide (ω-Conotoxin CVID; 5; Phase I). This molecule, a 27-residue peptide with three internal CYS-CYS bonds, similar in overall structure to ziconotide, had reached Phase I trials sponsored by Relevare Pharmaceuticals (Australia; previous name was CNSBio) for treatment of pain related to cancer. It is a calcium channel blocker and was originally identified by researchers at the University of Queensland. Although initial experiments used the intrathecal route (as with ziconotide) (Jayamanne et al. 2013), the protocol used systemic administration (Daly and Craik 2011). The company involved has now liquidated, so the current clinical status is unknown. No other natural conotoxins and/or analogues have yet reached Phase I clinical trials as of a search of the Integrity™ database in late December 2017 (Figure 1.1).

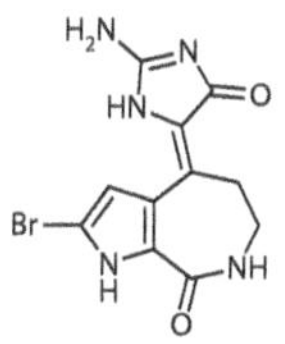

1. Hymenialdisine

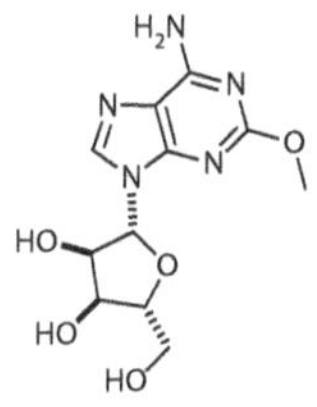

2. Spongosine

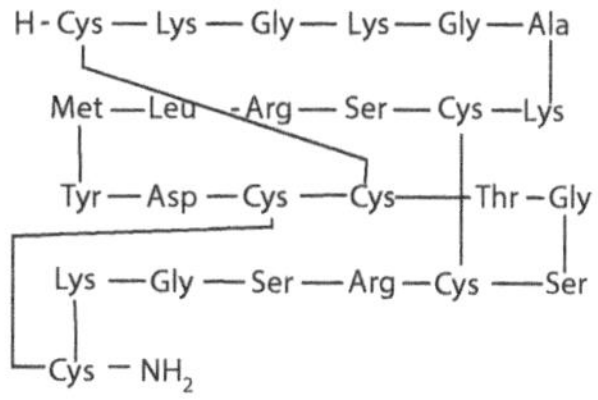

3. Ziconotide

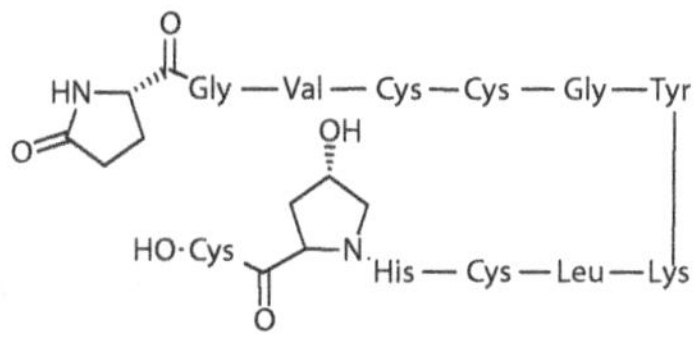

4. XEN-2174

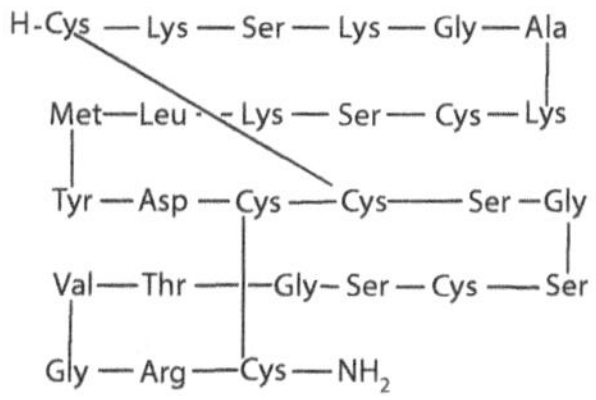

5. Leconotide

FIGURE 1.1 Compounds 1 to 5.

1.3 ECHINODERM MARINE TOXINS

Though we mentioned the recent review (Kudryavtsev et al. 2015) by the Stonik group earlier, there is one particular agent from the sea anemone *Stichodactyla helianthus* where a derivative has been in Phase I clinical trials. ShK (**6**) is a 35-residue peptide with 3 cystine bridges and was found to be the first peptide toxin from a sea anemone that blocked K+ channels, specifically the Kv1.3 potassium channel (Castañeda et al. 1995). The first ion channel blocker from these sources was a sodium-specific agent (ShN), a 48-residue peptide with three cystine links. Later, improved separation techniques permitted the isolation and purification of ShK (Kem et al. 1989, Pennington et al. 2015).

The derivative that has been in Phase I clinical trials is known as ShK-186 (**7**) or Dalazatide. The syntheses of this and other derivatives were described in 2006 by Beeton et al. (2005, 2006) with only subtle changes in the natural molecule (phosphotyrosine at one end, a replacement by lysine in the middle and an amidation of a terminal cysteine). The two trials were a safety trial in healthy volunteers (NCT02446340) and another Phase I covering tolerability and pharmacodynamics of the candidate drug in patients with active plaque psoriasis (NCT02435342). Both have now been completed with a result from the latter reported as an abstract in 2015 and then amplified in a formal report in 2017 demonstrating some efficacy (Tarcha et al. 2017).

In addition to psoriasis, the compound is in preclinical studies for treatment of type 1 diabetes (Upadhyay et al. 2013), inflammatory bowel disease, multiple sclerosis, lupus-induced nephritis, rheumatoid arthritis (Koshy et al. 2014), and vasculitis at the company, Kineta. There are other derivatives with improved activity in preparation and potential preclinical/clinical studies (Chang et al. 2015). It will be interesting to follow the clinical trials of these agents, particularly as these agents

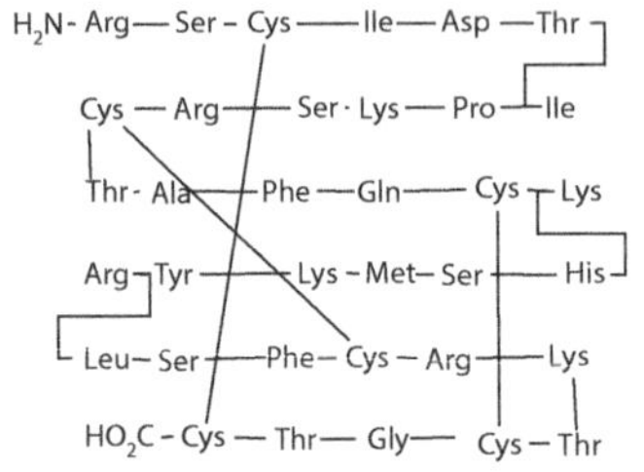

6. ShK Toxin 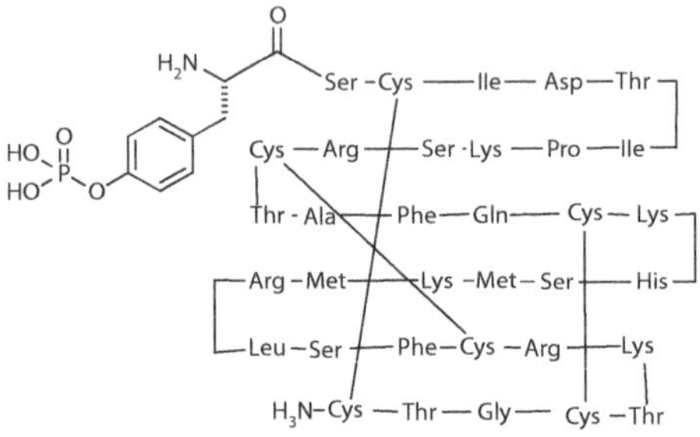   **7.** ShK-186; Dalazatide

FIGURE 1.2   Compounds 6 and 7.

appear to have interesting activities against T-lymphocytes (Matheu et al. 2008, Pennington et al. 2009, 2015). What is interesting about these trials is that the agent, although similar to conotoxins in terms of its cystine links, is a non-terminal linked compound and appears to be delivered either by injection or orally. Details are not available as to the methods used at the time of writing (Figure 1.2).

1.4   BROMINATED MARINE NATURAL PRODUCTS

1.4.1   INTRODUCTION

This section will be a combination of chemistry and pharmacological activities of certain classes of marine natural products that contain bromine in their structures. The pharmacological areas covered range from anti-fouling, anti-microbial, antiviral, and antiparasitic through neurological activities to antitumor agents, with some antifeedant activity as well. At times the same basic class may have different activities. This is usually due to different investigators using different bioactivity systems in their isolation schemes and is dependent in most cases upon who funded their research. Halogenated compounds are found in all forms of marine life but here we will not cover single-celled organisms or algae, simply marine invertebrates.

The numbers of halogenated marine natural products up through 2007 was quoted by Gribble (2015) as being 4714. Since then, Gribble in particular, has been collating and commenting on these compounds and the results are shown in a number of references in his 2015 review in Marine Drugs (Gribble 2015).

1.4.2   ANTIFOULING AND ANTIBACTERIAL AGENTS

In terms of brominated products and antifouling activity, the two simple tryptamine derivatives, 2-bromo-*N*-methyltryptamine (**8**) and its 6-bromo isomer (**9**), were isolated from the Mediterranean gorgonian *Paramuricea clavata* together with other analogues. Compound **8** demonstrated the highest activity of 10 compounds tested in preventing adhesion of *Pseudoalteromonas* sp. D41 and TC8 and *Paracoccus* sp. 4M6 (Pénez et al. 2011). Moving to colder waters, the ascidian *Synocium pulmonaria* contained a quartet of dibrominated compounds, synoxazolidinones A (**10**) and C (**11**), and pulmonarins A (**12**) and B (**13**). The synoxazolidinones had a chlorine

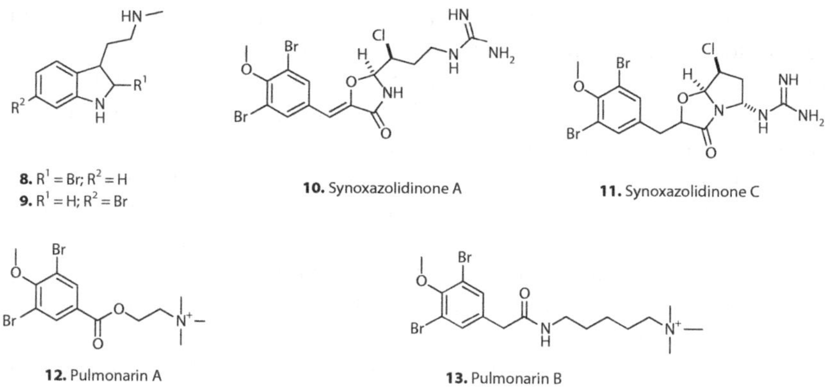

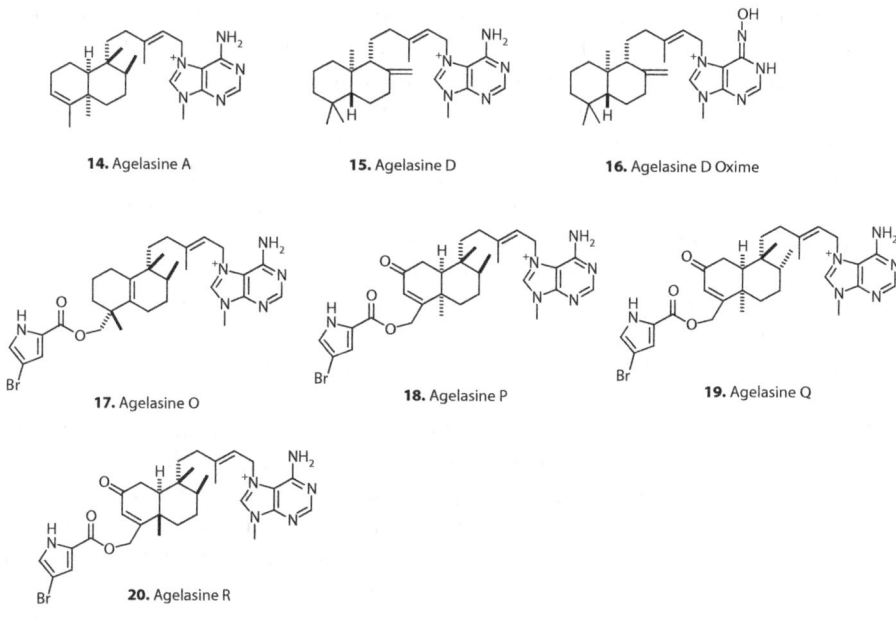

8. R^1 = Br; R^2 = H
9. R^1 = H; R^2 = Br

10. Synoxazolidinone A

11. Synoxazolidinone C

12. Pulmonarin A

13. Pulmonarin B

FIGURE 1.3 Compounds 8 to 13.

substituent with C being the ring-closed version of A. These two agents were very effective against fouling organisms, with C being equivalent to the most active commercial agent Sea-Nine-211. The other two compounds were solely brominated derivatives and only demonstrated antibacterial activity rather than activity against crustacean settlement (Trepos et al. 2014) (Figure 1.3).

A very interesting series of brominated compounds with antifouling and antimicrobial activity are the agelasines. These are formally 7,9-dialkylpurinium salts and usually have a diterpene substituent. The structure of agelasine A (**14**) is shown in Figure 1.4. Though it does not have a halogenated substituent, a fair number of these

14. Agelasine A

15. Agelasine D

16. Agelasine D Oxime

17. Agelasine O

18. Agelasine P

19. Agelasine Q

20. Agelasine R

FIGURE 1.4 Compounds 14 to 20.

compounds do and have antimicrobial activity. Interestingly, the non-halogenated natural product (+)-agelasine D (**15**), which had been synthesized by Utenova and Gundersen (2004), was isolated as the enantiomer (–) agelasine D and its oxime derivative (**16**) from an Indonesian sponge identified as an *Agelas* sp. (Hertiani et al. 2010). The very interesting aspect to these two compounds was that though both demonstrated cytotoxicity against mouse lymphoma cells, and both inhibited the settling of *Balanus improvisus*, the isolated molecules had different activities against biofilms. The (–) agelasine D epimer inhibited the growth of the planktonic forms of biofilm forming *S. epidermidis* but did not inhibit the formation of its biofilm; in contrast, the oxime derivative (**16**) inhibited the biofilm but had no effect upon the growth of the bacterium (Hertiani et al. 2010).

Kubota et al. (2012) reported the isolation of agelasines O to U with "O to R" (**17–20**) containing a brominated pyrrole side chain on the terpene. Of these, all had activity against *S. aureus* and *E. coli* together with activities against the human fungal pathogens *Candida albicans*, *Aspergillus niger*, *Trichophyton mentagrophytes*, and *Cryptococcus neoformans*. No details as to other activities were presented but extrapolating from other agelasines there is a strong possibility that these may also have some cytotoxic and anti-settling activities.

1.4.3 ANTIBACTERIAL AGENTS

As can be seen in the structures noted in the previous section, bromo-substituted pyrroles are a common substituent and there are significant numbers of bromopyrrole alkaloids isolated from marine organisms, predominately sponges. What is of significant import is the recent paper by Agarwal et al. (2017) demonstrating that a significant number of brominated compounds isolated from sponges are in fact the product(s) of as yet uncultured microbes. These findings may well alter the thought processes behind marine chemical ecology in the future.

From the same ascidian that eudistomins Y_2 to Y_7 were isolated from (a Korean *Synoicum* sp.) came a series of antimicrobial brominated furanones named as cadiolides E, G. H {Z}, H {E}, and I (**21–25**) and synolides A {Z}, A {E}, B {Z}, and B {E} (**29–32**) (Won et al. 2012). In the same time frame, cadiolide E (**21**) together with three others, cadiolide C (**26**), D (**27**), and F (**28**), were reported from another Korean ascidian, *Pseudodistoma antinboja* by Wang et al. (2012). Also in that same year, Sikorska et al. (2012) reported four new rubrolides (**33–36**) from a South African *Synoicum globosum* and all had some antimicrobial activities but they were not as potent as antibacterial agents as those isolated from the Korean sourced materials.

That these compounds can lead to some quite different activities can be seen from inspection of the recent paper by Pereira et al. (2015) where a large series of compounds based upon γ-alkylidene-γ-lactones were synthesized and tested as inhibitors of the photosynthetic pathway, led to two isomers (**37**; *Z* and *E*) that had reasonable activity to prove the point, even though this particular compound had a chloro rather than a bromo substituent. Other compounds in the same series with bromo substituents had activities around one third of the chloro derivatives (Figure 1.5).

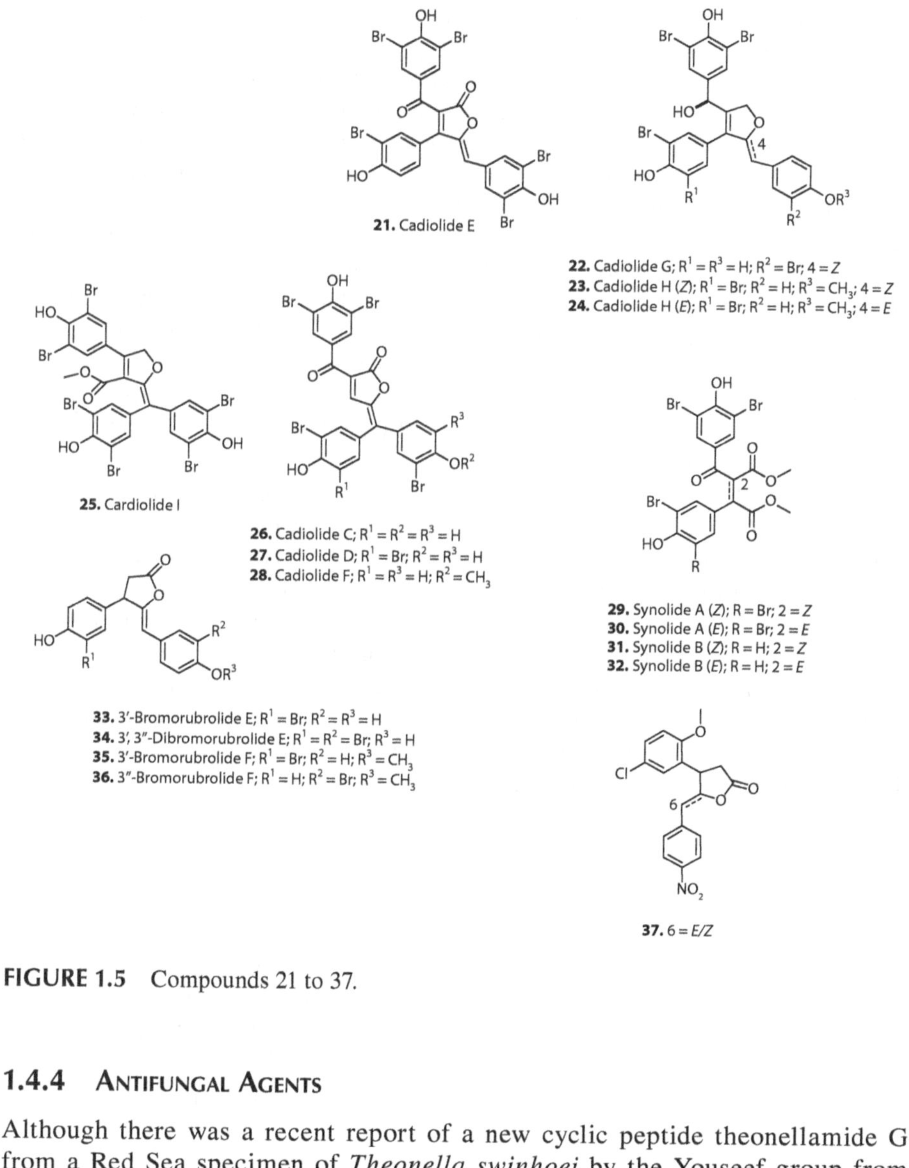

21. Cadiolide E

22. Cadiolide G; $R^1 = R^3 = H$; $R^2 = Br$; $4 = Z$
23. Cadiolide H (Z); $R^1 = Br$; $R^2 = H$; $R^3 = CH_3$; $4 = Z$
24. Cadiolide H (E); $R^1 = Br$; $R^2 = H$; $R^3 = CH_3$; $4 = E$

25. Cardiolide I

26. Cadiolide C; $R^1 = R^2 = R^3 = H$
27. Cadiolide D; $R^1 = Br$; $R^2 = R^3 = H$
28. Cadiolide F; $R^1 = R^3 = H$; $R^2 = CH_3$

29. Synolide A (Z); $R = Br$; $2 = Z$
30. Synolide A (E); $R = Br$; $2 = E$
31. Synolide B (Z); $R = H$; $2 = Z$
32. Synolide B (E); $R = H$; $2 = E$

33. 3′-Bromorubrolide E; $R^1 = Br$; $R^2 = R^3 = H$
34. 3′, 3″-Dibromorubrolide E; $R^1 = R^2 = Br$; $R^3 = H$
35. 3′-Bromorubrolide F; $R^1 = Br$; $R^2 = H$; $R^3 = CH_3$
36. 3″-Bromorubrolide F; $R^1 = H$; $R^2 = Br$; $R^3 = CH_3$

37. $6 = E/Z$

FIGURE 1.5   Compounds 21 to 37.

1.4.4 ANTIFUNGAL AGENTS

Although there was a recent report of a new cyclic peptide theonellamide G from a Red Sea specimen of *Theonella swinhoei* by the Youseef group from Saudi Arabia (Youssef et al. 2014), there is a strong possibility that this agent, which is very close to theonellamide A except for lacking both a methyl group on the para-bromophenylalanine and a hydroxyl group on these components, was produced by an as yet uncultured microbe in the sponge. The Piel group demonstrated that the *T. swinhoei* yellow variant in the Pacific produced almost all of its agents, including similar molecules, via a completely new mechanism starting from ribosomally mediated peptide production by the microbe, with the sponge apparently being a host (Wilson et al. 2014). Irrespective of the actual source, this agent demonstrated significant activity against amphotericin-resistant

Candida albicans. We should point out also that the MICs were lower against the resistant strain by a factor of 2.

In 2013, the Northcote group in New Zealand reported the isolation via NMR techniques of a number of closely related hamigerans from the sponge *Hamigera tarangaensis.* Only one of the eight compounds reported, however, hamigeran G (**38**), demonstrated any activity against *S. cerevisiae*, a non-pathogenic organism under normal conditions. Though the next series of compounds could have been split between antibacterial and antifungal categories, it was simpler to comment on them in this section. From the Okinawan sponge *Agelas* sp. (SS-162), the Kobayashi group reported the structures of agelamandins A–E (**39–43**) in two papers in Organic Letters in 2014 (Kusama et al. 2014a, 2014b). On testing against microbes (antibacterial activity against *B. subtilis* and *M. luteus*), only A (**39**) and B (**40**) demonstrated activity at ~10 μg/mL, but all bar D (**42**) had activity against *C. neoformans* with MICs of 8, 4, 32, and 32 μg/mL for A (**39**), B (**40**), C (**41**), and E (**43**), respectively.

Work with the same sponge and a closely related one, strain SS-156, led to isolation of mono-bromo and di-bromo-pyrrole derivatives, with the identification of nagelamides U–W in 2013 (Tanaka et al. 2013a) followed the same year by a report of nagelamides X–Z (Tanaka et al. 2013b). Then in 2015, came the report of the monobromo-substituted pyrrole-containing 2-debromomonagelamide, 2-debromomukanadin G, and 2-debromonagelamide P. When all of these compounds except for nagelamide V were screened against *C. albicans, T. mentagrophytes, C. neoformans,* and *Aspergillus niger,* only nagelamide Z (**44**) demonstrated good to excellent activities with MICs of 0.25, 4, 2, and 4, respectively, against these four pathogenic fungi (Figure 1.6).

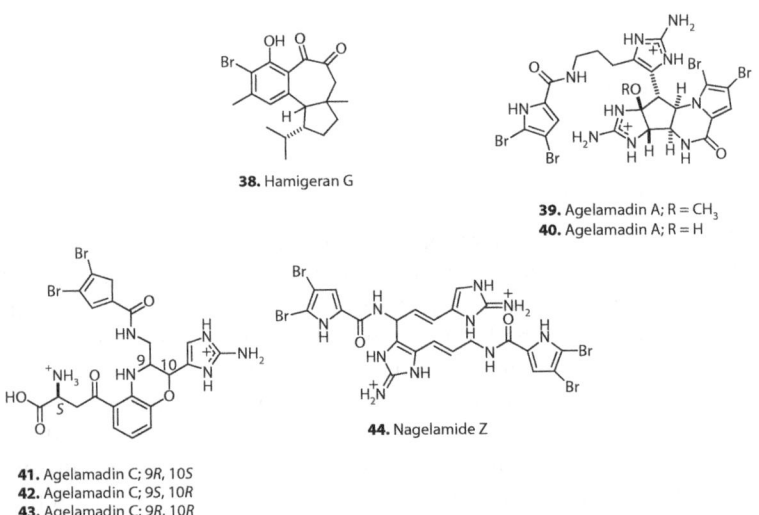

38. Hamigeran G

39. Agelamadin A; R = CH₃
40. Agelamadin A; R = H

41. Agelamadin C; 9R, 10S
42. Agelamadin C; 9S, 10R
43. Agelamadin C; 9R, 10R

44. Nagelamide Z

FIGURE 1.6 Compounds 38 to 44.

1.5 ANTIPARASITIC COMPOUNDS

1.5.1 INTRODUCTION

Perhaps the major problems in the developing world are parasitic infections with malaria probably causing the greatest mortality in the under 5 years of age population in Africa and the Far East. In Central and South America, one can also add examples such as Chagas disease and Leishmanial infections as major causes of death/significant impairment of young populations. We will discuss recent reports of both halogenated and non-halogenated compounds with significant activities in these areas. As mentioned previously, we will only comment on materials from invertebrates. Where a cultivatable microbe or other single-celled organisms are known to be involved, we will not cover these, though as mentioned earlier, the actual producer in a number of cases may be an as yet uncultured/unidentified microbe.

1.5.2 HALOGENATED COMPOUNDS

Recent examples are from three well-known classes of sponge-derived compounds. In 2011, a group at the Eskitis Institute in Brisbane, Australia reported the new psammaplysin derivative, psammaplysin H (**45**), which had good activity against the chloroquine-sensitive *Plasmodium falciparum* 3D7 line, with an IC_{50} value of 410 nM and minimal toxic activity (selectivity index above 96) against the mammalian lines HEK293 and HepG2. The following year, another Australian group, which included researchers from the University of Queensland but also utilized the assay and chemistry capabilities of the Eskitis Institute scientists, reported that the sponge *Aplysinella strongylata* from Bali contained 21 new psammaplysins and six previously known analogues, but none equaled the activity of the "H" analogue (Mudianta et al. 2012). The same year, Eskitis scientists reported the isolation of known tsitsikammamines and bioactivity of the new bispyrroleiminoquinone alkaloid tsitsikammamine C (**46**) (Davis et al. 2012). This material exhibited excellent in vitro activity against both the chloroquine-sensitive (3D7; IC_{50} 13nM) and chloroquine-resistant (Dd2; IC_{50} 18nM) strains. The selectivity index against HEK293 was >200. The known makaluvamines G, and J–L were also isolated from the same sponge but were somewhat less active than (**46**) but had selectivity indices 2–3 times lower.

Santos et al. (2015) reported the antiparasitic activities of a series of batzelladine and close derivatives isolated from the Brazilian sponge, *Monanchora arbuscula*. Of the compounds isolated and purified, batzelladines D, F, L, and nor-L (**47–50**) were assayed against *Trypanosoma cruzi* and *Leishmania infantum* since both parasites are major causes of disease in Brazil. Compound **47** was not significantly active against *T. cruzi* but the other three had similar activities against this parasite (IC_{50} 5–7 µM). Against *L. infantum*, the IC_{50} values for all were in the range 2–4 µM. Comparison of the *L. infantum* activities with their cytotoxicity activities gave ratios of 65 for **47**, 2.5 for **48**, 11 for **49**, and 43 for **50**. In fact, compound **50** was the best against both parasites as its ratio for *T. cruzi* was 11.5.

Recently, Avilés et al. (2015) published an interesting paper on the synthesis of hybrid molecules based on Ugi coupling methodologies to a pharmacophore-based upon the isocyanide (–)-DINCA (**51**). These types of compounds/activities are from an example of sponge metabolites known to show antimalarial activity that were initially reported in 1996 from the sponge *Cymbastela hooperi* by Konig et al. (1996). The isocyanide (**51**) was isolated from a Caribbean *Hymediacidon* sp. (Mayer et al. 2012) and coupled with acids, amines, and aldehydes under Ugi conditions. A representative example was compound **52**, but unfortunately none of the adducts were as potent as the isocyanide starting material. However, the ideas behind this type of reaction may well lead to more active compounds in the future (Figure 1.7).

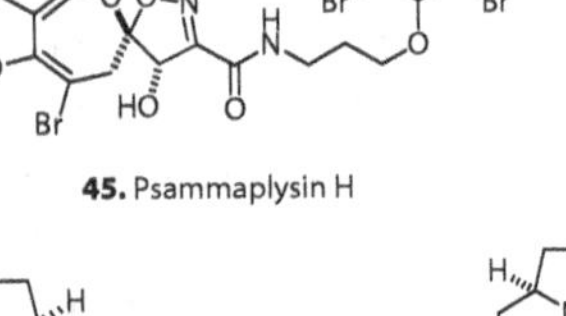

45. Psammaplysin H

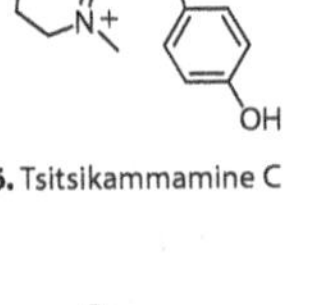

46. Tsitsikammamine C

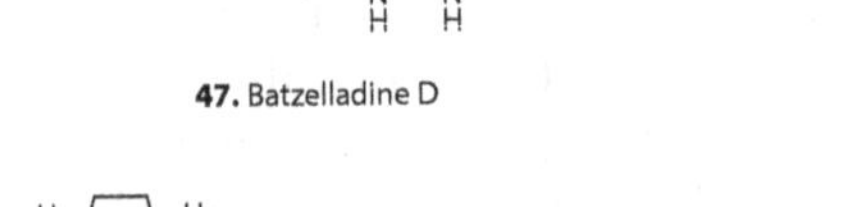

47. Batzelladine D

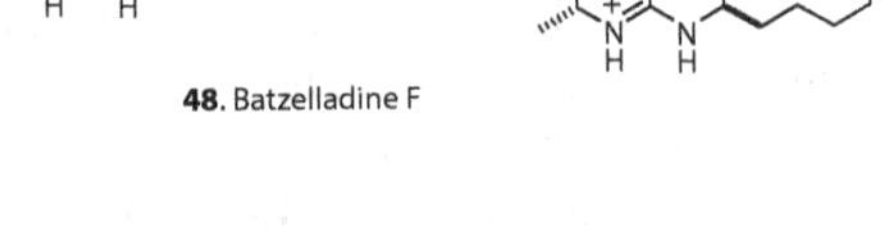

48. Batzelladine F

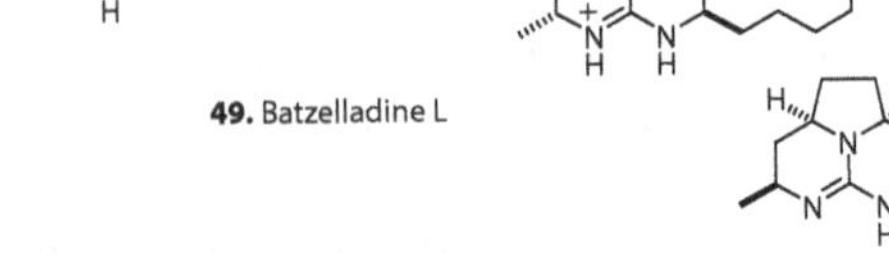

49. Batzelladine L

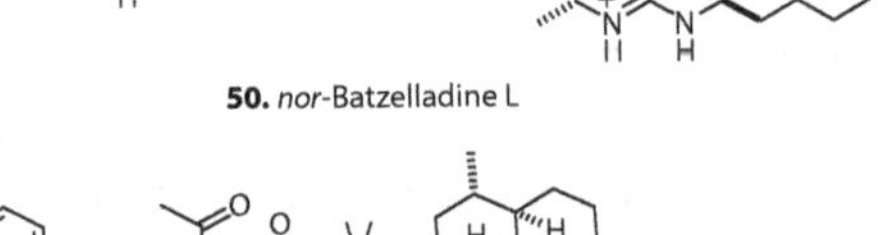

50. *nor*-Batzelladine L

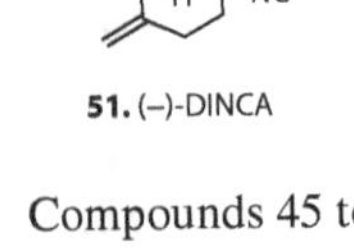

51. (–)-DINCA

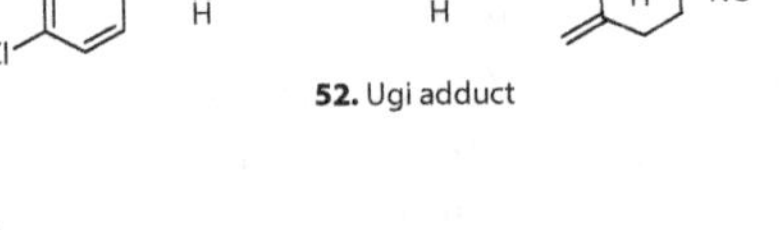

52. Ugi adduct

FIGURE 1.7 Compounds 45 to 52.

1.6 ANTIVIRAL COMPOUNDS FROM PORIFERA

1.6.1 INTRODUCTION

In the time from the start of the HIV epidemic and to other viral diseases, extending, in some cases such as herpetic infections, to before the recognition of HIV, the marine environment has been a source of potential agents for treatments. Examples such as the arabinose-containing nucleosides were referred to in the introduction, but it is now apparent that these are probably produced either solely by the endophytic bacterium or in conjunction with the host sponge (Bertin et al. 2015).

However, there are materials isolated from sponges that have been investigated for their ability to inhibit the life-cycles of viruses that infect humans. Obviously, the major impetus was searching for agents active against the various clades of HIV I and II and a large variety of different marine sponge-derived molecules have been and are still are being investigated. In 2015, an excellent review paper was published in *Chemical Reviews* by Gogineni et al. (2015) covering the genesis of antiviral agents from marine sources. It should be noted that the second author of that review was responsible to a large extent for the work that led to the approval of drugs d4T (stavudine), 3TC (lamivudine), FTC (emtricitabine/Emtriva), LdT (telbivudine), and recently sofosbuvir (Sovaldi). All are now approved antiviral drugs against either HIV or HCV, so rather than attempt to paraphrase this extensive article, we will select two or three recent examples for reference.

1.6.2 ANTIVIRAL ALKALOIDS AND CYCLIC PEPTIDES

Batzelladine C (**53**) was isolated along with other similar alkaloids from the Caribbean sponge *Monanchora ungelifera* and was shown to have activity against the gp120-CD4 interaction with an EC_{50} of 7.7 µM. Another alkaloid from the same sponge, crambescidin 800 (**54**) by contrast, had a similar potency but was directed against the *fusion* of the viron with the cell (Hua et al. 2007). The same year, Plaza et al. (2007) reported the cyclic depsipeptide, mirabamide A (**55**), which they isolated from the sponge *Siliquariaspongia mirabilis* collected in Chuuk lagoon as part of the NCI marine collection program, also inhibited fusion with an EC_{50} value of between 40 and 140 nM.

Moving to inhibitors of HCV from unusual places, Abbas et al. (2011) reported the isolation and anti-HCV activities of discorhabdins A (**56**) and C (**57**) and dihydrodiscorhabin C (**58**) all isolated from an as yet unspeciated *Latruncula* sp., collected in Arctic waters off the Aleutian Islands. These agents inhibited HCV with EC_{50} values below 10 µM, and **56** and **58** were also tested *in vivo* against a malarial infection but were found to cause weight loss and dehydration. However, this was the first report of any *in vivo* assay of materials from Porifera from Arctic environments.

We think that these examples are enough to show that the marine invertebrate environment does have the potential to discover agents active against viral diseases and if one included algae and microbes, then that prediction is certainly

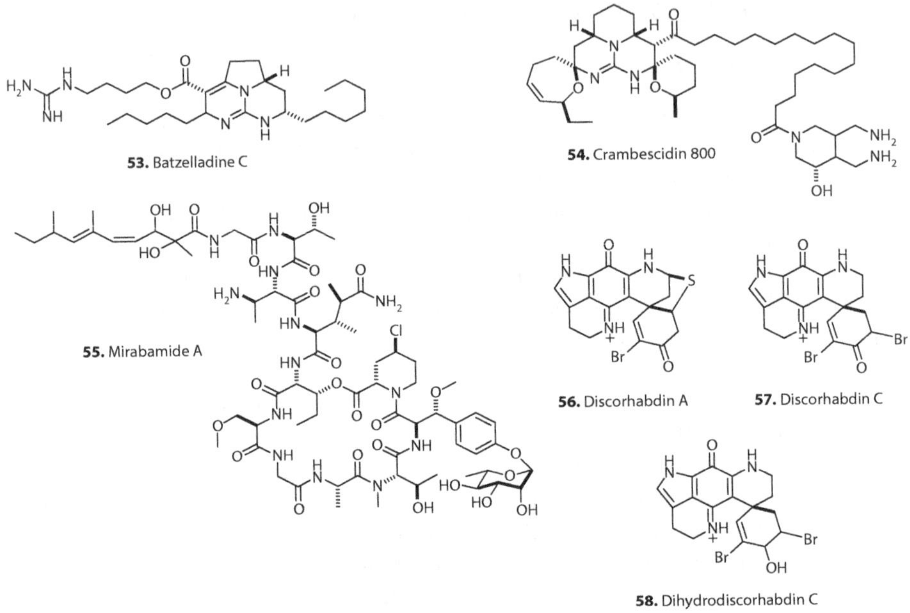

53. Batzelladine C

54. Crambescidin 800

55. Mirabamide A

56. Discorhabdin A   **57.** Discorhabdin C

58. Dihydrodiscorhabdin C

FIGURE 1.8   Compounds 53 to 58.

true, but in this chapter, we are only considering invertebrate sources (though these may well be simply hosts for as yet unrecognized or uncultured microbes) (Figure 1.8).

1.7   OTHER PHARMACOLOGICAL PROPERTIES OF MARINE INVERTEBRATE METABOLITES

1.7.1   INTRODUCTION

Over the last 13-plus years, professor Alejandro Mayer and colleagues have listed the metabolites that have been reported to show activities in a variety of pharmacologic areas. Though these reviews date from 2002 to 2013, they actually cover publications from 1999 to 2013, so reading these reviews will give a good idea as to the manifold areas that have been investigated (Mayer and Hamann 2002, 2004, 2005, Mayer et al. 2007, 2009, 2011, 2013, 2017). What was not covered in these reviews were antitumor agents but discussions of the anticancer agents approved and in late trials were published by Mayer et al. (2010). An update was published in early 2016 by the authors Newman and Cragg (2016). Both of these will give a better idea of where the compounds from marine invertebrates are in terms of approved and under clinical trials at the time of publication.

1.7.2 NEUROLOGICAL DISEASES

Although there have been some recent publications on activities of marine-derived agents as neuroprotective agents, a significant number have been isolated directly from single-celled organisms. As we mentioned earlier in this chapter, we were not going to discuss such compounds, though it is quite possible that a significant number of compounds, particularly from sponges, are in fact produced by single-celled endophytic organisms within the sponge (or other invertebrate). We will not discuss those compounds that are already identified from such sources. For a recent compendium of such agents we recommend reading the 2015 review by Choi and Choi (2015).

There is one particular class of molecules, the leucettines (Debdab et al. 2011), that have shown some interesting activities in the neurological area. These compounds, which are inhibitors of DYRKs (dual specificity, tyrosine phosphorylation regulated kinases) and CLKs (cdc2 like kinases), were derived from leucettamine B (**59**), which was originally isolated from the sponge *Leucetta microraphis* (Chan et al. 1993, Watanabe et al. 2000). The leucettamine B structure was chemically related to hymenialdisine (**1**) and other molecules that contain a 2-aminoimidazolone core. The synthetic routes to leucettamine B have been well covered by a number of investigators (Molina et al. 1994, Roué and Bergman 1999) and similar work leading to analogues was reported in 2009–2011 (Debdab et al. 2009, 2011).

One of the analogues, leucettine L_{41} (**60**), was found not only to interact very specifically with DYRKs and CLKs but also with GSK-3, PIMI, CK2, and the lipid kinase, PIKfyve. Co-crystal structures of this agent with the first three kinases, plus a similar crystal structure where the analogue leucettine L_4 (**61**) was co-crystallized with GSK-3β, demonstrated how these agents bound at the ATP site. There was definite interaction in cellulo with the L_{41} analogue in HT22 cells without any activity against GSK-3 in the same system, though both kinases and others were inhibited in in vitro assays. Inhibition of DYRKs in rat brain slices gave a neuroprotective effect, but efforts to link this action to effects on neurodegenerative diseases such as Alzheimer's disease had not succeeded when this paper was published (Tahtouh et al. 2012).

A very interesting paper from the same group was published in 2013 demonstrating how they were able to identify potential targets of the L_{41} analogue using immobilized versions of L_{41} on agarose beads (Burgy et al. 2013). Although this is a well-known technique for identifying potential kinase and other targets of agents (Knockaert et al. 2000), these investigators were meticulous in pointing out the problems involved with non-specific binding, etc. In addition, they deliberately used an inactive variant of L_{41} as a control agent for non-specific binding and found some interesting binding to kinases by this "inactive!" agent that indicated possible other binding sites within the kinome.

As mentioned previously, the Meijer group were not able to confirm their experimental hypothesis that inhibition of the DYRK1 series of kinases would be directly applicable to the treatment of Alzheimer's disease. However, Naert et al. (2015)

published an interesting paper demonstrating that L_{41} (**60**) delivered by intracranial injection to mice that had demonstrated SD-like symptoms following a 7-day treatment with oligomeric $A\beta_{25-35}$ peptide reversed the effect when measured by neuropharmacological methods such as maze memory and specific antibody western blots. The following quote from their conclusion is worth noting: "*L_{41} effects on $A\beta$ toxicity seem to be attributed to inhibition of DYRK1A in particular, as DYRK1A is the main target of L_{41} in mouse brain.*"

Leiros et al. (2015a) reported on the protective effect of the bromoalkaloids hymenialdisine (**1**) and hymenin (**62**) in an oxidative stress cellular model composed of cortical neurons preincubated with H_2O_2. Both compounds were active as protective agents at low nanomolar concentrations, with hymenialdisine being more protective. Further investigations were performed using assays that involved the principal sensor and modulator of the oxidative stress pathway, the nuclear factor erythroid 2-related factor 2 (Nrf2) antioxidant response element (ARE). If oxidative stress occurs, then Nrf2 relocates to the nucleus where it finally binds ARE and causes significant transcription of genes. Thus, these investigators looked at the influence of these two agents upon this pathway as mentioned earlier. They found that hymenialdisine (**1**) was the most protective, including modulation of the Nrf2-ARE pathway as an Nrf2 inducer, with a resulting down-regulation in the response to oxidative insults and an enhancement of GSH (reduced glutathione), which would promote a cellular antioxidant defense. This information where activities were seen at 1–10 nM, may well open a new research area for drugs against neuroinflammatory processes.

The same year, effectively the same group of investigators demonstrated that gracilin H (**63**), gracilin A (**64**), gracilin L (**65**), and tetrahydroaplysulphurin-1 (**66**) all isolated from sponges, with **63–65** from *Spongionella pulchella* (Rueda et al. 2006) and **66** from a *Spongiella* sp. (Rateb et al. 2009) exhibited some very interesting data on the effects of some of these agents upon processes associated with Alzheimer's disease. The work involved both in vitro and in vivo assays, giving the following results.

The BE(2)-M17 cell line was used as a model for APP metabolism studies, with SH-SY5Y-TMHT441 cells used for AD drugs by screening the targeting of *tau* phosphorylation. Using these lines, in vitro assays demonstrated that gracilins were able to inhibit BACE1, reduce *tau* hyper-phosphorylation, and inhibit ERK. The positive results led to the testing of gracilin H (**63**) and L (**65**) *in vivo* using chronic peritoneal injection in 3xTg-AD mice. Following these intraperitoneal treatments, a preliminary behavioral test pointed a positive trend on learning and spatial memory of mice treated with these compounds. In vivo assays on the treated mice confirmed the previous in vitro results showing decrease in Amyloid-b42 an hyperphosphorylated *tau* levels and inhibition of ERK was also seen (Leiros et al. 2015b).

A later paper from the same investigators also demonstrated that three of the four agents, gracilin A (**64**), gracilin L (**65**), and tetrahydroaplysulphurin-1 (**66**), significantly affected calcium ion fluxes in SH-SYSY neuroblastoma cells and probably interacted directly with cyclophilin D in a manner analogous to that shown by cyclosporin. Thus, these agents are potential sources of structures on which to design drugs against human diseases that involve mitochondrial alterations (Sanchez et al. 2015) (Figure 1.9).

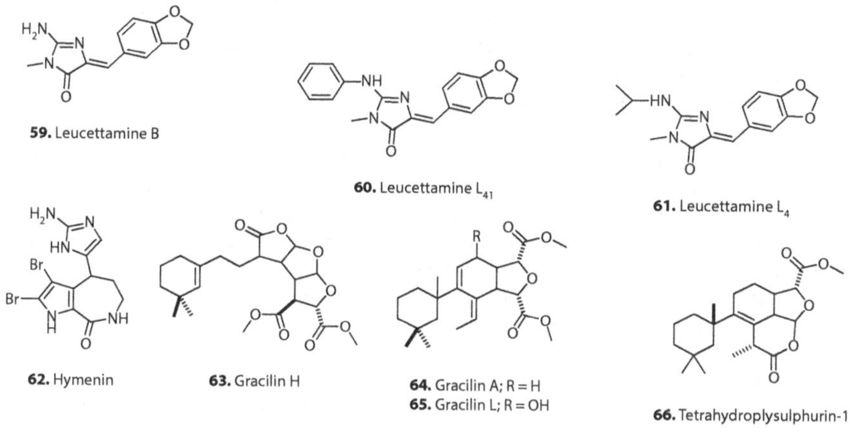

FIGURE 1.9 Compounds 59 to 66.

1.8 CONCLUSION

This chapter was not intended to be a full analysis of marine-derived compounds with known pharmacological activities as such an endeavor would be a multi-volume series. What we hope we have done is to lay a foundation of pharmacological activities shown by invertebrate-derived metabolites in areas that are not often covered, in particular antimicrobial, antiviral, and parasitic diseases and recent work on neurological diseases, specifically the areas related to Alzheimer's disease.

It will be noted that we have deliberately not included any significant discussion of antitumor agents. One major reason is that these have been thoroughly covered, as far as approved drugs and those in late clinical trials are concerned, by the authors in a current review in Planta Medica (Newman and Cragg 2016). In addition, there is now evidence that a significant proportion of the molecules involved in the agents referred to in that review are produced by either microbes within the invertebrate host or from interplay between the invertebrate and its endophytes. In some cases, free-living microbes are also a probable source. As a result, we have not covered these as such sources will be covered in other chapters in this book.

REFERENCES

Abbas, S., M. Kelly, J. Bowling, J. Sims, A. Waters, and M. T. Hamann. 2011. Advancement into the Arctic region for bioactive sponge secondary metabolites. *Mar. Drugs* 9(11):2423–2437.
Agarwal, V., J. M. Blanton, S. Podell, A. Taton, M. A. Schorn, J. Busch, Z. Lin et al. 2017. Metagenomic discovery of polybrominated diphenyl ether biosynthesis by marine sponges. *Nat. Chem. Biol.* 13:537–543.
Akondi, K. B., M. Muttenthaler, S. Dutertre, Q. Kaas, D. J. Craik, R. J. Lewis, and P. F. Alewood. 2014. Discovery, synthesis, and structure-activity relationships of conotoxins. *Chem. Rev.* 114:5815–5847.

Alicino, I., M. Giglio, F. Manca, F. Bruno, and F. Puntillo. 2012. Intrathecal combination of ziconotide and morphine for refractory cancer pain: A rapidly acting and effective choice. *Pain* 153(1):245–249.

Avilés, E., J. Prudhomme, K. G. Le Roch, S. G. Franzblau, K. Chandrasena, A. M. S. Mayer, and A. D. Rodríguez. 2015. Synthesis and preliminary biological evaluation of a small library of hybrid compounds based on Ugi isocyanide multicomponent reactions with a marine natural product scaffold. *Bioorg. Med. Chem. Lett.* 25:5339–5343.

Azam, L., A. Papakyriakou, M. Zouridakis, P. Giastas, S. J. Tzartos, and J. M. McIntosh. 2015. Molecular interaction of α-conotoxin RgIA with the rat α9α10 nicotinic acetylcholine receptor. *Mol. Pharmacol.* 87(5):855–864.

Beeton, C., M. W. Pennington, H. Wulff, S. Singh, D. Nugent, G. Crossley, I. Khaytin et al. 2005. Targeting effector memory T cells with a selective peptide inhibitor of Kv1.3 channels for therapy of autoimmune diseases. *Mol. Pharmacol.* 67(4):1369–1381.

Beeton, C., H. Wulff, N. E. Standifer, P. Azam, K. M. Mullen, M. W. Pennington, A. Kolski-Andreaco et al. 2006. Kv1.3 channels are a therapeutic target for T cell-mediated autoimmune diseases. *Proc. Natl. Acad. Sci. USA* 103(46):17414–17419.

Bergmann, W., and D. C. Burke. 1955. Marine products. XXXIX. The nucleosides of sponges. III. Spongothymidine and spongouridine. *J. Org. Chem.* 20:1501–1507.

Bergmann, W., and R. J. Feeney. 1950. Isolation of a new thymine pentoside from sponges. *J. Am. Chem. Soc.* 72:2809–2810.

Bergmann, W., and R. J. Feeney. 1951. Marine products. XXXII. The nucleosides of sponges. I. *J. Org. Chem.* 16:981–987.

Bertin, M. J., S. L. Schwartz, J. Lee, A. Korobeynikov, P. C. Dorrestein, L. Gerwick, and W. H. Gerwick. 2015. Spongosine production by a *Vibrio harveyi* strain associated with the sponge *Tectitethya crypta*. *J. Nat. Prod.* 78:493–499.

Brust, A., E. Palant, D. E. Croker, B. Colless, R. Drinkwater, B. Patterson, C. I. Schroeder et al. 2009. chi-Conopeptide pharmacophore development: Toward a novel class of norepinephrine transporter inhibitor (Xen2174) for pain. *J. Med. Chem.* 52(22):6991–7002.

Burgy, G., T. Tahtouh, E. Durieu, B. Foll-Josselin, E. Limanton, L. Meijer, F. Carreaux, and J.-P. Bazureau. 2013. Chemical synthesis and biological validation of immobilized protein kinase inhibitory leucettines. *Eur. J. Med. Chem.* 62:728–737.

Castañeda, O., V. Sotolongo, A. M. Amor, R. Stöcklin, A. J. Anderson, A. L. Harvey, Å. Engström, C. Wernstedt, and E. Karlsson. 1995. Characterization of a potassium channel toxin from the Caribbean Sea anemone *Stichodactyla helianthus*. *Toxicon* 33:603–613.

Chan, G. W., S. Mong, M. E. Hemling, A. J. Freyer, P. M. Offen, C. W. De Brosse, H. M. Sarau, and J. W. Westley. 1993. New leukotriene B4 receptor antagonist: Leucettamine A and related imidazole alkaloids from the marine sponge *Leucetta microraphis*. *J. Nat. Prod.* 56:116–121.

Chang, S. C., R. Huq, S. Chhabra, C. Beeton, M. W. Pennington, B. J. Smith, and R. S. Norton. 2015. N-Terminally extended analogues of the K⁺ channel toxin from *Stichodactyla helianthus* as potent and selective blockers of the voltage-gated potassium channel Kv1.3. *FEBS J.* 282(12):2247–2259.

Chang, Y. P., J. Banerjee, C. Dowell, J. Wu, R. Gyanda, R. A. Houghten, L. Toll, J. M. McIntosh, and C. J. Armishaw. 2014. Discovery of a potent and selective α3β4 nicotinic acetylcholine receptor antagonist from an α-conotoxin synthetic combinatorial library. *J. Med. Chem.* 57(8):3511–3521.

Choi, D.-Y., and H. Choi. 2015. Natural products from marine organisms with neuroprotective activity in the experimental models of Alzheimer's disease, Parkinson's disease and ischemic brain stroke: Their molecular targets and action mechanisms. *Arch. Pharm. Res.* 38(2):139–170.

Christensen, S. B., P. K. Bandyopadhyay, B. M. Olivera, and J. M. McIntosh. 2015. αS-conotoxin GVIIIB potently and selectively blocks α9α10 nicotinic acetylcholine receptors. *Biochem. Pharmacol.* 96:349–356.

Clark, R. J., J. Jensen, S. T. Nevin, B. P. Callaghan, D. J. Adams, and D. J. Craik. 2010. The engineering of an orally active conotoxin for the treatment of neuropathic pain. *Angew. Chem. Int. Ed.* 49:6545–6548.

Cruz, L. J., W. R. Gray, and B. M. Olivera. 1978. Purification and properties of a myotoxin from *Conus geographus* venom. *Arch. Biochem. Biophys.* 190(2):539–548.

Daly, N. L., and D. J. Craik. 2011. Conopeptides as novel options for pain management. *Drugs Fut.* 36(1):25–32.

Davis, R. A., M. S. Buchanan, S. Duffy, V. M. Avery, S. A. Charman, W. N. Charman, K. L. White et al. 2012. Antimalarial activity of pyrroloiminoquinones from the Australian marine sponge *Zyzzya* sp. *J. Med. Chem.* 55:5851–5858.

de Veer, S. J., J. Weidmann, and D. J. Craik. 2017. Cyclotides as tools in chemical biology. *Acc. Chem. Res.* 50:1557–1565.

Debdab, M., S. Renault, S. Eid, O. Lozach, L. Meijer, F. Carreaux, and J.-P. Bazureau. 2009. An efficient method for the preparation of new analogs of leucettamine B under solvent-free microwave irradiation. *Tetrahedron Lett.* 78:1191–1203.

Debdab, M., S. Renault, M. Soundararajan, O. Fedorov, P. Filippakopoulos, O. Lozach, L. Babault et al. 2011. Leucettines, a class of potent inhibitors of cdc2-like kinases and dual specificity, tyrosine phosphorylation regulated kinases derived from the marine sponge leucettamine B: Modulation of alternative pre-RNA splicing. *J. Med. Chem.* 54:4172–4186.

Durek, T., and D. J. Craik. 2015. Therapeutic conotoxins: A US patent literature survey. *Expert Opin. Ther. Pat.* 25(10):1159–1173.

Fedosov, A. E., S. A. Moshkovskii, K. G. Kuznetsova, and B. M. Olivera. 2012. Conotoxins: From the biodiversity of gastropods to new drugs. *Biochemistry (Moscow) Supp. Ser B, Biomed. Chem.* 6(2):107–122.

Gogineni, V., R. F. Schinazi, and M. T. Hamann. 2015. Role of marine natural products in the genesis of antiviral agents. *Chem. Revs.* 115(18):9655–9706.

Gribble, G. W. 2015. Biological activity of recently discovered halogenated marine natural products. *Mar. Drugs* 13(7):4044–4136. doi:10.3390/md13074044.

Henriques, S. T., Y.-H. Huang, S. Chaousis, M.-A. Sani, A. G. Poth, F. Separovic, and D. J. Craik. 2015. The prototypic cyclotide Kalata B1 has a unique mechanism of entering cells. *Chem. Biol.* 22:1087–1097.

Hertiani, T., R. Edrada-Ebel, S. Ortlepp, R. W. M. van Soest, N. J. de Voogd, V. Wray, U. Hentschel, S. Kozytsla, W. E. G. Müller, and P. Proksch. 2010. From anti-fouling to biofilm inhibition: New cytotoxic secondary metabolites from two Indonesian *Agelas* sponges. *Bioorg. Med. Chem.* 18:1297–1311.

Hua, H.-M., J. Peng, D. C. Dunbar, R. F. Schinazi, A. G. de Castro Andrews, C. Cuevas, L. F. Garcia-Fernandez, M. Kelly, and M. T. Hamann. 2007. Batzelladine alkaloids from the Caribbean sponge *Monanchora unguifera* and the significant activities against HIV-1 and AIDS opportunistic infectious pathogens. *Tetrahedron* 63:11179–11188.

Imperial, J. S., A. B. Cabang, J. Song, S. Raghuraman, J. Gajewiak, M. Watkins, P. Showers-Corneli et al. 2014. A family of excitatory peptide toxins from venomous crassispirine snails: Using constellation pharmacology to assess bioactivity. *Toxicon* 89:45–54.

Jayamanne, A., H. J. Jeong, C. J. Schroeder, R. J. Lewis, M. J. Christie, and C. W. Vaughan. 2013. Spinal actions of omega-conotoxins, CVID, MVIIA and related peptides in a rat neuropathic pain model. *Br. J. Pharmacol.* 170:245–254.

Kem, W. R., B. Parten, M. W. Pennington, D. A. Price, and B. M. Dunn. 1989. Isolation, characterization, and amino acid sequence of a polypeptide neurotoxin occurring in the sea anemone *Stichodactyla helianthus*. *Biochemistry* 28:3483–3489.

Knockaert, M., N. Gray, E. Damiens, Y. T. Chang, P. Grellier, K. Grant, D. Fergusson et al. 2000. Intracellular targets of cyclin-dependent kinase inhibitors: Identification by affinity chromatography using immobilised inhibitors. *Chem. Biol.* 7:411–422.

Konig, G. M., A. D. Wright, and C. K. Angerhofer. 1996. Novel potent antimalarial diterpene isocyanates, isothiocyanates, and isonitriles from the tropical marine sponge *Cymbastela hooperi. J. Org. Chem.* 61:3259–3267.

Koshy, S., R. Huq, M. R. Tanner, M. A. Atik, P. C. Porter, F. S. Khan, M. W. Pennington, N. A. Hanania, D. B. Corry, and C. Beeton. 2014. Blocking KV1.3 channels inhibits Th2 lymphocyte function and treats a rat model of asthma. *J. Biol. Chem.* 289(18):12623–12632.

Kubota, T., T. Iwai, A. Takahashi-Nakaguchi, J. Fromont, T. Gonoi, and J. Kobayashi. 2012. Agelasines O-U, new diterpene alkaloids with a 9-N-methyladenine unit from a marine sponge *Agelas* sp. *Tetrahedron* 68:9738–9744.

Kudryavtsev, D., I. Shelukhina, C. Vulfius, T. N. Makarieva, V. A. Stonik, M. Zhmak, I. Ivanov, I. Kasheverov, Y. Utkin, and V. Tsetlin. 2015. Natural compounds interacting with nicotinic acetylcholine receptors: From low-molecular weight ones to peptides and proteins. *Toxins* 7:1683–1701.

Kusama, T., N. Tanaka, K. Sakai, T. Gonoi, J. Fromont, Y. Kashiwada, and J. Kobayashi. 2014a. Agelamadins A and B, dimeric bromopyrrole alkaloids from a marine sponge *Agelas* sp. *Org. Lett.* 16:3916–3918.

Kusama, T., N. Tanaka, K. Sakai, T. Gonoi, J. Fromont, Y. Kashiwada, and J. Kobayashi. 2014b. Agelamadins C-E, bromopyrrole alkaloids comprising oroidin and 3-hydroxykynurenine from a marine sponge *Agelas* sp. *Org. Lett.* 16:5176–5179.

Lavergne, V., I. Harliwong, A. Jones, D. Miller, R. J. Taft, and P. F. Alewood. 2015. Optimized deep-targeted proteotranscriptomic profiling reveals unexplored *Conus* toxin diversity and novel cysteine frameworks. *Proc. Natl. Acad. Sci. USA* 112:E3782–E3791.

Leiros, M., E. Alonso, M. E. Rateb, W. E. Houssen, R. Ebel, M. Jaspars, A. Alfonso, and L. M. Botana. 2015a. Bromoalkaloids protect primary cortical neurons from induced oxidative stress. *ACS Chem. Neurosci.* 6:331–338.

Leiros, M., E. Alonso, M. E. Rateb, W. E. Houssen, R. Ebel, M. Jaspars, A. Alfonso, and L. M. Botana. 2015b. Gracilins: *Spongionella*-derived promising compounds for Alzheimer disease. *Neuropharmacol.* 93:285–293. doi:10.1016/j.neuropharm.2015.02.015.

Lewis, R. J. 2015. Case study 1: Development of the analgesic drugs Prialt® and Xen2174 from cone snail venoms. In *Venoms to Drugs: Venom as a Source for the Development of Human Therapeutics*, edited by G. F. King, pp. 245–254. Abington, UK: Marston.

Matheu, M. P., C. Beeton, A. Garcia, V. Chi, S. Rangaraju, O. Safrina, K. Monaghan et al. 2008. Imaging of effector memory T cells during a delayed-type hypersensitivity reaction and suppression by Kv1.3 channel block. *Immunity* 29(4):602–614.

Mayer, A. M. S., E. Avilés, and A. D. Rodríguez. 2012. Marine sponge *Hymeniacidon* sp. amphilectane metabolites potently inhibit rat brain microglia thromboxane B2 generation. *Bioorg. Med. Chem.* 20(1):279–282.

Mayer, A. M. S., K. B. Glaser, C. Cuevas, R. S. Jacobs, W. R. Kem, R. D. Little, J. M. McIntosh, D. J. Newman, B. C. Potts, and D. E. Shuster. 2010. The odyssey of marine pharmaceuticals: A current pipeline perspective. *Trends Pharmacol. Sci.* 31(6):255–265.

Mayer, A. M. S., and M. T. Hamann. 2002. Marine pharmacology in 1999: Compounds with antibacterial, anticoagulant, antifungal, anthelmintic, anti-inflammatory, antiplatelet, antiprotozoal and antiviral activities affecting the cardiovascular, endocrine, immune and nervous systems, and other miscellaneous mechanisms of action. *Comp. Biochem. Physiol. Part C* 132(3):315–339.

Mayer, A. M. S., and M. T. Hamann. 2004. Marine pharmacology in 2000: Marine compounds with antibacterial, anticoagulant, antifungal, anti-inflammatory, antimalarial, antiplatelet, antituberculosis, and antiviral activities; affecting the cardiovascular, immune, and nervous systems and other miscellaneous mechanisms of action. *Mar. Biotech.* 6(1):37–52.

Mayer, A. M. S., and M. T. Hamann. 2005. Marine pharmacology in 2001–2002: Marine compounds with anthelmintic, antibacterial, anticoagulant, antidiabetic, antifungal, antiinflammatory, antimalarial, antiplatelet, antiprotozoal, antituberculosis, and antiviral activities; affecting the cardiovascular, immune and nervous systems and other miscellaneous mechanisms of action. *Comp. Biochem. Physiol. Part C* 140(3–4):265–286.

Mayer, A. M. S., A. D. Rodríguez, R. G. S. Berlinck, and N. Fusetani. 2011. Marine pharmacology in 2007–8: Marine compounds with antibacterial, anticoagulant, antifungal, anti-inflammatory, antimalarial, antiprotozoal, antituberculosis, and antiviral activities; affecting the immune and nervous system, and other miscellaneous mechanisms of action. *Comp. Biochem. Physiol. Part C* 153:191–222.

Mayer, A. M. S., A. D. Rodríguez, R. G. S. Berlinck, and M. T. Hamann. 2007. Marine pharmacology in 2003–4: Marine compounds with anthelmintic antibacterial, anticoagulant, antifungal, anti-inflammatory, antimalarial, antiplatelet, antiprotozoal, antituberculosis, and antiviral activities; affecting the cardiovascular, immune and nervous systems, and other miscellaneous mechanisms of action. *Comp. Biochem. Physiol. Part C* 145(4):553–581.

Mayer, A. M. S., A. D. Rodríguez, R. G. S. Berlinck, and M. T. Hamann. 2009. Marine pharmacology in 2005–6: Marine compounds with anthelmintic, antibacterial, anticoagulant, antifungal, anti-inflammatory, antimalarial, antiprotozoal, antituberculosis, and antiviral activities; affecting the cardiovascular, immune and nervous systems, and other miscellaneous mechanisms of action. *Biochim. Biophys. Acta* 1790:283–308.

Mayer, A. M. S., A. D. Rodriguez, O. Taglialatela-Scafati, and N. Fusetani. 2017. Marine pharmacology in 2012–2013: Marine compounds with antibacterial, antidiabetic, antifungal, antiinflammatory, antiprotozoal, antituberculosis and antiviral Activities; affecting the immune and nervous system, and other miscellaneous mechanisms of action. *Mar. Drugs* 15(9):273. doi:10.3390/md15090273.

Mayer, A. M. S., A. D. Rodríguez, O. Taglialatela-Scafati, and N. Fusetani. 2013. Marine pharmacology in 2009–2011: Marine compounds with antibacterial, antidiabetic, antifungal, antiiInflammatory, antiprotozoal, antituberculosis, and antiviral activities; affecting the immune and nervous systems, and other miscellaneous mechanisms of action. *Mar. Drugs* 11:2510–2573.

Molina, P., P. Almendros, and P. M. Fresneda. 1994. An iminophosphorane-mediated efficient synthesis of the alkaloid leucettamine B of marine of origin. *Tetrahedron Lett.* 35:2235–2236.

Mudianta, I. W., T. Skinner-Adams, K. T. Andrews, R. A. Davis, T. A. Hadi, P. Y. Hayes, and M. J. Garson. 2012. Psammaplysin derivatives from the Balinese marine sponge *Aplysinella strongylata. J. Nat. Prod.* 75:2132–2143.

Munasinghe, N. R., and M. J. Christie. 2015. Conotoxins that could provide analgesia though voltage gated sodium channel inhibition. *Toxins* 7:5386–5407.

Naert, G., V. Ferré, J. Meunier, E. Keller, S. Malmström, L. Givalois, F. Carreaux, J.-P. Bazureau, and T. Maurice. 2015. Leucettine L41, a DYRK1A-preferential DYRKs/CLKs inhibitor, prevents memory impairments and neurotoxicity induced by oligomeric Aβ25-35 peptide administration in mice. *Eur. Neuropsychopharm.* 25(11):2170–2182.

Newman, D. J., and G. M. Cragg. 2016. Drugs and drug candidates from marine sources: An assessment of the current "state of play". *Planta Med* 82:775–789.

Northfield, S. E., C. K. Wang, C. L. Schroeder, T. Durek, M.-W. Kan, J. E. Swedberg, and D. J. Craik. 2014. Disulfide-rich macrocyclic peptides as templates in drug design. *Eur. J. Med. Chem.* 77:248–257.

Okkerse, P., J. L. Hay, E. Sitsen, A. Dahan, E. Klaassen, W. Houghton, and G. J. Groeneveld. 2017. Pharmacokinetics and pharmacodynamics of intrathecally administered Xen2174, a synthetic conopeptide with norepinephrine reuptake inhibitor and analgesic properties. *Br. J. Clin. Pharmacol.* 83(4):751–763.

Olivera, B. M., W. R. Gray, R. Zeikus, J. M. McIntosh, J. Varga, J. Rivier, V. de Santos, and L. J. Cruz. 1985. Peptide neurotoxins from fish-hunting cone snails. *Science* 230:1338–1343.

Pénez, N., G. Culioli, T. Pérez, J.-F. Briand, O. P. Thomas, and Y. Blache. 2011. Antifouling properties of simple indole and purine alkaloids from the Mediterranean gorgonian *Paramuricea clavata. J. Nat. Prod.* 74:2304–2308.

Pennington, M. W., C. Beeton, C. A. Galea, B. J. Smith, V. Chi, K. P. Monaghan, A. Garcia et al. 2009. Engineering a stable and selective peptide blocker of the Kv1.3 channel in T lymphocytes. *Mol. Pharmacol.* 75(4):762–773.

Pennington, M. W., S. C. Chang, S. Chauhan, R. Huq, R. B. Tajhya, S. Chhabra, R. S. Norton, and C. Beeton. 2015. Development of highly selective Kv1.3-blocking peptides based on the sea anemone peptide ShK. *Mar. Drugs* 13(1):529–542. doi:10.3390/md13010529.

Pereira, U. A., L. C. A. Barbosa, A. J. Demuner, A. A. Silva, M. Bertazzini, and G. Forlani. 2015. Rubrolides as model for the development of new lactones and their aza analogs as potential photosynthesis inhibitors. *Chem. Biodiv.* 12:987–1006.

Plaza, A., E. Gustchina, H. L. Baker, M. Kelly, and C. A. Bewley. 2007. Mirabamides A–D, depsipeptides from the sponge *Siliquariaspongia mirabilis* that inhibit HIV-1 fusion. *J. Nat. Prod.* 70:1753–1760.

Qu, H., B. J. Smithies, T. Durek, and D. J. Craik. 2017. Synthesis and protein engineering applications of cyclotides. *Aust. J. Chem.* 70:152–161.

Rateb, M. E., W. E. Houssen, M. Schumacher, W. T. A. Harrison, M. Diederich, R. Ebel, and M. Jaspars. 2009. Bioactive diterpene derivatives from the marine sponge *Spongionella* sp. *J. Nat. Prod.* 72(8):1471–1476.

Roué, N., and J. Bergman. 1999. Synthesis of the marine alkaloid leucettamine B. *Tetrahedron* 55:14729–14738.

Rueda, A., A. Losada, R. Fernandez, C. Cabanas, L. F. Garcia- Fernandez, F. Reyes, and C. Cuevas. 2006. Gracilins G-I, cytotoxic bisnorditerpenes from *Spongionella pulchella*, and the anti-adhesive properties of gracilin B. *Lett. Drug Des. Discov.* 3(10):753–760.

Sanchez, J. A., A. Alfonso, E. Leiros, E. Alonso, M. E. Rateb, M. Jaspars, W. E. Houssen, R. Ebel, and L. M. Botana. 2015. *Spongionella* secondary metabolites regulate store operated calcium entry modulating mitochondrial functioning in SH-SY5Y neuroblastoma cells. *Cell. Physiol. Biochem.* 37:779–792.

Santos, M. F. C., P. M. Harper, D. E. Williams, J. T. Mesquita, E. G. Pinto, T. A. da Costa-Silva, E. Hajdu et al. 2015. Anti-parasitic guanidine and pyrimidine alkaloids from the marine sponge *Monanchora arbuscula. J. Nat. Prod.* 78(5):1101–1112.

Schroeder, C. I., and D. J. Craik. 2012. Therapeutic potential of conopeptides. *Fut. Med. Chem.* 4(10):1243–1255.

Sharpe, I. A., E. Palant, C. I. Schroeder, D. M. Kaye, D. J. Adams, P. F. Alewood, and R. J. Lewis. 2003. Inhibition of the norepinephrine transporter by the venom peptide chi-MrIA. Site of Action, Na+ dependence, and structure-activity relationship. *J. Biol. Chem.* 278(41):40317–40322.

Sikorska, J., S. Parker-Nance, M. T. Davies-Coleman, O. B. Vining, A E. Sikora, and K. L. McPhail. 2012. Antimicrobial rubrolides from a South African species of *Synoicum* tunicate. *J. Nat. Prod.* 75:1824–1827.

Sun, D., Z. Ren, X. Zeng, Y. You, W. Pan, M. Zhou, L. Wang, and A. Xu. 2011. Structure-function relationship of conotoxin lt14a, a potential analgesic with low cytotoxicity. *Peptides* 32(2):300–305.

Tahtouh, T., J. M. Elkins, P. Filippakopoulos, M. Soundararajan, G. Burgy, E. Durieu, C. Cochet et al. Selectivity, cocrystal structures, and neuroprotective properties of leucettines, a family of protein kinase inhibitors derived from the marine sponge alkaloid leucettamine B. *J. Med. Chem.* 55:9312–9330.

Tanaka, N., T. Kusama, A. Takahashi-Nakaguchi, T. Gonoi, J. Fromont, and J. Kobayashi. 2013a. Nagelamides U–W, bromopyrrole alkaloids from a marine sponge *Agelas* sp. *Tetrahedron Lett.* 54:3794–3796.

Tanaka, N., T. Kusama, A. Takahashi-Nakaguchi, T. Gonoi, J. Fromont, and J. Kobayashi. 2013b. Nagelamides X–Z, dimeric bromopyrrole alkaloids from a marine sponge *Agelas* sp. *Org. Lett.* 15:3262–3265.

Tarcha, E. J., C. M. Olsen, P. Probst, D. Peckham, E. J. Muñoz-Elías, J. G. Kruger, and S. P. Iadonato. 2017. Safety and pharmacodynamics of dalazatide, a Kv1.3 channel inhibitor, in the treatment of plaque psoriasis: A randomized phase 1b trial. *PLoS One* 12(7):e0180762. doi:10.1371/journal.pone.0180762. eCollection 2017.

Teichert, R. W., and B. M. Olivera. 2010. Natural products and ion channel pharmacology. *Fut. Med. Chem.* 2(5):731–744.

Teichert, R. W., E. W. Schmidt, and B. M. Olivera. 2015. Constellation pharmacology: A new paradigm for drug discovery. *Ann. Rev. Pharmacol. Toxicol.* 55:573–589.

Trepos, R., G. Cervin, C. Hellio, H. Pavia, W. Stensen, K. Stensvåg, J.-S. Svendsen, T. Haug, and J. Svenson. 2014. Antifouling compounds from the sub-Arctic ascidian *Synoicum pulmonaria*: Synoxazolidinones A and C, pulmonarins A and B, and synthetic analogues. *J. Nat. Prod.* 77:2105–2113.

Upadhyay, S. K., K. L. Eckel-Mahan, M. R. Mirbolooki, I. Tjong, S. M. Griffey, G. Schmunk, A. Koehne et al. 2013. Selective Kv1.3 channel blocker as therapeutic for obesity and insulin resistance. *Proc. Natl. Acad. Sci. USA* 110(24):E2239–E2248.

Utenova, B. T., and L.-L. Gundersen. 2004. Synthesis of (+)-agelasine D from (+)-manool. *Tetrahedron Lett.* 45(22):4233–4235.

Wang, C. K., S. E. Northfield, B. Colless, S. Chaousis, I. Hamernig, R.-J. Lohman, D. S. Nielsen et al. 2014. Rational design and synthesis of an orally bioavailable peptide guided by NMR amide temperature coefficients. *Proc. Natl. Acad. Sci. USA* 111(49):17504–17509.

Wang, W., H. Kim, S.-J. Nam, B. J. Rho, and H. Kang. 2012. Antibacterial butenolides from the Korean tunicate *Pseudodistoma antinboja*. *J. Nat. Prod.* 75:2049–2054.

Watanabe, K., Y. Tsuda, M. Iwashima, and K. Iguchi. 2000. A new bioactive triene aldehyde from the marine sponge *Leucetta microraphis*. *J. Nat. Prod.* 63:258–260.

Wilson, M. C., T. Mori, C. Ruckert, A. R. Uria, M. J. Helf, K. Takada, C. Gernert et al. 2014. An environmental bacterial taxon with a large and distinct metabolic repertoire. *Nature* 506(7486):58–62.

Won, T. H., J. Jeon, S.-H. Kim, S.-H. Lee, B. J. Rho, D.-C. Oh, K.-B. Oh, and J. Shin. 2012. Brominated aromatic furanones and related esters from the ascidian *Synoicum* sp. *J. Nat. Prod.* 75:2055–2061.

Youssef, D. T. A., L. A. Shaala, G. A. Mohamed, J. M. Badr, F. H. Bamanie, and S. R. M. Ibrahim. 2014. Theonellamide G, a potent antifungal and cytotoxic bicyclic glycopeptide from the Red Sea marine sponge *Theonella swinhoei*. *Mar. Drugs* 12:1911–1923.

Yu, R., V. A. L. Seymour, G. Berecki, X. Jia, M. Akcan, D. J. Adams, Q. Kaas, and D. J. Craik. 2015. Less is more: Design of a highly stable disulfide-deleted mutant of analgesic cyclic alpha-conotoxin Vc1.1. *Sci. Rep.* 5:13264. doi:10.1038/srep13264.

2 The Status of Marine Chemical Ecology in Antarctica

Form and Function of Unique High-Latitude Chemistry

Jacqueline L. von Salm, Kathryn M. Schoenrock,
James B. McClintock, Charles D. Amsler,
and Bill J. Baker

CONTENTS

2.1   INTRODUCTION

As we move further into the twenty-first century, technological advances allow
access and discovery in difficult to reach environments to occur at a quickened
pace. This has been true for chemical ecology over the past few decades, including
in Antarctica where much of the continent and marine environment is difficult to
access. Marine chemical ecology is still behind its terrestrial counterpart in many
ways; however, the widespread use of SCUBA diving in this field has facilitated a
rapid increase in sampling effort and field research since the 1980s (Pawlik et al.
2013). In Antarctica, research has focused on the chemical interaction between
organisms, and between the organisms and their environment. This review aims
to provide an update on the progress of marine chemical ecology in the Southern
Ocean. Antarctic organisms have proven to be an interesting source of chemodiversity, both ecologically and pharmacologically, supporting the idea that high latitude
species can be well defended chemically (Avila et al. 2008, Lebar et al. 2007, Moles
et al. 2011, McClintock et al. 2010, Soldatou and Baker 2017).

2.1.1   Geography and Ecology of the Southern Ocean

The Southern Ocean is isolated from the rest of the world's oceans by the Antarctic
Circumpolar Current, which moves eastward driven by strong westerly winds
(Lüning 1990). Closer to the continent, the coastal topography and wind patterns
move water westward in the Antarctic Coastal Current. The Antarctic region
(Figure 2.1) occurs south of the Polar Front (50–60°S) and acts as a threshold for
environmental characteristics as well as species distributions. Antarctic seas are
characterized by consistently cold seawater temperatures that range from −1.8°C to
+4°C, extreme annual fluctuations in light level, and high nutrient content and dissolved oxygen levels. In addition, yearly formation of sea ice capable of doubling the
extent of the continent, high energy storm events, and large ice masses that disturb
and scour benthic communities contribute to this overall unique marine ecosystem
(Griffiths 2010). The bottom topography of the Antarctic region is characterized by
a very deep continental shelf (most >1000 m depth) and is primarily comprised of
glacial silt and high in silicates (Griffiths 2010). Though the seas are nutrient rich
(in ions such as nitrate and phosphate) they are very low in chlorophyll *a* (derived
from phytoplankton), largely due to low iron content in the water column (Korb et al.
2004). Still the best understood system in the Southern Ocean is the pelagic food
web that is rooted in phytoplankton production and subsequent secondary production of zooplankton (Hempel 1985). The benthic communities from the Antarctic are

FIGURE 2.1   Collection sites around the Antarctic continent marking location where chemical ecology studies were conducted. Subantarctic sites are not represented.

diverse, but under sampled, and much less is known about the ecological interactions in these communities than those at temperate or tropical latitudes.

The benthic marine environment is best understood in geographic regions that are en route or adjacent to research stations. These areas include the Antarctic Peninsula region as well as the Scotia, Bellingshausen, and Weddell Seas (Griffiths 2010). Communities range from deep water invertebrate aggregations to shallow peninsula habitats dominated by brown algal forests. Gutt et al. (2004) predicted that there could be as many as 17,000 species on the continental shelf alone but to date there are approximately 6,651 described marine animals (De Broyer et al. 2014, 2018) and 124 marine macroalgal (Wiencke et al. 2014) species known from the Antarctic marine environment (south of the Polar Front) based on sampling on SCUBA, using trawls or remotely operated vehicles, or identification on video (Griffiths 2010). While some species are well sampled others are represented by one individual, suggesting that there is very low resolution in our understanding of Antarctic biodiversity. Also, the application of molecular tools to differentiate between cryptic species has continually revealed new diversity in benthic communities (Griffiths 2010, Grant and Linse 2009), which may also impact the chemodiversity (Wilson et al. 2013, Engene et al. 2013) though genetic diversity and chemodiversity are not always correlated (Kremp et al. 2014).

Biological communities in the Antarctic region are regulated by yearly light cycles. With the onset of spring many macroalgae begin their annual growth season, and some even anticipate spring to maximize utilization of the yearly light hours

(Wiencke and Amsler 2012). Phytoplankton mimic this pattern and undergo seasonal blooms that deliver food not only to zooplankton in the upper photic zone, but also to benthic suspension feeders and microbial communities in the detrital food web (Griffiths 2010, Dayton 1975, Barnes et al. 1996). The distribution of communities in the benthic environment is patchy due to the complexity of the environments that the continent provides (Arntz et al. 1997) and the environmental and biological constraints on the continent (Griffiths 2010, Dayton 1975, Barnes et al. 1996). Benthic communities under sea ice are regulated by "biological accommodation" or structured by predatory and competitive interactions between organisms (Dayton et al. 1974). Benthic communities along the Antarctic Peninsula are dominated by chemically defended algal forests which structure the benthic community in the photic zone (Amsler et al. 2014), while deeper communities resemble the stable, ice-covered communities of higher latitudes (McClintock et al. 2010). All communities are susceptible to environmental conditions that could lead to their eradication (Miller and Pearse 1991, Peck et al. 1999).

Benthic Antarctic communities are devoid of certain taxa that are common worldwide, including most elasmobranch fishes and all decapod crustaceans (Aronson et al. 2007), and certain taxa are more diverse than others. For instance, the amphipod, sponge, and bryozoan taxa are rich in species while bivalve and gastropod mollusks are not (Griffiths 2010, De Broyer et al. 2018). Notably, the unique environmental conditions that shape this region are changing due to anthropogenic influences on the temperature, weather patterns, oceanic pH, and sea ice duration (Ducklow et al. 2013, Aronson et al. 2007, IPCC 2013, Clarke et al. 2007). This alters not only physical parameters of the habitats but the organisms within, as warmer temperatures and increased traffic in the Southern Ocean allow invasive species to settle on the continental shelf (Aronson et al. 2007). Specifically, the introduction of durophagous predators to the continental shelf could have a profound impact on the Antarctic benthos, which has not seen crushing predators for millions of years (Aronson et al. 2007). Also, the Antarctic region has shallow carbonate compensation depths (~300 m shelf, ~2,100–2,800 m deep; Feely et al. 2004) with low saturation states of carbonate species (Orr et al. 2005, Fabry et al. 2009). This would compound any predation stress on calcified organisms such as bivalves and coralline algae. Therefore, with the advent of changing climate conditions it is increasingly important to map biological diversity in many ways including genetic and chemical polymorphy within species.

2.1.2 Secondary Metabolites

Antarctic marine chemical ecology research began in the Ross Sea, McMurdo Sound where initial studies identified deterrent compounds in sponges, sea stars, and a pelagic pteropod (McClintock and Janssen 1990) that abducts a chemically defended amphipod as a shield against fish predation. Currently this field of research has expanded to the Antarctic Peninsula, very recently the deep sea (Moles et al. 2015), and encompasses many species including marine algae. The number of known Antarctic secondary metabolites has increased from ~200 in 2001 (Amsler et al. 2001a) to ~300 in 2008 (Avila et al. 2008), to ~600 present day (Koplovitz et al. 2011, McClintock et al. 2004).

The dramatic increase in metabolite isolation can be attributed to increased sampling effort and the technological advances mentioned previously.

The distribution of compounds within Antarctic taxa differs from patterns that characterize tropical and temperate reefs. Early work with Antarctic organisms found that algae, sponges, and echinoderms produce the majority of secondary metabolites; however, like their tropical counterparts, Antarctic soft corals and ascidians also contain unique secondary metabolites (Lebar et al. 2007, Soldatou and Baker 2017). Polar organisms contribute approximately 3% of marine natural products described as of 2008 (Lebar et al. 2007, Hu et al. 2011). In many cases it is not clear whether compounds are produced through de novo synthesis by the target organisms or whether they are acquired from microbial symbionts (Avila et al. 2008). Sequestration of active compounds from prey organisms has been shown to occur in a few Antarctic species and will be discussed later within this chapter. Both laboratory and field experiments have shown that secondary metabolite production in Antarctic marine organisms is critical in structuring marine benthic communities. What follows is a comprehensive discussion of known compounds and ecological relationships in the Antarctic benthos. Compounds are presented by taxonomic or functional group with any pharmacological uses discussed as well. Ecological interactions are discussed later, with a focus on more recent research.

2.2 ANTARCTIC MARINE SECONDARY METABOLITES AND THEIR ECOLOGICAL SIGNIFICANCE

Approximately 600 different secondary metabolites (Koplovitz et al. 2011, McClintock et al. 2004) have been isolated from Antarctic marine organisms to date, two-thirds of which were described in the 13 years since the last comprehensive Antarctic marine chemical ecology review by this group (Amsler et al. 2001a). Building on the 2001 review, recently published secondary metabolites were described from cnidarians and ascidians with new research done on Antarctic sea cucumbers (Figure 2.2). Many of the molecules described hereafter are discussed in a review by Lebar et al. (2007); however, much of that review focused on pharmacological rather than ecological significance.

2.2.1 ALGAE

Algae are a common component of the marine community in geographic regions of the Antarctic continent where light can reach the pelagic or benthic system. Many species are endemic and have a circumpolar distribution, but the highest biomass and diversity is found along the western Antarctic Peninsula (Wiencke et al. 2014). In this region, algal canopies function in providing structure for the benthic community (Clayton 1994) and mediate interactions between herbivores and their predators (Amsler et al. 2014) at biomass levels similar to that of kelp forests (Amsler et al. 2009a). Still, Antarctic macroalgae are less diverse than their temperate and tropical counterparts, though many regions of the continent are under-sampled (Wiencke et al. 2014).

There are considerably fewer known secondary metabolites in Antarctic algae when compared with tropical and temperate regions—approximately 64 without

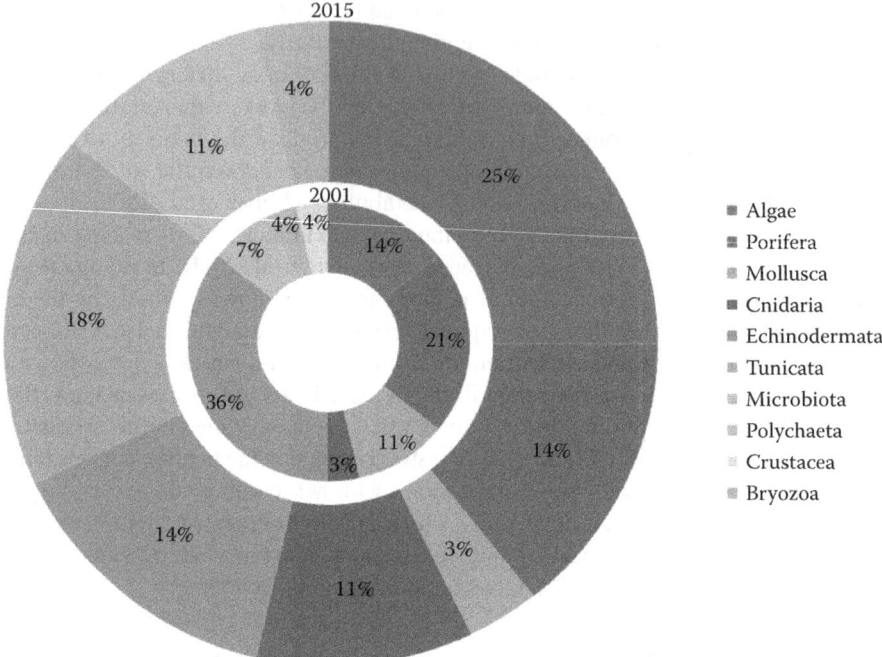

FIGURE 2.2 A comparison of the 2001 review (Amsler et al. 2001a) versus the current review (2002–2015). Shown is a percentage contribution from each group towards the total number of species investigated with ecologically relevant and identified secondary metabolites. The total number of species with full structure elucidation of published secondary metabolites in 2001 was 26 and in this review 28; however, many organisms from 2001 were further investigated in this review.

counting volatile halogenated organic compounds (VHOCs) and mycrosporine-like amino acids (MAAs) (Amsler et al. 2009a). Despite the minimal number of isolated metabolites, a large percentage of chemical extracts are deterrent to sympatric amphipod, sea star, and fish species indicating many species are chemically defended (Amsler et al. 2005a, Figuerola et al. 2012). In fact, all of the dominant brown algae and the majority of dominant red algae along the western Antarctic Peninsula have been shown to be unpalatable to these organisms (Amsler et al. 2009a). Along with anti-herbivore compounds many species elaborate compounds that defend against biofoulers and ultraviolet radiation (UVR) (Amsler et al. 2009a, McClintock and Karentz 1997), and some are utilized by higher trophic levels through sequestration (Amsler et al. 2013). Algae are a polyphyletic group of organisms and in the following we will discuss secondary metabolite production by subgroupings.

2.2.1.1 Microalgae

This group of algae encompasses ochrophytes such as the diatoms, the dinoflagellates in the phylum Alveolata, and other groups including the unicellular green algae (Chlorophyta). One unicellular green alga, *Chlamydomonas*-like strain (CCMP681), found in the intertidal zone, produces ice-binding proteins with strong recrystallization inhibition activity similar to antifreezes found in other organisms (Raymond et al. 2009).

The chemical ecology of this group is not well studied, though sea ice algae (microalgae) are at the base of the pelagic food web feeding krill, penguins, and marine mammals. Most of the compounds identified are MAAs and VHOCs; the first protect the organisms from harmful UVR while the later are halogenated alkanes that contribute to ozone depletion when broken down to C1–C2 alkyl halides (Laturnus et al. 1996). Specific species investigated to date include the Ochrophytes *Nitzchia stellata* and *Porosira pseudodenticulata* (Sturges et al. 1993). Phytoplankton such as *Phaeocystis antarctica* produce dimethyl sulfide (DMS) under stress like high levels of UVR and cold temperatures (Baumann et al. 1994, Hefu and Kirst 1997). DMS is also produced by a phyto-micro-zooplankton interaction in the Southern Ocean, which in turn has multi-trophic effects on the pelagic food web best outlined in Steinke et al. (2002).

2.2.1.2 Brown Algae (Phaeophyceae)

Brown algae are not very diverse in Antarctica, but ecologically very important in the Antarctic benthos and particularly along the western Antarctic Peninsula (Amsler and Fairhead 2006). There are 27 known species, 12 of which are endemic to the Southern Ocean, although a few large perennial species (*Desmarestia menziesii, Desmarestia anceps*, and *Himantothallus grandifolius*) usually dominate a given habitat (Wiencke et al. 2014). Interspersed throughout the canopy are other large browns *Ascoseira mirabilis*, *Cystosphaera jacquinotii*, and *Desmarestia antarctica*. All species are known to be unpalatable to common consumers such as amphipods, sea stars, and fish (Amsler et al. 2005a, Huang et al. 2006). Brown algae characteristically produce diterpenes (isoprenoids), phlorotannins (polyphenols), and acetogenins, which are rarely halogenated (Blunt et al. 2007). This group is often characterized by phlorotannins, which function in both primary and secondary metabolic roles.

Desmarestia menziesii and *D. anceps* are the most investigated of the Antarctic phaeophytes, particularly in terms of how their chemical defenses influence benthic ecology (Amsler et al, 2009a). Specifically, the unpalatability of these canopy species (Ankisetty et al. 2004b) protects mesograzers from predators, creating an associational refuge for the amphipod population in this ecosystem (Zamzow et al. 2010). Phlorotannin production varies across individuals, sites, years, and species in Antarctic phaeophytes (Amsler et al. 2008). Aside from phlorotannins, *D. menziesii* elaborates interesting plastoquinones that are bioactive against leukemia cells and inhibit mitosis of fertilized sea urchin eggs (Rivera 1996). This includes quinone derivative menzoquinone (**1**), which is active against *Staphylococcus aureus* (MRSA) and displays significant feeding deterrence against the sea star *Odontaster validus* (Ankisetty et al. 2004b). *D. anceps* can also produce 14.3 kg of bromoform a year per 10 kg of algae, which contributes to the breakdown of ozone in the atmosphere (Laturnus et al. 1996). In the first chemical investigation of *C. jacquinotii*, it was found to elaborate the steroid cystosphaerol (**2**), which resembles a compound used in female structures of plants to stimulate development of the anthers (Ankisetty et al. 2004b). *A. mirabilis* releases a burst of light-dependent reactive oxygen species (ROS) in response to wounding, which in turn affects grazing by a sympatric herbivore

(McDowell et al. 2014a, 2014b, 2016). The secondary metabolite composition of the remaining 21 species of Antarctic brown algae has not been thoroughly investigated to date.

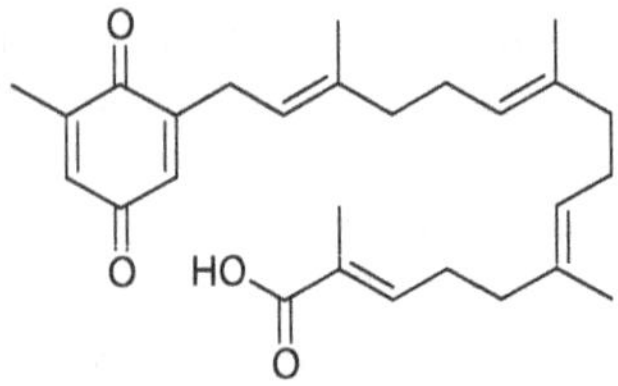

1

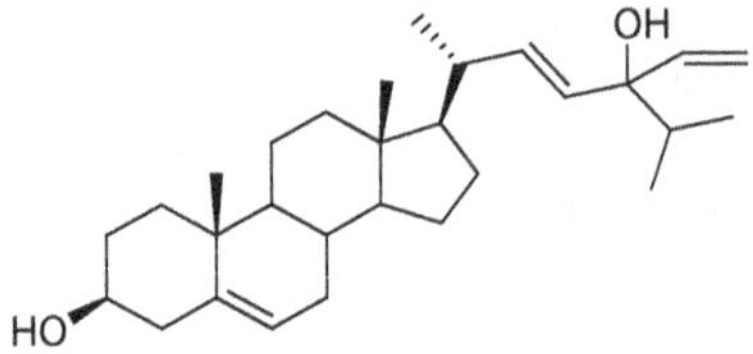

2

2.2.1.3   Red Algae (Rhodophyta)

There are 80 known species of red algae found in Antarctica (Wiencke et al. 2014). Red algae characteristically produce the largest array of secondary metabolites in all major classes excluding phlorotannins (Blunt et al. 2005), with over 90% of these compounds containing halogens (Maschek and Baker 2008). Along with VHOCs, a variety of halogenated terpenoids in Antarctic species show bioactivity (Avila et al. 2008). One of the best-studied species includes *Delisea pulchra* (cf. *fimbriata*), which has recently been shown to produce new halogenated furanones, pulchralides (3) and previously documented fimbrolides. Greatest biological activity was exhibited by the formerly reported acetoxyfimbrolide (4) and hydroxyfimbrolide (5) as antimicrobial agents (Ankisetty et al. 2004b). *Plocamium cartilagineum*, which produces a plethora of halogenated terpenes, was also found to produce an unreported halogenated monoterpene anverene (6) and the known *epi*-plocamene D (7) and pyranoid (8). Anverene and *epi*-plocamene D demonstrate antifeedant activity against the sympatric amphipod species, *Gondogeneia antarctica*, and all three structures exhibit antibiotic activity (Ankisetty et al. 2004b). The chemistry associated with these red algal species often plays unique ecological roles (Sections 2.3.1 and 2.3.5).

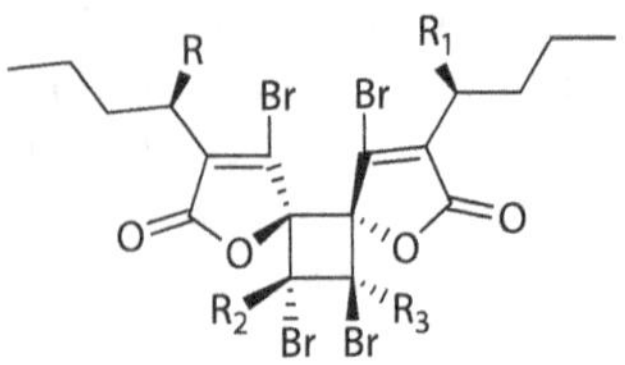

3

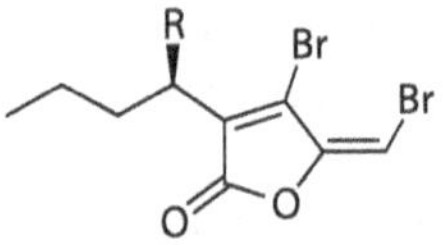

4: R = OAc
5: R = OH

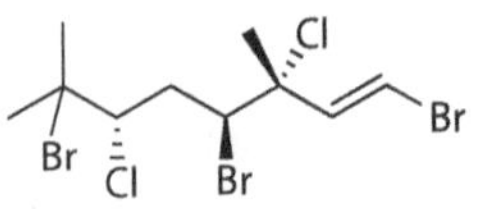

6

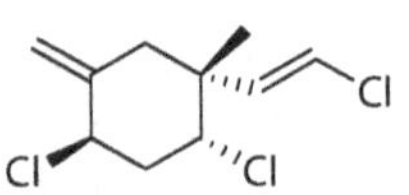

7

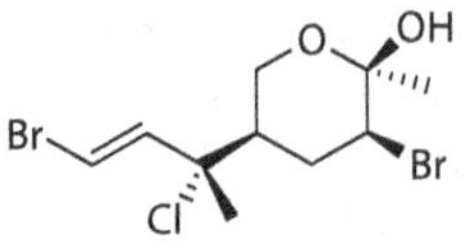

8

In feeding assays carried out by Amsler et al. (2005a), many of the common red macroalgae from the western Antarctic Peninsula were found to produce antiherbivore defenses. The two species found at the southernmost region for autotrophs,

Phyllophora antarctica and *Iridaea cordata*, also have defenses against sympatric sea urchins (Amsler et al. 1998), the latter of which may be related to cytotoxicity recently reported in its sulfated polysaccharides (Kim et al. 2017). Specific compounds, however, are known in only a few species. *Myriogramme smithii* elaborates the simple aromatics *p*-methoxyphenol (**9**) and *p*-hydroxybenzaldehyde (**10**), which are known to be a feeding deterrent in Antarctic sponges (Ankisetty et al. 2004b, Baker et al. 1995). Similar to *P. cartilagineum*, *Pantoneura plocamioides* produces a variety of terpenoids derivatives that can have antifeedant activity (Cueto et al. 1998, Argandona et al. 2002). *Audouinella purpurea*, *Delesseria lancifolia*, and *Phycodrys rubens* produce polyunsaturated fatty acids (PUFAs) that provide information on algal evolutionary history of the geographic regions and are key tools in investigating food web interactions (Graeve et al. 2002). Lastly, UV-absorbing pigments act as sunscreens in *Palmaria decipiens* (reported as *Leptosomia simplex*, Post and Larkum 1993) which also produces ROS in response to grazing, especially in high light conditions (McDowell et al. 2015, 2016). *P. decipiens* has also been previously reported to increase carotenoid content in the presence of UVR (Dohler 1998, Karentz and Bosch 2001).

9   **10**

2.2.1.4   Green Algae (Chlorophyta)

Among Antarctic algae, green algae produce the fewest and least diverse variety of secondary metabolites (Blunt et al. 2007). Most compounds in this group are terpenoids but lack the extensive halogenation found in the red algae (Maschek and Baker 2008). There are 17 known Antarctic green algal species, many of which produce VHOCs (Laturnus et al. 1996). *Ulva hookeriana* (reported as *Enteromorpha bulbosa*) has UV absorbing pigments (Post and Larkum 1993), and PUFAs in *Lambia antarctica* provide information on evolutionary history of the geographic regions and provide a substrate for investigating food web interactions (Graeve et al. 2002).

2.2.2   PORIFERA

Sponges have an integral role in the structure of Antarctic benthic communities. There are an estimated 436 species of sponges, including subspecies and undescribed species, in Antarctica, which provide nutrients and refuge for other species but also compete for space with many of the neighboring marine organisms (McClintock et al. 2005, Janussen and Tendal 2007). The presence of numerous ecological stressors has led to multiple investigations of high latitude sponge secondary metabolite production (Lebar et al. 2007, Avila et al. 2008, McClintock et al. 2005, Peters et al. 2005, Abbas et al. 2011, Turk et al. 2013). Some of the secondary metabolites responsible for chemical defense and pigmentation of sponges have been investigated among Antarctic sponge species, especially pertaining to the genera *Latrunculia*, which produces the cytotoxic discorhabdin alkaloids, *Suberites*, which is reported to produce suberitane sesterterpenes, *Isodictya*, which biosynthesizes the primary pigment eribusinone, and *Kirkpatrickia*, which elaborates the variolins pigments. Recent examinations of shallower benthic areas surrounding the western Antarctic Peninsula have shown 78% of 27 species tested had outer sponge tissue unpalatable to the predatory sea star *Odontaster validus* lending some support to optimal defense theory (ODT; Peters et al. 2009). In a follow up study, many of these same species were tested for in vitro chemical defenses against microorganisms. Ninety-six percent of the lipophilic extracts tested were found to cause mortality in the diatom *Syndroposis* sp.; however, few were shown to have antibacterial activity against ecologically relevant sympatric bacteria (Peters et al. 2010). About one third of a different suite of deeper water sponge species were found to exhibit antibacterial activity (Moles et al. 2015). The greatest levels of inhibition were detected against microbes found in Antarctic sea ice and seawater, providing evidence that microbial defense in sponges may be more obligatory than suggested by (Peters et al. 2005 and Turk et al. 2013). It is apparent from these data that although some Antarctic species of Porifera have been analyzed for antifouling and palatability (Nunez-Pons et al. 2012a, Taboada et al. 2013), many have yet to be chemically investigated to determine the compounds responsible for their ecological activity.

One such advancement involves the bright red, encrusting sponge *Crella* sp. Historically, sponges are known as prolific producers of a diverse assortment of steroids and sterols (Dauria et al. 1993, Baker and Kerr 1993), and *Crella* sp. follows this pattern by producing five oxygenated pregnane steroids, the norselic acids A–E (**11–15**). These molecules vary by levels of unsaturation in the A and B rings and oxidation of the steroid tail (Ma et al. 2009). Norselic acid A (**11**) deters predation by the sympatric amphipod *Gondogeneia antarctica* and displays broad spectrum antibiotic activity. Lipophilic crude extracts tested against both amphipod and sea star predators showed significant deterrence; however, the new steroids reported did not show deterrence against the sea star *Odontaster validus*, implying the presence of other unknown secondary metabolites responsible for this ecological activity. Continued investigations regarding the prominent demosponge *Dendrilla membranosa* have documented defensive secondary metabolites similar to the previously reported oxidized diterpenes membranolide, 9,11-dihydrogracilin, dendrillin, or dendrinolide (Baker et al. 1995, Molinski and Faulkner 1987, 1988, Fontana et al. 1997). The compounds have shown antibacterial activity and amphipod feeding deterrence; however, similar to norselic acid A, they did not inhibit feeding in sea stars (Molinski and Faulkner 1987, Baker et al. 1995). Due to the chemical instability of these molecules, methanolic artifacts have been reported in the literature and should be avoided when documenting ecological relevance; however, their bioactivity has proven intriguing in terms of their pharmacological properties (Diaz-Marrero et al. 2004, Ankisetty et al. 2004a, Maschek 2011). Recently, a diterpene scaffold isolated from *D. membranosa*, darwinolide, showed activity against methicillin resistant *Staphylococcus aureus* biofilms, which has great pharmacological implications (von Salm et al. 2016). Further ecological studies have shown this sponge to have habitat-mediated production of secondary metabolites (Section 2.3.5; Witowski 2015).

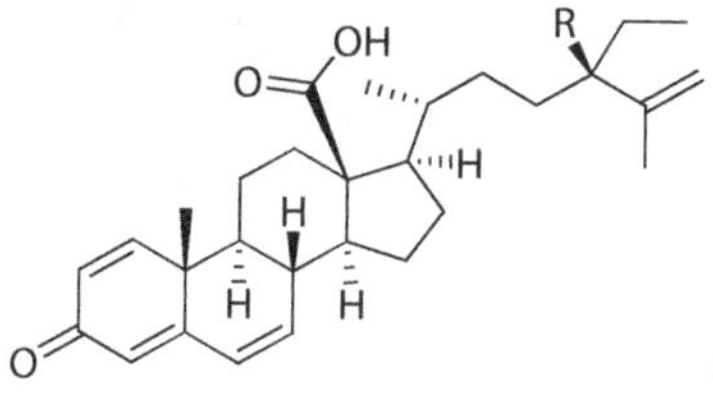

11: R = OH
15: R = OAc

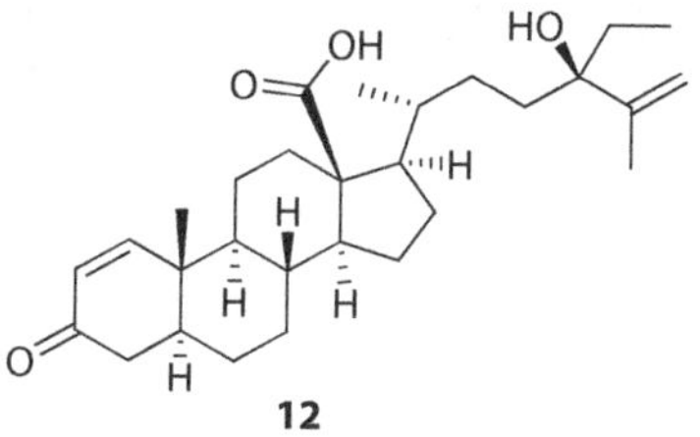

12

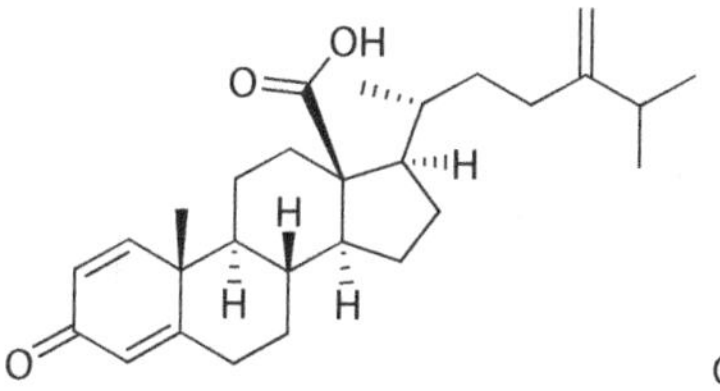

13

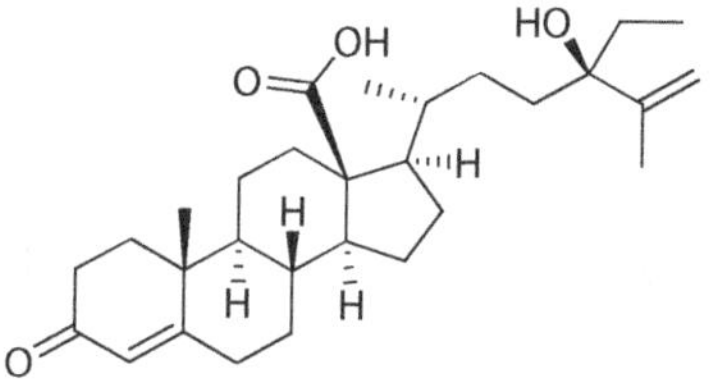

14

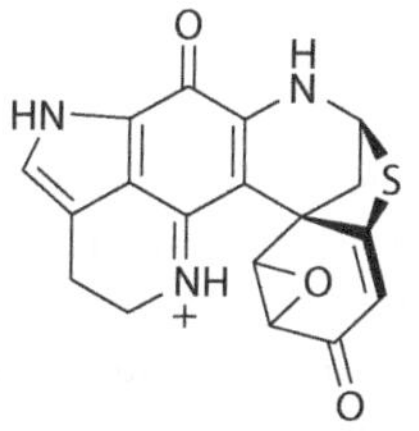

15

An antifreeze peptide has been reported from *Homaxinella balfourensis* collected from McMurdo Sound (Wilkins et al. 2002). Ice-binding proteins and peptides have been isolated from Antarctic bacteria and algae and represent chemistry unique to ecosystems with colder climates (Raymond et al. 2007, 2009, Janech et al. 2006, Raymond and Morgan-Kiss 2013, Davies 2014). The antifreeze peptide of *H. balfourensis* was partially sequenced, having a molecular weight of 2457.32; however, further analysis is needed for complete identification.

Latrunculia apicalis is commonly seen on the benthos as a dark green sponge. Extracts from the sponge were avoided by the spongivorous sea star *Perknaster fuscus*, providing initial evidence for defensive secondary metabolism (McClintock et al. 1993, Baker et al. 1994). These initial observations led to discoveries of a number of secondary metabolites produced by *L. apicalis*, most notably the highly cytotoxic discorhabdin C and other related alkaloid pigments. A new antibacterial addition to the discorhabdin family, discorhabdin R (**15**), was found from two species of the Latrunculiidae family, *Latrunculia* sp. collected near Prydz Bay, Antarctica, and Australian *Negombata* sp. (Ford and Capon 2000). While these compounds are not unique to Antarctic ecosystems, they appear to play important ecological roles in that environment (Section 2.3.4).

16   17   18

19   20   21

Analogous to other Poriferans known to produce ecologically relevant secondary metabolites, *Suberites* spp. are known for the production of unique polycyclic sester-terpenoids suberitenones A and B. Subsequent collections of *Suberites* sp. from King George Island, Antarctica resulted in the isolation of suberitenone C, D, and suberi-phenol (**16–18**; Lee et al. 2004). King George Island collections provided biomass for the further identification of sesterterpene metabolites from *Suberites caminatus*. The compounds were identified as caminatane caminatal (**19**; Diaz-Marrero et al. 2003) and suberitane-related sesterterpenes, oxaspirosuberitenone (**20**), and 19-episuberi-tenone (**21**; Diaz-Marrero et al. 2004). Ecologically focused biological assays were not performed for the majority of these new metabolites; however, suberitenones A and B are quintessential cytotoxic compounds and known to be potent sea star deterrents (Baker et al. 1997). The chemical similarities between these molecules in addition to their unique production from the genus *Suberites* has led researchers to propose their use as taxonomic markers for identification (Diaz-Marrero et al. 2004).

2.2.3 MOLLUSCA

Antarctic Molluscan secondary metabolite studies are limited to those relating to the class Gastropoda with the more recent research focused mostly on the order Nudibranchia (Davies-Coleman 2006). These shell-less marine organisms have adapted to resist pre-dation through the utilization of unique chemical defenses. A striking example is the relationship between the pteropod *Clione antarctica* and the amphipod *Hyperiella dila-tata* (McClintock and Janssen 1990). *C. antarctica* produces the polyketide secondary metabolite pteroenone de novo, a phenomenon seen in other species of Antarctic mollusks (Yoshida et al. 1995). Sequestration of secondary metabolites from their diet has also been shown in Antarctic nudibranchs such as *Tritoniella belli*. However, de novo biosynthesis is exemplified by the dorid nudibranchs *Bathydoris hodgsoni* (Avila et al. 2000), *Charcotia granulosa* (Moles et al. 2016), and *Austrodoris kerguelenensis* (Graziani et al. 1996, Iken et al. 2002), and much like *C. antarctica*, they produce defensive metabolites that deter predation from sympatric species of fish and the sea star *Odontaster validus* (McClintock et al. 2010, Davies-Coleman 2006, Amsler et al. 2001).

The secondary metabolites of *Austrodoris kerguelenensis* have been studied extensively (Davies-Coleman and Faulkner 1991, Gavagnin et al. 1995, McClintock and Baker 1997b). Diterpenoid glycerides continue to dominate the literature including a suite of natural products labeled the palmadorins (Gavagnin et al. 2003b, Diyabalanage et al. 2010, Maschek et al. 2012). Palmadorin A (**22**), M (**23**), and R (**24**) are representative of the three diterpene classes present in the nineteen palmadorins isolated thus far, clarodane, labdane, and halimane. The variability in diterpene secondary metabolism of *A. kerguelenensis* has led to multiple investigations of the biosynthetic routes responsible for these compounds and possible inter-speciation (Wilson et al. 2013, Cutignano et al. 2011). In addition to these compounds, nor-sesquiterpenes austrodoral (**25**) and austrodoric acid (**26**) were also isolated and identified as possible stress-induced secondary metabolites (Gavagnin et al. 2003a).

2.2.4 ECHINODERMATA

The Antarctic region is known for its high biodiversity of echinoderms (O'Loughlin et al. 2011). Studies of secondary metabolite chemistry associated with Antarctic echinoderms have reported steroids and their glycosides (McClintock et al. 2010, Amsler et al. 2001a), primarily from sea stars. For example, new compounds were found in *Diplasterias brucei* collected from Terra Nova Bay, Antarctica, include two new asterosaponins (**27** and **28**) and multiple known sea star steroids with promising biomedical activity (Ivanchina et al. 2006, 2011). Antimicrobial chemical defenses are produced by juveniles and embryos of the sea stars *Neosmilaster georgianus* and *Lysasterias perrierii*, although identification of the compounds responsible was not reported (McClintock et al. 2003).

Recent insight into the chemical ecology of representatives in the family Cucumariidae (Holothuroidea) collected in the South Georgia Islands shows similar saponins and other triterpene glycosides, such as the liouvillosides, achlioniceosides, and turquetoside A isolated from *Staurocucumis liouvillei*, *Achlionice violaecuspidata*, and *Staurocucumis turquetii* respectively (Maier et al. 2001, Antonov et al. 2008, 2009, 2011, Silchenko et al. 2013). Liouvillosides A2–A5 and turquetoside A have been suggested as possible chemotaxonomic markers of the *Staurocucumis* genus.

2.2.5 CNIDARIA

Within the context of the present review, the Antarctic cnidarian fauna consist largely of species within the order Alcyonacea (subclass Octocorallia). Octocorals, or soft corals, are sessile suspension feeders playing pivotal roles in the trilateral structure and ecology of benthic and deep-water communities (Roberts et al. 2006, Waller et al. 2011). Research associated with secondary metabolites of cold-water soft corals has greatly increased in the past decade, and the majority of polar specimens investigated have come from Antarctica. However, many of the compounds reported have been isolated for their biomedical potential (Lebar et al. 2007, Mellado et al. 2005). Accordingly, further chemical analysis of Antarctic, other high latitude, and deep-water soft corals, especially pertaining to the ecological relevance of secondary metabolites, is needed (Avila et al. 2008, Nunez-Pons et al. 2013, Skropeta 2008).

Similar to symbionts within the Porifera, the microbial populations associated with soft corals are attracting increased attention, including species exclusive to cold-water environments (Schottner et al. 2009). Zooxanthellae and other symbiotic microorganisms often produce secondary metabolites as part of their mutualistic relationship with their host coral (Coll 1992). As in other regions of the world, among Antarctic communities there are a variety of organisms that benefit from chemistry produced by soft corals. For example, the nudibranch *T. belli* sequesters chimyl alcohol from its prey, the stoloniferan coral *Clavularia frankliniana*, using it for its own chemical defense (McClintock et al. 1994a).

The secondary metabolite chemistry affiliated with soft corals is often similar to that of terrestrial plants, where terpene and terpenoid derivatives are dominant and mostly consist of sesqui-, di-, and triterpenes (Coll 1992). Early literature solely attributed this similarity to microscopic algal symbionts in host corals (Coll et al. 1985); however, the existence of terpene-producing octocorals that lack zooxanthellae, and further investigations of biosynthetic pathways, have proven otherwise in some species (Scheuer 1990). The cnidarian suborders Stolonifera and Holaxonia are no exception to this pattern, providing intriguing new cold-water terpene chemistry. The gorgonian coral *Ainigmaptilon antarcticus*, collected by trawl at 400 m depth in the Eastern Weddell Sea, was found to produce eudesmane sesquiterpenes ainigmaptilones A and B, with antibiotic properties as well as antifeedant activity against the sea star *O. validus* (Iken and Baker 2003).

	R_1	R_2			R_1	R_2	R_3	R_4			R_1	R_2
29 (A)	Cl	H		34 (C)	ONO_2	H	=O			39 (E)	ONO_2	H
30 (B)	ONO_2	H		35 (J)	ONO_2	OH	=O			40 (L)	Cl	OH
31 (D)	Cl	OH		36 (N)	OH	H	=O			41 (M)	ONO_2	OH
32 (G)	ONO_2	OH		37 (H)	ONO_2	H	OH	H				
33 (O)	OH	OH		38 (K)	Cl	OH	H	OH				

42 (F)   43 (I)

Initial investigations of organic extracts from the soft coral *Alcyonium paessleri* showed evidence of antifouling secondary metabolites (Slattery et al. 1995). Further collections of this octocoral near the South Georgian Islands resulted in the isolation of a suite of unique sesquiterpenoid structures. Fifteen alcyopterosin (Palermo et al. 2000) (29–43) and two paesslerin (Rodriguez Brasco et al. 2001) sesquiterpenoids that were not tested for ecological significance were isolated. The compounds revealed moderate cytotoxicity against cancer cell lines providing insight into the early reports of bioactivity (Slattery et al. 1995); however, geographical differences between the collections could complicate that relationship. The alcyopterosins represent the first illudalanes produced by a marine organism, some of which are the first nitrate esters seen in nature.

44 : R = COCH₂CH₂CH₃; R' = COCH₂CH₂CH₃
45 : R = COCH₃; R' = COCH₃
46 : R = COCH₃; R' = COCH₂CH₂CH₃

47 : R = COCH₃; R' = O-COCH₃
48 : R = COCH₃; R' = O-COCH₂CH₂CH₃
49 : R = COCH₃; R' = H
50 : R = COCH₂CH₂CH₃; R' = H
51 : R = H; R' = OH

52   **53**

The evidence for lypophilic illudalane sequiterpenes acting as ecologically relevant secondary metabolites produced by species of *Alcyonium* has been reported for *A. grandis* collected from the Weddell Sea. Nine further alcyopterosins (**44–52**) were isolated from hydrophobic extracts, all showing strong repellant activity against the sea star *Odontaster validus* (Carbone et al. 2009). Deep-water collections of dredged individuals of *Dasystenella acanthina* from Terra Nova Bay, Antarctica yielded two known chemical constituents and one new furanoeudesmane (**53**). The previously discovered isofuranodiene and the new sesquiterpene showed moderate ichthyotoxicity against the non-sympatric mosquito fish *Gambusia affinis* (Gavagnin et al. 2003c).

2.2.6 TUNICATA

As a subphylum of the Chordata, tunicates are the closest living invertebrate relatives of vertebrates (Delsuc et al. 2006). Studies of the chemical ecology of Antarctic tunicates were rare until the last decade. Benthic tunicates share a similar ecological niche as sponges on the Antarctic benthos. They are highly specialized sessile marine organisms with defensive properties limited to tough tunics, spicules, or chemical secretion. Like the Porifera, they support diverse microbial environments and have provided opportunities for studies of microbial symbioses and antifouling strategies (McClintock et al. 2004, Koplovitz et al. 2011). Chemical analyses of benthic tunicates in polar habitats have resulted in a wide variety of cytotoxic and ecologically significant secondary metabolites—some of which are, not surprisingly, antimicrobial in nature (Lebar et al. 2007, Koplovitz et al. 2011, Nunez-Pons et al. 2010, 2012b). A majority of ecologically relevant metabolites isolated from Antarctic tunicates originate from the members of the genera *Aplidium* and *Synoicum* and are distinguished by the presence of nitrogen in their structures. These and other families of tunicates also sequester rare metals including vanadium, nickel, and manganese (Lebar et al. 2011). Despite lower levels of these metals in Antarctic seas, representatives from the genera *Ascidia* and *Distaplia* that have been examined reveal a significant bioaccumulation of vanadium (Lebar et al. 2011). Although the biological necessity for tunicates to accumulate rare metals remains speculative, defensive properties of sequestered vanadium (III) and vanadium (IV) in tropical marine tunicates have been proposed (McHargue 1925, Odate and Pawlik 2007).

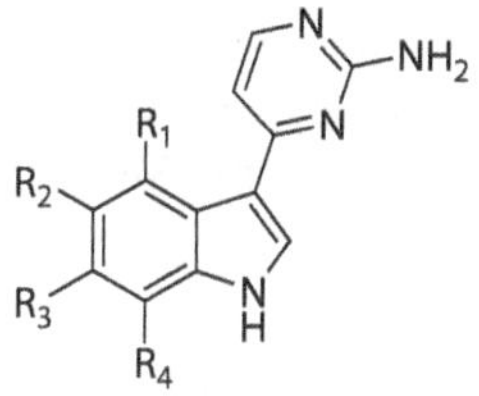

54 : R$_1$ = OH; R$_2$ = R$_3$ = R$_4$ = H
55 : R$_1$ = OH; R$_2$ = R$_4$ = H; R$_3$ = Br
56 : R$_1$ = R$_3$ = R$_4$ = H; R$_2$ = Br
57 : R$_1$ = R$_2$ = R$_4$ = H; R$_3$ = Br
58 : R$_1$ = OH; R$_2$ = R$_3$ = H; R$_4$ = Br
59 : R$_1$ = R$_4$ = H; R$_2$ = R$_3$ = Br
60 : R$_1$ = R$_2$ = R$_3$ = R$_4$ = H

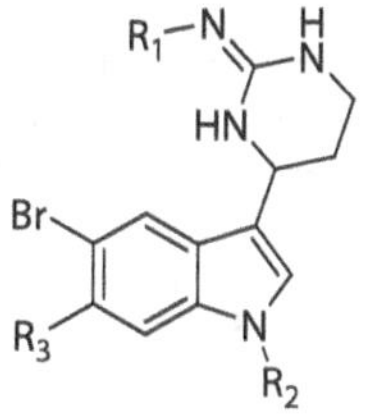

61 : R$_1$ = R$_2$ = R$_3$ = H
62 : R$_1$ = Ac; R$_2$ = R$_3$ = H
63 : R$_1$ = R$_3$ = H; R$_2$ = OCH$_3$
64 : R$_1$ = Ac; R$_2$ = OCH$_3$; R$_3$ = H
65 : R$_1$ = H; R$_2$ = OCH$_3$; R$_3$ = Br
66 : R$_1$ = Ac; R$_2$ = OCH$_3$; R$_3$ = Br

High levels of organic bioactive molecules are also produced by tunicates in the genera *Aplidium* and *Synoicum*. For example, *Aplidium meridianum* collected from South Georgia and *Synoicum* sp. collected from Anvers Island on the western Antarctic Peninsula produce potent indole alkaloids, the meridianins (Franco et al. 1998, Lebar and Baker 2010). These high nitrogen-content secondary metabolites are similar to variolin B (minus the heterocyclic ring), the defensive metabolite produced by the Antarctic sponge *Kirkpatrickia variolosa*. Meridianins A–G (**54–60**) have been assayed for their antifeedant properties toward the sea star *Odontaster validus* and amphipod *Cheirimedon femoratus*, in addition to antifouling properties against a sympatric bacterium (Nunez-Pons et al. 2012b). Other structurally related compounds, the aplicyanins A–F (**61–66**), have been isolated from lipophilic extracts of the congeneric tunicate *Aplidium cyaneum*. These brominated indole derivatives are considered to be biosynthetic precursors of the meridianins (Reyes et al. 2008).

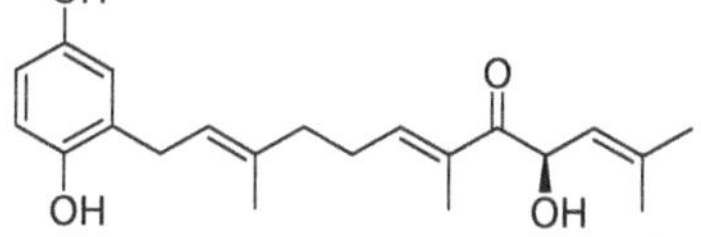

67

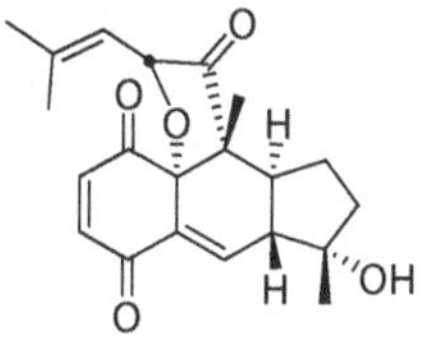

68

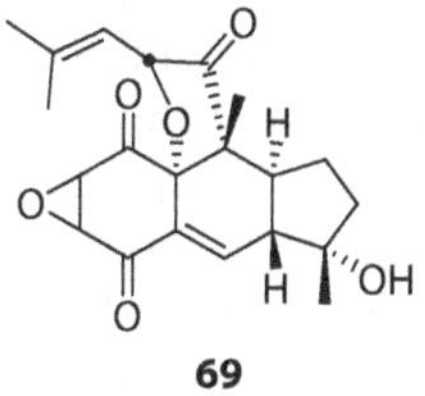

69

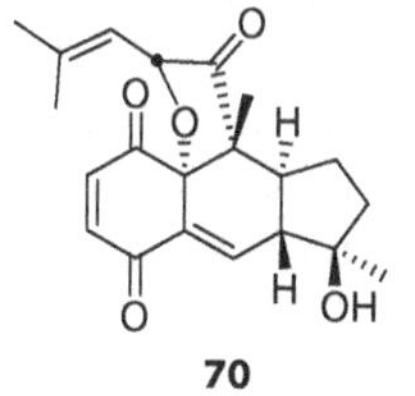

70

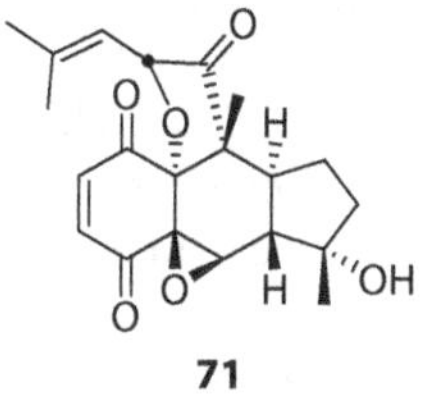

71

Non-nitrogenous meroterpenoids have also been reported from *Aplidium* spp. Rossinone A (**67**), B (**68**), and related terpenes (**69–71**) were isolated from *Aplidium* sp. and *Aplidium fuegiense* collected by dredging in the Ross and Weddell Seas, Antarctica (Appleton et al. 2009, Carbone et al. 2012). The ecological implications of these compounds do not appear to ascribe to the typical radical scavenging mechanisms of other meroterpenes, but widespread antimicrobial activity among this compound class may imply a role in tunicate-microbial interactions (Menna et al. 2013).

72 : R$_1$ = α-OH; R$_2$ = H
73 : R$_1$ = β-OH; R$_2$ = H
74 : R$_1$ = α-OH; R$_2$ = OH
75 : R$_1$ = β-OH; R$_2$ = OH

76

Representative of the family Polyclinidae, *Synoicum adareanum* produces a suite of unique secondary metabolites, the first of which were described as the enamide-bearing macrolides, the palmerolides (Diyabalanage et al. 2006, Noguez et al. 2011). Isolated in abundance from hydrophobic extracts of individuals collected by scuba near Palmer Station, Antarctica, these polyketides contain carbamate functionality and exhibit potent anticancer cytotoxicity. Although the bioactivity associated with these compounds is intriguing, the palmerolides have yet to be assigned ecological functionality. Subsequent studies of *S. adareanum* yielded new ecdysteroid feeding deterrents that may short-circuit molting in sympatric amphipods. Ecdysone-related steroids are known molting hormones in insects that have been widely isolated from terrestrial plants and algae and are known to possess insecticidal properties (Adler and Grebenok 1999). These compounds, hyousterones A–D (**72–75**) and abeohyousterone (**76**), represent the first report of ecdysteroids isolated from a tunicate. Hyousterones B and D display the rare 14β-hydroxy functionality (Miyata et al. 2007). The authors suggested the compounds may serve as deterrents to amphipod predation due to their structural similarities to phytoecdysteroids and the known amphipod predation pressure in Antarctic coastal ecosystems (Huang et al. 2006).

2.2.7 Microbiota

The first documentation of microbiota occurring in Antarctica was made by Johan Ekelöf in 1908 based on observations from the 1901 Swedish Antarctic Expedition (Ekelöf 1908). The field of Antarctic microbiology has progressed throughout the 1900s; however, even by the beginning of the twenty-first century, comparatively few studies had been carried out on marine microorganisms and their secondary metabolite chemistry. Recent technological advances in genetic sequencing and manipulation, as well as enhanced culturing techniques, have facilitated a dramatic expansion in the analyses of the marine microbiota (Schmidt 2008, Uria and Piel 2009). Marine microorganisms have been shown to produce unique chemistry involved in chemical defense, quorum sensing, biofilms, and other uses (Paul et al. 2011, Puglisi et al. 2014, Debbab et al. 2012, Engel et al. 2002); however, cold-water species have largely been cultivated for the purposes of drug discovery (Lebar et al. 2007, Su et al. 2013, Wu et al. 2012, 2013, Cutignano et al. 2013, Ren et al. 2009, Riesenfeld et al. 2008). Interests in symbiotic bacteria and fungi have expanded as well, and marine benthic invertebrates harboring diverse microorganisms such as sponges and tunicates may owe their chemical defense properties to many of these microscopic symbiotic microbes (Dubilier et al. 2008, Hochmuth and Piel 2009, Piel 2009, Balskus 2014). To date, studies have revealed that up to 50% of benthic filter feeder biomass can be attributed to symbiotic or pathogenic microbial organisms, and Antarctic invertebrates are no exception (Wang 2006, Xin et al. 2011). Although reports of secondary metabolites isolated from Antarctic host-associated marine microbes are rare, this is a rapidly expanding field in tropical oceans and a promising field of discovery in high-latitude seas.

Investigations of free-living Antarctic marine microorganisms have resulted in the discovery of a number of new secondary metabolites; however, few of these compounds have been investigated for their ecological significance. Of the few that have been investigated representative diketopiperazines, peptides, phenazine alkaloids, and diterpenoids (Avila et al. 2008, Amsler et al. 2001a) have been identified as extracellular metabolites used in quorum sensing or radical scavenging. Investigations of the Antarctic gram-negative psychrophile *Pseudoalteromonas haloplanktis* yielded two new diketopiperazines and two new linear oligopeptides (Mitova et al. 2005). This bacterium was isolated from seawater near the Antarctic station Dumont d'Urville, Antarctica, and was found additionally to produce a linear tetrapeptide Tyr-Val-Pro-Leu (**77**) with DPPH free-radical scavenging activity.

77   **78**

A strain of *Pseudoalteromonas* sp. (S-15-13) isolated from Antarctic sea ice and sediment samples produces an extracellular polysaccharide that has been identified to be a homopolymer of mannose (Li et al. 2006). Exopolysaccharides presumably benefit marine bacteria by offering physiological and chemical benefits in extreme environments where there are continuous freeze-thaw cycles (Nichols et al. 2005). A collection of sediments from shallow waters near Livingston Island, South Shetland Islands, yielded an isolate of the bacterium *Streptomyces* sp. that naturally produces a variety of natural products including one new metabolite (Ivanova et al. 2001), 2-amino-9,13-dimethyl heptadecanoic acid (**78**). All compounds displayed antimicrobial properties, suggesting a potential role in fouling inhibition or other chemical defenses.

79

2.2.8 OTHER ORGANISMS

New secondary metabolites have not been reported since 2001 from the phyla Annelida and Nemertea, nor the subphylum Crustacea (Figure 2.2); however, the phylum Bryozoa has contributed one new report, the discovery of tambjamine A (**79**) from the Antarctic bryozoan *Bugula longissimi* (Lebar et al. 2007). This alkaloid and its derivatives have been previously reported in tropical bryozoans as well as their nudibranch predators (Carte and Faulkner 1983, Blackman and Li 1994, Carbone et al. 2010). Tambjamine A serves as a defensive metabolite when in high concentration and may also serve as an attractant in nudibranch mucus trails. These are also the compounds responsible for the characteristic blue coloration of the bryozoan.

2.3 ECOLOGICAL RELEVANCE OF CHEMICAL INTERACTIONS

Chemical interactions between organisms, or secondary metabolite production, are derived from interactions between organism(s) and their environment, often structuring ecological interactions in specific habitats. As illustrated previously, Antarctic marine chemical ecology encompasses a wide variety of compounds that marine organisms exploit for a range of functions within their environment. Many marine organisms, ranging from microbiota to tunicates, produce fouling or feeding deterrents, and some capitalize on metabolite production of other organisms by associating with or sequestering bioactive compounds within their tissues. The roles of these secondary metabolites vary across organism and compound class. Some species simply produce metabolites that make the organism unpalatable to common predators, while others influence predator fitness (e.g., regulation of crustacean molting; Moon et al. 2000). Fouling deterrence is especially important in sessile species because space and resources are at a premium for Antarctic marine invertebrates and algae. Marine organisms often have chemical or mechanical modes of removing fouling organisms (Steinberg et al. 1997) to prevent infection or better access basic resources such as light and nutrients. Few studies to date have examined the prospective chemosensory (pheromones) or allelopathic functions of Antarctic compounds, though recent studies have revealed that Antarctic marine invertebrates display chemo-attraction to fish blood (Kidawa et al. 2008), that amphipods seek refuge in macroalgae in the presence of fish cues (Zamzow et al. 2010), or that extracts of host macroalga chemotactically attract an obligate filamentous algal endo/epiphyte (Bucolo et al. 2012). In another study, the ciliate *Euplotes focardii* produces allelopathic compounds that have repellent activity toward a sympatric ciliate competitor (Guella et al. 1996). The major functions of marine secondary metabolite chemistry in Antarctica are described in the next section.

2.3.1 FEEDING DETERRENTS

In contrast to tropical reefs that are dominated by fish predators, Antarctic benthic communities are subject to predation by omnivorous invertebrates including sea stars, nudibranchs, and amphipods (Amsler et al. 2001b). Model omnivorous consumers used in chemical feeding deterrent assays typically include the sea star *Odontaster validus* (Figure 2.3) the fish *Notothenia coriiceps,* and the amphipod *Gondogeneia antarctica.* Palatability studies found that the fresh tissue and organic extracts from 35 common species of macroalgae from the western Antarctic Peninsula are unpalatable to sea stars and fish (Amsler et al. 2005a) while many are also unpalatable to amphipods (Aumack et al. 2010, Huang et al. 2006). Many species of sponges from the western peninsula are also unpalatable to these model predators (Peters et al. 2005, Nunez-Pons et al. 2012a); and specific compounds have been isolated from *Crella* sp. (Ma et al. 2009), *Dendrilla membranosa* (Ankisetty et al. 2004a), *Suberites* sp. (Lebar et al. 2007), *Isodictya erinacea* (Figure 2.6; Moon et al. 2000), *Leucetta leptoraphis* (Amsler et al. 2001a), and *Latrunculia apicalis* (Furrow et al. 2003). Other taxa with feeding or predator deterrent compounds include mollusks (*Austrodoris kerguelenensis*; Figure 2.5), *Bathydoris hodgsoni, Tritoniella belli,* and *Marseniopsis mollis* (Amsler et al. 2001a), cnidarians (*Clavularia frankliniana*; Figure 2.4), *Ainigmaptilon antarcticus, Dasystenella acanthina,* and *Alcyonium paessleri*; Avila et al. 2008), brachiopods (*Liothyrella uva*; Mahon et al. 2003), and tunicates (*Cnemidocarpa verrucosa* and symbionts, and *Aplidium* spp.; Koplovitz et al. 2011, Nunez-Pons et al. 2010, Amsler et al. 2001b). Many of the compounds described earlier in this review are employed against predation (Taboada et al. 2013, Aumack et al. 2010), but they are outliers. For most unpalatable species, the compounds responsible for deterrence are still unknown.

In some Antarctic marine invertebrates, secondary metabolites are localized to certain tissues in order to optimize defense in those regions most vulnerable to

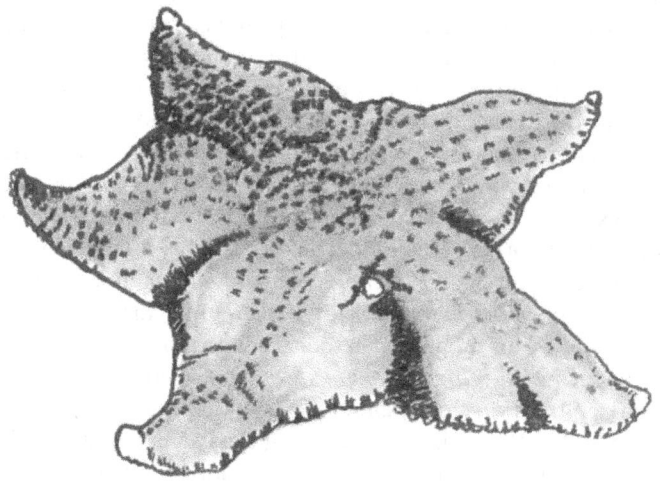

FIGURE 2.3 The omnivorous predatory sea star *Odontaster validus.*

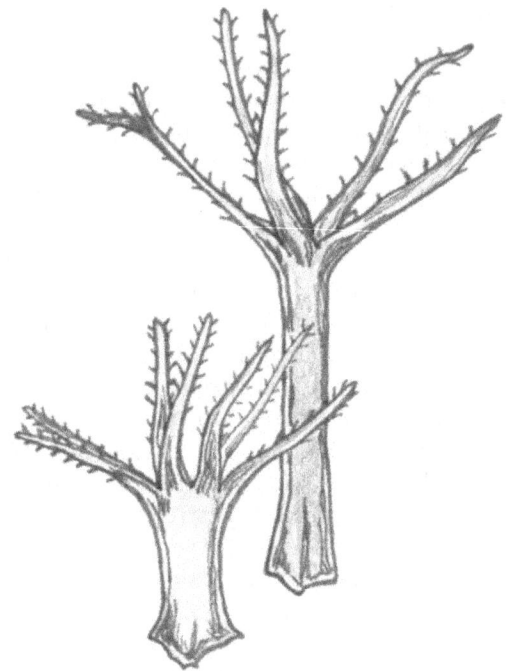

FIGURE 2.4   Polyps of stoloniferan coral *Clavularia frankliniana*.

predators. For example, in the nudibranchs *Bathydoris hodgsoni* and *Austrodoris kerguelenensis*, hodgsonal and mono-acylglycerides, respectively, are localized in the mantle tissues (the mantle is the outermost body tissue in nudibranchs) where prospective fish predators will first bite (Amsler et al. 2001a, Davies-Coleman 2006). Furthermore, in some sponges (e.g., *Suberites* sp. and *Latrunculia apicalis*), feeding deterrent natural products occur at their highest concentration at the sponge surface (pinacoderm) where the cardiac stomach of a prospective sea star predator is extruded during feeding (Furrow et al. 2003, Baker et al. 1997).

Examples of feeding deterrents sequestered from the diet include the aforementioned *Tritoniella belli*, which feeds on *Clavularia frankliniana* (Section 2.2.5; Figure 2.4), then sequesters chemical defenses into its own tissues as well as its egg masses to protect its offspring (McClintock and Baker 1997a). Recent investigations into the Antarctic red macroalga *Plocamium cartilagineum* and its primary grazer, the amphipod *Paradexamine fissicauda* (Figure 2.7) indicate similar sequestration of chemical deterrents. This amphipod associates with, and seeks refuge within, the filamentous branches of *P. cartilagineum* in order to avoid fish predation, while at the same time consuming this chemically defended host alga and sequestering the defensive compounds for its own defense (Amsler et al. 2013). Studies have demonstrated that the sequestered halogenated monoterpenes within the tissues of *P. fissicauda* deter predation by a sympatric omnivorous fish, *Notothenia coriiceps* (Amsler et al. 2013).

Associational chemical defenses also occur in the shell-less pteropod *Clione antarctica*, which is captured and carried by the hyperiid amphipod *Hyperiella dilatata* as a mechanism of avoiding fish predation (McClintock and Janssen 1990). In addition, behavioral chemical sequestration is seen in the urchin *Sterechinus neumayeri*, which utilizes the chemically defended red macroalgae *Iridaea cordata* and *Phyllophora antarctica* as physical cover (pieces of macroalgae are held by the tube feet to cover the test in a covering response) against one of its major predators, the common sea anemone *Isotealia antarctica* (Amsler et al. 1999). Although the defensive compounds in both macroalgae do not increase the effectiveness of the algae as a physical defense against the sea anemone, they do prevent the sea urchins from consuming their defensive cover (Amsler et al. 1998, 1999).

2.3.2 FOULING DETERRENTS

Fouling inhibition is important for marine benthic macroalgae and invertebrates with a sessile life history. Fouling organisms in Antarctica include microbiota such as bacteria and microalgae, as well as filamentous algae, and small (e.g., serpulid worms) and large (e.g., soft corals and sponges) invertebrates. Not only does fouling impede host access to essential resources such as light or nutrients, but fouling organisms can increase drag on the host organism or increase attractiveness to consumers (Karentz and Bosch 2001, Witman and Suchanek 1984, Anderson and Martone 2014). Sessile marine organisms have the potential to utilize two antifouling techniques: mechanical and chemical. Mechanical defenses include the production of mucus or the sloughing of outer tissue layers (Steinberg et al. 2002). Effective chemical fouling inhibitors are sequestered at the surfaces of organisms and may be either lipophilic or hydrophilic compounds that repel the settlement of early life history stages or deter the further proliferation of once-established fouling organisms. Potential antifouling compounds and crude organic extracts that deter fouling in in vitro assays have been detected in a variety of Antarctic marine organisms (Turk et al. 2013, Silchenko et al. 2013, Amsler et al. 1999, Witman and Suchanek 1984, Anderson and Martone 2014, McClintock et al. 1994b); however, more recently, in situ field experiments have provided ecologically relevant evidence for fouling deterrence against microorganisms (Angulo-Preckler et al. 2015).

Specifically, two common macroalgal species in the Ross Sea and 22 species from the western Antarctic Peninsula have been found to have either lipophilic or hydrophilic extracts that are toxic to sympatric diatoms in in vitro assays (Amsler et al. 2005b). Phlorotannins from two common species of Desmarestiales in addition to *Cystosphaera jacquinotii* were found to possess in vitro antibacterial activity, and are also toxic to sympatric diatoms; however, this fouling deterrent activity cannot be generalized to all phlorotannins that occur in Antarctic phaeophytes (Iken et al. 2009). One additional study revealed that the spores of a host-dependent epiphyte, *Elachista antarctica*, were repelled by extracts of sympatric non-host macroalgae (Bucolo et al. 2012).

Among Antarctic sponges, 60% of lipophilic and hydrophilic extracts of 25 species examined were found to have toxicity against sympatric diatoms (Peters et al. 2010). A similar pattern was detected in extracts from the colonial tunicate *Distaplia cylindrica* in in vitro assays (McClintock et al. 2004). In contrast, most sponges and tunicates from the western Antarctic Peninsula have revealed little or no antibacterial activity against sympatric species (McClintock et al. 2010). Moderate antibacterial activity has been detected in some sponges from McMurdo Sound as well as bioactivity against the fungus *Candida tropicalis* (McClintock and Gauthier 1992). Norselic acid A (**11**) isolated from the sponge *Crella* sp. revealed broad antimicrobial activity (Ma et al. 2009). In contrast, extracts from discrete tissues of the brachiopod *Liothyrella uva* had potent antimicrobial activity, particularly those extracts prepared from tissues of the lophophore and gut (Mahon et al. 2003). The cnidarian *Ainigmaptilon antarcticus* was found to produce a sesquiterpene that shows strong antibacterial activity (Iken and Baker 2003). To summarize, fouling deterrents are exhibited in many of the organisms that make up the Antarctic marine community and future investigations are likely to reveal more metabolites with antibiotic properties. It remains uncertain whether symbiont or host is responsible for the production of active metabolites in some species (Konig et al. 2006, McFall-Ngai et al. 2013, Lopanik 2014). Sessile marine organisms live in the presence of the marine microbiome, and some species may permit or even promote bacterial colonization of their surfaces. Such investigations may prove fruitful in the future.

2.3.3 PATTERNS OF CHEMICAL DEFENSE

Marine organisms exploit a variety of strategies to ensure defense against predation, fouling, and overgrowth including the differential allocation of chemical resources to vulnerable tissues (Carbone et al. 2013). Theories of defense allocation are often predicated on the assumption that there is a significant metabolic cost to defense that is likely to detract from investments in to other metabolic processes. One such theory is the ODT (Rhoades 1979), and a variety of studies of chemical defenses in benthic marine invertebrate species in Antarctica have evaluated the allocation of secondary metabolites in this context. Not all species investigated have provided support for the ODT, but enough have done so to lend support to the value of ODT for framing investigations. For example, the dominant canopy-forming macroalgae *Desmarestia menziesii* and *D. anceps* have higher concentrations of defensive secondary metabolites in regions with less mechanical protection (Fairhead et al. 2005), though this may be specific to individual, year, or geographic location (Amsler et al. 2009a). Sessile marine invertebrates can sequester feeding deterrents at their surface (e.g., sponges; Furrow et al. 2003), which is their first line of defense against predators. In contrast, some colonial tunicates produce the defensive secondary metabolite meridianin (**54–60**; Franco et al. 1998) and distribute it throughout the tunic and inner tissues with no specific locational allocation (Nunez-Pons et al. 2010), similar to hexactinellid glass sponges (Nunez-Pons et al. 2012a). The absence of chemical deterrent localization in glass sponges is hypothesized to be a result of predation pressure coming largely from small mesograzers that have equal access to inner as much as outer tissues (Nunez-Pons et al. 2012a).

FIGURE 2.5   The dorid nudibranch *Austrodoris kerguelenensis.*

Other prospective mechanisms of chemical defense in Antarctic marine organisms include induced defenses, which to date have only been demonstrated to occur in macroalgae from the western Antarctic Peninsula. In this instance, innate immune response ROS is produced upon wounding in light (McDowell et al. 2014a). The production of phlorotannins found in brown macroalgae can be induced in algae from other biogeographic regions, but apparently not in Antarctic species (Fairhead et al. 2006). In addition to macroalgae, austrodoral and austrodoric acid in the common Antarctic nudibranch *Austrodoris kerguelenensis* (Figure 2.5) may actually be stress-induced metabolites induced by environmental conditions rather than a response to predation (Gavagnin et al. 2003a).

Another defense mechanism documented in Antarctica is the behavioral association with chemically defended species. For example, in coastal waters along the western Antarctic Peninsula mesograzers (amphipods) affiliate with chemically defended macroalgae as a refuge from fish predation (Zamzow et al. 2010). This behavioral interaction plays an important role in structuring the benthic community as unprecedented high-density populations of mesograzers find refuge in chemically defended macroalgae while grazing potentially filamentous algal epiphytes that are potentially detrimental to their macroalgal hosts (Amsler et al. 2014). Mesograzers are also known to seek refuge within chemically defended invertebrates, for example the sponge *Dendrilla membranosa*, where they also may consume sponge tissue (Amsler et al. 2009b).

Although other defense theories exist (e.g., Resource-Allocation Model, Growth-Differentiation Balance Model, etc.) few have been used to frame studies to the extent of the ODT in Antarctica. Because the Southern Ocean is not nutrient limited as are many other oceans (Holm-Hansen 1985), Amsler et al. examined the production of nitrogen-based compounds in macroalgae from the WAP in the context of the Carbon-Nutrient Balance hypothesis. The basic assumption under this model was that the suite of Antarctic macroalgae examined would be found to be replete in nitrogenous compounds. Upon analysis, they were not nitrogen replete; nitrogenous secondary metabolites were just as rare as they are in macroalgae from temperate and tropical latitudes (Peters et al. 2005).

2.3.4   THE ROLE OF PIGMENTATION

Many benthic marine species inhabiting the Southern Ocean encircling the Antarctic continent possess chemical compounds that protect them from UVR in incident light. MAAs have been detected in Antarctic red algae, porifera, cnidarians, nemerteans, mollusks, bryozoans, arthropods, echinoderms, tunicates, and fish (McClintock and Karentz 1997), while green algae have been found to produce UV-protective pigments including carotenoids (Post and Larkum 1993, Karentz and Bosch 2001). Despite their known role as UV-absorbing compounds, the function of MAAs as sunscreens has never been experimentally demonstrated in Antarctica, and their presence may be vestigial. Brown algae also have photo-protective phlorotannins (Iken et al. 1997, 1999, 2007, 2009, Rivera L 1996, Amsler et al. 2001a, Fairhead et al. 2005) that have been shown experimentally to increase in concentration with increased exposure to UVR, although this pattern is not consistent across all common brown canopy species for the western Antarctic Peninsula (Fairhead et al. 2006, Rautenberger et al. 2013). A recent study found an increase in phlorotannins and radical scavenging in the brown alga *Desmarestia anceps* that may protect photosynthetic efficiency at high temperature, but not high UVR environments (Flores-Molina et al. 2016).

The colorful pigments found in many species of sessile Antarctic marine invertebrates are unlikely to play any role in capturing light energy or warding off visual predators (e.g., fish, sea turtles) as those in marine organisms that occur in tropical and terrestrial marine environments. However, the bioactivity of these pigments seems to serve a different purpose. For example, pigments from some Antarctic sponges deter predation by sea stars or inhibit biofouling (Amsler et al. 2001b). Specific examples include a purple pigment in *Kirkpatrickia variolosa* (Baker et al. 1994, Jayatilake et al. 1995), a quinolone pigment in *Dendrilla membranosa* (Molinski and Faulkner 1988), discorhabdins from *Latruncula apicalis* (Yang et al. 1995), eribusinone from *Isodictya erinacea* (Figure 2.6; Moon et al. 2000, Vankayala et al. 2017) and suberitenones from *Suberites* sp. (Amsler et al. 2001a, Baker et al. 1997). Notably, these pigments occur in representatives of the Porifera.

2.3.5   HABITAT SPECIFICITY AND SEQUESTRATION

The continent of Antarctica separated from South America 50 to 35 mya, followed by seafloor spreading that widened the Drake Passage and led to a large cooling event that formed Antarctic ice sheets (Livermore et al. 2005). Tandem analyses of genetics and secondary metabolite production have shown that divergent phylogroups parallel divergent metabolomes (Young et al. 2013). This partnership between chemo- and phylo-group provides insights into how marine organisms interact with their environment and their own evolution. Organisms such as the common dorid nudibranch *Austrodoris kerguelenensis* (Figure 2.5), which produces the palmadorins (Section 2.2.3), provide evidence of cryptic speciation events driven by glaciation or predation on the Antarctic Peninsula (Wilson et al. 2013). Similar metabolomic analyses carried out with individuals of the yellow demosponge *Dendrilla membranosa* reveal the production of habitat-specific secondary metabolites, where sponges that live below the algal canopy elaborate a more diverse array of secondary

FIGURE 2.6   The sponge *Isodictya erinacea*, known to produce the pigment eribusinone that may short circuit molting in certain mesocrustacea.

metabolites than their counterparts in deeper algal-free benthos (Witowski 2015). Additional putative defenses likely driven by geographical constraints have recently been observed in representatives of Antarctic bryozoans (Figuerola et al. 2013).

Among marine algae, the red alga *Plocamium cartilagineum* is known for producing a wide variety of halogenated monoterpenes (Section 2.2.1) and shows a distinct site specificity in its production of these compounds across a geographic scale of a few kilometers or less (Young et al. 2013). Gene sequencing and analysis of the halogenated metabolome verified site specificity on the WAP indicates, that much like *A. kerguelenensis*, this intraspecific diversity of chemistry is likely a result of episodic glaciation and qualitative or quantitative changes in predation pressure (Young et al. 2013). Furthermore, the amphipod *Paradexamine fissicauda* (Figure 2.7) sequesters halogenated compounds from *P. cartilagineum* and uses them for its own chemical defense from fish predators (Amsler et al. 2013). This remarkable co-evolution between two species and other similar interactions observed in Antarctica and yet to be discovered (McClintock and Janssen 1990, McClintock et al. 1994a) will provide further insights into the unique chemistry associated with this extreme ecosystem.

2.3.6   Climate Change Impacts

While few studies worldwide have examined the effects of climate change on the secondary metabolite chemistry of marine organisms, it is clear that environmental shifts associated with the anthropogenic production of atmospheric carbon dioxide are already affecting biological processes that range from genetic expression to ecosystem function (Brierley and Kingsford 2009). The western portion of the

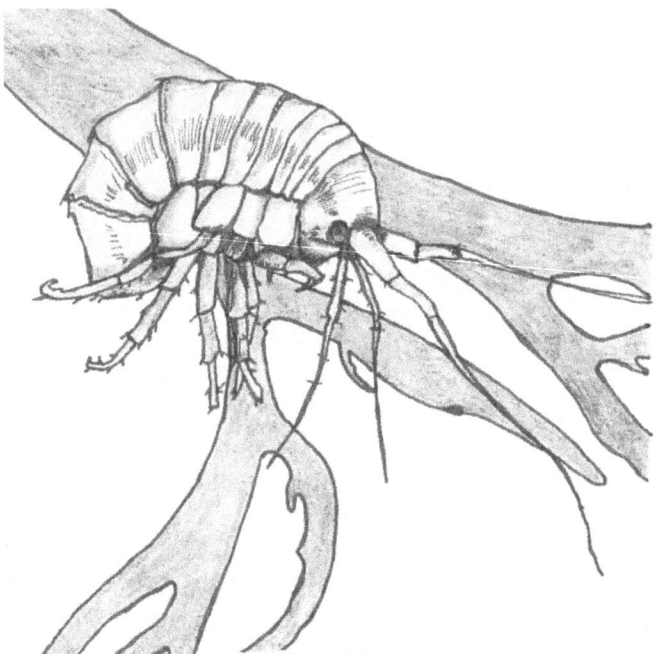

FIGURE 2.7   The amphipod *Paradexamine fissicauda* perched on the red alga *Plocamium cartilagineum*.

Antarctic continent and in particular, the WAP, is experiencing rapid changes in climate that encompass both increases in air and sea temperature (Vaughan et al. 2003, Meredith and King 2005, Steig et al. 2009), the melting of large ice forma- tions and early breakup of sea ice (Ducklow et al. 2013), high UVR and a reduced ozone layer (from chlorofluorocarbons; McKenzie et al. 2011), increases in primary productivity and habitat availability (Quartino et al. 2013, Moreau et al. 2015), and more (IPCC 2013). Importantly, it is now well established that environments can influence gene expression through epigenetic modification (Jaenisch and Bird 2003), and that climate change can alter regulation of genetic information (Hoffmann and Willi 2008).

Few studies have characterized climate change effects on secondary metabolite production. The most in-depth study examined the Australasian red alga *Delisea pulchra* in a region where a rapidly warming ocean was linked to a reduction in anti- bacterial compounds, which concurrently caused bleaching (Campbell et al. 2011). In Antarctica, the common brown alga *Desmarestia menziesii* was found to increase production of phlorotannins at a lowered seawater pH based on near future predic- tions of ocean acidification (Schoenrock et al. 2015). This amplification could be the result of either increased demand for a primary or secondary metabolic function. Amphipods living in the species assemblage associated with WAP macroalgal com- munities were found to shift their feeding preferences at increased seawater tem- peratures (+2°C), a pattern that could significantly impact community structure in the near future (Schram et al. 2015). Given that Antarctica is a model for the study of rapid global climate change, there is considerable latitude to expand studies of the potential impacts of rising seawater temperature and ocean acidification on secondary metabolites and their diverse roles in marine ecology.

2.4 FUTURE DIRECTIONS

High-latitude regions are rapidly changing environments regardless of added effects due to climate change. The limited scope of the present review is indicative that there remains still much to learn about the chemical ecology of Antarctic marine organisms. In addition to a geographic remoteness that limits access, the physical and logistical constraints and challenges of carrying out collections, and in particular, in situ manipulative experimentation, has historically constrained the field of Antarctic marine chemical ecology. Nevertheless, field studies are becoming more common in Antarctic seas (Angulo-Preckler et al. 2015, Amsler et al. 2012, Schoenrock et al. 2013), and technological advances now allow for more rapid and informative analyses of chemical structures (Bingol et al. 2015). Combined efforts of metabolic and genetic analysis using metabolomics (Young et al. 2013, Viant and Sommer 2013), proteomics (Williams et al. 2012), and genomics (Gianoulis et al. 2009, Patel et al. 2010, Riesgo et al. 2015), will expand our understanding of chemically-mediated interactions and co-evolution of species that deeply influence Antarctic marine ecosystems (Desjardins 2008, Raguso et al. 2015). In this sense, there are areas ripe for investigation. For example, studies to date on the production and role of secondary metabolites among Antarctic symbiotic microorganisms are few and could very well provide a major new research area. In summary, the rich biodiversity of Antarctic marine organisms that comprise communities largely under biotic selective pressures promises a wealth of chemical diversity of considerable ecological and biomedical interest (Nunez-Pons and Avila 2015).

ACKNOWLEDGMENTS

Marine chemical ecology is a logistically challenging field of study requiring considerable effort and collaboration. We are grateful for the contributions of other Antarctic researchers in the field, particularly Dr. Conxita Avila and past and current members of her marine chemical ecology laboratory at the Universitat de Barcelona. We are also indebted to our own field team members during many field seasons in Antarctica, as well as the outstanding logistical support of the employees and subcontractors of Antarctic Support Associates, Raytheon Polar Services Company, and Antarctic Support Contract. Our Antarctic research has been generously supported by the Antarctic Organisms and Ecosystems Program in the Office of Polar Programs of the National Science Foundation. Recent and current awards include the grants ANT-0838773, ANT-1041022, and PLR-1341333 (CDA, JBM) and ANT-0838776 and PLR-1341339 (BJB). JBM acknowledges support from an Endowed Professorship provided by the University of Alabama at Birmingham. Figures 2.3 through 2.7 are original artwork by KMS.

REFERENCES

Abbas, S., M. Kelly, J. Bowling, J. Sims, A. Waters, and M. Hamann. 2011. Advancement into the Arctic region for bioactive sponge secondary metabolites. *Mar. Drugs* 9(11):2423–2437. doi:10.3390/Md9112423.

Adler, J. H., and R. J. Grebenok. 1999. Occurrence, biosynthesis, and putative role of ecdysteroids in plants. *Crit. Rev. Biochem. Mol. Biol.* 34(4):253–264. doi:10.1080/10409239991209282.

Amsler, C. D., and V. A. Fairhead. 2006. Defensive and sensory chemical ecology of brown algae. *Adv. Bot. Res.* 43:1–91. doi:10.1016/S0065-2296(05)43001-3.

Amsler, C. D., K. Iken, J. B. McClintock, M. O. Amsler, K. J. Peters, J. M. Hubbard, F. B. Furrow, and B. J. Baker. 2005a. Comprehensive evaluation of the palatability and chemical defenses of subtidal macroalgae from the Antarctic Peninsula. *Mar. Ecol. Prog. Ser.* 294:141–159. doi:10.3354/Meps294141.

Amsler, C. D., K. Iken, J. B. McClintock, and B. J. Baker. 2009a. Defenses of polar macroalgae against herbivores and biofoulers. *Bot. Mar.* 52(6):535–545.

Amsler, C. D., J. B. McClintock, and B. J. Baker. 1998. Chemical defense against herbivory in the Antarctic marine macroalgae *Iridaea cordata* and *Phyllophora antarctica* (Rhodophyceae). *J. Phycol.* 34(1):53–59. doi:10.1046/j.1529-8817.1998.340053.x.

Amsler, C. D., J. B. McClintock, and B. J. Baker. 1999. An Antarctic feeding triangle: Defensive interactions between macroalgae, sea urchins, and sea anemones. *Mar. Ecol. Prog. Ser.* 183:105–114.

Amsler, C. D., J. B. McClintock, and B. J. Baker. 2001b. Secondary metabolites as mediators of trophic interactions among Antarctic marine organisms. *Am. Zool.* 41(1):17–26.

Amsler, C. D., J. B. McClintock, and B. J. Baker. 2012. Amphipods exclude filamentous algae from the western Antarctic Peninsula: Experimental evidence. *Polar Biol.* 35(2):171–177.

Amsler, C. D., J. B. McClintock, and B. J. Baker. 2014. Chemical mediation of mutualistic interactions between macroalgae and mesograzers structure unique coastal communities along the western Antarctic Peninsula. *J. Phycol.* 50(1):1–10.

Amsler, C. D., J. McClintock, and B. J. Baker. 2008. Macroalgal chemical defenses in polar marine communities. In *Algal Chemical Ecology*, C. D. Amsler (Ed.), pp. 91–103. Berlin, Germany: Springer-Verlag.

Amsler, C. D., I. N. Okogbue, D. M. Landry, M. O. Amsler, J. B. McClintock, and B. J. Baker. 2005b. Potential chemical defenses against diatom fouling in Antarctic macroalgae. *Bot. Mar.* 48(4):318–322. doi:10.1515/Bot.2005.041.

Amsler, C. D., K. B. Iken, J. B. McClintock, and B. J. Baker. 2001a. Secondary metabolites from Antarctic marine organisms and their ecological implications. In *Marine Chemical Ecology*, J. B. McClintock and B. J. Baker (Eds.), pp. 267–300. Boca Raton, FL: CRC Press.

Amsler, M. O., C. D. Amsler, J. L. von Salm, C. F. Aumack, J. B. McClintock, R. M. Young, and B. J. Baker. 2013. Tolerance and sequestration of macroalgal chemical defenses by an Antarctic amphipod: A 'cheater' among mutualists. *Mar. Ecol. Prog. Ser.* 490:79–90. doi:10.3354/Meps10446.

Amsler, M. O., J. B. McClintock, C. D. Amsler, R. A. Angus, and B. J. Baker. 2009b. An evaluation of sponge-associated amphipods from the Antarctic Peninsula. *Antarct. Sci.* 21(6):579–589. doi:10.1017/s0954102009990356.

Anderson, L. M., and P. T. Martone. 2014. Biomechanical consequences of epiphytism in intertidal macroalgae. *J. Exp. Biol.* 217:1167–1174.

Angulo-Preckler, C., C. Cid, F. Oliva, and C. Avila. 2015. Antifouling activity in some benthic Antarctic invertebrates by "in situ" experiments at Deception Island, Antarctica. *Mar. Environ. Res.* 105:30–38.

Ankisetty, S., S. Nandiraju, H. Win, Y. C. Park, C. D. Amsler, J. B. McClintock, J. A. Baker et al. 2004b. Chemical investigation of predator-deterred macroalgae from the Antarctic Peninsula. *J. Nat. Prod.* 67(8):1295–1302. doi:10.1021/Np049965c.

Ankisetty, S., C. D. Amsler, J. B. McClintock, and B. J. Baker. 2004a. Further membranolide diterpenes from the Antarctic sponge *Dendrilla membranosa*. *J. Nat. Prod.* 67(7):1172–1174.

Antonov, A. S., S. A. Avilov, A. I. Kalinovsky, S. D. Anastyuk, P. S. Dmitrenok, E. V. Evtushenko, V. I. Kalinin et al. 2008. Triterpene glycosides from Antarctic sea cucumbers. 1. Structure of liouvillosides A(1), A(2), A(3), B-1, and B-2 from the sea cucumber *Staurocucumis liouvillei*: New procedure for separation of highly polar glycoside fractions and taxonomic revision. *J. Nat. Prod.* 71(10):1677–1685. doi:10.1021/Np800173c.

Antonov, A. S., S. A. Avilov, A. I. Kalinovsky, S. D. Anastyuk, P. S. Dmitrenok, V. I. Kalinin, S. Taboada, A. Bosh, C. Avila, and V. A. Stonik. 2009. Triterpene glycosides from Antarctic sea cucumbers. 2. Structure of achlioniceosides A(1), A(2), and A(3) from the sea cucumber *Achlionice violaecuspidata* (=*Rhipidothuria racowitzai*). *J. Nat. Prod.* 72(1):33–38. doi:10.1021/Np800469v.

Antonov, A. S., S. A. Avilov, A. I. Kalinovsky, P. S. Dmitrenok, V. I. Kalinin, S. Taboada, M. Ballesteros, and C. Avila. 2011. Triterpene glycosides from Antarctic sea cucumbers III. Structures of liouvillosides A(4) and A(5), two minor disulphated tetraosides containing 3-O-methylquinovose as terminal monosaccharide units from the sea cucumber *Staurocucumis liouvillei* (Vaney). *Nat. Prod. Res.* 25(14):1324–1333. doi:10.1080/1478 6419.2010.531017.

Appleton, D. R., C. S. Chuen, M. V. Berridge, V. L. Webb, and B. R. Copp. 2009. Rossinones A and B, biologically active meroterpenoids from the Antarctic ascidian, *Aplidium* species. *J. Org. Chem.* 74(23):9195–9198. doi:10.1021/Jo901846j.

Argandona, V. H., J. Rovirosa, A. San-Martin, A. Riquelme, A. R. Diaz-Marrero, M. Cueto, J. Darias, O. Santana, A. Guadano, and A. Gonzalez-Coloma. 2002. Antifeedant effects of marine halogenated monoterpenes. *J. Agric. Food Chem.* 50(24):7029–7033.

Arntz, W. E., J. Gutt, and M. Klages (Eds.). 1997. Antarctic marine biodiversity: An overview. In *Antarctic Communities: Species, Structure, and Survival*, B. Battaglia and J. Valencia (Eds.). Cambridge, UK: Cambridge University Press.

Aronson, R. B., S. Thatje, A. Clarke, L. S. Peck, D. B. Blake, C. D. Wilga, and B. A. Seibel. 2007. Climate change and invasibility of the Antarctic benthos. *Annu. Rev. Ecol. Evol. Syst.* 38:129–154.

Aumack, C. F., C. D. Amsler, J. B. McClintock, and B. J. Baker. 2010. Chemically mediated resistance to mesoherbivory in finely branched macroalgae along the western Antarctic Peninsula. *Eur. J. Phycol.* 45(1):19–26. doi:10.1080/09670260903171668.

Avila, C., K. Iken, A. Fontana, and G. Cimino. 2000. Chemical ecology of the Antarctic nudibranch *Bathydoris hodgsoni* Eliot, 1907: Defensive role and origin of its natural products. *J. Exp. Mar. Biol. Ecol.* 252(1):27–44. doi:10.1016/S0022-0981(00)00227-6.

Avila, C., S. Taboada, and L. Nunez-Pons. 2008. Antarctic marine chemical ecology: What is next? *Mar. Ecol.-Evol. Persp.* 29(1):1–71. doi:10.1111/j.1439-0485.2007.00215.x.

Baker, B. J., T. L. Barlow, and J. B. McClintock. 1997. Evaluation of the functional role of suberitenones A and B from the sponge *Suberites* sp. found in McMurdo Sound, Antarctica. *Antarct. J.* 32:90–92.

Baker, B. J., R. W. Kopitzke, W. Y. Yoshida, and J. B. Mcclintock. 1995. Chemical and ecological studies of the Antarctic sponge *Dendrilla membranosa*. *J. Nat. Prod.* 58(9):1459–1462. doi:10.1021/Np50123a020.

Baker, B. J., W. Y. Yoshida, and J. B. McClintock. 1994. Chemical constituents of four antarctic sponges in McMurdo Sound, Antarctica. *Antarct. J.* 29(5):153–155.

Baker, B. J., and R. G. Kerr. 1993. Biosynthesis of marine sterols. In *Topics in Current Chemistry*, P. J. Scheuer (Ed.), pp. 1–31. Berlin, Germany: Springer-Verlag.

Balskus, E. P. 2014. Natural products: Sponge symbionts play defense. *Nat. Chem. Biol.* 10(8):611–612. doi:10.1038/nchembio.1588.

Barnes, D. K. A., P. Rothery, and A. Clarke. 1996. Colonization and development in encrusting communities from the Antarctic intertidal and sublittoral. *J. Exp. Mar. Biol. Ecol.* 196(1):251–265.

Baumann, M. E. M., F. P. Brandini, and R. Staubes. 1994. The influence of light and temperature on carbon-specific DMS release by cultures of *Phaeocystis antarctica* and three Antarctic diatoms. *Mar. Chem.* 45(1):129–136.

Bingol, K., L. Bruschweiler-Li, C. Yu, A. Somogy, F. Zhang, and R. Bruschweiler. 2015. Metabolomics beyond spectroscopic databases: A combined MS/NMR strategy for the rapid identification of new metabolites in complex mixtures. *Anal. Chem.* 87(7):3864–3870. doi:10.1021/ac504633z.

Blackman, A. J., and C. P. Li. 1994. New tambjamine alkaloids from the marine bryozoan *Bugula dentata. Aust. J. Chem.* 47(8):1625–1629.

Blunt, J. W., B. R. Copp, W. Hu, M. H. G. Munro, P. T. Northcote, and M. R. Prinsep. 2007. Marine natural products. *Nat. Prod. Rep.* 24:31–86.

Blunt, J. W., B. R. Copp, M. H. G. Munro, P. T. Northcote, and M. R. Prinsep. 2005. Marine natural products. *Nat. Prod. Rep.* 23:26–78.

Brierley, A. S., and M. J. Kingsford. 2009. Impacts of climate change on marine organisms and ecosystems. *Curr. Biol.* 19(14):R602–R614. doi:10.1016/j.cub.2009.05.046.

Bucolo, P., C. D. Amsler, J. B. McClintock, and B. J. Baker. 2012. Effects of macroalgal chemical extracts on spore behavior of the Antarctic epiphyte *Elachista antarctica* Phaeophyceae. *J. Phycol.* 48(6):1403–1410.

Campbell, A. H., T. Harder, S. Neilsen, S. Kjelleberg, and P. D. Steinberg. 2011. Climate change and disease: Bleaching of a chemically defended seaweed. *Global Change Biol.* 17:2958–2970.

Carbone, M., M. Gavagnin, M. Haber, Y. W. Guo, A. Fontana, E. Manzo, G. Genta-Jouve et al. 2013. Packaging and delivery of chemical weapons: A defensive trojan horse stratagem in chromodorid nudibranchs. *PLoS One* 8(4):e62075. doi:10.1371/journal.pone.0062075.

Carbone, M., C. Irace, F. Costagliola, F. Castelluccio, G. Villani, G. Calado, V. Padula et al. 2010. A new cytotoxic tambjamine alkaloid from the Azorean nudibranch *Tambja ceutae. Bioorg. Med. Chem. Lett.* 20(8):2668–2670. doi:10.1016/j.bmcl.2010.02.020.

Carbone, M., L. Nunez-Pons, F. Castelluccio, C. Avila, and M. Gavagnin. 2009. Illudalane sesquiterpenoids of the alcyopterosin series from the Antarctic marine soft coral *Alcyonium grandis. J. Nat. Prod.* 72(7):1357–1360. doi:10.1021/np900162t.

Carbone, M., L. Nunez-Pons, M. Paone, F. Castelluccio, C. Avila, and M. Gavagnin. 2012. Rossinone-related meroterpenes from the Antarctic ascidian *Aplidium fuegiense. Tetrahedron* 68(18):3541–3544. doi:10.1016/j.tet.2012.03.013.

Carte, B., and D. J. Faulkner. 1983. Defensive metabolites from three nembrothid nudibranchs. *J. Org. Chem.* 48(14):2314–2318. doi:10.1021/Jo00162a003.

Clarke, A., E. J. Murphy, M. P. Meredith, J. C. King, L. S. Peck, D. K. A. Barnes, and R. C. Smith. 2007. Climate change and the marine ecosystem of the western Antarctic Peninsula. *Philos. Trans. R. Soc. B: Biol. Sci.* 367(1477):149–166.

Clayton, M. N. 1994. Evolution of the Antarctic marine benthic algal flora. *J. Phycol.* 30(6):897–904.

Coll, J. C. 1992. The chemistry and chemical ecology of Octocorals (Coelenterata, Anthozoa, Octocorallia). *Chem. Rev.* 92(4):613–631. doi:10.1021/Cr00012a006.

Coll, J. C., B. F. Bowden, D. M. Tapiolas, R. H. Willis, P. Djura, M. Streamer, and L. Trott. 1985. Studies of Australian soft corals. 35. The terpenoid chemistry of soft corals and its implications. *Tetrahedron* 41(6):1085–1092. doi:10.1016/S0040-4020(01) 96476-2.

Cueto, M., J. Darias, J. Rovirosa, and A. San-Martin. 1998. Pantoneurotriols: Probable bioge-netic precursors of oxygenated monoterpenes from Antarctic *Pantoneura plocamioi-des. Tetrahedron* 54(14):3575–3580.

Cutignano, A., G. Nuzzo, D. D'Angelo, E. Borbone, A. Fusco, and A. Fontana. 2013. Mycalol: A natural lipid with promising cytotoxic properties against human anaplas-tic thyroid carcinoma cells. *Angew. Chem. Int. Ed.* 52(35):9256–9260. doi:10.1002/anie.201303039.

Cutignano, A., W. Zhang, C. Avila, G. Cimino, and A. Fontana. 2011. Intrapopulation vari-ability in the terpene metabolism of the Antarctic opisthobranch mollusc *Austrodoris kerguelenensis. Eur. J. Org. Chem.* 27:5383–5389. doi:10.1002/ejoc.201100552.

Dauria, M. V., L. Minale, and R. Riccio. 1993. Polyoxygenated steroids of marine origin. *Chem. Rev.* 93(5):1839–1895. doi:10.1021/Cr00021a010.

Davies-Coleman, M. T. 2006. Secondary metabolites from the marine gastropod molluscs of Antarctica, southern Africa and South America. *Prog. Mol. Subcell. Biol.* 43:133–157.

Davies-Coleman, M. T., and D. J. Faulkner. 1991. New diterpenoic acid glycerides from the Antarctic nudibranch *Austrodoris kerguelensis. Tetrahedron* 47(47):9743–9750. doi:10.1016/S0040-4020(01)80714-6.

Davies, P. L. 2014. Ice-binding proteins: A remarkable diversity of structures for stop-ping and starting ice growth. *Trends Biochem. Sci.* 39(11):548–555. doi:10.1016/j.tibs.2014.09.005.

Dayton, P. K. 1975. Experimental evaluation of ecological dominance in a rocky intertidal algal community. *Ecol. Monogr.* 45(2):137–159.

Dayton, P. K., G. A. Robilliard, R. T. Paine, and L. B. Dayton. 1974. Biological accom-modation in the benthic community at McMurdo Sound, Antarctica. *Ecol. Monogr.* 44(1):105–128.

De Broyer, C., A. Clarke, P. Koubbi, F. Scott, E. Vanden Berghe, and B. Danis. 2018. Register of Antarctic Marine Species. http://www.marinespecies.org/rams.

De Broyer, C., P. Koubbi, H. J. Griffiths, B. Raymond, C. d'Udekem d'Acoz, A. P. Van de Putte, B. Danis et al. 2014. *Biogeographic Atlas of the Southern Ocean.* Cambridge, UK: Scientific Committee on Antarctic Research.

Debbab, A., A. H. Aly, and P. Proksch. 2012. Endophytes and associated marine derived fungi-ecological and chemical perspectives. *Fungal Divers.* 57(1):45–83. doi:10.1007/s13225-012-0191-8.

Delsuc, F., H. Brinkmann, D. Chourrout, and H. Philippe. 2006. Tunicates and not cepha-lochordates are the closest living relatives of vertebrates. *Nature* 439(7079):965–968. doi:10.1038/nature04336.

Desjardins, A. E. 2008. Natural product chemistry meets genetics: When is a genotype a chemotype? *J. Agric. Food Chem.* 56(17):7587–7592. doi:10.1021/Jf801239j.

Diaz-Marrero, A. R., I. Brito, M. Cueto, A. San-Martin, and J. Darias. 2004. Suberitane net-work, a taxonomical marker for Antarctic sponges of the genus Suberites? Novel sester-terpenes from *Suberites caminatus. Tetrahedron Lett.* 45(24):4707–4710. doi:10.1016/j.tetlet.2004.04.091.

Diaz-Marrero, A. R., I. Brito, E. Dorta, M. Cueto, A. San-Martin, and J. Darias. 2003. Caminatal, an aldehyde sesterterpene with a novel carbon skeleton from the Antarctic sponge *Suberites caminatus. Tetrahedron Lett.* 44(31):5939–5942. doi:10.1016/S0040-4039(03)01426-6.

Diyabalanage, T., C. D. Amsler, J. B. McClintock, and B. J. Baker. 2006. Palmerolide A, a cytotoxic macrolide from the Antarctic tunicate *Synoicum adareanum. J. Am. Chem. Soc.* 128(17):5630–5631. doi:10.1021/Ja0568508.

Diyabalanage, T., K. B. Iken, J. B. McClintock, C. D. Amsler, and B. J. Baker. 2010. Palmadorins A-C, diterpene glycerides from the Antarctic nudibranch *Austrodoris kerguelenensis. J. Nat. Prod.* 73(3):416–421. doi:10.1021/Np900617m.

Dohler, G. 1998. Effect of UV radiation on pigments of the Antarctic macroalga *Leptosomia simplex* L. *Photosynthetica* 35(3):473–476.

Dubilier, N., C. Bergin, and C. Lott. 2008. Symbiotic diversity in marine animals: The art of harnessing chemosynthesis. *Nat. Rev. Microbiol.* 6(10):725–740.

Ducklow, H. W., W. R. Fraser, M. P. Meredith, S. E. Stammerjohn, S. C. Doney, D. G. Martinson, S. F. Sailley et al. 2013. West Antarctic Peninsula: An ice-dependent coastal marine ecosystem in transition. *Oceanography* 26(3):190–203.

Ekelöf, E. 1908. Bakteriologische Studien während der Schwedischen Südpolar-Expedition 1901–1903. In *Wissenschaftliche Ergebnisse der Schwedischen Süpolar-Expedition 1901–1903*, O. Nordenskjöld (Ed.). Stockholm, Sweden: Generalstabs Lithograph Institut.

Engel, S., P. R. Jensen, and W. Fenical. 2002. Chemical ecology of marine microbial defense. *J. Chem. Ecol.* 28(10):1971–1985. doi:10.1023/A:1020793726898.

Engene, N., V. J. Paul, T. Byrum, W. H. Gerwick, A. Thor, and M. H. Ellisman. 2013. Five chemically rich species of tropical marine cyanobacteria of the genus *Okeania* gen. nov. (Oscillatoriales, Cyanoprokaryota). *J. Phycol.* 49(6):1095–1106.

Fabry, V. J., J. B. McClintock, J. T. Mathis, and J. M. Grebmeier. 2009. Ocean acidification at high latitudes: The bellweather. *Oceanography* 22(4):160–171.

Fairhead, V. A., C. D. Amsler, J. B. McClintock, and B. J. Baker. 2006. Lack of defense or phloro-tannin induction by UV radiation or mesograzers in *Desmarestia anceps* and *D. menziesii* (Phaeophyceae). *J. Phycol.* 42(6):1174–1183. doi:10.1111/j.1529-8817.2006.00283.x.

Fairhead, V. A., C. D. Amsler, J. B. McClintock, and B. J. Baker. 2005. Variation in phlo-rotannin content within two species of brown macroalgae (*Desmarestia anceps* and *D. menziesii*) from the Antarctic Peninsula. *Polar Biol.* 28:680–686.

Feely, R. A., C. L. Sabine, K. Lee, W. Berelson, J. Kleypas, V. J. Fabry, and F. J. Millero. 2004. Impact of anthropogenic CO_2 on the $CaCO_3$ system in the oceans. *Science* 305(5682):362–366.

Figuerola, B., L. Nunez-Pons, J. Moles, and C. Avila. 2013. Feeding repellence in Antarctic bryozoans. *Naturwissenschaften* 100(11):1069–1081. doi:10.1007/s00114-013-1112-8.

Figuerola, B., L. Nunez-Pons, J. Vazquez, S. Taboada, J. Cristobo, M. Ballesteros, and C. Avila. 2012. Chemical interactions in Antarctic marine benthic ecosystems. In *Marine Ecosystems*, A. Cruzado (Ed.), pp. 105–126. Rijeka, Croatia: InTech Open Access Publisher.

Flores-Molina, M. R., R. Rautenberger, P. Munoz, P. Huovinen, and I. Gomez. 2016. Stress tolerance of the endemic Antarctic brown alga *Desmarestia anceps* to UV radiation and temperature is mediated by high concentrations of phlorotannins. *Photochem. Photobiol.* 92(3):455–466.

Fontana, A., G. Scognamiglio, and G. Cimino. 1997. Dendrinolide, a new degraded diterpe-noid from the Antarctic sponge *Dendrilla membranosa*. *J. Nat. Prod.* 60(5):475–477. doi:10.1021/Np960712w.

Ford, J., and R. J. Capon. 2000. Discorhabdin R: A new antibacterial Pyrroloiminoquinone from two latrunculiid marine sponges, *Latrunculia* sp. and *Negombata* sp. *J. Nat. Prod.* 63(11):1527–1528. doi:10.1021/Np000220q.

Franco, L. H., E. B. D. Joffe, L. Puricelli, M. Tatian, A. M. Seldes, and J. A. Palermo. 1998. Indole alkaloids from the tunicate *Aplidium meridianum*. *J. Nat. Prod.* 61(9):1130–1132. doi:10.1021/Np970493u.

Furrow, F. B., C. D. Amsler, J. B. McClintock, and B. J. Baker. 2003. Surface sequestration of chemical feeding deterrents in the Antarctic sponge *Latrunculia apicalis* as an optimal defense against sea star spongivory. *Mar. Biol.* 143:443–449.

Gavagnin, M., M. Carbone, E. Mollo, and G. Cimino. 2003a. Austrodoral and austro-doric acid: Nor-sesquiterpenes with a new carbon skeleton from the Antarctic nudi-branch *Austrodoris kerguelenensis*. *Tetrahedron Lett.* 44(7):1495–1498. doi:10.1016/S0040-4039(02)02849-6.

Gavagnin, M., M. Carbone, E. Mollo, and G. Cimino. 2003b. Further chemical studies on the Antarctic nudibranch *Austrodoris kerguelenensis*: New terpenoid acylglycerols and revision of the previous stereochemistry. *Tetrahedron* 59(29):5579–5583. doi:10.1016/S0040-4020(03)00775-0.

Gavagnin, M., E. Mollo, F. Castelluccio, A. Crispino, and G. Cimino. 2003c. Sesquiterpene metabolites of the Antarctic gorgonian *Dasystenella acanthina*. *J. Nat. Prod.* 66(11):1517–1519.

Gavagnin, M., E. Trivellone, F. Castelluccio, G. Cimino, and R. Cattaneovietti. 1995. Glyceryl ester of a new halimane diterpenoic acid from the skin of the Antarctic nudibranch *Austrodris kerguelenensis*. *Tetrahedron Lett.* 36(40):7319–7322. doi:10.1016/0040-4039(95)01476-X.

Gianoulis, T. A., J. Raes, P. V. Patel, R. Bjornson, J. O. Korbel, I. Letunic, T. Yamada et al. 2009. Quantifying environmental adaptation of metabolic pathways in metagenomics. *Proc. Natl. Acad. Sci. USA* 106(5):1374–1379. doi:10.1073/pnas.0808022106.

Graeve, M., G. Kattner, C. Wiencke, and U. Karsten. 2002. Fatty acid composition of Arctic and Antarctic macroalgae: Indicator of phylogenetic and trophic relationships. *Mar. Ecol. Prog. Ser.* 231:67–74.

Grant, R. A., and K. Linse. 2009. Barcoding Antarctic biodiversity: Current status and the CAML initiative, a case study of marine invertebrates. *Polar Biol.* 32(11):1629–1637.

Graziani, E. I., R. J. Andersen, P. J. Krug, and D. J. Faulkner. 1996. Stable isotope incorporation evidence for the de novo biosynthesis of terpenoic acid glycerides by dorid nudibranchs. *Tetrahedron* 52(20):6869–6878. doi:10.1016/0040-4020(96)00327-4.

Griffiths, H. J. 2010. Antarctica marine biodiversity-what do we know about the distribution of life in the Southern Ocean. *PLoS One* 5(8):e11683.

Guella, G., F. Dini, and F. Pietra. 1996. Epoxyfocardin and its putative biogenetic precursor, focardin, bioactive, new-skeleton diterpenoids of the marine ciliate *Euplotes focardii* from Antarctica. *Helv. Chim. Acta* 79(2):439–448. doi:10.1002/hlca.19960790211.

Gutt, J., B. I. Sirenko, I. S. Smirnov, and W. E. Arntz. 2004. How many macrozoobenthic species might inhabit the Antarctic shelf. *Antarct. Sci.* 16(1):11–16.

Hefu, Y., and G. O. Kirst. 1997. Effect of UV-radiation on DMSP content and DMS formation of *Phaeocystis antarctica*. *Polar Biol.* 18(6):402–409.

Hempel, G. 1985. *Antarctic Marine Food Webs, Antarctic Nutrient Cycles and Food Webs*. Berlin, Germany: Springer.

Hochmuth, T., and J. Piel. 2009. Polyketide synthases of bacterial symbionts in sponges–Evolution-based applications in natural products research. *Phytochemistry* 70(15–16):1841–1849.

Hoffmann, A. A., and Y. Willi. 2008. Detecting genetic responses to environmental change. *Nat. Rev. Genet.* 9(6):421–432.

Holm-Hansen, O. 1985. Nutrient cycles in Antarctic marine ecosystems. In *Antarctic Nutrient Cycles and Food Webs*, W. R. Siegfried, P. R. Condy, and R. M. Laws (Eds.). Berlin, Germany: Springer.

Hu, G. P., J. Yuan, L. Sun, Z. G. She, J. H. Wu, X. J. Lan, X. Zhu, Y. C. Lin, and S. P. Chen. 2011. Statistical research on marine natural products based on data obtained between 1985 and 2008. *Mar. Drugs* 9(4):514–525.

Huang, Y. M., J. B. McClintock, C. D. Amsler, K. J. Peters, and B. J. Baker. 2006. Feeding rates of common Antarctic gammarid amphipods on ecologically important sympatric macroalgae. *J. Exp. Mar. Biol. Ecol.* 329:55–65.

Iken, K., C. D. Amsler, M. O. Amsler, J. B. McClintock, and B. J. Baker. 2009. Field studies on deterrent properties of phlorotannins in Antarctic brown algae. *Bot. Mar.* 52(6):547–557. doi:10.1515/Bot.2009.071.

Iken, K., C. D. Amsler, J. M. Hubbard, J. B. McClintock, and B. J. Baker. 2007. Allocation patterns of phlorotannins in Antarctic brown algae. *Phycologia* 46(4):386–395. doi:10.2216/06-67.1.

Iken, K., C. Avila, A. Fontana, and M. Gavagnin. 2002. Chemical ecology and origin of defensive compounds in the Antarctic nudibranch *Austrodoris kerguelenensis* (Opisthobranchia: Gastropoda). *Mar. Biol.* 141(1):101–109. doi:10.1007/s00227-002-0816-7.

Iken, K., E. R. Barrera-Oro, M. L. Quartino, R. J. Casaux, and T. Brey. 1997. Grazing by the Antarctic fish *Notothenia coriiceps*: Evidence for selective feeding on macroalgae. *Antarct. Sci.* 9(4):386–391.

Iken, K., M. L. Quartino, and C. Wiencke. 1999. Histological identification of macroalgae from stomach contents of the Antarctic fish *Notothenia coriiceps* using semi-thin sections. *Mar. Ecol.* 20(1):11–17.

Iken, K. B., and B. J. Baker. 2003. Ainigmaptilones, sesquiterpenes from the Antarctic gorgonian coral *Ainigmaptilon antarcticus*. *J. Nat. Prod.* 66(6):888–890.

IPCC. 2013. Summary for policymakers. In *Climate Change 2013: The Physical Science Basis. Contribution of Working Group I to the Fifth Assessment Report of the Intergovernmental Panel on Climate Change*, T.F. Stocker, D. Qin, G.-K. Plattner, M. Tignor, S.K. Allen, J. Boschung, A. Nauels, Y. Xia, V. Bex and P.M. Midgley (Eds.). New York: Cambridge University Press.

Ivanchina, N. V., T. V. Malvarenko, A. A. Kicha, A. I. Kalinovskii, P. S. Dmitrenok, and E. Mollo. 2006. Polar steroidal compounds from the Antarctic starfish *Diplasterias brucei*. *Chem. Nat. Compd.* 42(5):621–622. doi:10.1007/s10600-006-0235-y.

Ivanchina, N. V., T. V. Malyarenko, A. A. Kicha, A. I. Kalinovsky, P. S. Dmitrenok, and S. P. Ermakova. 2011. Structures and cytotoxic activities of two new asterosaponins from the Antarctic starfish *Diplasterias brucei*. *Russ. J. Bioorg. Chem.* 37(4):499–506. doi:10.1134/S1068162011030083.

Ivanova, V., M. Oriol, M. J. Montes, A. Garcia, and J. Guinea. 2001. Secondary metabolites from a *Streptomyces* strain isolated from Livingston Island, Antarctica. *Z. Naturforsch. C J. Biosci.* 56(1–2):1–5.

Jaenisch, R., and A. Bird. 2003. Epigenetic regulation of gene expression: How the genome integrates intrinsic and environmental signals. *Nat. Genet.* 33:245–254.

Janech, M. G., A. Krell, T. Mock, J. S. Kang, and J. A. Raymond. 2006. Ice-binding proteins from sea ice diatoms (Bacillariophyceae). *J. Phycol.* 42(2):410–416. doi:10.1111/j.1529-8817.2006.00208.x.

Janussen, D., and O. S. Tendal. 2007. Diversity and distribution of Porifera in the bathyal and abyssal Weddell Sea and adjacent areas. *Deep Sea Res. Part II Top. Stud. Oceanogr.* 54(16–17):1864–1875. doi:10.1016/j.dsr2.2007.07.012.

Jayatilake, G. S., B. J. Baker, and J. B. McClintock. 1995. Isolation and identification of a stilbene derivative from the Antarctic sponge *Kirkpatrickia variolosa*. *J. Nat. Prod.* 58(12):1958–1960.

Karentz, D., and I. Bosch. 2001. Influence of ozone-related increases in utraviolet radiation on Antarctic marine organisms. *Am. Zool.* 41(1):3–16.

Kidawa, A., K. Stepanowska, M. Markowska, and S. Rakusa-Suszczewski. 2008. Fish blood as a chemical signal for Antarctic marine invertebrates. *Polar Biol.* 31(4):519–525.

Kim, H. J., W. J. Kim, B.-W. Koo, D.-W. Kim, J. H. Lee, and W. S. K. Nugroho. 2017. Anticancer activity of sulfated polysaccharides isolated from the Antarctic red seaweed *Iridaea cordata*. *Ocean Polar Res.* 38(2):129–137.

Konig, G. M., S. Kehraus, S. F. Seibert, A. Abdel-Lateff, and D. Muller. 2006. Natural products from marine organisms and their associated microbes. *Chembiochem* 7:229–238.

Koplovitz, G., J. B. McClintock, C. D. Amsler, and B. J. Baker. 2011. A comprehensive evaluation of the potential chemical defenses of Antarctic ascidians against sympatric fouling microorganisms. *Mar. Biol.* 158(12):2661–2671. doi:10.1007/s00227-011-1764-x.

Korb, R. E., M. J. Whitehouse, and P. Ward. 2004. SeaWiFS in the southern ocean: Spatial and temporal variability in phytoplankton biomass around South Georgia. *Deep Sea Res. Part II: Trop. Stud. Oceanogr.* 51(1):99–116.

Kremp, A., P. Tahvanainen, W. Litaker, B. Krock, S. Suikkanen, C. P. Leaw, and C. Tomas. 2014. Phylogenetic relationships, morphological variation, and toxin patterns in the *Alexandrium ostenfeldii* (Dinophyceae) complex: Implications for species boundaries and identities. *J. Phycol.* 50(1):81–100. doi:10.1111/Jpy.12134.

Laturnus, F., C. Wiencke, and H. Kloser. 1996. Antarctic macroalgae—Sources of volatile halogenated organic compounds. *Mar. Environ. Res.* 41(2):169–181.

Lebar, M. D., and B. J. Baker. 2010. Synthesis and structure reassessment of psammopemmin A. *Aust. J. Chem.* 63(6):862–866. doi:10.1071/CH10042.

Lebar, M. D., J. L. Heimbegner, and B. J. Baker. 2007. Cold-water marine natural products. *Nat. Prod. Rep.* 24(4):774–797. doi:10.1039/b516240h.

Lebar, M. D., L. Luttenton, J. B. McClintock, C. D. Amsler, and B. J. Baker. 2011. Accumulation of vanadium, manganese, and nickel in Antarctic tunicates. *Polar Biol.* 34(4):587–590. doi:10.1007/s00300-010-0902-0.

Lee, H. S., J. W. Ahn, Y. H. Lee, J. R. Rho, and J. Shin. 2004. New sesterterpenes from the Antarctic sponge *Suberites* sp. *J. Nat. Prod.* 67(4):672–674. doi:10.1021/np030342t.

Li, J., K. H. Chen, X. H. Lin, P. Q. He, and G. Y. Li. 2006. Production and characterization of an extracellular polysaccharide of Antarctic marine bacteria *Pseudoalteromonas* sp. S-15-13. *Acta Oceanol. Sin.* 25(6):106–115.

Livermore, R., A. Nankivell, G. Eagles, and P. Morris. 2005. Paleogene opening of Drake passage. *Earth Planet. Sci. Lett.* 236(1–2):459–470.

Lopanik, N. B. 2014. Chemical defensive symbioses in the marine environment. *Funct. Ecol.* 28(2):328–340.

Lüning, K. 1990. *Seaweeds: Their Environment, Biogeography, and Ecophysiology.* New York: John Wiley & Sons.

Ma, W. S., T. Mutka, B. Vesley, M. O. Amsler, J. B. McClintock, C. D. Amsler, J. A. Perman et al. 2009. Norselic acids A-E, highly oxidized anti-infective steroids that deter mesograzer predation, from the Antarctic sponge *Crella* sp. *J. Nat. Prod.* 72(10):1842–1846. doi:10.1021/Np900382x.

Mahon, A. R., C. D. Amsler, J. B. McClintock, M. O. Amsler, and B. J. Baker. 2003. Tissue-specific palatability and chemical defenses against macropredators and pathogens in the common articulate brachiopod *Liothyrella uva* from the Antarctic Peninsula. *J. Exp. Mar. Biol. Ecol.* 290(2):197–210.

Maier, M. S., A. J. Roccatagliata, A. Kuriss, H. Chludil, A. M. Seldes, C. A. Pujol, and E. B. Damonte. 2001. Two new cytotoxic and virucidal trisulfated triterpene glycosides from the Antarctic sea cucumber *Staurocucumis liouvillei*. *J. Nat. Prod.* 64(6):732–736. doi:10.1021/Np000584i.

Maschek, J. A. 2011. Chemical investigation of the Antarctic marine invertebrates *Austrodoris kerguelenensis* and *Dendrilla membranosa* and the Antarctic Red Alga *Gigartina skottsbergii*. Dissertation, Chemistry, University of South Florida, Tampa, FL.

Maschek, J. A., and B. J. Baker. 2008. The chemistry of algal secondary metabolism. In *Algal Chemical Ecology*, C. D. Amsler (Ed.), pp. 1–24. Berlin, Germany: Springer-Verlag.

Maschek, J. A., E. M. Mevers, T. Diyabalanage, L. Chen, Y. Ren, C. D. Amsler, J. B. McClintock, J. Wu, and B. J. Baker. 2012. Palmadorin chemodiversity from the Antarctic nudibranch *Austrodoris kerguelenensis* and inhibition of Jak2/STAT5-dependent HEL leukemia cells. *Tetrahedron* 68:9095–9104.

McClintock, J. B., C. D. Amsler, and B. J. Baker. 2010. Overview of the chemical ecology of benthic marine invertebrates along the western Antarctic peninsula. *Integr. Comp. Biol.* 50(6):967–980. doi:10.1093/icb/icq035.

McClintock, J. B., C. D. Amsler, B. J. Baker, and R. W. M. van Soest. 2005. Ecology of Antarctic marine sponges: An overview. *Integr. Comp. Biol.* 45(2):359–368. doi:10.1093/Icb/45.2.359.

McClintock, J. B., M. O. Amsler, C. D. Amsler, K. J. Southworth, C. Petrie, and B. J. Baker. 2004. Biochemical composition, energy content and chemical antifeedant and anti-foulant defenses of the colonial Antarctic ascidian *Distaplia cylindrica*. *Mar. Biol.* 145(5):885–894. doi:10.1007/s00227-004-1388-5.

McClintock, J. B., and B. J. Baker. 1997a. Palatability and chemical defense of eggs, embryos and larvae of shallow-water Antarctic marine invertebrates. *Mar. Ecol. Prog. Ser.* 154:121–131. doi:10.3354/Meps154121.

McClintock, J. B., and B. J. Baker. 1997b. A review of the chemical ecology of Antarctic marine invertebrates. *Am. Zool.* 37(4):329–342.

McClintock, J. B., B. J. Baker, M. Slattery, J. N. Heine, P. J. Bryan, W. Yoshida, M. T. Davies-Coleman, and D. J. Faulkner. 1994a. Chemical defense of common Antarctic shallow-water nudibranch *Tritoniella belli* eliot (Mollusca: Tritonidae) and its prey, *Clavularia frankliniana* rouel (Cnidaria: Octocorallia). *J. Chem. Ecol.* 20(12):3361–3372. doi:10.1007/BF02033732.

McClintock, J. B., and J. J. Gauthier. 1992. Antimicrobial activities of Antarctic sponges. *Antarct. Sci. US* 4(2):179–183.

McClintock, J. B., and J. Janssen. 1990. Pteropod abduction as a chemical defense in a pelagic Antarctic amphipod. *Nature* 346(6283):462–464. doi:10.1038/346462a0.

McClintock, J. B., and D. Karentz. 1997. Mycosporine-like amino acids in 38 species of subtidal marine organisms from McMurdo Sound, Antarctica. *Antarct. Sci.* 9(4):392–398.

McClintock, J. B., A. R. Mahon, K. J. Peters, C. D. Amsler, and B. J. Baker. 2003. Chemical defenses in embryos and juveniles of two common Antarctic sea stars and an isopod. *Antarct. Sci.* 15(3):339–344. doi:10.1017/S0954102003001354.

McClintock, J. B., M. Slattery, B. J. Baker, and J. N. Heine. 1993. Chemical ecology of antarctic sponges from McMurdo Sound, Antarctica: Ecological aspects. *Antarct. J.* 28(5):134–135.

McClintock, J. B., B. J. Baker, M. Slattery, J. N. Heine, P. J. Bryan, W. Yoshida, M. T. Davies-Coleman, and D. John Faulkner. 1994b. Chemical defense of the common Antarctic nudibranch *Tritonella belli* Eliot (Mollusca: Tritonidae) and its prey, *Clavularia frankliniana* Rouel (Cnidaria: Octocorallia). *J. Chem. Ecol.* 20:3361–3372.

McDowell, R. E., C. D. Amsler, D. A. Dickson, J. B. McClintock, and B. J. Baker. 2014a. Reactive oxygen species and the Antarctic macroalgal wound response. *J. Phycol.* 50(1):71–80. doi:10.1111/jpy.12127.

McDowell, R. E., C. D. Amsler, J. B. McClintock, and B. J. Baker. 2014b. Reactive oxygen species as a grazing defense in marine systems: H_2O_2 and wounded *Ascoseira mirabilis* both inhibit feeding by an amphipod grazer. *J. Exp. Mar. Biol. Ecol.* 458:34–38.

McDowell, R. E., M. O. Amsler, Q. Li, J. R. Lancaster, and C. D. Amsler. 2015. The immediate wound-induced oxidative burst of *Saccharina latissima* depends on light via photo-synthetic electron transport. *J. Phycol.* 51:431–441.

McDowell, R. E., C. D. Amsler, M. O. Amsler, Q. Li, and J. R. Lancaster Jr. 2016. Control of grazing by light availability via light-dependent, wound-induced metabolites: The role of reactive oxygen species. *J. Exp. Mar. Biol. Ecol.* 477:86–91.

McFall-Ngai, M., M. G. Hadfield, T. C. G. Bosch, H. V. Carey, T. Domazet-Loso, A. E. Douglas, N. Dubilier et al. 2013. Animals in a bacterial world, a new imperative for the life sciences. *Proc. Natl. Acad. Sci. USA* 110(9):3229–3236. doi:10.1073/pnas.1218525110.

McHargue, J. S. 1925. The occurrence of copper, manganese, zinc, nickel, and cobalt in soils, plants, and animals, and their possible function as vital factors. *J. Agric. Res.* XXX(2):193–196.

McKenzie, R. L., P. J. Aucamp, A. F. Bais, L. O. Bjorn, M. Ilyas, and S. Madronich. 2011. Ozone depletion and climate change: Impacts on UV radiation. *Photochem. Photobiol. Sci.* 10(2):182–198. doi:10.1039/C0pp90034f.

Mellado, G. G., E. Zubia, M. J. Ortega, and P. J. Lopez-Gonzalez. 2005. Steroids from the Antarctic octocoral *Anthomastus bathyproctus. J. Nat. Prod.* 68(7):1111–1115. doi:10.1021/Np050080q.

Menna, M., C. Imperatore, F. D'Aniello, and A. Aiello. 2013. Meroterpenes from marine invertebrates: Structures, occurrence, and ecological implications. *Mar. Drugs* 11(5):1602–1643. doi:10.3390/Md11051602.

Meredith, M., and J. C. King. 2005. Rapid climate change in the ocean west of the Antarctic Peninsula during the second half of the 20th century. *Geophys. Res. Lett.* 32(19):L19604.

Miller, K. A., and J. S. Pearse. 1991. Ecological studies of seaweeds in McMurdo Sound, Antarctica. *Am. Zool.* 31(1):35–48.

Mitova, M., M. L. Tutino, G. Infusini, G. Marino, and S. De Rosa. 2005. Exocellular peptides from Antarctic psychrophile *Pseudoalteromonas haloplanktis. Mar. Biotechnol. (NY)* 7(5):523–531. doi:10.1007/s10126-004-5098-2.

Miyata, Y., T. Diyabalanage, C. D. Amsler, J. B. McClintock, F. A. Valeriote, and B. J. Baker. 2007. Ecdysteroids from the Antarctic tunicate *Synoicum adareanum. J. Nat. Prod.* 70:1859–1864.

Moles, A. T., I. R. Wallis, W. J. Foley, D. I. Warton, J. C. Stegen, A. J. Bisigato, L. Cella-Pizarro et al. 2011. Putting plant resistance traits on the map: A test of the idea that plants are better defended at lower latitudes. *New Phytol.* 191:777–788.

Moles, J., L. Nunez-Pons, S. Taboada, B. Figueroa, J. Cristobo, and C. Avila. 2015. Antipredatory chemical defences in Antarctic benthic fauna. *Mar. Biol.* 162:1813–1821.

Moles, J., H. Wagele, A. Cutignano, A. Fontana, and C. Avila. 2016. Distribution of granuloside in the Antarctic nudibranch *Charcotia granulosa* (Gastrobod: Heterobranchia: Charcotiidae). *Mar. Biol.* 163(3):1–11. doi:10.1007/s00227-016-2831-0.

Molinski, T. F., and D. J. Faulkner. 1987. Metabolites of the Antarctic sponge *Dendrilla membranosa. J. Org. Chem.* 52(2):296–298. doi:10.1021/Jo00378a031.

Molinski, T. F., and D. J. Faulkner. 1988. An antibacterial pigment from the sponge *Dendrilla membranosa. Tetrahedron Lett.* 29(18):2137–2138. doi:10.1016/S0040-4039(00)86692-7.

Moon, B., Y. C. Park, J. B. McClintock, and B. J. Baker. 2000. Structure and bioactivity of eribusinone, a pigment from the Antarctic sponge *Isodictya erinacea. Tetrahedron* 56(46):9057–9062.

Moreau, S., B. Mostajir, S. Belanger, I. R. Schloss, M. Vancoppenolle, S. Demers, and G. A. Ferreyra. 2015. Climate change enhances primary production in the western antarctic peninsula. *Global Change Biol.* 21(6):2191–2205.

Nichols, C. A. M., J. Guezennec, and J. P. Bowman. 2005. Bacterial exopolysaccharides from extreme marine environments with special consideration of the Southern Ocean, sea ice, and deep-sea hydrothermal vents: A review. *Mar. Biotechnol. (NY)* 7(4):253–271. doi:10.1007/s10126-004-5118-2.

Noguez, J. H., T. K. K. Diyabalanage, Y. Miyata, X. S. Xie, F. A. Valeriote, C. D. Amsler, J. B. McClintock, and B. J. Baker. 2011. Palmerolide macrolides from the Antarctic tunicate *Synoicum adareanum. Bioorg. Med. Chem.* 19(22):6608–6614. doi:10.1016/j.bmc.2011.06.004.

Nunez-Pons, L., and C. Avila. 2015. Natural products mediating ecological interactions in Antarctic benthic communities: A mini-review of the known molecules. *Nat. Prod. Rep.* 32(7):1114–1130.

Nunez-Pons, L., M. Carbone, D. Paris, D. Melck, P. Rios, J. Cristobo, F. Castelluccio, M. Gavagnin, and C. Avila. 2012a. Chemo-ecological studies on hexactinellid sponges from the Southern Ocean. *Naturwissenschaften* 99(5):353–368. doi:10.1007/s00114-012-0907-3.

Nunez-Pons, L., M. Carbone, J. Vazquez, M. Gavagnin, and C. Avila. 2013. Lipophilic defenses from alcyonium soft corals of Antarctica. *J. Chem. Ecol.* 39(5):675–685. doi:10.1007/s10886-013-0276-1.

Nunez-Pons, L., M. Carbone, J. Vazquez, J. Rodriguez, R. M. Nieto, M. M. Varela, M. Gavagnin, and C. Avila. 2012b. Natural products from Antarctic colonial ascidians of the genera *Aplidium* and *Synoicum*: Variability and defensive role. *Mar. Drugs* 10(8):1741–1764. doi:10.3390/md10081741.

Nunez-Pons, L., R. Forestieri, R. M. Nieto, M. Varela, M. Nappo, J. Rodriguez, C. Jimenez et al. 2010. Chemical defenses of tunicates of the genus *Aplidium* from the Weddell Sea (Antarctica). *Polar Biol.* 33(10):1319–1329. doi:10.1007/s00300-010-0819-7.

O'Loughlin, P. M., G. Paulay, N. Davey, and F. Michonneau. 2011. The Antarctic region as a marine biodiversity hotspot for echinoderms: Diversity and diversification of sea cucumbers. *Deep Sea Res. Part II Top. Stud. Oceanogr.* 58(1–2):264–275. doi:10.1016/j. dsr2.2010.10.011.

Odate, S., and J. R. Pawlik. 2007. The role of vanadium in the chemical defense of the solitary tunicate, *Phallusia nigra*. *J. Chem. Ecol.* 33(3):643–654. doi:10.1007/s10886-007-9251-z.

Orr, J. C., V. J. Fabry, O. Aumont, L. Bopp, S. C. Doney, R. A. Feely, A. Gnanadesikan et al. 2005. Anthropogenic ocean acidification over the twenty-first century and its impact on calcifying organisms. *Nature* 437:681–686.

Palermo, J. A., M. F. Brasco, C. Spagnuolo, and A. M. Seldes. 2000. Illudalane sesquiterpenoids from the soft coral *Alcyonium paessleri*: The first natural nitrate esters. *J. Org. Chem.* 65(15):4482–4486.

Patel, P. V., T. A. Gianoulis, R. D. Bjornson, K. Y. Yip, D. M. Engelman, and M. B. Gerstein. 2010. Analysis of membrane proteins in metagenomics: Networks of correlated environmental features and protein families. *Genome Res.* 20(7):960–971. doi:10.1101/gr.102814.109.

Paul, V. J., R. Ritson-Williams, and K. Sharp. 2011. Marine chemical ecology in benthic environments. *Nat. Prod. Rep.* 28(2):345–387. doi:10.1039/C0np00040j.

Pawlik, J. R., C. D. Amsler, R. Ritson-Williams, J. B. McClintock, B. J. Baker, and V. J. Paul. 2013. Marine chemical ecology: A science born of scuba. In *Research Discoveries: The Revolution of Science through Scuba*, M. A. Lang, R. L. Marinelli, S. J. Roberts, and P. R. Taylor (Eds.), pp. 53–69. Washington, DC: Smithsonian Institution Scholarly Press.

Peck, L. S., S. Brockington, S. Vanhove, and M. Beghyn. 1999. Community recovery following catastrophic iceberg impacts in a soft-sediment shallow-water site at Signy Island, Antarctica. *Mar. Ecol. Prog. Ser.* 186:1–8. doi:10.3354/Meps186001.

Peters, K. J., C. D. Amsler, J. B. McClintock, and B. J. Baker. 2005. Palatability and chemical defenses of Antarctic Peninsula sponges. *Integr. Comp. Biol.* 45(6):1056.

Peters, K. J., C. D. Amsler, J. B. McClintock, and B. J. Baker. 2010. Potential chemical defenses of Antarctic sponges against sympatric microorganisms. *Polar Biol.* 33(5):649–658. doi:10.1007/s00300-009-0741-z.

Peters, K. J., C. D. Amsler, J. B. McClintock, R. W. M. van Soest, and B. J. Baker. 2009. Palatability and chemical defenses of sponges from the western Antarctic Peninsula. *Mar. Ecol. Prog. Ser.* 385:77 85. doi:10.3354/Meps08026.

Piel, J. 2009. Metabolites from symbiotic bacteria. *Nat. Prod. Rep.* 26(3):338–362.

Post, A., and A. W. D. Larkum. 1993. UV-absorbing pigments, photosynthesis and UV exposure in Antarctica: Comparison of terrestrial and marine algae. *Aquat. Biol.* 45(2):231–243.

Puglisi, M. P., J. M. Sneed, K. H. Sharp, R. Ritson-Williams, and V. J. Paul. 2014. Marine chemical ecology in benthic environments. *Nat. Prod. Rep.* 31(11):1510–1553. doi:10.1039/C4np00017j.

Quartino, M. L., D. Deregibus, G. L. Campana, G. E. J. Latorre, and F. Momo. 2013. Evidence of macroalgal colonization on newly ice-free areas following glacial retreat in Potter Cove (South Shetland Islands), Antarctica. *PLoS One* 8(3):e58223.

Raguso, R. A., A. A. Agrawal, A. E. Douglas, G. Jander, A. Kessler, K. Poveda, and J. S. Thaler. 2015. The raison d'être of chemical ecology. *Ecology* 96(3):617–630.

Rautenberger, R., C. Wiencke, and K. Bischof. 2013. Acclimation to UV radiation and antioxidative defense in the endemic Antarctic brown macroalga *Desmarestia anceps* along a depth gradient. *Polar Biol.* 36:1779–1789.

Raymond, J. A., C. Fritsen, and K. Shen. 2007. An ice-binding protein from an Antarctic sea ice bacterium. *FEMS Microbiol. Ecol.* 61(2):214–221. doi:10.1111/j.1574-6941.2007.00345.x.

Raymond, J. A., M. G. Janech, and C. H. Fritsen. 2009. Novel ice-binding proteins from a psychrophilic Antarctic alga (Chlamydomonadaceae, Chlorophyceae). *J. Phycol.* 45(1):130–136. doi:10.1111/j.1529-8817.2008.00623.x.

Raymond, J. A., and R. Morgan-Kiss. 2013. Separate origins of ice-binding proteins in Antarctic *Chlamydomonas* species. *PLoS One* 8(3). doi:10.1371/journal.pone.0059186.

Ren, J. W., C. M. Xue, L. Tian, M. J. Xu, J. Chen, Z. W. Deng, P. Proksch, and W. H. Lin. 2009. Asperelines A-F, peptaibols from the marine-derived fungus *Trichoderma asperellum. J. Nat. Prod.* 72(6):1036–1044. doi:10.1021/Np900190w.

Reyes, F., R. Fernandez, A. Rodriguez, A. Francesch, S. Taboada, C. Avila, and C. Cuevas. 2008. Aplicyanins A-F, new cytotoxic bromoindole derivatives from the marine tunicate *Aplidium cyaneum. Tetrahedron* 64(22):5119–5123. doi:10.1016/j.tet.2008.03.060.

Rhoades, D. F. 1979. Evolution of plant chemical defense against herbivores. In *Herbivores: Their Interaction with Secondary Plant Metabolites,* G. A. Rosenthal and D. H. Janzen (Eds.), pp. 3–54. New York: Academic Press.

Riesenfeld, C. S., A. E. Murray, and B. J. Baker. 2008. Characterization of the microbial consortia and polyketide synthetic potential in the Palmerolide-producing tunicate, *Synoicum adareanum. J. Nat. Prod.* 71:1812–1818.

Riesgo, A., S. Taboada, and C. Avila. 2015. Evolutionary patterns in Antarctic marine invertebrates: An update on molecular studies. *Mar. Genomics* 23:1–13.

Rivera, P. L., 1996. Plastoquinones and a chromene isolated from the Antarctic brown alga *Desmarestia menziesii. Bol. Soc. Chil. Quim.* 41(1):103–105.

Roberts, J. M., A. J. Wheeler, and A. Freiwald. 2006. Reefs of the deep: The biology and geology of cold-water coral ecosystems. *Science* 312(5773):543–547. doi:10.1126/science.1119861.

Rodriguez Brasco, M. F., A. M. Seldes, and J. A. Palermo. 2001. Paesslerins A and B: Novel tricyclic sesquiterpenoids from the soft coral *Alcyonium paessleri. Org. Lett.* 3(10):1415–1417.

Scheuer, P. J. 1990. Some marine ecological phenomena: Chemical basis and biomedical potential. *Science* 248(4952):173–177.

Schmidt, E. W. 2008. Trading molecules and tracking targets in symbiotic interactions. *Nat. Chem. Biol.* 4(8):466–473. doi:10.1038/Nchembio.101.

Schoenrock, K. M., J. B. Schram, C. D. Amsler, J. B. McClintock, and R. A. Angus. 2015. Climate change impacts on overstory *Desmarestia* spp. from the western Antarctic Peninsula. *Mar. Biol.* 162(2):377–389.

Schoenrock, K. M., C. D. Amsler, J. B. McClintock, and B. J. Baker. 2013. Endophyte presence as a potential stressor on growth and survival in Antarctic macroalgal hosts. *Phycologia* 52(6):595–599.

Schottner, S., F. Hoffmann, C. Wild, H. T. Rapp, A. Boetius, and A. Ramette. 2009. Inter- and intrahabitat bacterial diversity associated with cold-water corals. *ISME J.* 3(6):756–759. doi:10.1038/ismej.2009.15.

Schram, J. B., J. B. McClintock, C. D. Amsler, and B. J. Baker. 2015. Impacts of acute elevated seawater temperature on the feeding preferences of an Antarctic amphipod toward chemically deterrent macroalgae. *Mar. Biol.* 162:425–433.

Silchenko, A. S., A. I. Kalinovsky, S. A. Avilov, P. V. Andryjashchenko, P. S. Dmitrenok, V. I. Kalinin, S. Taboada, and C. Avila. 2013. Triterpene glycosides from Antarctic sea cucumbers IV. Turquetoside A, a 3-O-methylquinovose containing disulfated tetraoside from the sea cucumber *Staurocucumis turqueti* (Vaney, 1906) (=*Cucumaria spatha*). *Biochem. Syst. Ecol.* 51:45–49. doi:10.1016/j.bse.2013.08.012.

Skropeta, D. 2008. Deep-sea natural products. *Nat. Prod. Rep.* 25(6):1131–1166. doi:10.1039/B808743a.

Slattery, M., J. B. McClintock, and J. N. Heine. 1995. Chemical defenses in Antarctic soft corals—Evidence for antifouling compounds. *J. Exp. Mar. Biol. Ecol.* 190(1):61–77. doi:10.1016/0022-0981(95)00032-M.

Soldatou, S., and B. J. Baker. 2017. Cold-water marine natural products, 2006 to 2016. *Nat. Prod. Rep.* 34(6):585–626. doi:10.1039/c6np00127k.

Steig, E. J., D. P. Schneider, S. D. Rutherford, M. E. Mann, J. C. Comiso, and D. T. Shindell. 2009. Warming of the Antarctic ice-sheet surface since the 1957 International Geophysical Year. *Nature* 457(7228):459–462.

Steinberg, P. D., R. DeNys, and S. Kjellberg. 2002. Chemical cues for surface colonization. *J. Chem. Ecol.* 28(10):1935–1951.

Steinberg, P. D., R. Schneider, and S. Kjelleberg. 1997. Chemical defenses of seaweeds against microbial colonization. *Biodegradation* 8(3):211–220.

Steinke, M., G. Malin, S. D. Archer, P. H. Burkill, and P. S. Liss. 2002. DMS production in a coccolithophorid bloom: Evidence for the importance of dinoflagellate DMSP lysases. *Aquat. Microb. Ecol.* 26:259–270.

Sturges, W. T., C. W. Sullivan, R. C. Schnell, L. E. Heidt, and W. H. Pollock. 1993. Bromoalkane production by Antarctic ice algae. *Tellus B* 45(2):120–126.

Su, S. S., L. Tian, G. Chen, Z. Q. Li, W. F. Xu, and Y. H. Pei. 2013. Two new compounds from the metabolites of a marine-derived actinomycete *Streptomyces cavourensis* YY01-17. *J. Asian Nat. Prod. Res.* 15(3):265–269. doi:10.1080/10286020.2012.762764.

Taboada, S., L. Nunez-Pons, and C. Avila. 2013. Feeding repellence of Antarctic and sub-Antarctic benthic invertebrates against the omnivorous sea star *Odontaster validus*. *Polar Biol.* 36(1):13–25. doi:10.1007/s00300-012-1234-z.

Turk, T., J. A. Avgustin, U. Batista, G. Strugar, R. Kosmina, S. Civovic, D. Janussen, S. Kauferstein, D. Mebs, and K. Sepcic. 2013. Biological activities of ethanolic extracts from deep-sea Antarctic marine sponges. *Mar. Drugs* 11(4):1126–1139. doi:10.3390/Md11041126.

Uria, A., and J. Piel. 2009. Cultivation-independent approaches to investigate the chemistry of marine symbiotic bacteria. *Phytochem. Rev.* 8(2):401–414. doi:10.1007/s11101-009-9127-7.

Vankayala, S. L., F. L. Kearns, B. J. Baker, J. D. Larkin, and H. L. Woodcock. 2017. Elucidating a chemical defense mechanism of Antarctic sponges: A computational study. *J. Mol. Graph. Model.* 71:104–115.

Vaughan, D. G., G. J. Marshall, W. M. Connolley, C. Parkinson, R. Mulvaney, D. A. Hodgson, J. C. King, C. J. Pudsey, and J. Turner. 2003. Recent rapid regional climate warming on the Antarctic Peninsula. *Clim. Change* 60(3):243–274.

Viant, M. R., and U. Sommer. 2013. Mass spectrometry based environmental metabolomics: A primer and review. *Metabolomics* 9(1):S144–S158. doi:10.1007/s11306-012-0412-x.

von Salm, J. L., C. G. Witowski, R. M. Fleeman, J. B. McClintock, C. D. Amsler, L. N. Shaw, and B. J. Baker. 2016. Darwinolide, a new diterpene scaffold that inhibits methicillin-resistant *Staphylococcus aureus* biofilm from the Antarctic sponge *Dendrilla membranosa*. *Org. Lett.* 18(11):2596–2599. doi:10.1021/acs.orglett.6b00979.

Waller, R. G., K. M. Scanlon, and L. F. Robinson. 2011. Cold-water coral distributions in the drake passage area from towed camera observations–Initial interpretations. *PLoS One* 6(1):e16153. doi:10.1371/journal.pone.0016153.

Wang, G. Y. 2006. Diversity and biotechnological potential of the sponge-associated microbial consortia. *J. Ind. Microbiol. Biotechnol.* 33(7):545–551. doi:10.1007/s10295-006-0123-2.

Wiencke, C., and C. D. Amsler. 2012. Seaweeds and their communities in Polar Regions. In *Seaweed Biology*, C. Wiencke and K. Bischof (Eds.), pp. 265–294. Berlin, Germany: Springer-Verlag.

Wiencke, C., C. D. Amsler, and M. N. Clayton. 2014. Macroalgae. In *Biogeographic Atlas of the Southern Ocean*, C. De Broyer, P. Koubbi, H. J. Griffiths, B. Raymond, C. d'Udekem d'Acoz, A. P. Van de Putte, B. Danis et al., pp. 66–73. Cambridge, UK: Scientific Committee on Antarctic Research.

Wilkins, S. P., A. J. Blum, D. E. Burkepile, T. J. Rutland, A. Wierzbicki, M. Kelly, and M. T. Hamann. 2002. Isolation of an antifreeze peptide from the Antarctic sponge *Homaxinella balfourensis*. *Cell. Mol. Life Sci.* 59(12):2210–2215. doi:10.1007/s000180200020.

Williams, T. J., E. Long, F. Evans, M. Z. DeMaere, F. M. Lauro, M. J. Raftery, H. Ducklow, J. J. Grzymski, A. E. Murray, and R. Cavicchioli. 2012. A metaproteomic assessment of winter and summer bacterioplankton from Antarctic Peninsula coastal surface waters. *ISME J.* 6(10):1883–1900. doi:10.1038/ismej.2012.28.

Wilson, N. G., J. A. Maschek, and B. J. Baker. 2013. A species flock driven by predation? Secondary metabolites support diversification of slugs in Antarctica. *PLoS One* 8(11):e80277. doi:10.1371/journal.pone.0080277.

Witman, J. D., and T. H. Suchanek. 1984. Mussles in flow: Drag and dislodgement by epizoans. *Mar. Ecol. Prog. Ser.* 16:259–268.

Witowski, C. W. 2015. Investigation of bioactive metabolites from the Antarctic sponge *Dendrilla membranosa* and marine microorganisms. Dissertation, Chemistry, University of South Florida, Tampa, FL.

Wu, G. W., A. Q. Lin, Q. Q. Gu, T. J. Zhu, and D. H. Li. 2013. Four new chloro-eremophilane sesquiterpenes from an Antarctic deep-sea derived fungus, *Penicillium* sp. PR19N-1. *Mar. Drugs* 11(4):1399–1408. doi:10.3390/Md11041399.

Wu, G. W., H. Y. Ma, T. J. Zhu, J. Li, Q. Q. Gu, and D. H. Li. 2012. Penilactones A and B, two novel polyketides from Antarctic deep-sea derived fungus *Penicillium crustosum* PRB-2. *Tetrahedron* 68(47):9745–9749. doi:10.1016/j.tet.2012.09.038.

Xin, Y. J., M. Kanagasabhapathy, D. Janussen, S. Xue, and W. Zhang. 2011. Phylogenetic diversity of Gram-positive bacteria cultured from Antarctic deep-sea sponges. *Polar Biol.* 34(10):1501–1512. doi:10.1007/s00300-011-1009-y.

Yang, A., B. J. Baker, J. Grimwade, A. Leonard, and J. B. McClintock. 1995. Discorhabdin alkaloids from the Antarctic sponge *Latrunculia apicalis*. *J. Nat. Prod.* 58(10):1596–1599.

Yoshida, W. Y., P. J. Bryan, B. J. Baker, and J. B. McClintock. 1995. Pteroenone–A defensive metabolite of the abducted Antarctic pteropod *Clione antarctica*. *J. Org. Chem.* 60(3):780–782. doi:10.1021/Jo00108a057.

Young, R. M., J. L. von Salm, M. O. Amsler, J. Lopez-Bautista, C. D. Amsler, J. B. McClintock, and B. J. Baker. 2013. Site-specific variability in the chemical diversity of the Antarctic red alga *Plocamium cartilagineum*. *Mar. Drugs* 11(6):2126–2139. doi:10.3390/Md11062126.

Zamzow, J. P., C. D. Amsler, J. B. McClintock, and B. J. Baker. 2010. Habitat choice and predator avoidance by Antarctic amphipods: The roles of algal chemistry and morphology. *Mar. Ecol. Prog. Ser.* 400:155–163. doi:10.3354/Meps08399.

3 From the Tropics to the Poles

Chemical Defense Strategies in Sea Slugs (Mollusca: Heterobranchia)

Conxita Avila, Laura Núñez-Pons, and Juan Moles

CONTENTS

The opisthobranchs are to the molluscs what the orchids are to the angiosperms, or the butterflies to the arthropods.

T.E. Thompson, 1976

3.1 INTRODUCTION

Marine organisms produce a wide variety of molecules, often unique and critical for their survival in terms of feeding, reproduction, or protection. These natural products are at the basis of ecological specialization because they may affect species distribution, feeding patterns, community structure, and biodiversity (McClintock and Baker 2001). However, much remains to be explored on how chemical ecology regulates marine ecosystems (Ianora et al. 2006, Paul et al. 2007). "Opisthobranch" mollusks, comprising the commonly known sea slugs and sea hares, are no exception to this, and they present a wide variety of bioactive compounds protecting them against potential predators and competitors, and enhancing their ecological performance. Other molecules may be involved in reproduction and feeding, although these are not reviewed here. Sea slug defenses include chemicals obtained directly from their prey, transformed dietary metabolites, or even de novo biosynthesized bioactive compounds. Over the last 30 years, a great deal of advances was achieved on chemical defensive strategies in sea slugs throughout different latitudes, from the tropics to the poles. Since our previous review on opisthobranch chemical ecology (Avila 1995), a number of reviews regarding mollusk chemistry have been published, covering Australia and New Zealand (Garson 2006), Antarctica, South Africa, and South America (Davies-Coleman 2006), the Indo-Pacific (Wahidulla et al. 2006), Japan (Miyamoto 2006), and China (Wang et al. 2013), and the chemistry of nudibranchs from the west coast of North America (Andersen et al. 2006). Other reviews covered the chemical ecology of selected opisthobranch groups (Cimino and Ghiselin 1998, 1999, 2001, 2009, Kamiya et al. 2006), and the chemistry of pulmonate gastropods (Darias et al. 2006). Cimino and Gavagnin (2006) provided an overview of marine molluscan chemistry and biotechnology, while Benkendorff (2010) discussed the chemical diversity of mollusks focusing on the taxonomical distribution of compounds within the phylum, as well as their medicinal properties. Wider general reviews covered marine chemical ecology (McClintock and Baker 2001, Ianora et al. 2006, Puglisi et al. 2014), the ecological roles of sterols (Kandyuk 2006), or the role of marine natural products as antifeedants (Garson 2010), among others. Chemically mediated ecological interactions in polar organisms have also been reviewed recently (Avila et al. 2008, McClintock et al. 2010, Núñez-Pons and Avila 2015). Blunt and collaborators (2015) (and their previous reports) have been providing complete reviews of the natural products described yearly. In this chapter we describe what is known about the chemical ecology of marine slugs ("opistobranchs") from a biological perspective, and the molecules involved, focusing on the origins, bioactivities, and ecological roles in combination with other defensive strategies (Table 3.1). Biogeographical and evolutionary aspects are also considered.

Classically, gastropod mollusks were divided into three subclasses: Prosobranchia, Opisthobranchia, and Pulmonata, on the basis of the position and type of their respiratory organs. This classification, however, has been a matter of great controversy among malacologists for decades. Part of the problem is that Pulmonata and Opisthobranchia are very closely related, and are often placed together into

TABLE 3.1

Summary of Heterobranchia Data Reviewed in This Chapter, Regarding Species Number, Geographical Distribution, Natural Compounds, Origin and Function, According to WoRMS

Group	Total Number of Species Described[a]	Number of Genera Reviewed Here	Number of Species Reviewed Here	Number of Species by Region: Tropical, Temperate, or Polar			Number of Natural Products Found	Number of Species vs Origin: Biosynthetic or Biotransformed, from Diet, or Symbionts			Number of Species vs Role: Defensive in situ, or Laboratory Assays		% of Species with Anatomically Localized Compounds
				TRO	TEM	POL		BIO	DIE	SYM	DEF	LAB	
Nudibranchia	~2500	56	116	48	53	7	~250	12	59	0	41	26	21
Pleurobranchoidea	~80	5	16	4	11	1	25	9	6	1	10	5	69
Tylodinoidea	~10	2	3	0	3	0	6	0	3	0	3	2	100
Cephalaspidea	~900	11	24	7	14	2	40	11	10	0	13	1	29
Anaspidea	~80	5	17	8	10	0	~200	2	11	0	6	6	76
Pteropoda	~140	1	1	0	0	1	5	1	1	0	1	1	0
Sacoglossa	~350	15	38	25	13	0	~120	26	25	1	29	6	42
Pulmonata (marine)	~500	5	26	11	10	0	~75	2	0	0	6	7	8
Total	~8000	100	241	103	114	11	~725	63	115	2	109	54	Mean = 32

Note: www.marinespecies.org.

a Approximate numbers; the total includes unassigned taxa.

Heterobranchia, because of their detorsioned nervous system and their modified respiratory organs. After recent advances in mollusk phylogeny, Heterobranchia together with its closest prosobranch relatives are considered to form a monophyletic group, named Heterogastropoda, with Prosobranchia being paraphyletic. The consequence of all these ambiguities is that the taxonomy of opisthobranchs still remains highly debated (Ponder and Lindberg 1997, Dinapoli and Klussmann-Kolb 2010, Göbbeler and Klussmann-Kolb 2011, Schrödl et al. 2011, Medina et al. 2011, Wägele et al. 2014, Zapata et al. 2014). In this chapter we use the term "sea slug" *sensu lato*, with no phylogenetic implications, to include all the Heterobranchia, that is, considering together evolutionary lineages of "modified" marine gastropods (including opisthobranch gastropods as well as their marine pulmonate relatives) regardless the reduction or not of the adult shell. In some cases, the shell remains vestigial, or only present during larval stages, while in many groups it is overgrown by soft tissue and becomes internal, no longer providing mechanical protection, but likely providing some skeletal support. Heterobranchia comprises more than 8000 marine species, from which less than 300 have been chemically studied (Avila 1995, 2006, Blunt et al. 2015) (Table 3.1). Since they have been on Earth already since the Paleozoic, a great deal of morphological and ecological diversification has occurred, resulting in amazing trophic, reproductive, and defensive adaptations (Medina et al. 2011).

Sea slugs play a crucial role within the benthos, occupying many different ecological niches and displaying a wide array of trophic relationships with organisms from many different phyla, such as Chlorophyta, Ochrophyta, Rhodophyta (green, red, and brown algae, respectively), Porifera, Cnidaria, Bryozoa, Chordata (tunicates), other Mollusca, and so on. Such interactions, therefore, comprise macroalgal or plant herbivory, as well as carnivore prey–predator relationships, and occasional cannibalism. Among the key innovations behind the evolutionary success of sea slugs are the abilities to steal functional structures or chemical products from other organisms through cleptoplasty and cleptochemistry, respectively. In cleptoplasty, slugs retain certain functional structures from their prey, such as chloroplasts from algae to obtain energy and camouflage (Händeler et al. 2009), or nematocysts (cleptocnides) from cnidarians to use as protective devices (Putz et al. 2010). Cleptochemistry, instead, is the incorporation of natural products from the diet (cleptochemicals), being called cleptochemodefenses when used for their own defensive means (Avila 1992, 1993, 1995, Avila and Durfort 1996). The astonishing biological adaptations of sea slugs have also turned several species into important model organisms in neurobiology (e.g., *Aplysia* [Kandel 1979]; *Hermissenda* [Alkon et al. 2005]; *Tritonia* [Wyeth and Willows 2006]; pleurobranchs [Jing and Gillette 1999]), and pharmaceutical research (Blunt et al. 2015, Newman and Cragg 2004, 2014, Faircloth and Cuevas 2006).

One well-known characteristic of sea slugs is their ability to avoid major generalist predation, even if lacking obvious physical protection, and this is thanks to their unpleasant taste, combined with other mechanisms (see below). It is currently accepted that chemical defense originated in sea slugs' ancestors when these were still protected by shells, since shelled as well as shell-less forms possess defensive chemicals (Cimino and Ghiselin 1998, 1999, Faulkner and Ghiselin 1983, Wägele and Klussmann-Kolb 2005). Nonetheless, shelled mollusks are usually poorer sources of natural products (Benkendorff 2010). Interestingly, there are some extraordinary exceptions, such as cone snails and turrids (Terlau and Olivera 2004, López-Vera et al. 2004), which do synthesize amazingly bioactive genome-derived peptides for both predation and defense (Olivera 2002, Lin et al. 2013). Shell reduction is in any case a common phenomenon in heterobranchs, extended paraphyletically (independently) to nearly all lineages (Wägele and Klussmann-Kolb 2005). From an evolutionary perspective, the loss of the shell represents an economical advantage in terms of energy savings, which would otherwise be used for shell production and transportation, as well as other respiratory and excretory advantages. However, it simultaneously entails the investment in alternative defense strategies to survive in front of putative predators. Shell-less groups were probably favored to develop escape reactions, including swimming in some groups, but mostly resulted in an increase of specialization and allowed the colonization of new ecological niches. In fact, most heterobranchs are soft-bodied and shell-less animals, which possess an extremely wide array of defensive traits ranging from behavioral and morphological (mechanical or physical) to chemical strategies (Todd 1981, Faulkner and Ghiselin 1983, Vermeij 1987, 1994, Perrone 1989, Avila 1995, Putz et al. 2010). These are common defensive strategies shared in most animal groups, electrical defense being the only type missing in sea slugs so far. As in other animals, the different co-existing defensive strategies act as a whole system, thus combining behavioral, morphological, and chemical protection for the species survival. Some species are active at night when predators are inactive or are more difficult to be seen, or live hidden under rocks. Some sea slugs possess an indistinguishable coloration (homochromy), shape (homomorphy), or texture from the surrounding environment, using cryptic camouflage to hide from predators. Others may present escape reactions involving detachment from the substrate and writhing to be carried away by currents, especially pelagic forms inhabiting the water column (e.g., Pteropoda), or even swimming by body undulations (e.g., *Bornella*, *Hexabranchus*), or flapping parapodia (e.g., *Aplysia*). Potential enemies can also be kept away through deimatic behavior, like exhibiting sudden bright color patterns (e.g., *Hexabranchus sanguineus* [Edlinger 1982]). In some cases, ink along with toxic metabolites is secreted as dense clouds that distract enemies, as reported in several aplysiomorphs (Derby 2007). Pleurobranchs, cylichnid cephalaspideans, as well as some dorids, produce highly acidic secretions (pH~1) by specialized dermal glands to deter predators (Thompson 1960, 1988, Avila 1995, Wägele et al. 2006). Mechanical defenses other than the shell, if present, may include calcareous spicules, or nematocysts with their toxins derived from preyed cnidarians (Thompson 1976, Edmunds 2009). Autotomy is also a quite extended strategy by which body parts that will later be regenerated are detached, distracting predators. This, however, represents an ultimate and quite expensive strategy. Usually, the autotomized structures are not essential for survival, are the most exposed, and/or contain the most powerful, or the highest concentrations of deterrents (Stasek 1967, Ros 1976, Todd 1981, Di Marzo et al. 1991a, 1991b, Carbone et al. 2013), but not always (Avila 1996). Thus, in some sacoglossans and nudibranchs detachment of the tail, mantle rim, or cerata is effective, especially if accompanied by toxic mucus secretions, and particularly when it is followed by behavioral and physiological traits that enhance the surprising effect.

The use of chemical weapons is, with no doubt, the most interesting and noteworthy strategy among sea slugs (Thompson 1960, Avila 1995, Cimino and Ghiselin 2009). Bioactive metabolites often derive from the diet, and they may be used in several ways, transferred, and accumulated in particular sites, glands, or exposed parts, or they may also be released within mucous secretions. Some species are able to biotransform the dietary metabolites in order to make them less toxic for the slug itself, or more noxious and deterrent towards predators (Avila 1995, 2006, Cimino and Ghiselin 2009). Finally, some sea slugs may completely de novo biosynthesize some chemicals from simple precursors (Cimino and Ghiselin 1999, Cimino et al. 2001). Often, chemical defenses are associated to warning (aposematic) colorations, allowing species to survive in exposed habitats, where predators learn to associate bright colorations to their bad taste (Margalef 1977, 1982, Guilford and Cuthill 1991, Avila 1995, Gamberale 1998, Härlin and Härlin 2003, Bandaranayake 2006, Cortesi and Cheney 2010). Many pigments, in fact, possess bioactive properties themselves (e.g., alkaloids), while being part of photosynthetic systems, or may act as sunscreens protecting from UV light (Czeczuga 1984, Whitehead et al. 2001, Bandaranayake 2006, Kamio et al. 2011). As in other animals, several strategies of using warning colorations appeared during evolution. Batesian mimicry is used by several taxa (not only sea slugs) as a false warning appearance to discourage potential predators by imitating chemically defended species, without real toxic chemistry. Müllerian mimicry, instead, involves species that possess defensive metabolites and use similar coloration patterns, making the warning effect more effective against potential predators (Margalef 1977, 1982, Avila 1995, Begon et al. 2006). Typical colors with warning effect in the sea often include: dark blue-black background, combined with yellow, white, light blue, red, and orange lines or dots. These are common chromatic patterns of phyllidid and chromodoridid nudibranchs, and *Thuridilla* sacoglossans, which may be considered well-known examples of Müllerian or Batesian mimetic groups (Endler 1988, Guilford and Cuthill 1991, Avila 1995, Cortesi and Cheney 2010, Cheney et al. 2014).

In general, the chemicals behind ecological bioactivities are organic compounds, named natural products or secondary metabolites, to distinguish them from primary metabolites (sugars, lipids, proteins, and nucleic acids) (Cimino and Ghiselin 2009). Those are called "secondary" even if they may be crucial for survival, because they are synthesized from primary metabolites, although their role is not always clear. The biochemical pathways to build natural products are usually complex and metabolically expensive, even when they come from the diet, since the compounds must be processed, transferred, or stored in specific sites. Resources allocated to natural products, thus, are not available for growth or reproduction, and therefore, it is generally assumed that a benefit is obtained, which often is not so evident (Cronin 2001). Why and how an organism invests in defense is obviously a key question in chemical ecology, and several theories and models have been proposed around the topic (Ianora et al. 2006). Among them, the resource availability model (RAM) predicts that species that evolved in a nutrient-rich environment are inherently fast growing species, with low investment in defenses, since they can easily replace lost tissues (Coley et al. 1985). Species evolved in low-resource environments, instead, should tend to be slow growers, investing more in defense. Organisms under environmental stress (temperature or salinity fluctuations, high UV radiation, etc.) have probably more difficulties to acquire resources. Thus, the environmental stress theory (EST) predicts that in extreme conditions the levels of chemical defenses would be lower, and therefore, the organisms should be more vulnerable to predation (Rhoades 1979). According to the growth differentiation balance hypothesis (GDBH) there is a trade-off between resources devoted to differentiation processes (production of chemical defenses, cell specialization, etc.) and growth, with differentiation occurring only after growth (Herms and Mattson 1992). This means that young, actively growing tissues should contain low levels of defenses, and thus juveniles should be less defended than adults. In contrast, the optimal defense theory (ODT) predicts that younger individuals and younger reproductive parts of organisms should be better defended, because these are under a higher risk of predation, since they usually are more nutritious (McKey 1974, Feeny 1976, Rhoades and Gates 1976, Rhoades 1979). This assumes that organisms should defend themselves maximizing fitness, i.e. protecting preferentially their tissues in function of the vulnerability or ecological value of that tissue or body part. In all, it is clear that different models lead to completely different predictions. In part, this is related to the fact that completely different organisms were used when making these models, whether plants or animals. Many of the predictions of the ODT and GDBH relate to well-developed, complex organisms that can invest differentially in their different organs or body parts. Chemical defense allocation is particularly common in sea slugs, with extensive literature reporting bioactive products stored in exposed, vulnerable areas, such as the mantle, foot, gills, and rhinophores; within mucus or ink secretions; in specialized glands, like mantle dermal formations (MDFs); and also occasionally in eggs, embryos, and larval stages (Avila 1995, Wägele et al. 2006, Cimino and Ghiselin 2009).

Although not a major subject of this chapter, many bioactive compounds found in nature are pharmacologically interesting as drugs to fight diseases. Oceans are the source of many unique natural compounds produced or accumulated both by micro- and macroorganisms, including sea slugs, as a result of evolutionary and ecological pressures, such as predation and competition (Paul 1992, McClintock and Baker 2001, Avila et al. 2008, Puglisi et al. 2014, Blunt et al. 2015). These marine molecules, shaped into a wide range of structurally diverse compounds, display specific biological activities, and they are rarely found in terrestrial organisms (Carte 1996, Haefner 2003, Clardy and Walsh 2004, Newman and Cragg 2004, Molinski et al. 2009, Cragg et al. 2014). Obviously, marine organisms are much less explored than terrestrial sources (Radjasa et al. 2011). Only 1% of the recorded species has been chemically analyzed, while just about 10% of the natural compounds described to date come from marine organisms (Hu et al. 2011, Blunt et al. 2015). Also, less than 3% of the described marine natural products originate from high latitudes (Núñez-Pons and Avila 2015), due to the geographically uneven research efforts done so far. The discovery potentialities of novel marine compounds, thus, remain wide open. In particular, the pharmacological potential of sea slugs in drug discovery is quite relevant. Remarkably, a few drugs from the sea are commercially in use. For instance, in marine mollusks, ziconotide (Prialt®), a synthetic peptide derived from *Conus* venom, is used as pain-killer. Compounds such as kahalalide F from sacoglossans, and aplyronines and dolastatins from anaspideans, are known to be highly effective against tumors, but they are still undergoing clinical trials

(Faircloth and Cuevas 2006, Molinski et al. 2009, Newman and Cragg 2014, Kigoshi and Kita 2015). Promising sea slug compounds, however, include several dolastatins from *Dolabella auricularia*, bursatellanins from *Bursatella leachii*, kahalalides from *Elysia rufescens*, jorumycin from *Jorunna funebris*, ulapualides from *Hexabranchus sanguineus*, and kabiramide from *Hexabranchus* sp.

3.2 CHEMICAL DEFENSE MECHANISMS OF SEA SLUGS

As already mentioned, the taxonomy and phylogeny of Gastropoda are still under intense debate, and it is not the topic of this chapter to discuss these in detail. We will use the most recent general classification of the groups reviewed here, the Heterobranchia, named "sea slugs" *sensu lato* (Figure 3.1). In the following sections we describe what is known about their chemical ecology, in terms of the chemistry involved, metabolic origins, bioactivities, and ecological roles, in combination with other defensive traits (Table 3.1). All the information compiled has been interpreted from a biological perspective, in order to provide a general scheme about the trends followed by the eight major taxa, namely: Nudibranchia, Pleurobranchoidea, Tylodinoidea, Cephalaspidea, Anaspidea, Pteropoda, Sacoglossa, and Pulmonata. This review does not cover data regarding synthesis nor pharmacological activity of the compounds and includes mostly data published after 1995 (see Avila [1995] for previous studies).

3.2.1 NUDIBRANCHIA

Nudibranchs include four major taxa: Doridacea, Dendronotida, Euarminida, and Aeolidida, which have been unequally studied over the years for their chemical ecology. These four taxa are not equally considered taxonomically, since Doridacea is currently an Infraorder, while the other three belong to the Infraorder Cladobranchia. All nudibranchs are carnivorous and shell-less in the adult form. This group is the most diverse within Heterobranchia, displaying an amazing variety of biological and chemical defensive strategies.

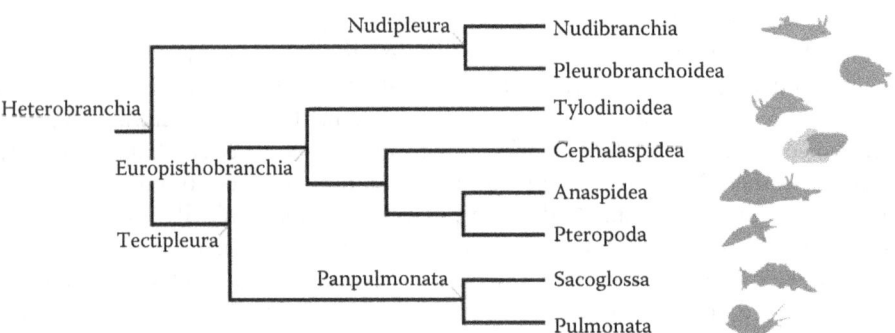

FIGURE 3.1 Schematic tree of the phylogeny of Heterobranchia used in this chapter. (Simplified from Wägele, H. et al., *Org. Divers. Evol.*, 14, 133–149, 2014; Zapata, F. et al., *Proc. R Soc. B: Biol. Sci.*, 281, 20141739, 2014.)

3.2.1.1 Doridacea

Species in this group are generally predators of sponges, bryozoans, tunicates or other opisthobranchs. They possess mainly terpenes, which may be dietary or de novo biosynthesized, but also alkaloids and other compounds (Avila 1995). The Antarctic *Bathydoris hodgsoni* produces the feeding deterrent drimane sesquiterpene, hodgsonal (Iken et al. 1998, Avila et al. 2000). Hodgsonal (Figure 3.2) is found in the mantle and dorsal papillae, the most exposed body parts, and seems to be produced de novo by the slug. It is a very effective deterrent against sympatric predators, such as the sea star *Odontaster validus* and the anemone *Epiactis* sp. (Avila et al. 2000) Egg masses of this species are physically protected against predators (Moles et al. 2017a). The Antarctic *Prodoris (=Bathydoris) clavigera* is also chemically protected from *O. validus*, although the responsible compound has not yet been identified (C. Avila and K. Iken, unpublished results).

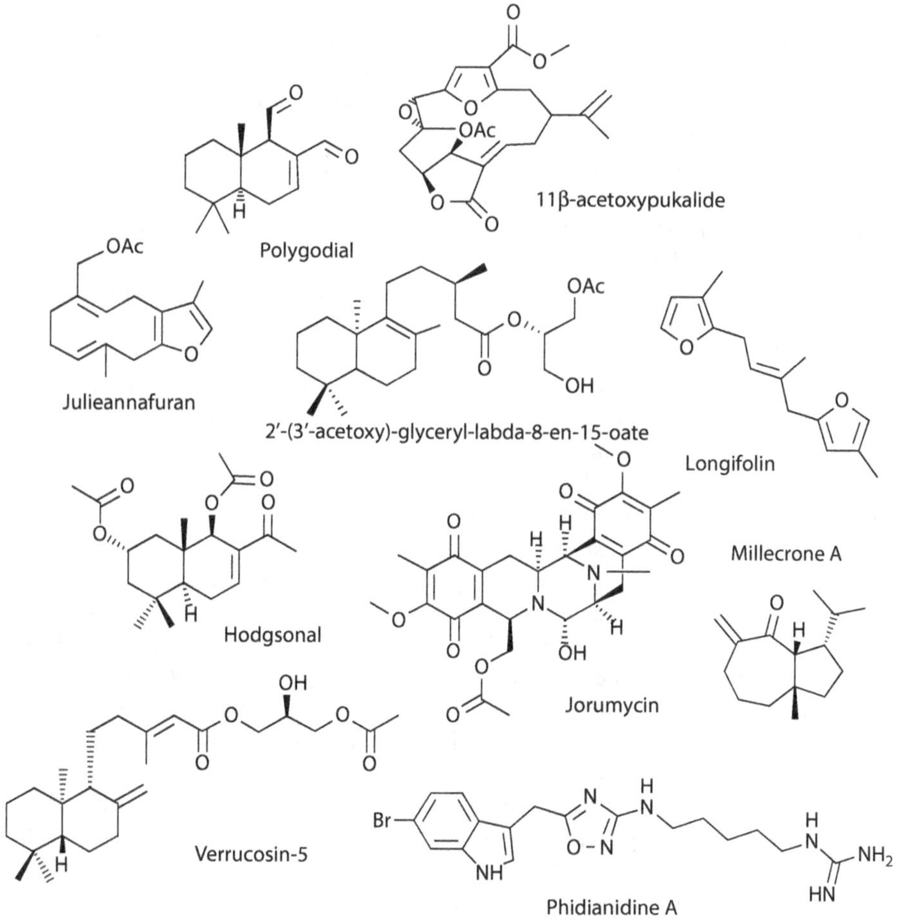

FIGURE 3.2 Structures of selected defensive compounds from nudibranchs.

Following previous studies on *Dendrodoris* slugs (Avila 1995), drimane sesquiterpenes were widely found allocated in different body parts. Drimane esters are usually
associated to reproductive organs and egg masses, while drimane sesquiterpenes are
present in the mantle. The known sesquiterpene 7-deacetoxyolepupuane was isolated
from *D. arborescens* (Fontana et al. 1999), while *D. carbunculosa* possessed several
cytotoxic drimane sesquiterpenes, dendrocarbins A–N (Sakio et al. 2001). *D. krebsi*
from Mexico presented also drimane sesquiterpenes and esters (Gavagnin et al. 2001),
distributed similarly to what was previously reported in other *Dendrodoris* species
(Avila et al. 1991). Also *D. denisoni* from New Zealand contains in its mantle cin-
namolide, olepupuane, and polygodial (Grkovic et al. 2005) (Figure 3.2). The genus
Doriopsilla continued to yield interesting chemistry, related to that of *Dendrodoris*.
Furanosesquiterpene alcohols, pelseneeriols-1 and -2, were described from the
mantle of *Doriopsilla pelseneeri* from Portugal (Gaspar et al. 2005). Stable-isotope
studies allowed to confirm that the drimane esters, sesquiterpenes, and 15-acetoxy-
ent-pallescensin were de novo biosynthesized via mevalonic acid in *D. areolata* and
in *Doriopsilla* sp. (Spinella et al. 1994, Gavagnin et al. 2001, Fontana et al. 2003,
Gaspar et al. 2008); also, two additional diastereomeric acetates of pelseneeriol-1
and -2 were described there. *D. albopunctata* from California and *D. areolata* from
Portugal also contained drimane sesquiterpenes and *ent*-pallescensin A (Gavagnin
et al. 2001), distributed as reported for other *Dendrodoris* species (Avila et al. 1991).
Chemical ecology studies found that two fish, the blenny *Chasmodes bosquianus*
and the mummichog *Fundulus heteroclitus*, learned to avoid food items that contained extracts from *Doriopsilla pharpa* (Long and Hay 2006). The extracts were
similarly refused by the crabs *Callinectes similus* and *Panopeus herbstii* in the field.
The compound behind this deterrence was the well-known polygodial (Figure 3.2).

The colorful phyllidids have been reported to contain abundant isocyanate compounds (Avila 1995) with antifouling, antibiotic, antifungal, and antitumor properties, extensively investigated over the years (Fusetani et al. 1991, Okino et al.
1996, Hirota et al. 1998, Cimino et al. 1999, Garson and Simpson 2004). This is
a particularly difficult group to study since many species and genus are similar
in shape and color and this has produced many misidentifications over the years
(Brunckhorst 1993). Studies on the antifouling substances from *Phyllidia ocelata*,
P. varicosa, *Phyllidiella pustulosa* and *Phillidiopsis krempfi*, led to the isolation of
three new sesquiterpene isonitriles: 10-epi-axisonitrile-3, 10-isocyano-4-cadinene,
and 2-isocyanotrachyopsane, as well as the peroxide, 1,7-epidioxy-5-cadinene, and
several known sesquiterpene isonitriles (Fusetani et al. 1996, Okino et al. 1996).
The dietary transfer of the terpenes from the sponge *Acanthella cavernosa* to
P. pustulosa was further demonstrated (Dumdei et al. 1997). A new sesquiterpene
isonitrile was also isolated as an antifouling agent from the Japanese *P. pustulosa*
(Hirota et al. 1998). *P. pustulosa* from China resulted in the first finding of isocyanide diterpenoids in a phyllidiid, along with isocyanide sesquiterpenoids (Manzo
et al. 2004), some of them previously identified from *Acanthella* sponges, thus
confirming the prey–predator relationship (Dumdei et al. 1997, Shimomura et al.
1999). In contrast, 3-isocyano-theonellin, previously found in *Phyllidia* sp. from
Sri Lanka, is closely related to a cyanide from an *Axinyssa* sponge (Iwashima et al.
2002). *P. pustulosa* from Vietnam contained sesquiterpenoids also, and a new

isothiocyanate, as well as some sterols (Lyakhova et al. 2010). *P. pustulosa* from Fiji contained the known axisonitrile-3, a new isothiocyanate, and several minor related sesquiterpenes (Wright 2003). A moderated antiplasmodial activity was shown for the new compound. Several nitrogenous bisabolene sesquiterpenes from *Phyllidia* sp. from Sri Lanka exhibited a potent in vitro antifouling activity towards barnacle larvae (Gulavita et al. 1986, Kitano et al. 2002). Two thiocyanatopupuke-anane sesquiterpenes were isolated as an epimeric mixture from *P. varicosa* and its sponge prey, *Axinyssa aculeata* (Yasman et al. 2003). While both compounds were isolated from its digestive gland, one of them did accumulate also in the man-tle, suggesting a role in chemical defense. Both compounds exhibited mild toxicity towards brine shrimp and some antimicrobial activity against *Candida albicans* and *Bacillus subtilis*. *P. coelestis* from Thailand possessed two cytotoxic pupuke-anane sesquiterpenoids (Jaisamut et al. 2013). A series of experiments on Guam, using agar-based food combined with different color patterns, tested how phyl-lidiids were defended from fish predators (Ritson-Williams and Paul 2007). Intact living *Phyllidiella granulatus* offered to fish in the field were never consumed. Crude lipophilic extracts of three species of phyllidiid were deterrent to fish in the field; however, *P. pustulosa* from Palau had deterrent extracts whereas *P. pustu-losa* from Guam did not. The results of this study suggest that visual and chemi-cal cues together are more effective defenses than either cue alone. All these data support the specific chemotaxonomic features of the family Phyllidiidae, based on the presence of dietary sesquiterpene isocyanides, suggesting that the nudibranchs may sequester the compounds from sponges belonging to different genera, thus showing a broader feeding variability than previously described (Wahidulla et al. 2006). The fact that usually different species, and even genera of phyllidiids, are found commonly coexisting in the same habitats, displaying similar morpholo-gies, colors, and defensive chemicals, indicates the presence of Müllerian mimicry groups, as observed for example in the Indo-Pacific and in Guam (C. Avila, unpub-lished results). In Guam, the extracts of several species of *Phyllidia*, *Phyllidiella*, *Phyllidiopsis*, and *Fryeria* were deterrent to sympatric crabs, and those from the mantle were more deterrent than those from the viscera (C. Avila and V. J. Paul, unpublished results). In fact, we detected a quick transformation of the secreted metabolites, and the deterrence disappeared in a matter of minutes. As an excep-tion in the family, the Okinawan *Reticulidia fungia* contained two sesquiterpenes of a rare class of sponge metabolites, the cytotoxic carbonimidic dichlorides, reti-culidins A and B (Tanaka and Higa 1999).

Among bryozoan-feeders, the North Sea *Adalaria loveni* contained a cyto-toxic degraded triterpenoid, lovenone, suggested to come from an unidentified prey, while *Adalaria* sp. from the Pacific did not present this compound (Graziani et al. 1995). *Polycera* and *Triopha* species were previously reviewed (Avila 1995). *Polycera atra* concentrated bryostatins from its diet, *Bugula neritina*, on which it is cryptic, and transferred them to its conspicuous egg masses (Paul et al. 1997, Davidson 1999). Bryostatins are in fact synthesized by a microbial symbiont of the bryozoan, *Endobugula sertula*. Contrastingly, median buccal glands that exude sul-furic acid were identified in the species *Plocamopherus ceylonicus* (Polyceridae) (Wägele et al. 2017). Limaciamine is a diacylguanidine isolated from the external part of the North Sea *Limacia clavigera* (Graziani and Andersen 1998). Not surprisingly, *Tambja* species continue affording new tambjamines (Avila 1995). *T. capensis* from South Africa feeds on *Bugula dentata*, and both contained 4-methoxypyrrolic metabolites, and tambjamines A and E, indicating a prey–predator relationship (Rapson 2004, Davies-Coleman 2006). *T. eliora* from Brazil afforded tambjamines A and D, and their imino salts (Berlinck et al. 2004), in agreement with a study on the same species from Mexico (Carte and Faulkner 1983). *T. verconis* and *T. morosa* possessed also tambjamines (Grkovic et al. 2005), as well as *T. ceutae* and its prey *Bugula dentata* from Azores, including the cytotoxic tambjamine K (Carbone et al. 2010). *T. stegosauriformis* from Brazil possessed tambjamines C and K, probably obtained from its bryozoan prey, *Bugula dentata* (Pereira et al. 2012). Some tambjamines have been reported to cause DNA damage and induce apoptosis (Cavalcanti et al. 2008). *Roboastra* species feeding on *Tambja* species were detected to capture tambjamines (Avila 1995). *Okenia zoobotryon* from Brazil yielded also an alkaloid from its bryozoan prey, *Zoobotryon verticillatum* (Pereira et al. 2012). Instead, *Nembrotha kubaryana* from Pohnpei (Micronesia) possess a blue tetrapyrrole with antimicrobial properties, presumably derived from a diet on ascidians (Avila 1995, Karuso and Scheuer 2002).

Aegires species (previously *Notodoris*) sequestered iminoimidazole alkaloids from calcareous *Leucetta* sponges (Avila 1995), while *Acanthodoris* species de novo biosynthesized sesquiterpenoids (Avila 1995, Graziani and Andersen 1996). *A. hudsoni* actually contained these same metabolites (Andersen et al. 2006). Within other sponge-feeders, a progesterone homologue was isolated from the external tissues of the dorid *Aldisa smaragdina* from Spain (Gavagnin et al. 2002). Previously, *A. sanguinea* was reported to have similar steroids used against predators (Avila 1995). *A. andersoni* from India yielded two phorboxazoles, mainly in the external part of the slugs, with feeding deterrent activity (Nuzzo et al. 2012). These compounds, along with the previously known phorbazoles A, B, and D were found mainly in external tissues, suggesting a defensive role. The two new compounds and phorbazole A were tested for feeding deterrence against the shrimp, *Palaemon elegans*, were they significantly reduced feeding at a concentration of 1.0 mg mL^{-1}; however, the natural concentrations within the nudibranch were unknown. *Hexabranchus sanguineus* from the Indo-Pacific presented a series of kabiramides, sanguinamides, and halichondriamides, probably most of them from *Halichondria* sponges, and some with antitumoral and antifungal activity (Matsunaga et al. 1989, Dalisay et al. 2009). The same species from China possessed sesquiterpenes and diterpenes (Zhang et al. 2007), some of them previously isolated from *Axinella* and *Dysidea* sponges, among which ulapualide A is an actin-capping toxin (Vincent et al. 2007). *H. sanguineus* from the Red Sea presented oxazole containing macrolides, but also an unusual esterified carotenoid pigment, named hurghadin (Guo et al. 1998), structurally related to actinioerythrin from a sea anemone (Hertzberg et al. 1969). The geographical variations observed in the chemistry of *H. sanguineus* from China and Egypt significantly increase the number of chemicals isolated from the species and strongly support the dietary origin of the metabolites, which appears to be a generalist sponge feeder (Wahidulla et al. 2006).

The Mediterranean *Doris verrucosa* yielded nine diterpene glycerides, verrucosins 1–9 (Avila 1995, Gavagnin et al. 1997) (Figure 3.2), and a further series of diterpenoid glycerides, among which de novo biosynthesis was proven for verrucosin A (Cimino et al. 1988, Gavagnin 1990) in experiments using ^{13}C- and ^{14}C-labelled precursors (Fontana et al. 2003). This seems to be a common ability in the group. Granato and coauthors (2000) found a variety of common sterols in the mantle of *Doris* aff. *verrucosa* from Brazil, along with the previously known xylosil-MTA (Cimino and Sodano 1993), which was absent in its sponge prey, *Hymeniacidon* aff. *heliophila*. Other *Doris* (*Archidoris*) species also possess similar glycerids, some of them de novo biosynthesized (Avila 1995, Graziani et al. 1996). Anisodorins 1–5, for instance, are likely de novo biosynthetized in the Patagonian dorid *Anisodoris fontaini* (Gavagnin et al. 1999). Tanyolides A and B from the dorsal mantle of *Sclerodoris tanya* from California were effective fish feeding deterrents (Krug et al. 1995). Ichtyotoxic diterpene glycerides were also described in mantle and egg masses of *Archidoris pseudoargus* from UK (Young and Baker 2015). In Antarctica, *Doris* (*Austrodoris*) *kerguelenensis* also presented a series of diterpene diacylglycerides (Figure 3.2), accumulated in the mantle, which effectively protect them from sea star predators and the anemone *Epiactis* sp., along with the corresponding monoacylglycerides, and monoacylglycerides of regular fatty acids (Avila 1995, Gavagnin et al. 1999a, 1999b, Iken et al. 2002). Austrodorins A and B, and the two nor-sesquiterpenes austrodoral and austrodoric acid, were also found in different collections, although no ecological role was described for them yet (Davies-Coleman and Faulkner 1991, Gavagnin et al. 1995, 1999a, 2003a, 2003b). More recently, additional diterpene glycerides and clerodane diterpenes from *D. kerguelenensis* were described at different Antarctic locations (Diyabalanage et al. 2010, Maschek et al. 2012). This wide array of terpene glycerides in *D. kerguelenensis* indicates the existence of different chemotypes, with terpene synthase variants involved in their biosynthesis (Cutignano et al. 2011). Wilson and collaborators (2013) actually linked the secondary metabolome with genotype variability, suggesting that cryptic speciation exists in *D. kerguelenensis*. Remarkably, high chemodiversity in this species exists also within populations (Cutignano et al. 2011, Maschek et al. 2012). From an ecological perspective, most of the wide chemical arsenal of this species remains to be ecologically tested, as compounds occur in complex mixtures where there is an enormous variability.

Diaulula and *Jorunna* are the only two genera chemically studied within the caryophyllidid doridoidea (Avila 1995). Diaulusterol A is, at least in part, biosynthesized by *Diaulula sandieguensis* (Kubanek and Andersen 1999). *Jorunna funebris* from India, found on a sponge of the genus *Oceanapia*, contained, among other related compounds, the cytotoxic isoquinoline alkaloid jorumycin (Fontana et al. 2000, 2001) (Figure 3.2), which is similar to renieramycin E from the sponge *Reniera* sp. (He and Faulkner, 1989) Jorumycin is antimicrobial against *Bacillus subtilis* and *Staphylococcus aureus*, and cytotoxic at very low concentrations against different tumor cell lines (Cimino et al. 2001). Since jorumycin was found in the mucus secretion, the authors proposed a defensive role (Fontana et al. 2001). *J. funebris* from Thailand yielded the cytotoxic jorunnamycins A–C (Charupant et al. 2007), whereas *J. funebris* from Sri Lanka presented a series of dietary isoquinoline–quinone metabolites from the sponge *Xestospongia* sp., but it did not seem to contain jorumycins (Wahidulla et al. 2006). Further renieramycin-like isoquinoline-quinone alkaloids, fennebricins A and B, and others, were found in *J. funebris* from South China, probably from a *Xestospongia* sp. sponge (He et al. 2014, Huang et al. 2016).

Paradoris (=*Discodoris*) *indecora* from the Mediterranean incorporated sesterterpenes with fish-deterrent activity from its sponge diet, *Ircinia variabilis* and *I. fasciculata* (Marín et al. 1997). Interestingly, the specialist-feeder *Peltodoris atromaculata*, predator of the sponge *Petrosia ficiformis*, was observed feeding also on *Haliclona fulva*, both sponges contain cytotoxic fulvinol-like polyacetilenes (Gemballa and Schermutzki 2004). These authors suggested that this highly specific diet preference for cytotoxic sponges could be related to their ability to chemically detect their defended prey. In fact, further studies showed long-chain fulvinol-related polyacetylenes from *H. fulva* found in the slug (Ciavatta et al. 2014). A chlorinated pyrrolidone and some halogenated metabolites, such as dibromophenols, were isolated from extracts of the viscera of *Asteronotus cespitosus* from Australia and the Philippines, probably deriving from its diet, the sponge *Dysidea herbacea*, with similar metabolites (Fahey and Garson 2002). *Halgerda aurantiomaculata* from Australia presented a new tryptophan derivative, halgerdamine (Fahey and Carroll 2007). This species was collected from three sites in Australia, and contained also the known compounds trigonellin, esmodil, zooanemonin, and C2-a-D-mannosylpyranosyl-L-tryptophan in all localities. Five different *Halgerda* species were collected in that study, but only those with contrasting warning color patterns, *H. aurantiomaculata* and *H. gunnessi*, contained characteristic natural products. The other three species were cryptic and did not contain novel compounds. However, crude extracts were not tested against potential predators. Finally, the mantle of the dorid *Actinocyclus papillatus* from South China presented the mildly cytotoxic compound, actisonitrile (Manzo et al. 2011, Guo et al. 2012), as well as a terpenoid diacylguanidine, named actinofide (Carbone et al. 2017).

The world of "chromodoridids" (as used in the classical nomenclature; see below) continued to show an astonishing variety of chemicals obtained from sponges, usually located within specific defensive glands (e.g., MDFs) (Avila 1995). The Atlantic species *Cadlina laevis* and *C. pellucida* possessed several sesquiterpenes (laevidiene, albicanol, and derivates), some of them dietary from the sponge *Spongia agaricina* (Hellou et al. 1982, Avila 1995, Barrero et al. 1995, Fontana et al. 1995, Nakano et al. 1995). The egg masses of *C. luteomarginata* from British Columbia contained a drimane sesquiterpenoid, while five additional diterpenoids, cadlinaldehyde, spongian, secospongian, 20-acetoxy-12-marginatone, and lutenolide were obtained from the external extracts (Dumdei et al. 1997). This slug is able to biosynthesize some of its terpenoids (Kubanek et al. 1997). *C. luteomarginata* from Canada and its sponge prey, *Phorbas* sp., shared the sesterterpenoid ansellone A, still with no described ecological role (Daoust et al. 2010). Other *Cadlina* species selectively accumulate some dietary compounds (Thompson et al. 1982), while *Glossodoris* species possess different kinds of terpenoids (Avila 1995). Two scalaranes from MDF-like structures were described in Indian *Glossodoris atromarginata* and its dietary sponge, *Spongia* sp. (Fontana et al. 1999) *G. atromarginata* from India was also found crawling on two other potential preys, the sponge *Hyattella cribriformis* containing pentacyclic scalaranes, and an unidentified sponge, probably *Spongia,* containing heteronemin and other scalarens (Fontana et al. 1999, 2001, Wahidulla et al. 2006). The sesterterpene deoxoscalarin, previously found in a Mediterranean sponge, was isolated in *G. atromarginata* and the sponge *H. cribriformis,* again supporting a dietary origin. The nudibranchs found on *Spongia,* instead, contained the sesterterpene heteronemin, known from the sponge *Heteronema erecta* and two new scalaranes, one of them also reported from an unidentified sponge. These findings also support a biotransformation of the dietary scalaranes (Fontana et al. 2001, Wahidulla et al. 2006). Similar to previous approaches in chromodoridids (Avila 1995), an anatomically guided chemical analysis of the Australian *G. atromarginata* led to the isolation of two diterpenes in different body parts (Somerville et al. 2006). Scalarane metabolites from *G. rufomarginata* from China seem to come from feeding on an unidentified sponge and being later biotransformed into several scalaradial derivatives. The sponge contains scalaradial and its 12-deacetyl derivative as main constituents (Gavagnin et al. 2004), which are potent anti-inflammatory compounds (Jacobs et al. 1993). The prey–predator relationship was confirmed by the presence of 12-deacetylscalaradial in both mollusk and sponge, while the absence of scalaradial in the nudibranch, suggested the ability of the mollusk to transform this toxic compound into related scalaranes (Gavagnin et al. 2004). Natural products of *G. pallida* from China and Guam were compared: specimens from China, as well as *G. vespa* and *G. averni* from Australia, contained 12-deacetoxy-12-oxoscalaradial (Manzo et al. 2007). *G. pallida* from Guam contained different sesquiterpenes including scalaradial, deacetylscalaradial, and deoxoscalarin, located in the mantle border and acting as feeding deterrents against crabs and reef fish (Rogers and Paul 1991, Avila and Paul 1997). Scalaradial and deacetylscalaradial were also found in the sponge *Cacospongia* sp. on which *G. pallida* preys upon. Probably *Glossodoris* species biotransform dietary scalaranes into related molecules in a detoxification process. Actually, injection of scalaradial in the viscera of *G. pallida* was not toxic for the slug and resulted in a fast transformation of the compound in less than 24 h (C. Avila and V. J. Paul, unpublished results). Costa Rican specimens of *G. sedna* afforded new sesterterpenes, one of them an inhibitor of phospholipase A2, while specimens of *G. dalli* from the same location contained other known sesterterpenes (Fontana et al. 2000). Five spongian diterpenes were obtained from the digestive gland and dorsal mantle of *G. cincta* (reported as *G. atromarginata*) from Egypt (Fontana et al. 1997, 1999, 2001), as well as in specimens from Sri Lanka (Wahidulla et al. 2006, De Silva and Scheuer 1982). Some of these compounds are cytotoxic and antiviral in the laboratory (Kohmoto et al. 1987, Cambie et al. 1988).

Chromodoris species (now synonymized to *Felimida* for Eastern Pacific, Atlantic, and Mediterranean species, and other genera for other areas) (Johnson and Gosliner 2012), typically contain sponge-derived diterpenes located in MDFs along the mantle border (Avila 1995). Many new compounds have been reported recently in this genus. Four unusual chlorinated homoditerpenes, hamiltonins A–D, and the sesterterpene hamiltonin E were isolated from the South African *C. hamiltoni* together with latrunculins A and B (Pika and Faulkner 1995), while specimens from Mozambique had two new spongian diterpene lactones, along with latrunculin B (McPhail and Davies-Coleman 1997). In the Japanese *C. obsoleta,* seven cytotoxic sponge diterpenoids, dorisenones A–D, and related compounds were found (Miyamoto et al. 1996). *C. africana* from the Red Sea presents the furanoterpene kurospongin and the 14-membered macrolide with an attached 2-thiazolidinone unit, latrunculin B (Guo 1997), also reported in *C. quadricolor* (Mebs 1985) (as *Glossodoris quadricolor*). Kurospongin was obtained from an Okinawan *Spongia* sp., showing ichthyotoxicity and feeding deterrency (Tanaka and Higa 1988), while the ichthyotoxic latrunculin B, which may also induce changes in cell morphology, was found in the sponge *Latrunculia magnifica* (Kashman et al. 1980, 1985). *C. inornata* from Japan contained three cytotoxic sesterterpenes, inorolides A, B, and C, and five new scalaranes (Miyamoto et al. 1999). *C. petechialis* from Hawaii yielded three sponge diterpenes, probably from *Chelonaplysilla* sp. (Karuso and Scheuer 2002), one of them cytotoxic (Miyamoto et al. 1996). *C. mandapamensis* from India contained spongiadiol, also found in *G. cincta* from Egypt and previously isolated from Australian sponges, within a mixture of related spongiane compounds (Fontana et al. 2001). The unidentified sponge on which this slug was found also presented spongiadiol, thus supporting a dietary origin. Spongiadiol presents antiviral activity against Herpes simplex virus (Kohmoto et al. 1987). *C. reticulata* from China yielded some known cytotoxic furanoditerpenoids (Wahidulla et al. 2006). Instead, specimens from Australia presented a series of diterpenes, some of them previously reported in dendroceratid sponges and in other nudibranchs, with characteristic body allocations (Suciati et al. 2011). This suggests a defensive role for some of them. Also from Australia, an unidentified *Chromodoris* species was reported to present mildly cytotoxic diterpenes (Yong et al. 2008) while *C. willani* contained two deoxy analogues of manoalide with antimicrobial activity (Uddin et al. 2009), and *C. albopunctata* yielded oxygenated diterpenes from its sponge prey, again differently distributed in their anatomy (Katavic et al. 2012). *C. kunei* from Okinawa contained a cytotoxic spongian diterpene, probably from the sponge *Dysidea* cf. *arenaria* (Agena et al. 2009). Finally, *Goniobranchus splendidus* from Australia presented a cytotoxic spiroepoxide lactone, suspected to derive from its dietary sponge prey (Forster et al. 2017).

The genus *Hypselodoris* (now synonymized to *Felimida* for Eastern Pacific, Atlantic, and Mediterranean species, and other genera for other areas) (Johnson and Gosliner 2012) continued providing examples of sponge-derived furanosesquiterpenes, usually accumulated into MDFs as well. The South African *H. capensis* and the sponges *Fasciospongia* sp. and *Dysidea* sp., on which it was probably feeding, yielded a sesterterpene, 22-deoxy-23-hydroxymethyl-variabilin, together with several known sesquiterpenes, the antifeedant nakafurans 8 and 9, and several other sesterterpenes (McPhail et al. 1998). *H. kanga* from India, and the associated sponge *Dysidea* sp., contained the sesquiterpenoid furodysinin, already known from an unidentified *Dysidea*, suggesting again a trophic relationship (Fontana et al. 2001). *H. infucata* from Hawaii again possessed nakafurans-8 and -9, probably from *Dysidea fragilis* (Karuso and Scheuer 2002), while *H. lajensis* from Brazil contained furodisinin lactone also from *Dysidea* species (Pereira et al. 2012). Crude extracts from *H. cantabrica* showed higher bioactivity in palatability assays using *Palaemon elegans* compared to its prey, *Dysidea fragilis*, indicating a possible cumulative strategy for defense (Da Cruz et al. 2012). Both extracts contained mainly nakafuran-9, thus the authors hypothesized a defensive escalation during coevolution in the nudi-branch (discussed in the following section). A new furan was described from the Australian *H. jacksoni* (Mudianta et al. 2013). As previously described (Avila 1993, 1995), Mediterranean and North Atlantic *Hypselodoris* form Müllerian mimicry groups. These species have distinctive blue, white and yellow aposematic color-ations and are chemically protected by furanosesquiterpenes accumulated in MDFs. However, *H. fontandraui* lacks MDFs, which could indicate Batesian rather than Müllerian mimicry (Haber et al. 2010). Despite the absence of MDFs, *H. fontandraui* contained the furanosesquiterpenoid tavacpallescensin, mainly along the mantle border (Avila 1993, 1995, Avila and Durfort 1996). This metabolite was deterrent to the generalist shrimp *Palaemon elegans* at 1.0 mg mL^{-1}, which is a much lower concentration than that found in the mantle (25.98 ± 1.41 mg mL^{-1}) (Haber et al. 2010). Thus, as expected, the presence of defensive metabolites in *H. fontandraui* supported the fact that it is also part of the Müllerian mimicry complex.

Some chromodorids also display "trail-following" behavior (*e.g. Hypselodoris (Risbecia) tryoni* [Koehler 1999]), although the chemicals involved in this behavior have not been described yet. An unusual seco-spongiane diterpene, tyrinnal, was isolated together with some known sesquiterpenes (dendrolasin, pallescensin A, dehydropallescensin-2) from the Patagonian *Tyrinna nobilis* (Fontana et al. 1998). *Ceratosoma amoena* from New Zealand contained allolaurinterol, an algal metabolite (Grkovic et al. 2005) probably ingested accidentally along with the sponge prey or by sample contamination. More natural products were described from two species of *Ceratosoma* from China (Mollo et al. 2005, Wahidulla et al. 2006). *C. trilobatum* and *C. gracillimum* contained four furanosesquiterpenes, pallescensin B, (-)-furodysinin, (-)-dehydroherbadysidolide, and (-)-herbadysidolide, previously reported from *Dysidea* spp. sponges, and thus probably from dietary origin. These dorids possess dorsal horns with MDFs covering their gills, and containing MDFs with high concentrations of such chemicals, suggesting a defensive role. However, among the products mentioned, only (-)-furodysinin deterred feeding by goldfish, while the remaining compounds tested were neither deterrent nor toxic to mosquito fish. Their effect on ecologically relevant predators remains to be tested.

3.2.1.2 Dendronotida

These species feed either on cnidarians (octocorallia or hydrozoa) or on a variety of small animals (crustaceans and turbellarians). Some sesquiterpenes were isolated from *Tochuina tetraquetra* and its dietary soft coral *Gersemia rubiformis* (Williams and Andersen 1987), but no ecological role has been assigned for them yet. The Antarctic *Tritoniella belli* obtains defense against potential sym-patric predators by sequestering chimyl alcohol (McClintock et al. 1994a, 1994b, Bryan et al. 1998) and other unidentified compounds of unknown origin that deter sea star predators (Avila and Iken, unpublished results). Also, its egg masses seem to be chemically protected, but only against some potential predators (McClintock and Baker 1997, Bryan et al. 1998). Chimyl alcohol in particular is obtained from its prey, the soft coral *Clavularia frankliniana* (McClintock et al. 1994a, 1994b, Bryan et al. 1998). *Tritonia hamnerorum* from Florida sequesters a furano-germacrene, julieannafuran (Figure 3.2), from the sea fan *Gorgonia ventalina*, which is deterrent against reef fish in one of the few reliable field assays done for sea slugs (Cronin et al. 1995). *Tritonia* sp. feeding on the octocoral *Telesto riisei* accumulates its cytotoxic prostaglandins, punaglandins (Baker and Scheuer 1994). The Antarctic *T. challengeriana* is also chemically protected against sympatric sea star predators, although the compounds have not been identified yet (C. Avila and K. Iken, unpublished results). Homarine, a well-known and widely common antifoulant and deterrent, was found in *Marionia blainvillea* from the Mediterranean, although no role has been assigned in the slug (Affeld et al. 2006). *Tritoniopsis elegans* from China contains the tritoniopsins A–D from its coral prey, *Cladiella krempfi* (Ciavatta et al. 2011, Guo et al. 2012). *Doto pinnatifida* from the Atlantic presents dotofide, a terpenoid not found in its hydrozoan prey and with unknown function (Putz et al. 2011). De novo biosynthesis of 2,6-dimethyl-5-heptenal, a volatile component of the external extracts of *Melibe leonina,* was proved using stable-isotope experiments (Ayer and Andersen 1983, Barsby et al. 2002). *Melibe viridis*, an invasive species in the Mediterranean, contained an ichthyotoxic (against mosquito fish) prostaglandin lactone in mucus and cerata (Mollo et al. 2008), which was previously reported from *Tethys fimbria* (Di Marzo et al. 1991b). *T. fimbria*, in fact, contains a variety of de novo synthesized prostaglandins with diverse functions (Avila 1995).

3.2.1.3 Euarminida

Members of this group may feed on octocorals or on bryozoans. *Dermatobranchus ornatus* from China presents four known diterpenoids of the eunicellin class in its mantle extract (Zhang et al. unpublished data, cited in Wahidulla et al. 2006). Two of the compounds could be of dietary origin, from the cnidarian *Muricella sinensis*, and another one was previously isolated from an unidentified Pacific soft coral (Hochlowski and Faulkner 1980). No role as chemical defense has been attributed to any of them, although activities such as moderate cytotoxicity and inhibition of cell division of fertilized starfish eggs were cited (Wahidulla et al. 2006). The authors suggested that *D. ornatus* feeds selectively on cnidarians and accumulate their metabolites (Wahidulla et al. 2006). *D. otome*, instead, presented sesquiterpenoids of unknown origin and activity (Ishibashi et al. 2006). Similar diterpenoids as those of *D. ornatus* were previously isolated in the Mediterranean *Armina maculata* and its prey, the pennatulacean octocoral *Veretillum cynomorium* (Guerriero et al. 1987, 1988, 1990). *A. babai* possessed ceramide (Ishibashi et al. 2006), a compound found in the gorgonian *Acabaria undulata* (Shin and Seo 1995). In India, *A. babai* extracts showed antimicrobial activity (Ramya et al. 2014). Contrastingly, *Janolus cristatus* possessed a toxic tripeptide, janolusimide, of unknown origin (Cimino et al. 1986, Sodano and Spinella 1986).

Chemical studies of the South African *Leminda millecra* yielded several new sesquiterpenes, probably from its dietary octocorals (*Alcyonium foliatum, A. valdivae, A. fauri, Capnella thyrsoidea*) (Pika and Faulkner 1994, McPhail et al. 2001). Millecrone A (Figure 3.2) inhibited the growth of *Candida albicans,* and millecrone B was active against *Staphylococcus aureus* and *Bacillus subtilis*, while millecrol B was active only against the latter (Davies-Coleman 2006). Specimens from a different location contained also quinones; here some of the metabolites were traced to the gorgonian *Leptogorgia palma* (McPhail et al. 2001). Some of these quinones and hydroquinones induced apoptosis in esophageal tumor cells via generation of reactive oxygen moieties (McPhail et al. 2001, Whibley et al. 2007), but no role in chemical defense has been described so far. Surprisingly enough, three specimens of *Pseudotritonia quadrangularis* from Antarctica presented two polyketide structures, impossible to further identify due to the small amount of material collected (C. Avila, K. Iken, G. Cimino and A. Fontana, unpublished results). Recently, the Antarctic *Charcotia granulosa* was found to possess a unique linear homosesterterpene, granuloside (Cutignano et al. 2015). The authors suggested granuloside is de novo biosynthesized and stored in MDF-like structures, probably being released as deterrent (Moles et al. 2016).

3.2.1.4 Aeolidida

These generally cnidarian-feeder slugs are usually well protected by nematocysts from their prey relocated into their dorsal cerata. This is considered a key feature for the radiation in this group. However, they also contain interesting chemistry (Avila 1995). Several hydroxy and acetoxysterols from *Eudendrium* hydorids were found in the Mediterranean *Cratena peregrina*, *Flabellina affinis*, and *Flabellina* (*Coryphella*) *lineata* (Cimino et al. 1980). Later, two prenyl chromanols and an acid were isolated from *C. peregrina* feeding on the hydroid *Eudendrium racemosum* (Ciavatta et al. 1996). Homarine was found in *Cratena pilata* and *Cuthona gymnota* from the Atlantic (Woods Hole, MA), and *Hermissenda crassicornis* from California (location for these species was mistaken in [Affeld et al. 2006]), as well as in *Cuthona coerulea* from the Mediterranean and *Phestilla lugubris* from Australia (Affeld et al. 2006). No specific role has been proven for homarine in these species yet (Section 3.4). *Phestilla melanobrachia* contained indole alkaloids, probably derived from *Tubastrea* corals (Okuda et al. 1982). The aeolid *Phidiana militaris* from South China possessed the cytotoxic phidianidines A and B (Carbone et al. 2011, Guo et al. 2012) (Figure 3.2). Interestingly, Stachowicz and Lindquist compared the efficiency of chemistry versus nematocysts (2000), and more recently, an experimental study using agar-based food intended to test whether fish learn to avoid warning contrasting color patterns due to the unpleasant experience, by adding nematocysts to artificial models (Aguado and Marin 2007). All this has generated some debate that will be discussed in the following section.

Phyllodesmium species are particularly interesting because they do not possess functional nematocysts and they mimic the color and morphology of their soft coral hosts. Bogdanov and collaborators (2014) reviewed the defenses in this genus. In Australia, *P. longicirrum* presented cembranoid diterpenes from its prey, the soft coral *Sarcophytum trochelioforum* (Coll et al. 1985). *P. lizardensis* also from Australia accumulated muurolene sesquiterpenes from its coral prey, *Bayerxenia* (*Heteroxenia*) sp. (Affeld et al. 2009). *P. magnum* from China possessed a rare asteriscane sesquiterpene, together with other sesquiterpenes (Mao et al. 2011), related to the previously described defensive compounds of *P. guamensis* (11β-acetoxypukalide, Figure 3.2) (Slattery et al. 1998), leading the authors to suggest that they were obtained from *Sinularia* soft corals. *P. briarieum*, *P. longicirrum*, and *P. magnum* yielded new diterpenes and some sesquiterpenes, probably also from their octocoral diets (Bogdanov et al. 2014). Recent studies showed four new polycyclic diterpenes in *P. longicirrum* from Australia, as well as several other fish deterrent compounds (Bogdanov et al. 2016, 2017).

All in all, nudibranchs are continuously yielding interesting chemicals, although the defensive role in their natural environment is still scarcely understood due to the low number of field, reliable experimental data (Table 3.1). Evolutionary implications of these chemical strategies are discussed in the following section.

3.2.2 Pleurobranchoidea

The order Pleurobranchomorpha (previously known as Notaspidea, side-gill slugs, together with Tylodinoidea) has one child clade superfamily, Pleurobranchoidea, and two families, Pleurobranchaeidae and Pleurobranchidae. Adult pleurobranchomorphs are carnivorous or scavengers, characterized by a single, external gill on the right side, rolled rhinophores, prominent mantle, and an internal, reduced juvenile shell lost in adults. Many species produce defensive secretions against predators (Table 3.1), and the group is further known for the possession of defensive epidermal, sub epidermal and median buccal glands that exude sulfuric acid (Thompson 1988, Wägele et al. 2006, 2017). Among the species with copious acidic secretions are *Pleurobranchaea californica*, *Berthellina citrina*, and *Pleurobranchus strongi* from the Pacific (California), and *Berthella plumula* and *Pleurobranchus membranaceus* from the North Atlantic, with pHs as low as 1.2. Furthermore, *Berthella* sp. 1 from the Mediterranean and *Berthella* sp. 2 from Antarctica also display pH~1 (C. Avila, unpublished results). *P. californica* and *P. membranaceus*, in addition possess buccal acid glands (Gillette et al. 1991). Within Pleurobranchaeidae there are generalist scavengers, but certain species have preference for sea anemones and other cnidarians, sponges, nematodes, polychaetes, amphipods, opisthobranchs, ophiuroids, squid, or fish (Cattaneo-Vietti et al. 1993). In the first family, the Mediterranean *Pleurobranchaea meckelii* contained labdane aldehydes in the external part (Ciavatta et al. 1995), and the grey side-gill sea slug *P. maculata* from New Zealand presented dietary or symbiotic tetrodotoxin (TTX; Figure 3.3) in adult tissues, gonads, and eggs, suggesting a defensive function in adult individuals, who then transfer this to their progeny for protection (McNabb et al. 2010, Salvitti et al. 2015). Tetrodotoxin is a very potent neurotoxin that inhibits action potentials in nerves, and was previously found in many groups of poisonous animals, including fish (pufferfish, porcupinefish, sunfish, angelfish, triggerfish), toads, blue-ringed octopus, amphibians, sea stars, chaetognaths (arrow worms), nemerteans (ribbon worms), xanthid crabs, predator sea snails, and a polyclad flatworm. This toxin is produced by symbiotic bacteria (*Pseudoalteromonas*, *Pseudomonas*, *Vibrio*, and other strains) and is bioaccumulated along the food chain. Nonetheless, there is still a lot of debate as to whether the bacteria are truly the source of TTX (Chau et al. 2011, Salvitti et al. 2016). *Pleurobranchaea* is a model for neurobiology research due to the

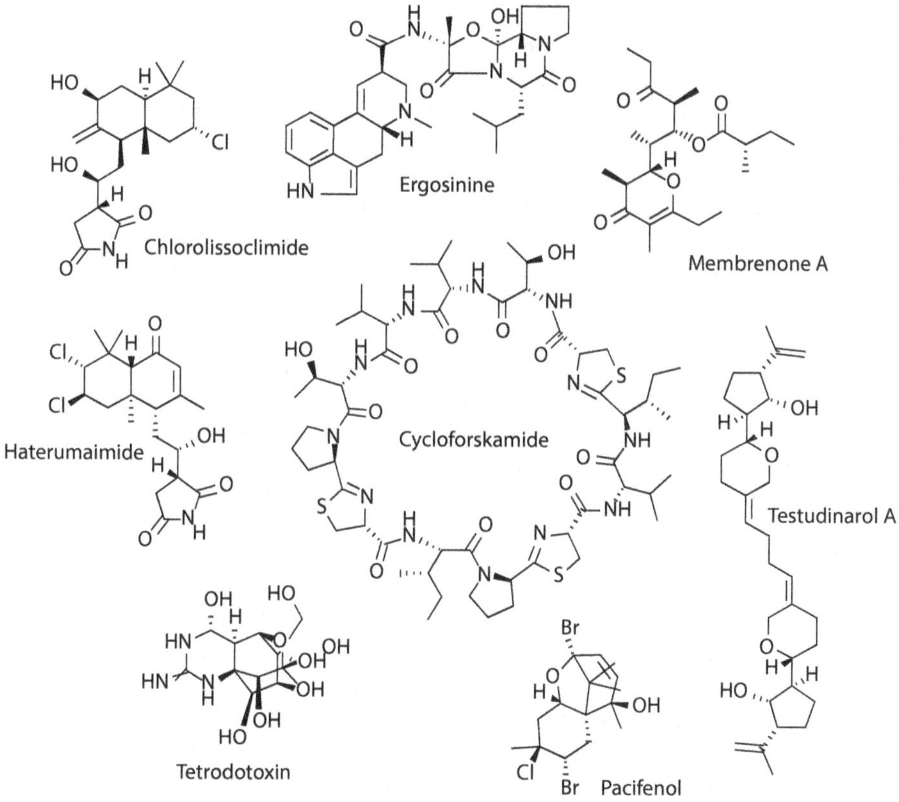

FIGURE 3.3 Structures of representative defensive compounds from Pleurobranchoidea.

escape swimming behavior, achieved by alternating dorsal and ventral body flexions (Jing and Gillette 1999).

In the second family, Pleurobranchidae, several representatives feed on ascidians, including species with specialized tunic bladder cells filled with acidic vesicles, which some slugs sequester for defense. *Pleurobranchus albiguttatus* and *P. forskalii* from Philippines yielded several cytotoxic chlorinated diterpenes (chlorolissoclimide, dichlorolissoclimide, haterumaimide D, H, L, M, 3ß-hydroxylissoclimide; Figure 3.3) derived from direct bioaccumulation and biotransformation of metabolites from their ascidian prey, *Lissoclinum* (Fu et al. 2004). *P. forskalii* from Indonesia additionally afforded a bioactive cyclic peptide, keenamide A, probably derived from another tunicate source (Wesson and Hamann 1996). Also recently, an ergot alkaloid peptide, ergosinine (Figure 3.3), and a cytotoxic macrocylic dodecapeptide, cycloforskamide (the latter probably from a diet on *Didemnum* ascidians; Figure 3.3), were found in specimens from Japan and Okinawa, respectively. Ergot alkaloids are sometimes psicoactive and are found in terrestrial higher plants and fungi. This was the first report of an ergopeptine from the sea (Tan et al. 2013, Wakimoto et al. 2013).

The predator *P. membranaceus*, feeding upon the ascidians *Botryllus schlosseri* and *Ascidia mentula*, contained membrenones (polypropionates; Figure 3.3) in the external part, likely biosynthesized de novo by the slug (Thompson and Slinn 1959, Ciavatta et al. 1993). In the Mediterranean, *P. testudinarius* presented in the mantle and mucus two triterpenoids, testudinariol A and B (Figure 3.3), similar to those of sponges (Spinella et al. 1997) and limpets (Pawlik et al. 1986). An unidentified species of *Pleurobranchus* from the South China Sea contains the halogenated algal sesquiterpene pacifenol (Figure 3.3), along with testudinariol B (Carbone 2007). The sesquiterpenoid was detected in the digestive gland, suggesting an incidental or indirect ingestion of items containing this algal metabolite (as it happens in the umbracullid *Tylodina perversa* feeding on cyanobacteria-rich tissues from the sponge *Aplysina aerophoba*) (Becerro et al. 2003). A Brazilian population of *P. areolatus* possessed diketopiperazines in the form of oxidized rodriguesin A derivatives (rodriguesic acids), and the respective esters (rodriguesic acid hydroxamates), probably accumulated when feeding on the ascidian *Didemnum* sp. (Pereira et al. 2014). On the contrary, the triterpenoids testudinariols, located in the mantle, are of probable de novo synthesis. In the pleurobranchs *Berthella* and *Berthellina,* mainly consumers of demosponges and occasionally of calcareous sponges and corals (Willan 1984), no chemical defense was recorded besides the acid secretions noted previously. The acid skin secretions of *B. citrina* were seen to cause immunotoxicity in mice spleen (Awaad and Moustafa 2016). Acid secretions with anti-predatory bioactivity were also reported in polar extracts of the mantle of the Antarctic *Bathyberthella antarctica* (Taboada et al. 2013, Moles et al. 2015). It is worth to mention that some metabolites of pleurobranchs share similarities with those of intertidal prosobranch and pulmonate limpets (Díaz-Marrero et al. 2003). For example, two diterpenes from *Pleurobranchaea meckelii* are close to that of the pulmonate *Trimusculus reticulatus.* Also, membrenone-C from *P. membranaceus* resembles the polypropionate vallartanone-D from the limpet *Siphonaria maura*. Moreover, testudinariols of *P. testudinarius* are chemically related to limatulone of *Lottia limatula*. The basis for this pattern though is not yet understood.

3.2.3 Tylodinoidea

Members in this group are sometimes referred to as false limpets, as they show large limpet-like, cap shaped external shell and a small mantle. Recent molecular phylogenetic reconstructions consider them a sister group to the Cephalaspidea, together with Anaspidea and Pteropoda (Figure 3.1). The clade comprises a single superfamily, Umbraculoidea, which includes Tylodinidae and Umbraculidae, previously placed in Notaspidea. Animals of both families feed upon sponges, and this was proposed to be the primitive condition for the previous Notaspidea and for Nudibranchia. In fact, it was speculated (with not much support) that nudibranchs could be polyphyletic and would include old spongivorous notaspideans (Valdés and Lozouet 2000).

Tylodinidae comprises two genera, of which *Tylodina* is the only one investigated from chemical and ecological perspectives (Table 3.1). In California, *T. fungina* was found to contain an ester derivative of the brominated alkaloid, 3,5-dibromotyrosine, which is used as feeding deterrent in *Aplysina* sponges (Andersen and Faulkner 1972). Similar strategies seem to exist in other congeners from different geographic areas. In the Mediterranean, *T. perversa* derives analogous metabolites from *Aplysina (Verongia) aerophoba* (Teeyapant et al. 1993). While *T. corticalis* from Australia selectively accumulates a few bromotyrosine-derived alkaloids with antibiotic activity from its diet, the sponge *Pseudoceratina purpurea* containing a larger variety of these products (Gotsbacher and Karuso 2015). The compounds are sequestered by the slugs and then stored in the mantle, mucus, reproductive organs and egg masses (Ebel et al. 1999, Thoms et al. 2004). These sponges are loaded with associated cyanobacteria, and *T. perversa* was observed to feed preferentially on the symbiotic tissues (Becerro et al. 2003). The mimetic yellow color of the slug and its eggs when they are on their *Aplysina* substrata is due to uranidine (Figure 3.4), a phenolic pigment that darkens upon exposure to the air, and which is also derived from the sponge (Cimino et al. 1984, Cimino and Sodano 1994). In the second family, the Umbraculidae, only one species has been chemically investigated. This is *Umbraculum mediterraneum* from the Mediterranean, which contains fatty acid esters with toxic properties towards mosquito fish (Cimino et al. 1988, Gavagnin et al. 1990). These compounds are the diacylglycerols umbraculumins A, B, and C (Figure 3.4), with B being an ester (Cimino et al. 1989). A transfer of such metabolites from the underlying prey-and-substrate sponge to the slug for its own defense was proposed (Cimino and Sodano 1994).

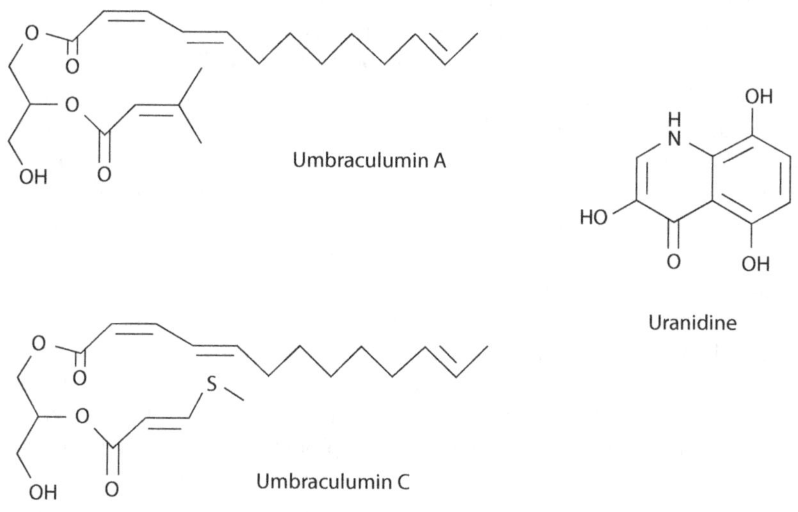

Umbraculumin A

Uranidine

Umbraculumin C

FIGURE 3.4 Structures of representative compounds from Tylodinoidea.

3.2.4 CEPHALASPIDEA

Cephalaspideans, known as head-shielded slugs and snails, are characterized by a broad cephalic shield used to burrow in the sand. Most of them possess a shell, although it may be reduced or internal. These can be large external bubble-like shells, where the animal can hide, or thin, internal, vestigial shells without any protective function. Basal, shelled cephalaspideans are protected against sympatric predators even when removing their shell in experimental bioassays, suggesting that shell reduction does not strictly correlate with a progressive increment of chemical defenses (Neves et al. 2009). This group was split up into Architectibranchia, Runcinacea, and Cephalaspidea *sensu stricto* (Malaquias et al. 2009, Oskars et al. 2015). Cephalaspidea *sensu lato* is phylogenetically related to Anaspidea and Pteropoda (Medina et al. 2011, Klussmann-Kolb and Dinapoli 2006). Among the synapomorphies shared, the Blochmann's gland is actually involved in chemical defense (Cutignano et al. 2012). and it morphologically resembles the ink gland of anaspideans (Wägele et al. 2006). Cephalaspideans are well known to possess polyketides and polyacetates, some of which are de novo biosynthesized, biomodified or just accumulated from their prey. Although most of them feed upon algae, some may also prey actively upon other heterobranchs (including other cephalaspideans), annelids, and sponges (Avila 1995). Among the Architectibranchia, the only species chemically analyzed is the acteonoid *Micromelo undatus* from the Canary Islands, which feeds on cirratulid annelids. Two polypropionates (micromelones A and B; Figure 3.5) with untested defensive properties were suggested to be de novo biosynthesized by the snail (Napolitano et al. 2008).

Several families have been chemically analyzed among Cephalaspidea *s.s.* The Mediterranean and amphi-Atlantic *Bulla striata* is a generalist algal feeder, able to synthesize its own defensive metabolites (Fontana et al. 2004, Neves et al. 2009). The cephalaspidean *Philinopsis depicta* (Aglajidae) feeds upon *B. striata*, and sequesters the polypropionates, aglajnes 1–3 (Figure 3.5), for its own defense (Cimino et al. 1985, 1987, Marín et al. 1999). In Hawaii, *Philinopsis speciosa* possessed the polypropionates niuhinone A and B (Figure 3.5), as well as a pyridine derivate, pulo'upone (Coval and Scheuer 1985, Coval et al. 1985), of unknown origin. *Bulla gouldiana* presented an isomer of pulo'upone (Figure 3.5) also found in its predator *Navanax inermis* (Spinella et al. 1993). These facts support that other cephalaspideans are probably the source of the metabolites in *P. speciosa*. In this species, also a depsipeptide, kulolide-1, a linear tetrapeptide, pupukeamide, additional peptides, and the macrolide tolytoxin-23-acetate were found (Reese et al. 1996, Nakao et al. 1996, 1998). Their molecular structures suggest an ultimate cyanobacterial origin for them, which could be further grazed by anaspideans and other opisthobranchs. *B. occidentalis* from the Gulf of Mexico led to the known niuhinones A and B, along with an acyclic polypropionate, niuhinone C (Cutignano et al. 2011). The occurrence of structurally related polypropionates in Bullidae species from different geographical areas suggests a de novo origin and adds further evidence on prey–predator relationships between species belonging to herbivorous *Bulla* and carnivorous *Philinopsis* genera. The aglajid *N. inermis* from the Pacific swallows whole *B. gouldiana* specimens, obtaining ichthyotoxic and shrimp-toxic polypropionates (niuhinone-B, isopulo'upone, and 5,6-dehydroagajne-3) (Spinella et al.

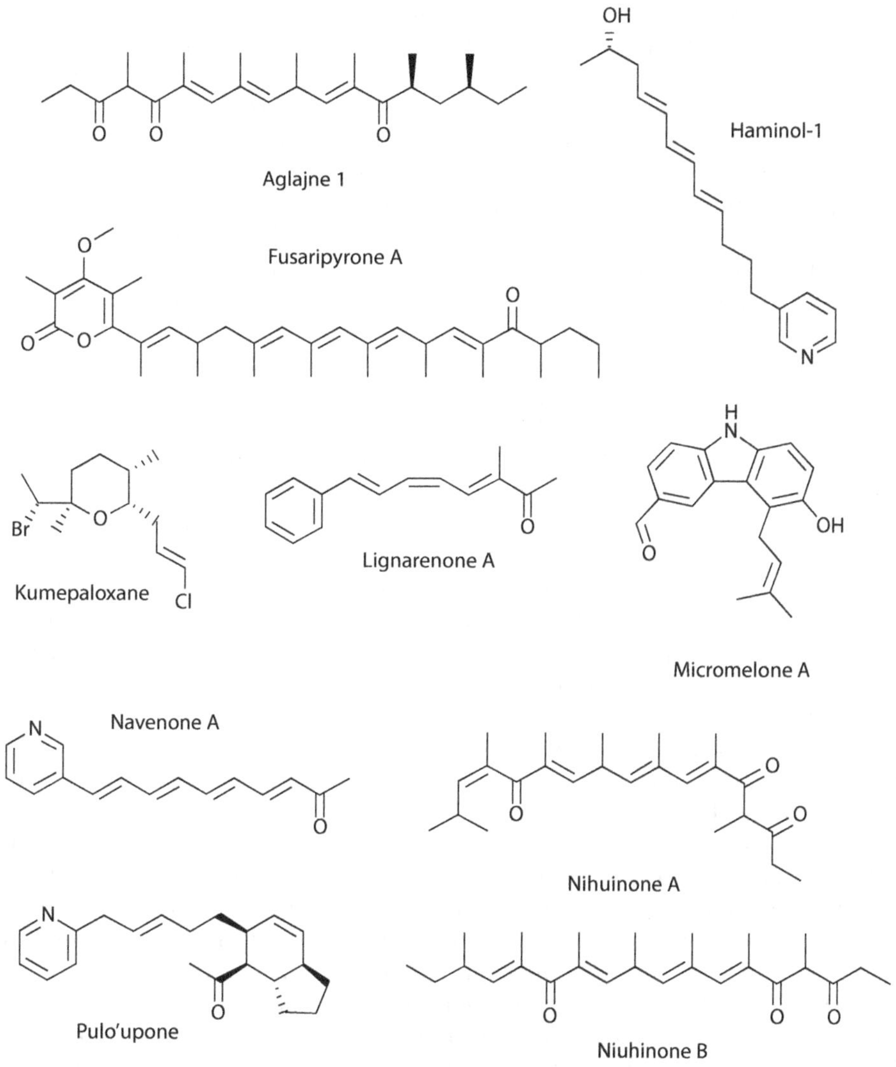

FIGURE 3.5 Structures of representative compounds known from Cephalaspidea.

1993). *N. inermis* also contained alarm pheromones, navenones A–C (Figure 3.5) in a specialized ventral gland, which are secreted in a bright yellow slime trail when disturbed, and which induce an avoidance-alarm response in trail-following conspecifics (Sleeper and Fenical 1977, Fenical et al. 1979, Coval and Scheuer 1985). These slugs are cannibalistic but, interestingly, secretion is not released when a small slug is attacked by a larger conspecific (Sleeper et al. 1980). Although synthesis experiments did not rigorously exclude the possibility of dietary origin, it was suggested that *N. inermis* may produce its own pheromones (Fenical et al. 1979). Homarine was found in *Aglaja tricolorata* from the Mediterranean, probably from predation on other sea slugs (Affeld et al. 2006). *Nakamigawaia spiralis* from Guam is chemically protected against sympatric reef fish, although the compounds have not yet been identified (Cruz-Rivera 2011).

Mediterranean haminoeids of the genus *Haminoea* (*H. exigua*, *H. fusari*, *H. orbignyana*, *H. orteai*, *H. navicula*) are chemically characterized by the presence of oxygenated 3-alkylpyridines, commonly known as haminols (Figure 3.5) (Cimino and Ghiselin 2009). Haminols secreted in the mucus act as alarm pheromones inducing escape reactions in conspecifics (Cimino 1991, Marín 1999), as well as in the laboratory nematode model *Caenorhabditis elegans* (Storvick et al. 2011). Haminoeids synthesize these polyacetates from nicotinic acid by sequential addition of acetic acid units. The de novo biosynthesis of haminol-1 and -2 in *H. orbignyana* was proved using stable-isotope labelled precursors (Cutignano et al. 2003, 2004), thus supporting the pioneering data mentioned previously for *Navanax inermis* (Fenical et al. 1979). Similarly, *H. japonica* contained a series of alkylphenols, closely related to alarm pheromones like navenone-C (Spinella et al. 1998, Marín et al. 1999). In addition to haminols, the Mediterranean *H. fusari* and *H. exigua* possess linear polypropionates in external and internal tissues: fusaripyrones A and B (Figure 3.5), and exiguapyrone and exiguaone, respectively (Cutignano et al. 2007, Nuzzo et al. 2015). The biosynthetic relationships among polypropionates to nalodi-onol, aglajne-3, and dehydroaglajne-3 from *Smaragdinella calyculata*, *Bulla striata*, and *B. gouldiana*, respectively, was compared; the structural analogies suggested a closely phylogenetic relationship (Cutignano et al. 2007). Indeed, the haminoeid *S. calyculata* from Hawaii contained the 2-alkylpyridine naloamine and the propionate nalodionol (Szabo et al. 1996), very similar to *Bulla* metabolites. Usually cephalaspideans contain either polyacetates or polypropionates, but the only two genera from the family Haminoeidae chemically analyzed to date (i.e., *Haminoea* and *Smaragdinella*) contain both. Their location in different body parts suggests different ecological functions. *H. cymbalum* from Guam presented an halogenated polyacetate, kumepaloxane (Figure 3.5), exuded when disturbed and acting as deterrent for porcupine fish (Poiner et al. 1989). Conspecific specimens from India as well as the Mediterranean *H. cyanomarginata* also presented kumepaloxane (Fontana et al. 2001, Mollo et al. 2008). Although the origin of kumepaloxane is unknown, its structure reminds that of some *Laurencia* algal metabolites (e.g., obtusenol) (Imre et al. 1981). These two *Haminoea* species are brightly colored and conspicuous, suggesting a warning coloration (Cimino and Ghiselin 2009). Indeed, it was proposed that aposematism, together with the wide distribution of its feeding source (i.e., *Laurencia* spp.), as predisposing factors to migration as an invasive species (Mollo et al. 2008). Therefore, low predation in the new geographical range could be directly attributed to a specific chemical defense against evolutionarily naïve, native generalist predators (Enge et al. 2012). Moreover, egg masses of the Eastern Pacific *H. virescens* present deterrent activity against decapods (Chang 2014). Finally, *H. templadoi* contained an unusual fatty acid, 10,15-eicosadienoic acid with unknown ecological properties (Carballeira et al. 1992).

Several species of the philinoid genus *Philine* exude sulfuric acid secretions from subepithelial notal glands (Thompson 1960, Wägele et al. 2006). *P. quadripartita* from the Mediterranean, East Atlantic, South Africa, and Indo-Pacific presented sulfuric and hydrochloric acid (Thompson 1986) in acidic glands. Likewise, other Pleurobranchomorpha and Nudibranchia species often present these glands (Wägele et al. 2006). However, two unidentified *Philine* species from the Mediterranean and Antarctica did not present acid secretions (C. Avila, unpublished results). The Scaphandridae species *Scaphander lignarius,* with a remarkable geographical range (Domènech et al. 2006, Eilertsen and Malaquias 2015), feeds mostly upon foraminiferans. Populations from muddy bottoms in the Eastern Atlantic and Mediterranean exude ω-arylmethylketones, named lignarenones (Figure 3.5), from the Blochmann's gland (Cutignano et al. 2012). Analogously to haminols from *Haminoea, S. lignarius* biosynthesizes the lignarenones through a polyketide pathway. This involves benzoic acid elongation with two acetate units, one propionate, and a final acetate, for lignarenone-C, or the loss of one C in lignarenones A and B (Cutignano et al. 2008, 2012) (Figure 3.10). These compounds could act as alarm pheromones, similarly to navenones from *Navanax* species (Cimino and Ghiselin 2009). This would further support the hypothesis that these species acquired the genes to produce the chemicals they use for survival, as assumed in other biosynthetic slugs. A series of minor metabolites correlated to lignarenones were constantly found among populations of *S. lignarius* from the Italian and Spanish Mediterranean coasts (Della Sala et al. 2007). Additionally, several polyunsaturated fatty acids with cytotoxic activity against a range of human cancer cells lines were described from Arctic specimens of *S. lignarius* (Vasskog et al. 2012).

Finally, two gastropterid species from Guam, *Sagaminopteron nigropunctatum* and *S. psychedelicum,* feeding on the sponge *Dysidea granulosa,* seem to rely on the same chemistry while exhibiting contrasting functional coloration patterns for defense. *S. psychedelicum* is aposematic, whereas *S. nigropunctatum* is highly cryptic on the sponge. Both species sequester polybrominated diphenyl ethers from their host sponge and accumulate them in the mantle and parapodia for defense (Becerro et al. 2006).

The overview of the chemical defenses in cephalaspideans, with different dietary habits, offers extraordinary models for investigation, as they produce many of their compounds, including polypropionates, but their defensive role, however, has not always been demonstrated (Table 3.1).

3.2.5 ANASPIDEA

Anaspideans or aplysiomorphs are commonly known as sea hares. They range from less than 2 cm to over 70 cm in length, and usually possess an internal, thin, laminar shell (Carefoot 1987). Their head presents a pair of oral tentacles and a pair of rolled rhinophores, which are sensory structures, derived by modification of the cephalaspidean head-shield. They present a well-developed pair of lateral parapodia, used for swimming in some species. Parapodia are sometimes completely fused into an anterior syphon, also used for movement (e.g., *Notarchus*). The gill is usually present inside the mantle cavity, protected by the vestigial shell. Sea hares produce ink and secretions from the opaline glands inside the mantle cavity. Most species are restricted to temperate and warm waters, except for *Aplysia parvula*, distributed globally, extending into the Arctic in Norway, and making it the most polar Anaspidean. However, no species are known from Antarctica (Carefoot 1987). The monogeneric family Akeridae is the more basal family within Anaspidea. *Akera* feeds on green algae, it has a big external bubble-shell, and lacks rhinophores and oral tentacles. This genus can swim using the parapodial lobes and secretes purple ink for defense (Thompson and Seaward 1989), but its natural products have not been investigated so far.

The internal shell of Aplysiidae can be thick and whorled as in *Dolabella*, thin and plate-like as in *Aplysia* and *Syphonota*, calcareous wedge-like in *Dolabrifera*, *Petalifera*, and *Phyllaplysia*, or be also completely absent as in *Stylocheilus*, *Bursatella*, and *Notarchus* (Rudman and Willan 1998). All anaspideans are herbivorous, feeding on algae, sea grass, or cyanobacteria, from which they obtain a rich arsenal of compounds. However, most of their secondary metabolites, mainly polyketides, terpenes, peptides, and depsipeptides, are obtained from red algae (Rhodophyta). This is clearly different from other herbivore sea slugs, i.e. Sacoglossa, which rely mostly on green algae (Chlorophyta; see below). There are very few cases of predation on sea hares reported in the literature. Unusual experiments conducted with sharks showed how specimens of *Aplysia brasiliana* (currently a junior synonym of *A. fasciata*) were avoided, even when offered camouflaged into fish fillets. The sharks rejected all the slug's parts, except for the buccal mass, likely depleted of metabolites (Kinnel et al. 1979). Pennings (1990) performed deterrence assays against kelp bass specimens forced to feed on *A. californica* fed on *Ulva* (chlorophyte) and on *Plocamium* (rhodophyte). The results showed significant deterrence, especially when slugs were fed on red alga, due to a richer terpene content. Plenty of algal defensive metabolites have been found in anaspideans' digestive gland, not much as a strategic location, since the animal should have to be fully ingested to yield deterrence, but likely because this is the site where detoxification takes place (see below).

Sea hares possess many glandular structures loaded with metabolites to deter predators, which are either exuded within secretions (opaline and ink glands) or accumulated in cells along the parapodial edges and mantle. The opaline and ink secretions provide defense against crabs in *A. juliana* (Kamiya et al. 1989). The ink gland of *A. californica*, *A. dactylomela*, and *A. parvula* possesses two effective deterrents against blue crabs: aplysioviolin (Figure 3.6) and phycoerythrobilin produced by modifying photosynthetic algal proteins in the digestive gland (Kamio et al. 2010a, 2010b). Additionally, mixing opaline and ink secretions enzymatically produces hydrogen peroxide, which also deters crabs, providing an apparent redundant protection. The emission of millimolar quantities of amino acids and the enzyme escapin was described in the ink-opaline secretions of *A. californica* when attacked by spiny lobsters (Kicklighter et al. 2005). These compounds stimulate sensory appetitive, ingestive behavior, and sensory disruption on the predator, which attends to a false food stimulus (phagomimicry) and grooming, while the sea hare can escape (Derby 2007, Sheybani et al. 2009). An internal compartmentalization of substances, in both the ink and opaline glands, seems to exist. In fact, escapin is only accumulated in the amber vesicles of the ink gland and not in the red-purple vesicles, which contain algal-derived chromophores, which give the distinctive violet color (Johnson et al. 2006). Ink secretions containing the enzyme escapin elicited deterrence to several fishes (Nusnbaum and Derby 2010), and tentacle shriveling and/or retraction in sea anemones, while opaline secretions elicited a feeding response on them (Kicklighter and Derby 2006). Escapin is a L-amino acid oxidase that quickly forms reaction

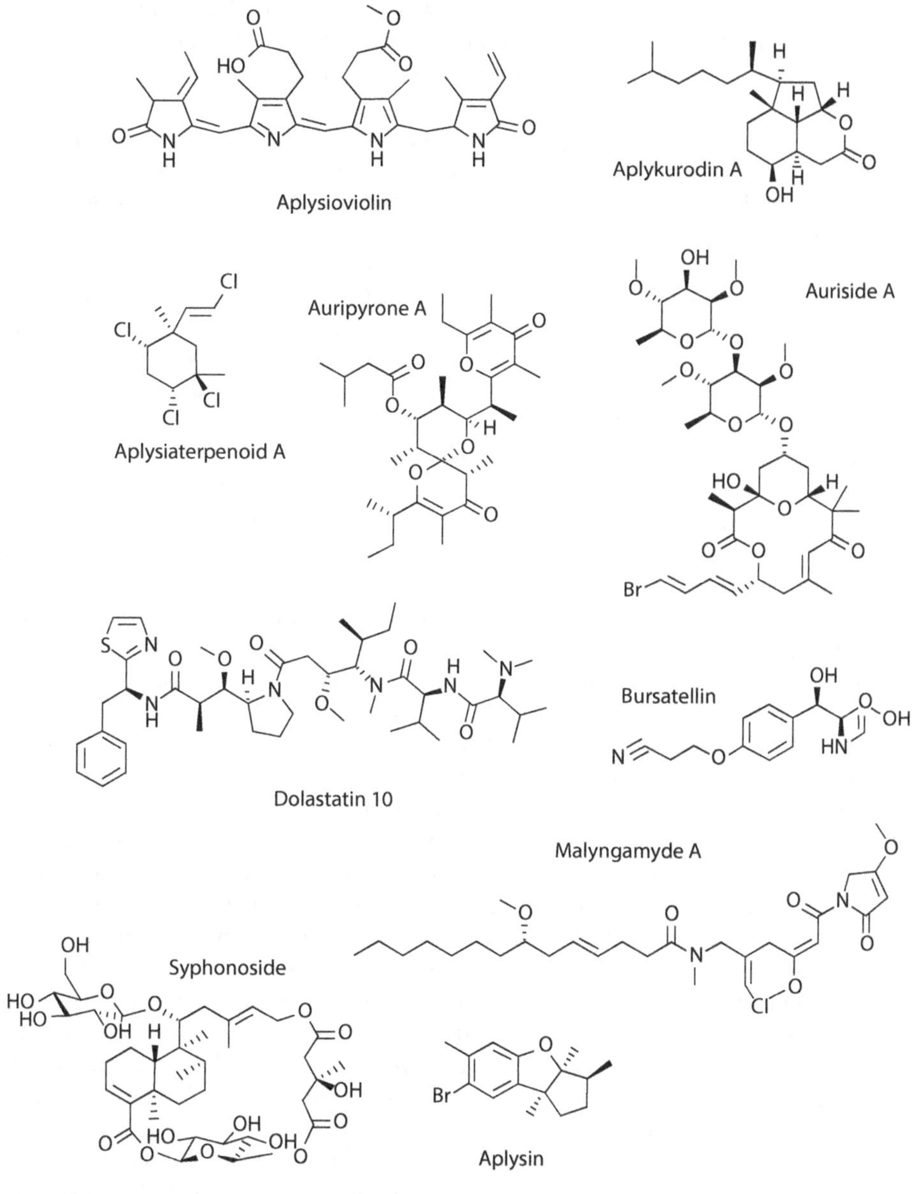

FIGURE 3.6 Structures of representative compounds from Anaspidea.

products when mixed with the opaline amino acids L-lysine and L-arginine (Kamio et al. 2009), when ink and opaline secretions are released into the sea hare mantle cavity. Thus, by packaging escapin and its substrate, lysine, into two separated glands the slug obtains defensive compounds from innocuous precursors, just when a predator attack happens and those products are released and mixed (Johnson et al. 2006).

The genus *Aplysia*, probably due to its wide distribution, large dimensions, and near-shore habitats is by far the most studied anaspidean with a plethora of studies involving natural products (see Avila 1995, Pereira et al. 2016 and references therein). They are generalist herbivores and contain many halogenated terpenoids as well as occasionally carotenoids, which provide UV protection (Czeczuga 1984). *A. californica* also presents mycosporine-like amino acids acting as alarm cues and/or sunscreens (Kamio et al. 2011). Most metabolites are diet-derived, and only a few seem to play a defensive role. Cryptic speciation might be behind the chemical differences found in some species of the genus, especially in widely distributed species (Ellingson and Krug 2006, Hunt et al. 2008, Jörger et al. 2012). *Aplysia* generally obtain and biomodify secondary metabolites from red algae. Specimens of *A. fasciata* from the Mediterranean possessed polyhalogenated monoterpenes related to *Plocamium* red algae (Imperato et al. 1977); however, a different population possessed ichthyotoxic, degraded sterols (i.e., 4-acetylaplykurodin-B, aplykurodinone B, and 3-*epi*-aplykurodinone B) (Spinella et al. 1992) in the external tissues, similarly to the steroids present in the Atlantic *A. fasciata* (Ortega et al. 1997). These steroids are closely related to aplykurodin B from the Pacific *A. kurodai* (see below) (Miyamoto et al. 1986). Recently, three new sesquiterpenes, 6-hydroxy-1-brasilene, epibrasilenol acetate, and 6-*epi*-β-snyderol, one acetogenin, (3Z, 9Z)-7-chloro-6-hydroxy-12-oxo-pentadeca-3,9-dien-1-yne, and one diterpene, 16-acetoxy-15-bromo-7-hydroxy-9(11)-parguerene, were further characterized from Mediterranean *A. fasciata* (Ioannou et al. 2009). Although *A. fasciata* has been synonymized to the Western Atlantic *A. brasiliana* (Medina et al. 2005), completely different metabolites were found in both populations (Kinnel et al. 1977, 1979, Stallard and Fenical 1978, Dieter et al. 1979), supporting bioaccumulation of bioactive molecules from diverse local diets. Sri Lankan *A. oculifera* possessed metabolites from *Laurencia*, for example, srilankenyne (De Silva et al. 1983), while the digestive gland of Indic and West Pacific specimens presented two brominated, isomeric acetylenes (Schulte et al. 1981). Mucus secretions in NE Atlantic and Mediterranean *A. punctata*, contained *Plocamium coccineum* halogenated monoterpenes (Quiñoá et al. 1989, Ortega et al. 1997), and epidioxy sterols, some of them identical to those of *A. depilans*, in its digestive gland (Jiménez et al. 1986). New Zealand specimens of the ubiquitous *A. parvula* obtained brominated and chlorinated terpenoids from *Plocamium costatum*, with costatone being fourteen times more concentrated in the slug (Grkovic et al. 2005). Similar to *A. dactylomela*, *A. parvula* also acetylates two phenolic compounds in the digestive gland, allolaurinterol and isolaurinesol, after ingesting the algae *Laurencia distichophylla* and *Hymenea variolosa*. Interestingly, as mentioned previously, allolaurinterol was found in the spongivore and dorid-eater nudibranch *Ceratosoma amoena* (Carefoot 1987), and suggested possible transference from *A. parvula* and/or *A. dactylomela* egg masses as a result of accidental ingestion (McDonald and Nybakken 1997). *A. parvula* from Guam feeds upon the red alga *Portieria hornemanii*, sequestering the halogenated monoterpenes, apakaochtodene A and B for fish deterrence (Ginsburg and Paul 2001). Japanese *A. parvula* accumulates the ichthyotoxic brominated acetogenin and a dicyclic ether, aplyparvunin (Miyamoto et al. 1995); whereas specimens from SE Africa derive a halogenated cyclic acetogenin, (3Z)-bromofucin, from *Laurencia implicata* (McPhail and Davies-Coleman 2005). Three new brominated diterpenes, glandulaurencianols A–C, from *Laurencia*

glandulifera and *A. punctata* (Kladi et al. 2014) as well as punctatol also from
A. punctata (Findlay and Li 2002) shared the skeleton of laurencianol, an antibacterial
diterpene in *Laurencia obtusa* (Caccamese et al. 1982). The widely studied, NE
Atlantic *A. californica* transforms, through acid catalysis, laurinterol and pacifenol
from *Laurencia* and *Plocamium* (Faulkner and Stallard 1973, Faulkner et al. 1973,
Stallard and Faulkner 1974, Ireland et al. 1976) into the halogenated terpenoids: aplysin
(Figure 3.6) and pacifidiene, within the digestive gland. The Atlantic and Mediterranean
A. depilans specializes in feeding upon brown algae of the family Dictyotaceae. This
species presents guanidine diterpenes, dictyol-A and B, from *Dictyota coriacea*
(Minale and Riccio 1976), as well as steroids (Di Prisco et al. 1973) and peroxy sterols
(Jiménez et al. 1986) in Atlantic specimens, instead of the halogenated monoterpenes
associated to red algae (Finer et al. 1979, Quiñoá et al. 1989). *A. depilans* also yielded
ichthyotoxic fatty acid lactones, aplyolides A–E (Spinella et al. 1997), as well as
nine brominated diterpenes featuring the rare dactylomelane skeleton, previously
described in *A. dactylomela* (Petraki et al. 2015). In the Eastern Pacific, *A. vaccaria*
also contained ichthyotoxic non-halogenated diterpenoids, crenulides, in the digestive
gland, derived from *Dictyota crenulata* (Midland et al. 1983, Sun et al. 1983). The
cosmopolitan anaspidean *A. juliana*, instead, exclusively consumes green algae, and
presents two toxic chlorophyll derivatives, pyropheophorbides *a* and *b* (Kobayashi et al.
1991) and an halogenated diterpenoid lactone (Alvi et al. 1991). The purple secretion of
this sea hare includes also an antibacterial and cytotoxic peptide, julianin-S (Kamiya
et al. 1989), and its egg mass is protected from microbial infections by unsaturated
fatty acids (Benkendorff et al. 2005). Twenty-two polyunsaturated fatty acids and
eight carotenoids with anti-inflammatory potential were identified in the digestive
gland of *A. depilans* (Oliveira et al. 2015).

The extensively studied anaspidean *Aplysia kurodai* from the NW Pacific, contains cytotoxic polyketide macrolides, aplyronines A–H (Yamada et al. 1993, Ojika
et al. 2012), and the cytotoxic aplaminal-1 (Kuroda and Kigoshi 2008). Some of
these compounds are in clinical trials as antitumor drugs (Kigoshi and Kita 2015).
Additionally, *A. kurodai* displays halogenated and brominated mono- and diterpenes
distributed in different body parts, such as kurodainol (Katayama et al. 1982), aplysi-
aterpenoids A–D (Figure 3.6) (Kusumi et al. 1987, Miyamoto et al. 1988), aplysin-20
(Yamamura and Hirata 1971), isoaplysin-20 (Yamamura and Terada 1977), aplysiadiol (Ojika et al. 1990), *epi*-aplysin-20, and *ent*-isoconcinndiol, arguably derived
from isoconncindiol in *Laurencia snyderae* (Ojika et al. 1992). Two brominated sesquiterpenoids, aplysin (Figure 3.6) and aplysinol (Yamamura and Hirata 1963), and
aplykurodin A and B (Miyamoto et al. 1986) (Figure 3.6) were also found in *A. kurodai*. This species is also source of several alkaloids, like the cytotoxic aplaminone,
neoaplaminone, and neoaplaminone sulfate (Kigoshi et al. 1990), as well as other
compounds including aplydilactone (Ojika et al. 1990), a dieicosanoid lactone, aply-
sepsine (Ojika et al. 1993), and a 1,4-benzoidiasepine alkaloid. The egg masses and
albumen gland moreover contain cytotoxic peptides (aplysianin A) (Iijima et al. 1995,
Johnson et al. 2006) with antimicrobial and antifungal activity to reduce biofouling
(Iijima et al. 1995, 2003, Johnson et al. 2006, Kaviarasan et al. 2012). Finally, *A. dactylomela* is one of the most prolific sources of natural products, such as polyketides,
diterpenes, triterpenes, and sesquiterpenes, which are mostly biotransformed from

red algal compounds. Curiously, porcupine fish displays identical annulated pattern as *A. dactylomela* (Heck and Weinstein 1978), in a possible case of Batesian mimicry. Thereby, these harmless fish can imitate the warning signals of defended slugs. The species is presently divided into two siblings based on both molecular and morphological data: *A. dactylomela* and *A. argus*, from the Atlantic and Indo-Pacific, respectively (Alexander and Valdés 2013). Specimens of *A. argus* from Bahamas presented cyclic polyketides and acetylenic ethers: dibromochloro dactylyne (McDonald et al. 1975) and its isomer isodactylyne in the digestive gland (Vanderah and Schmitz 1976, Gopichand et al. 1981); whereas whole body extracts had sesquiterpene ethers: dactyloxene-B and dihydroxydeodactol monoacetate (Schmitz and McDonald 1974, Schmitz et al. 1978, 1980). Manzo and collaborators (2005) found a series of ethers in specimens of the South China Sea, some of which described from *Laurencia pinnatifida* and *L. obtusa*, as well as other enantiomers. In addition, these authors found the sesquiterpene ether (+)-brasilenol, known from *L. obtusa* and *Aplysia brasiliana*. The mantle of *A. argus* from China also possessed the triterpene polyethers similar to *Laurencia*, the aplysiols A and B (Manzo et al. 2007), and other studies characterized the halogenated bicyclic C_{15} ether dactylallene from Atlantic specimens (Ciavatta et al. 1997). Further evidence of the biotransformation of *Laurencia* metabolites was found in Australian specimens, containing aplysiastatin (Pettit et al. 1977). Japanese (Ichiba and Higa 1986), Brazilian (Kaiser et al. 1998), and South African specimens (McPhail et al. 1999) presented further halogenated sesquiterpenes, while East Caribbean specimens contained brominated diterpenes: parguerol, deoxyparguerol, and isoparguerol (Schmitz et al. 1982). Moreover, *A. argus* from the Western Caribbean biotransforms and bioaccumulates ichthyotoxic metabolites from the brown alga *Stypopodium* zonale (Gerwick et al. 1979); contrastingly, specimens feeding on *Cladophora vagabunda*, present the bioactive diphenyl ether 2-(2′,4′dibromophenoxy)-dibromoanisole from this green alga in the digestive gland (Kuniyoshi et al. 1985). Several species of *Laurencia* and *A. argus* from the Indo-Pacific contained twenty halogenated metabolites related to: syndrean, chamigrane, non-chamigrane sesquiterpene, cuparane, bromoindole, and a C_{15} acetogenin (Palaniveloo and Vairappan 2013). A new sesquiterpene, dactylomelatriol, was found *A. dactylomela* from the Canary Islands, derived from an omphalane skeleton, and previously described only in terrestrial fungi (Díaz-Marrero et al. 2012). The authors suggested a modification from a precursor obtained from *Laurencia*. To add more evidence to the biotransformation of metabolites in *A. dactylomela*, a new diterpene with significant in vitro anti-neuroinflammatory activity, dactyloditerpenol acetate (Jiménez-Romero et al. 2014), resulted structurally related to laurenditerpenol from *Laurencia intricata* (Mohammed et al. 2004). Moreover, evidence of biomodification due to photochemical addition was suggested for aplydactone, a brominated ladderane sesquiterpenoid likely formed from the chamigranes dactylone (Matsuura et al. 2016). *A. argus* from China presents derived *Laurencia* products, cyclopropane and cyclobutane rings, in five brominated primaranes diterpenoids, all of them biotransformed by adding acetoxy groups (Bian et al. 2014). In the Mediterranean-invasive *Syphonota geographica,* two degraded sterols, aplykurodinone-1 and -2, were isolated from the mantle (Gavagnin et al. 2005); whereas a macrocyclic glycoterpenoid, syphonoside (Figure 3.6), derived from the sea grass *Halophila stipulacea* was accumulated in the digestive gland (Carbone et al. 2008). This might be the precursor of other terpenoids found exclusively in *S. geographica*, thus suggesting again biotransformation of the main algal metabolite, syphonoside.

Dolabella auricularia (*D. californica*) lives camouflaged between algae, in shallow bottoms from tropical and subtropical waters of the Indic and Pacific Oceans. Adults of *D. auricularia* are sold to acuariophils, and the egg masses (named "Lokot") are consumed as a delicacy in the Philippines. In the drug industry, antitumor and antineoplastic compounds from this species are currently in clinical trials (Kigoshi and Kita 2015). *D. auricularia* feeds on brown and red algae and modifies their secondary metabolites, while being able as well to synthesize de novo polypropionates and peptides (Pennings et al. 1999). Among the metabolites derived from red algae three polyketides, the macrolide glycosides, aurisides A and B (Figure 3.6) (Sone et al. 1996) and the polyether bromotriterpene aurilol (Suenaga et al. 1998), display cytotoxicity against HeLa tumor cell line. Doliculols A and B are non-halogenated, acetylenic cyclic ethers similar to esthers in *Laurencia* (Ojika et al. 1993), and Dolabelides A–D are moderate cytotoxic macrolides (Ojika et al. 1995, Suenaga et al. 1997, Yamada et al. 2010). Specimens from the Gulf of California presented diterpenes, the dolabellanes (Ireland et al. 1976, Ireland and Faulkner 1977) from brown Dictyotaceae algae. In the Indian Ocean, these slugs yield a similar diterpene, dolatriol, together with the monoterpenoid (-)–loliolide (Pettit et al. 1976, 1980), that is probably a degraded carotenoid from algae. The internal organs of *D. auricularia* presented biosynthetic polypropionates, such as auripyrone-A and -B (Figure 3.6), in small quantities (Suenaga et al. 1996). These resemble the non-contiguous polypropionates, dolabriferols A–C, from *Dolabrifera dolabrifera* (Ciavatta et al. 1996, Jiménez-Romero et al. 2012). This latter species is supposed to synthesize polypropionates de novo, although this has not been confirmed experimentally. It has been suggested that the lack of defensive metabolites due to a diet change (*D. dolabrifera* feeds on diatoms instead of macroalgae) could be promoting a de novo biosynthesis of defenses (Cimino and Ghiselin 2009). Although *Dolabella auricularia* and *D. dolabrifera* are not closely related species, both are unique in possessing polypropionates within Anaspidea. *Dolabella* is noteworthy for various antitumoral linear and cyclic peptides, especially dolastatins (Figure 3.6) (Kigoshi and Kita 2015). A modified form of dolastatin 10 (Figure 3.6), the monomethyl auristatin E, is an antimitotic agent that blocks the polymerization of tubulin (Verdier-Pinard et al. 2000, Doronina et al. 2003). In 2011, brentuximab vedotin (Adcetris®), an antibody-dolastatin 10 conjugate, was approved by the FDA for the treatment of Hodgkin's lymphoma (Newman and Cragg 2014). Pettit and co-authors (1982, 1993) described the cyclic peptide dolastatin 3, together with dolastatins 10–15 from *D. auricularia*, all of them with mild to strong biological activities (Yamada and Kigoshi 1997). Californian *D. auricularia* possess a new cancer cell inhibitor, the macrocyclic lactone dolastatin 19, obtained together with the macrolides debromoaplysiatoxin and anhydrodebromoaplysiatoxin. Plenty of other cytotoxic peptides have been characterized from this species: dolastatins C, D (Sone et al. 1993a, 1993b), dolastatin H and isodolastatin H (Sone et al. 1996), dolastatin G and nordolastatin G (Mutou et al. 1996), and aurilide (Suenaga et al. 1996, 2004). Dolastatin G and nordolastatin G, with moderate cytotoxicity (Yamada et al. 2010), are analogs of lyngbyastatin 2 and nor-lyngbyastatin 2 isolated from the cyanobacterium *Lyngbya majuscula* from Guam (Luesch et al. 1999). Dolabellin, a moderate cytotoxic bisthiazole metabolite, was found in a Japanese population (Sone et al. 1995); this molecule strongly resembles hectochlorin, isolated from Jamaican populations of the cyanobacterium *L. majuscula* (Márquez et al. 2002). These findings keep supporting that many compounds originally isolated from sea hares are most likely of cyanobacterial origin. Nonetheless, the glycoprotein, dolabellanin A, found in the albumen gland is probably biosynthesized de novo by the mollusk (Kisugi et al. 1992). This compound shows antibacterial and antineoplastic properties, which may serve to protect the eggs from bacterial pathogens.

The last two genera, *Stylocheilus* and *Bursatella*, belong to Notarchinae, and are both cyanobacterial feeders. The tropical *Stylocheilus* presents two chemically analyzed species: *S. striatus* distributed mainly in the Caribbean and Cape Verde Islands, and *S. longicauda* from the Indo-Pacific Ocean (West Africa to Australia) (OBIS 2014). Both species names are mixed indistinctly in the literature, since they were formerly synonymized. A chemo-taxonomical review is needed to confirm which chemistry belongs to each one. Kato and Scheuer (1974, 1975), characterized aplysiatoxin and debromoaplysiatoxin from the Hawaiian *S. longicauda*. Further studies described the non-toxic amide, stylocheilamide from other Hawaiian specimens, feeding on a red alga and tuft mats of the cyanobacterium *Lyngbya majuscula* (Rose et al. 1978). A similar chemical structure, with ichthyotoxic properties, was revised for being identical to acetyl malyngamide I from the Hawaiian *Lyngbya majuscula* (Todd and Gerwick 1995). Different Hawaiian populations of *S. longicauda*, further afforded complex proline esters, makalika and makalikone, together with lyngbyatoxin A with antitumor properties (Gallimore et al. 2000). Two alkaloids, malyngamides O and P, again structurally correlated to the compounds of *L. majuscula* were also described (Gallimore and Scheuer 2000). The first malyngamides, A and B (Figure 3.6), were described from the marine cyanobacterium *Microcoleus lyngbyaceus* (supposedly *L. majuscula*) (Pennings and Paul 1993). *S. longicauda* from Guam bioaccumulate malyngamydes and biotransform malyngamyde B into an acetate. Feeding preference assays showed that *S. longicauda* had preference for cyanobacteria, and its metabolites were then used for deterrence against fish, amphipods, crabs, and even the herbivorous cephalaspidean *Diniatys dentifer* (Paul and Pennings 1991, Capper et al. 2006). Recently, the filamentous cyanobacterial genus *Moorea* emerged from molecular evidence on various cosmopolitan and pan-tropical samples (Engene et al. 2012). Due to the morphological similarity to the genus "*Lyngbya*," this group has often been incorrectly cited in the literature. The extraordinarily rich production of bioactive secondary metabolites of these cyanobacteria seemed to be well profited by *S. longicauda* (Paul et al. 2007). Similarly, *Bursatella leachii* feeds on "*Lyngbya*" and accumulates lyngbyatoxin A and debromoaplysiatoxin in its digestive gland (Capper et al. 2005). This species inhabits the Indo-West Pacific, the Caribbean, and the Mediterranean Sea. Although several subspecies are presently described, some metabolites seem to be ubiquitously present in populations worldwide. Bursatellin, a diol nitrile alkaloid, (Figure 3.6), was found in a population from Puerto Rico (*B. leachii plei*), which is structurally related to chloramphenicol (Gopichand and Schmitz 1980). The subspecies from the Tyrrhenian Sea (*B. leachii leachii*) and the Adriatic Sea (*B. leachii savignyana*) presented the + and − isomer forms of bursatellin, respectively, in external extracts (Cimino et al. 1987). Specimens from New Zealand revealed another alkaloid, derived from cyanobacteria, malyngamide S, with cytotoxic and antiinflammatory properties (Appleton et al. 2002). In addition, Thai specimens possessed the cytotoxic hectochlorin and deacetylhectochlorin, previously isolated from *Lyngbya majuscula* and structurally related to dolabellin from *Dolabella auricularia* (Suntornchashwej et al. 2005).

In general, anaspidean chemical protection relies on a wide variety of natural products mostly halogenated terpenes, but also macrolides, sterols, and alkaloids, mostly derived from algal prey (Table 3.1), and often displaying cytotoxic, antibacterial, antifungal, antiviral, and antifeedant activities (Pereira et al. 2016). Different species of sea hares seem to have similar mechanisms for sequestering algal metabolites rather than processes tightly linked to particular compounds. The generalist feeding habits of sea hares correlated with the panoply of natural products harbored in each algal species (especially red) undoubtedly contribute to the amount of metabolites found in anaspideans. For instance, species from the red alga *Portiera*, which is ingested by *Aplysia parvula*, possess ~300 compounds only in the Philippines (Payo et al. 2011). Additionally, cyanobacteria and probably other microorganisms commonly fouling algae (e.g., dinoflagellates and diatoms) are intentionally or accidentally ingested by anaspideans, and are furnishing with a vast array of natural products. Undoubtedly, the high ecological success of this widely distributed group can be attributed to their ability to feed upon a wide variety of toxic seaweeds and cyanobacteria, and using the consumed metabolites for their own benefit, either by storing or transforming them. These factors may also facilitate their invasion into perturbated areas, such as the Mediterranean Sea (Mollo et al. 2008, Moles et al. 2017b).

3.2.6 PTEROPODA

Pteropods comprise two taxa of pelagic heterobranchs: the shelled Thecosomata and the shell-less Gymnosomata. They are phylogenetically related to anaspideans, and also cephalaspideans, sharing some characters with them (Malaquias et al. 2009, Klussmann-Kolb and Dinapoli 2006, Dayrat and Tillier 2002). Both groups present wing-like extensions of the foot used to swim in the water column. Thecosomes feed on phytoplankton by using mucous nets, while gymnosomes are active predators, usually preying upon thecosomes. If marine benthic organisms have been only partially studied, the planktonic ones are still much less known. Although they are found worldwide, only one species of gymnosome has been chemically characterized (Table 3.1). The Antarctic *Clione limacina* provided a polypropionate-derived natural product, pteroenone (Figure 3.7), with strong feeding deterrence against fish predators

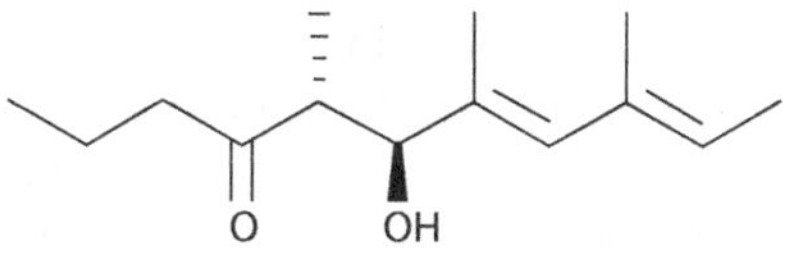

Pteroenone

FIGURE 3.7 Structure of pteroenone from the pteropod *Clione limacina*.

(Yoshida et al. 1995). A pelagic hyperiid crustacean, *Hyperiella dilatata,* increases its chances of survival by carrying the chemical-defended pteropod on its back, in an unusual symbiosis based on noxiousness (McClintock and Janssen 1990). Pteroenone does not occur in the prey of *C. limacina,* the thecosome *Limacina helicina* (Bryan et al. 1995), suggesting a de novo biosynthesis. Once again, the use of defensive poly-propionates accompanies a major shift in diet and habitat. Additionally, *L. helicina* and *C. limacina* present mycosporine-like amino acids obtained from phytoplankton, conferring them UV protection (Whitehead et al. 2001). *C. limacina* further contains odd-chain fatty acids, absent in its prey (Kattner et al. 1998).

3.2.7 SACOGLOSSA

Sacoglossans are usually cryptic and possess a modified radula with a single row of teeth, used to pierce algal cells and feed suctorially. The worn-out teeth are depos-ited in a "sac," giving the name to the group (Jensen 1980, 1993). This characteristic "sac" is only missing in the most primitive genus, *Cylindrobulla.* Sacoglossa includes primitive shelled forms, as well as shell-less forms, both being monophyletic sister groups. Their absence in the Poles is remarkable. This seems to be related to the nar-row trophic requirements of early forms, correlated to the absence of photosynthetic prey items and reduced photoperiod, avoiding the group to radiate in high latitudes (Jensen 1996, 1997).

These slugs are specialized herbivores, arisen from a shelled ancestry that, as the living shelled lineages, fed by suction on siphonaceous algae, probably *Caulerpa* or *Udotea.* These algae contain bioactive sesquiterpenoids and diterpenoids (often acyclic) toxic towards microorganisms, invertebrate eggs, sperm, larvae, fishes, and mollusks. Thus, the ancestral sacoglossan would have fed upon algal substrata, rich in deterrents against generalist herbivores, by developing mechanisms of detoxifi-cation (Paul and Fenical 1983, 1984, Marín and Ros 2004). Among the shell-less Plakobranchidae (=Elysiidae), the most primitive species feed on siphonales (fami-lies Derbesiaceae, Caulerpaceae, Bryopsidaceae, and Codiaceae) and have adopted the ability to retain algal plastids (cleptoplasty) and keep them functional for months (Laetz et al. 2017). The slugs may digest them when needed, and thus, they work as food reserves, with no translocation of photosynthates to the animal (Christa et al. 2014). Actually, sacoglossans have even been considered ectoparasites of the endo-symbiotic chloroplasts (Muscatine and Greene 1973, Hinde and Smith 1974, Clark and Busacca 1978, Ros and Marín 1991, Rumpho et al. 2000, Wägele and Johnsen 2001, Christa et al. 2014). More evolved shell-less species within Plakobranchidae and Limapontioidae have a broad dietary range, but are instead unable to retain chlo-roplasts, being able to de novo biosynthesize their natural products. Primitive shelled sacoglossans specialized on *Caulerpa* live in close association, and even adopt simi-lar shapes as the algae, but they do not maintain photosynthesizing chloroplasts. They instead release algal noxious products within defensive mucous secretions (Cimino and Ghiselin 2009). In certain primitive shelled and some cerata-bearing shell-less sacoglossans, autotomy is also used to deter predators. This has been observed in the shelled *Oxynoe* and *Lobiger* with the detachment of the tail and/or parapodia, and in some species of Limapontioidea that shed the cerata.

The use of toxic natural products from caulerpalean algae by sacoglossans was first reported in 1970 (Doty and Aguilar-Santos 1970) for the Californian *Oxynoe panamensis*. This species accumulated caulerpicin and caulerpin (Figure 3.8) from *Caulerpa sertularioides*. Comparative studies have shown striking chemical and dietary parallelisms for shelled congeneric species from tropical and temperate

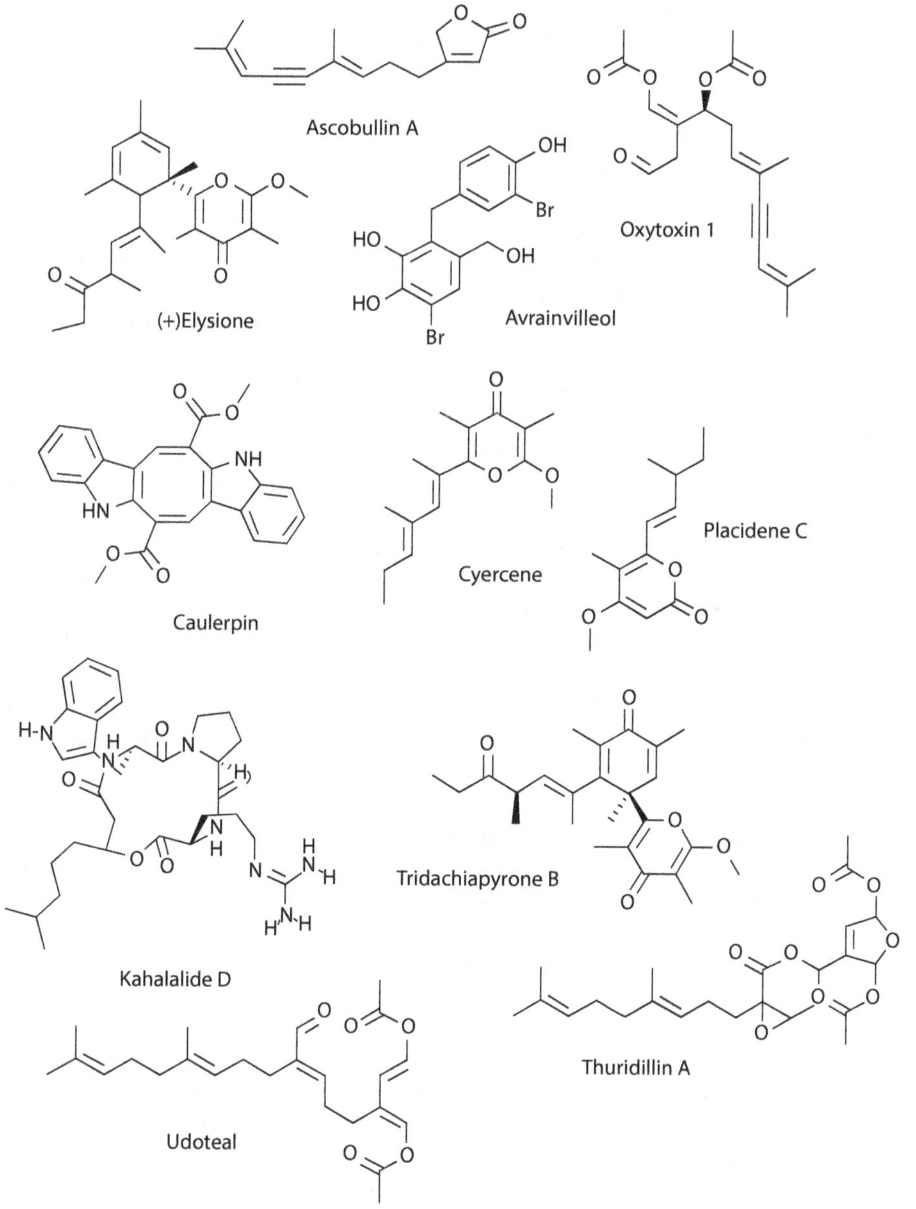

FIGURE 3.8 Structures of representative defensive compounds from Sacoglossa.

seas, suggesting that these are plesiomorphic, retained features (Gavagnin et al. 2000). In the Mediterranean, *Oxynoe olivacea*, *Lobiger serradifalci*, and *Ascobulla* (=*Cylindrobulla*) *fragilis* feed on different parts of *Caulerpa prolifera*, which contains the sesquiterpenoid caulerpenyne (Gavagnin et al. 1994). *O. olivacea* and *A. fragilis* are able to transform caulerpenyne, present in the digestive gland, into the ichthyotoxic aldehydes, oxytoxin-1 and -2 (Figure 3.8), and translocate them to the mucus and mantle. *Ascobulla fragilis* further retains unmodified caulerpenyne in the digestive gland, likely as primitive character. *L. serradifalci* only contained oxytoxin-1 in the parapodial lobes and in the defensive mucus (Cimino et al. 1990). The Caribbean *Ascobulla ulla* (which feeds on *Caulerpa fastigiata*), *Oxynoe antillarum* (feeding on *Caulerpa* sp.), and *Lobiger souberveii* (which feeds on *Caulerpa racemosa*), showed all the same basic pattern of chemicals as their Mediterranean counterparts: caulerpenyne accumulation (only this compound in *L. souberveii*) and transformation into oxytoxins (Gavagnin et al. 2000). In *A. ulla,* oxytoxins have been replaced by the structurally related, less reactive (*i.e.* detoxification), ascobullin A and B (Figure 3.8). Caulerpenyne, along with the modified sesquiterpenoid, volvatellin, was also found in an Indian *Volvatella* sp. (Fontana et al. 1999).

The rest of Sacoglossa are all shell-less in adult stages. There are two groups separated on the basis of morphology: the family Plakobranchidae with leaf-shaped bodies and foot edges forming parapodia (similar to those of anaspideans), often green in color due to the retained algal chloroplasts; and a second group comprising three families: Caliphyllidae, Limapontiidae and Hermaeidae, displaying dorsal ceratal processes in two series (similar, but not the same as those of nudibranchs) (Jensen 1996). Plakobranchidae experienced important shifts from one food item to another during evolution, but usually within the same order of algae (Bryopsidales=Caulerpales=Codiales=Siphonales) (Gosliner 1995). *Bosellia* is a highly modified genus. In the Mediterranean, *B. mimetica* feeds on *Halimeda tuna* (Gavagnin et al. 1994), from which it acquires the diterpenoid halimedatrial, from its inactivated form halimedatetraacetate. Halimedatetraacetate is quite similar to caulerpenyne of *Caulerpa* except for having 20 rather than 15 C atoms (Paul and Fenical 1983, 1984, Paul and van Alstyne 1988, 1992). *Elysia* species generally use *Caulerpa* and closely related algae as diet (Jensen 1980, 1992, 1997), and as major source of defensive compounds. Several Caribbean species show analogies with shelled sacoglossans and contain caulerpenyne and oxytoxin-1, indicating that they all modify algal metabolites similarly. This was observed in *E. subornata* (feeding on *Caulerpa prolifera*), and *E. patina* and *E. nisbeti* (*Caulerpa* sp.) (Gavagnin et al. 2000). Moreover, *Elysia* cf. *expunsa* from India was reported to possess caulerpenyne, together with minor amounts of the reduced derivatives, dihydrocaulerpenyne, and expansinol (Ciavatta et al. 2006), suggesting similarities with the Mexican *Ascobulla ulla*. There are also parallelisms in species feeding on *Halimeda* in the Caribbean with the aforementioned Mediterranean *Bosellia*: *E. pusilla* (formerly *E. halimedae*) uses halimedatetraacetate and halimedatrial from *Halimeda macroloba* (Paul and van Alstyne 1988), whereas *E. tuca* obtains halimedatetraacetate from *Halimeda incrassata* (Gavagnin et al. 2000). Similarly, *E. translucens* from Bermuda and Florida uses a closely related metabolite to halimedatetraacetate, the linear diterpenoid, udoteal

(Figure 3.8) from *Udotea* (Paul et al. 1982, Paul and Fenical 1984, Gavagnin et al. 1994). In contrast, *Elysia* species that feed upon other types of algae possessed quite different chemical defenses. In Hawaiian waters, *E. rufescens* obtains toxic polypeptides, kahalalides A–F (Figure 3.8) (Hamann and Scheuer 1993, Hamann et al. 1996, Becerro et al. 2001), and two acyclic kahalalides H–J from *Bryopsis* (Goetz et al. 1997), while *E. ornata* and its *Bryopsis* sp. prey contain the depsipeptide kahalalide O (Horgen et al. 2000). Further kahalalides A–F, K, O, and G–J have been found in *E. grandifolia* from India (Ashour et al. 2006), although not all them were found in its diet, *Bryopsis plumosa* (Tilvi and Naik 2007). Symbiotic bacteria, like *Vibrio*, produce kahalalides in *E. rufescens* (Davis et al. 2014). Rather than deriving chemicals from food, other sacoglossans de novo biosynthesize polypropionates. Polypropionates are rare in nature, but rather common in marine mollusks, where they have been found in cephalaspideans, anaspideans, pulmonates, and sacoglossans (see above). The Mediterranean *E. timida* feeds exclusively upon *Acetabularia acetabulum*, and contains three propionates, none of which occurs in the alga (Gavagnin et al. 1994). *E. viridis* feeds upon *Codium vermiliara* and biosynthesizes the polypropionate (+) elysione (Gavagnin et al. 1994, Cutignano et al. 2009) (Figure 3.8), which is also produced by *E. chlorotica* feeding on *Cladophora* (Dawe and Wright 1986). Related compounds were obtained from *E. (Tridachia) crispata* and *E. (Tridachiella) diomedea* (Ireland et al. 1978, Ireland and Faulkner 1981, Ksebati and Schmitz 1985, Gosliner 1995) in the Caribbean and Pacific respectively, which produce the unusual propionate-derived γ-pyrones, tridachiapyrones A–F (Figure 3.8), including tridachiahydropyrone and 9,10-deoxytridachione, respectively. Tridachiapyrone A has been suggested to be the enantiomer of (+) elysione, as well as tridachiapyrone-C and crispatene (Dawe and Wright 1986). These polyene γ-pyrone metabolites were recently localized at cell membranes of sacoglossans, where they may act as sunscreens, protecting the photosynthetic apparatus of the chloroplasts (Ireland and Scheuer 1979, Powell et al. 2017). In this particular case, such compounds seem to represent protected forms of elysione under light exposition (Ireland and Scheuer 1979). Elysiapyrones were also found in *E. diomedea* from the Pacific (Cueto et al. 2005, Díaz-Marrero et al. 2008), whereas a population of *E. crispata* from Venezuela was found to contain, in addition to typical algal metabolites and polypropionates, the sesquiterpenoids, crispatenine and onchidal, the latter also found in the pulmonate *Onchidella* (Gavagnin et al. 1996, 1997, 2000). The chemical profile of *E. crispata* is indeed intriguing for including all types of strategies: bioaccumulation, biotransformation, and biosynthesis. Recently, *E. patagonica,* the first sacoglossan described from Argentina (Patagonia) and also the Southern-most species, was observed to accumulate chloroplasts from *Bryopsis* cf. *plumosa* and to synthesize phototridachiapyrone J, with UV protective properties, as other polypropionates (Carbone et al. 2013). These γ-pyrone polypropionates have been proposed to be chemical markers for a selected group of *Elysia* sacoglossans including: *E. crispata, E. diomedea, E. timida, E. chlorotica, E. viridis,* and *E. patagonica.* Interestingly, the sacoglossan *Placobranchus ocellatus* is also featured by the same chemical pattern (Cimino and Ghiselin 2009, Carbone et al. 2013).

Placobranchus ocellatus and *Placobranchus* sp. from Hawaii possess propionate-derived γ-pyrones (e.g., 9,10-deoxy-tridachione, photodeoxytridachione, tridachiahydropyrone B and C, iso-9,10-deoxy-tridachione), similar to those of *E. crispata* and *E. diomedea*, probably used as sunscreens (Ireland and Scheuer 1979, Manzo et al. 2005). Recently, related propionates were reported in *P. ocellatus* from Philippines (including 9,10-deoxytridachione) (Fu et al. 2000). These compounds were also found along with the elysiapyrone-related ocellapyrones in specimens from India (Manzo et al. 2005). The genus *Thuridilla* comprises mostly tropical species, having evolved towards brilliantly colored forms rather than cryptic, with only basal members being cryptic (Gosliner 1995). The presence of functional chloroplasts in this genus (e.g., *T. gracilis, T. ratna, T. hopei*) and the consequent need of sunlight (Wägele and Johnsen 2001) gives the animals further reasons for acquiring chemical protection and use aposematism. *T. hopei* from the Mediterranean is a brightly colored species, exhibiting a dark green mantle with blue, red, and yellow lines, which contained three diterpenoids, thuridillins A–C (Gavagnin et al. 1993) (Figure 3.8), and three nor-thuridillonals (nor-thuridillonal, dihydronor-thuridillonal, deacetyl-dihydro-nor-thuridillonal), all believed to derive from a common dietary precursor, epoxylactone (Carbone et al. 2014). *T. hopei* feeds and lives in association with the small caulerpal alga *Derbesia tenuissima* (Gavagnin et al. 1994), which contains the highly active diterpenoid epoxylactone 1/A (with a conjugated dienolacetate moiety). Thuridillins D–F were found in *T. splendens* (Somerville et al. 2012). Thuridillins, as well as nor-thuridillonals, might derive by oxydation/reduction processes from algal precursors. The conversion of these metabolites is thought to be a detoxification mechanism (Gavagnin et al. 1993, Carbone et al. 2014).

Sacoglossans possessing cerata, or descending from ancestors possessing them, are divided into the families Limapontiidae, Caliphyllidae, and Hermaeidae, and form a monophyletic sister group of the Plakobranchidae. Similar to primitive conchoid sacoglossans, some species use autotomy of their cerata as a defensive mechanism (Jensen 1996, Somerville et al. 2012). Among the Limapontiidae, the Caribbean *Costasiella ocellifera* (*C. lilianae*), feeding on *Avrainvillea longicaulis* contains avrainvilleol (Figure 3.8), a brominated diphenylmethane dietary algal derivative, with deterrent properties against fish (Hay et al. 1990, Gavagnin et al. 2000). In the Mediterranean, *Placida dendritica* feeds upon both *Bryopsis* and *Codium* making use of the chloroplasts for camouflage. This slug has no evident capacity to shed appendages for defense, yet it has a rich secondary metabolite arsenal. It produces polypropionates γ-pyrones, which have a pyrone chromophore, and are closely related to cyercene (Vardaro et al. 1992). These are biosynthesized following a pathway similar to bacteria (mixed acetate-propionate), but different from fungi (methylation of polyacetate) (Cutignano et al. 2009). In addition, *P. dendritica* contains α-pyrones (placidenes C–F; Figure 3.8) and placidene A, a hydroperoxide 10-hydroperoxyl (Cutignano et al. 2003). *Ercolania viridis* (*E. funerea*) from the Mediterranean, which easily undergoes autotomy of cerata, feeds on *Chaetomorpha linum* and retains non-active chloroplasts. It possesses similar γ- and α-pyrone polypropionates as those reported in *P. dendritica* (Vardaro et al. 1992), including 7-methyl-cyercene-1, also found in the phytopathogenic fungus *Phoma tracheiphila* (Tringali et al. 1993). This finding has raised the hypothesis of a symbiotic origin for these metabolites in mollusks. Cyercenes however exhibit a wide occurrence and variability, and according to what was mentioned previously, the slugs seem to really build them from propionate, whereas fungi biosynthesize propionic units from acetate followed by methylation (Gavagnin et al. 1996).

Among the Caliphyllidae, cyercene polypropionates (Figure 3.8) were found in the Mediterranean *Cyerce cristallina*, of unknown feeding habits and with autotomy behavior (Di Marzo et al. 1991a, Vardaro et al. 1991). They were also found in *C. nigricans* from the Great Barrier Reef, which feeds upon *Chlorodesmis* and further contains the algal diterpenoid chlorodesmin (Roussis et al. 1990). *Mourgona germaineae* from the Caribbean readily sheds cerata when disturbed, while it secretes a toxic mucus (Jensen 1984). This slug retains active chloroplasts, and accumulates prenylated bromohydroquinones, including cyclocymopol, from the calcareous green alga *Cymopolia barbata* (Högberg et al. 1976). Cyclocymopol is similar to the brominated diphenylmethane derivative, avrainvilleol, found in *Costasiella ocellifera* and *Avrainvillea longicaulis* (Hay et al. 1990). The isolation of aplysiopsenes A–D in *Aplysiopsis formosa* from Azores (Atlantic) further confirmed the existence of α-pyrones in Hermaeidae (Ciavatta et al. 2009). Production of propionates by cerata-bearing sacoglossans is often associated with defensive autotomy and regenerative stimulation of the lost cerata, as proven in regeneration experiments with *Hydra* (Di Marzo et al. 1993). According to this, *Caliphylla mediterranea*, which exhibits no autotomy, has been reported to lack propionates or any other type of protective chemistry. This slug seems to rely on a defensive cryptic behavior to avoid predators, by capturing algal chloroplasts from *Bryopsis plumula* for camouflage (Hamann et al. 1996). In *Placida dendritica* instead, polypropionate γ-pyrones are likely used only for deterrence, since this species does not autotomize. During evolution, polypropionate compounds could originate as means of defense, and be secondarily used for regeneration, or vice versa.

Early sacoglossans seem to be very specialized grazers that became able to feed on siphonaceous algae, stealing functional chloroplasts and bioaccumulating chemical defenses. The defensive use of the algal metabolites and its biotransformed compounds allowed sacoglossans to extend their ecological niches (Table 3.1). They then diversified towards wider diets on different algal species. When the dietary shift meant ingestion on poorly defended algae and, thus, a loss of protection from dietary metabolites, the slugs may have adopted de novo mechanisms for biosynthesis (e.g., polypropionates [Cimino and Ghiselin 2009]). The role of polypropionates in regeneration suggests a dual functional origin, either as defenses, or as signaling molecules for healing, or both. Photoactive polypropionates are also important as sunscreens, as defensive chemicals, or probably as both, due to the use of chloroplasts, and their need of living in photophylic environments for photosynthesis (Cimino and Ghiselin 2009).

3.2.8 Pulmonata

Pulmonates inhabit terrestrial, fresh-water, and marine environments. Their mantle cavity is transformed into a "lung" to breath. Pulmonates are divided into Stylommatophora, with stalked eyes, and Basommatophora, with unstalked eyes. The few marine examples belong to the Basommatophora. These are limpets restricted to the intertidal and possess a wide array of propionates and terpenoids (Table 3.1). *Siphonaria* possess two classes of polypropionates, some of them ichthyotoxic and located in the mucus and mantle border (Darias et al. 2006). Type I includes acyclic compounds with a 2-pyrone and furanone rings, such as siphonaienolone (Figure 3.9), structurally related to polypropionates of cephalaspideans. Type I polypropionates are present in several species from Australia, West and East Atlantic, and South Africa; some of them displaying antimicrobial and cytotoxic activity (Biskupiak and Ireland 1983, Hochlowski and Faulkner 1983, 1984, Norte et al. 1990, 1994, Paul et al. 1997, Beukes et al. 1999). Type II polypropionates vary in the length of the alkyl chain, yielding a profuse polyoxygenated network that frequently

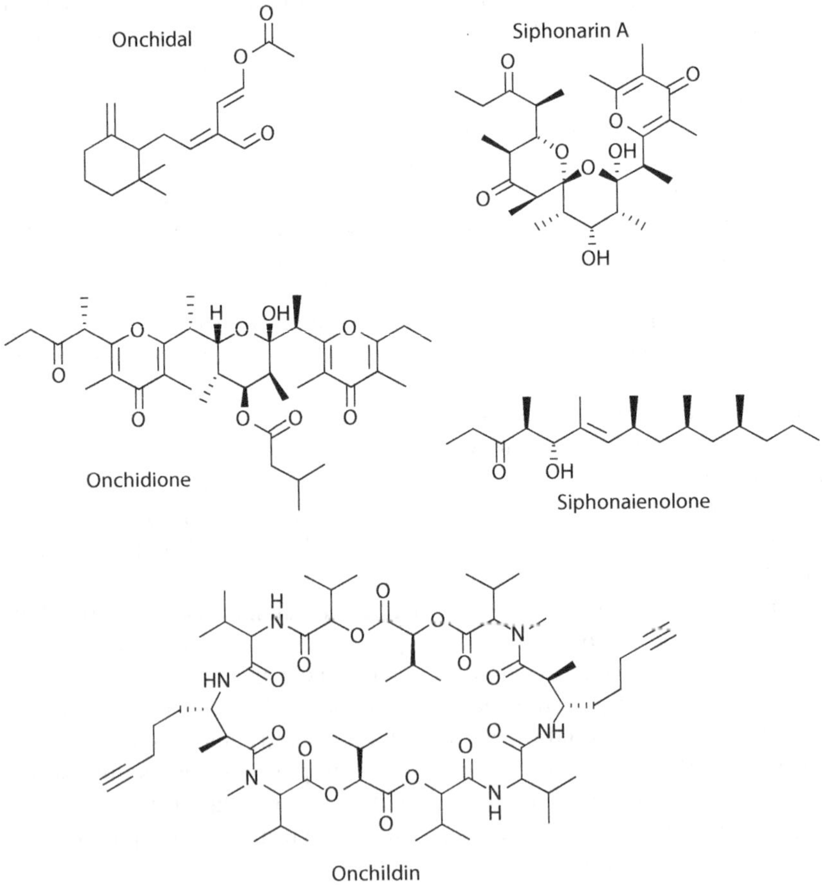

FIGURE 3.9 Structures of representative compounds from Pulmonata.

cyclizes. Type II polypropionates, such as siphonarin A (Figure 3.9), are similar to those of actinomycetes. These metabolites are found in *Siphonaria* species from Australia, New Zealand, North-East Pacific, Pacific Islands, and South Africa (Hochlowski et al. 1983, 1984, Roll et al. 1986, Manker and Faulkner 1989, Faulkner et al. 1989, Brecknell et al. 2000). The studied species include *S. capensis*, *S. concinna*, *S. cristatus*, and *S. serrata*. Some of the siphonariid polypropionates displayed deterrence against fish (Hochlowski et al. 1983) and are considered to be de novo synthesized by the mollusks (Manker et al. 1988). In contrast to *Siphonaria*, the genus *Trimusculus* lacks propionates, but presents a single type of labdane diterpenoids. Some of these metabolites are similar to those of the pleurobranchomorph *Pleurobranchaea meckelii* (Ciavatta et al. 1995) (see above). *T. reticulatus* from New Zealand displayed deterrence against sea stars by secreting several diterpenes located in the mantle and foot (Manker and Faulkner 1996). Other species from Chile and South Africa, including *T. costatus* and *T. peruvianus*, also displayed cytotoxic and antifeeding activities (Rovirosa et al. 1992, Gray et al. 1998, Díaz-Marrero et al. 2003, Van et al. 2008). Finally, representatives of the family Onchidiidae present repugnatorial glands containing defensive metabolites, such as sesquiterpenoids, depsipeptide acetates, and propionates. From *Onchidella binneyi*, onchidal (Figure 3.9) is secreted in its active form, ancistrodial, to elicit deterrence against predators (Ireland and Faulkner 1978). Several species of the genus *Onchidella* from different latitudes may have uneven amounts of secondary metabolites (Abramson et al. 1989), but this does not seem to alter their effectivity in deterring sea stars (Young et al. 1986). Additional cytotoxic depsipeptides, such as onchidin (Figure 3.9), have been characterized from *Onchidium* spp. (Rodríguez et al. 1994, Fernández et al. 1996). *Onchidium* spp., as well as *Peronia peronii*, also possessed similar propionates, as those of *Siphonaria* (Biskupiak and Ireland 1985, Carbone et al. 2013). Finally, *Onchidium* sp. from China possessed onchidione (Figure 3.9) in mucus and mantle (Carbone et al. 2009), as well as onchidiol and 4-*epi*-onchidiol (Guo et al. 2012, Wang et al. 2012). Ilikonapyrone esters were isolated in *O. verruculatum* (Ireland et al. 1984), while the tropical *Onchidium* sp. contained cytotoxic acetates and propionates (Rodríguez et al. 1992).

3.3 ORIGIN AND ANATOMICAL ALLOCATION OF CHEMICAL DEFENSES

As mentioned in previous sections, sea slugs have an astonishing variety of trophic strategies, which often correlate to their chemical defenses. They may accumulate natural products directly from their diet on marine algae and other invertebrates, biotransform them, and/or de novo biosynthesize them (Table 3.1). In fact, there may be several mechanisms coexisting in the same species for different compounds and roles. Moreover, the possibility of symbiont microorganisms producing some of the natural compounds should not be disregarded (see below). Examples of direct dietary derived defense metabolites include many anaspideans, some chromodorids, and sacoglossans. In general, anaspideans do not discern among dietary chemicals, however, other groups are able to sort and sequester specific compounds from their prey, such as terpenes from sponges in chromodorid nudibranchs. The compounds then may be redistributed towards specific body regions, particularly the mantle, and further they may be stored in functional defensive structures, for example, MDFs

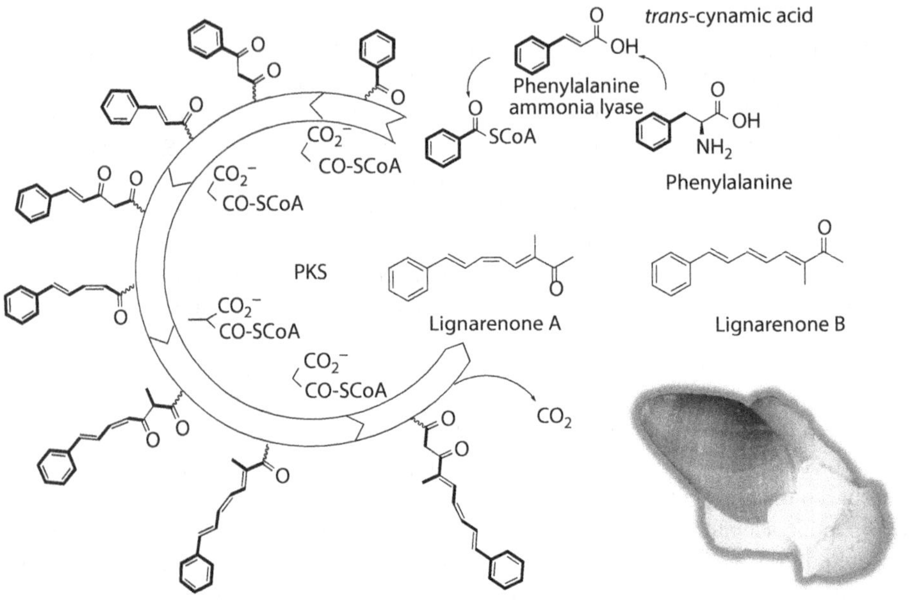

FIGURE 3.10 Proposed biosynthetic pathway of lignarenone A in the Blochmann's glands of the cephalaspidean *Scaphander lignarius*, as reported in Cutignano et al. (2012). The putative PKS and corresponding intermediates assume a minimal polyketide process. Bold bonds indicate units added at each round of chain extension, as determined by labelling experiments.

(Figure 3.10). The accumulation of bioactive products in characteristic glandular bodies seems to serve a dual role, as a strategy against predators, and also, simultaneously preventing autotoxicity in the slug (see below). In fact, Wägele (2004) argued that storing toxic dietary chemicals in MDFs could have fostered the radiation in chromodorids, by allowing them to benefit by feeding upon toxic sponges and expand their dietary resources. Biotransformation consists on modifications of dietary compounds, either by detoxification when the original products are highly toxic, or by transformation to obtain more stable or more toxic metabolites (Avila 1995). Usually, any of such processes entails energy costs. Within *Ascobulla* sacoglossans, for example, *A. fragilis* transforms the algal sesquiterpene caulerpenyne into the more toxic oxytoxins, while *A. ulla* detoxifies the algal compound to the less active ascobullins (Gavagnin et al. 2000). Algal lipases were initially suggested to be involved in the biotransformation of caulerpenyne (Jung and Pohnert 2001). However, the Mediterranean *Oxynoe olivacea* transforms caulerpenyne to its defensive allomone, oxytoxin-2, by using two lipases, LIP-1 and LIP-2 (Cutignano et al. 2004). There is also extensive evidence that *Aplysia* spp. are able to biomodify polyketides, monoterpenes, diterpenes, triterpenes, and sesquiterpenes from *Laurencia* and *Plocamium* red algae (see above). The anaspidean *Syphonota geographica* similarly transforms terpenoids from a macrocyclic glycoterpenoid from the sea grass *Halophila stipulacea* (Carbone et al. 2008). *Stylocheilus longicauda* also transforms malyngamyde B obtained from cyanobacteria into an acetate (Pennings and Paul 1993).

The acquisition of defensive chemicals in slugs co-evolved with the selection of living host substrata, serving as habitat and food source, similarly to what happened in insects (Opitz and Muller 2009). Some prey–predator pairs sharing the same natural products are, for example, *Bugula* bryozoans-polycerids, *Dysidea* sponges-*Hypselodoris*, cyanobacteria-*Stylocheilus*, Caulerpales-elysiids, to name a few. Indeed, there are many natural sources from which sea slugs could obtain their chemical defenses (or the precursors to build them), these include: cyanobacteria, algae (green, red, and brown) for some cephalaspidean, anaspideans, sacoglossans; sponges (mostly demosponges) for some dorids, phyllidids, umbraculids; cnidarians (anthozoans and hydrozoans) for aeolidid nudibranchs and pleurobranchoids; bryozoans for dorid and dendronotacean nudibranchs; ascidians for some dorid nudibranchs and pleurobranchs; and even annelids are the prey of some cephalaspidea taxa, as well as nematodes, polychaetes, amphipods, other opisthobranchs, ophiuroids, squid, and fish are among the diet items of pleurobrachomorphids and dendronotaceans (Todd 1981, McDonald and Nybakken 1997). Chemoreception and mechanisms for prey detection have been rarely investigated, with only a few studies reporting chemotaxis towards prey, and sea slug behavior with ingestion conditioning (Avila 1998, Iken et al. 2002, Wyeth and Willows 2006, Paul et al. 2007). Interesting studies tried to link evolution with diet specialization, chemical patterns, and biogeography (e.g., sacoglossans [Gavagnin et al. 2000]). Remarkably, some sponge-eating dorids may not be as specialized as previously thought, and they may show some diet variation according to their patchy or scarce sponge preys at each different locality (Penney 2013). This of course, may also generate interspecific chemical variability. In general, animals that incorporate defensive compounds from food tend to vary their metabolite profiles geographically (Avila 1993, Faulkner et al. 1990). Contrastingly, if the organisms synthesize their own protective chemicals, they are not as dependent upon a certain food type, and their chemical profile is likely to be more constant over a wider geographical range (e.g., several pleurobrachomorphs, some dorids). Some groups have never colonized certain areas in the planet, probably limited by their narrow feeding habits. A clear example is the absence of anaspideans and sacoglossans in polar and subpolar regions (with very few exceptions, see below [Carefoot 1987, Linse 1999]).

Some sea slugs are clear examples of defensive escalation, in the sense that predators often acquire better protection than their defended prey because those sequester dietary deterrents, which accumulate in higher concentrations with respect to the original source (Vermeij 1987, 1994). This has been proved for some aeolidid species of *Phyllodesmium* (Slattery et al. 1998, Affeld et al. 2009, Bogdanov et al. 2014), chromodoridids (Da Cruz et al. 2012, Carbone et al. 2013), as well as in sacoglossans (Cutignano et al. 2004), and anaspideans (Grkovic et al. 2005). The accumulation of defensive molecules in specific glands has been described in chromodorids, which possess the typical MDFs, but also in other groups harboring different structures. Dorid nudibranchs are a clear example for defense allocation (Avila 1996, Wägele et al. 2006), and also many cladobranch dendronotids possess special cells, and many arminids present marginal sacs suspected to be involved in chemical defense (Wägele et al. 2006, Putz et al. 2010). Aeolidids do not present glandular devices but are the only animals known to store nematocysts in cnidosacs, with few exceptions (e.g., the dendronotid

Hancockia [Putz et al. 2010]). Besides a few examples mentioned previously, published data about the concentrations in which the defensive compounds appear in specialized anatomical sites (e.g., MDFs [Wägele et al. 2006]) are scarce. We found amounts ranging 3%–9% of defensive chemicals per total dry mass in the cerata of the tropical *Phyllodesmium guamensis,* up to 22% in the mantle border of the tropical *Glossodoris pallida,* and 50%–95% in the MDFs of the Mediterranean *Hypselodoris picta* (C. Avila, unpublished results). These findings agree with the ODT, which predicts allocation of bioactivity towards the most exposed or vulnerable parts to reduce predators' attacks (Rhoades and Gates 1976), while reducing metabolic costs. As we said previously, this is also safer for the slug to avoid autotoxicity. Defenses in sea slugs are thus regularly placed in the mantle and papilla or particular glands, as well as in some autotomized body parts, and mucus secretions. A curious case is the biosynthesized polygodial, which is toxic for *Dendrodoris,* and it is only transformed once secreted, from macrovacuolated epidermal cells containing the precursor (Cimino et al. 1985, Avila et al. 1991). Contrastingly, sponge furanosesquiterpenes are not toxic to *Hypselodoris* nudibranchs, which store them into MDFs (Fontana et al. 1994, Avila 1995). A particular case is that of Chromodoridids (see below), where the most primitive forms (*Cadlina, Glossodoris*) tend to contain mixtures formed mainly by sponge sesterterpenes, while the genus *Chromodoris* contain mostly diterpenes and *Hypselodoris* contain sesquiterpenes. This is obviously related to their specific sponge diet and the diversification of the taxon, which co-evolved with poriferans (Rudman and Bergquist 2007). However, *Cadlina* also can biosynthesize some of its compounds too (Kubanek and Andersen 1997, Kubanek et al. 2000), and there are some implications of the dietary specificity in the ecology of this species: *Cadlina* may feed upon several species of sponges, while *Chromodoris* and *Hypselodoris* are constrained to certain sponge preys.

Over and above, it is still a matter of debate whether the accumulated compounds really act as chemical defenses. Quite often the defensive role is never proved or just assumed by homology with the few tested examples. For instance, most algal compounds have not been tested for deterrence, but are assumed to be deterrent, or even to possess multiple functions: allelopathic, antimicrobial, antifouling, UV-screening agents, and herbivore deterrents (Hay and Fenical 1988, Potin et al. 2002, Amsler and Fairhead 2005). Furthermore, a large variability exists in the concentration and effectiveness of algal compounds towards different herbivore species (Hay and Steinberg 1992, Paul et al. 2001, Ianora et al. 2006). Regarding induction of chemical defenses, this has been rarely described in sea slugs prey. An example is the arminoidea *Dirona,* which induces the production of compounds in its bryozoan prey (Harvell 1984). Induction of defenses in macroalgae by herbivory is also an interesting topic in which sea slugs may play an important role (Hay 1996, Cronin and Hay 1996, Pavia and Toth 2000, Toth et al. 2005).

3.4   BIOSYNTHESIS AND THE ROLE OF SYMBIONTS

Biosynthesis is a multi-step, enzyme-catalyzed process, where simple precursor molecules (primary metabolites) are transformed into usually more complex products. Generally, "newly" formed defensive products are considered to be the most metabolically expensive defenses (Paul 1992). By using precursors labeled with stable isotopes, the de novo biosynthesis of the sesquiterpenoids nanaimoal and isoacan-thodoral in the British Columbian *Acanthodoris nanaimoensis* was proved (Graziani and Andersen 1996). Cimino and collaborators reviewed the use of stable isotopes for biosynthesis studies (Cimino et al. 2004). Actually, biosynthesis in sea slugs has been reviewed several times and evidence has been accumulated in several taxa (Table 3.1) proving that this is a relevant process in the group (see reviews [Garson 1993, Cimino and Ghiselin 1999, Gavagnin and Fontana 2000, 2001, Cimino et al. 2001, 2004, Fontana 2006, Moore 2006]). For example, compounds following the propionate route, which are otherwise unusual in Eukaryota, are quite extended among Pleurobranchoidea, Cephalaspidea, Anaspidea, Sacoglossa, and Pulmonata (Ireland and Scheuer 1979, Gavagnin et al. 1994, Cimino et al. 2004, Cutignano et al. 2009). Enzymes implicated in biosynthetic processes are known for *Aplysia fasciata* (glycosidases) (Giordano et al. 2004), *Oxynoe olivacea* (lipases) (Cutignano et al. 2004) and *Tethys fimbria* (cyclooxigenase) (Cimino and Ghiselin 2009). Examples of de novo biosynthesis by *in vivo* experiments and molecular labeling include poly-propionates in elysioidean sacoglossans and haminoean cephalaspideans. The ori-gin of polypropionates of *Bulla striata* was investigated by experiments with [1-^{14}C] propionate (Fontana et al. 2004). A very low (~0.01%) but consistent incorporation was measured in aglajne-1 and aglajne-3, thus suggesting that polypropionates from other cephalaspidean mollusks may also be de novo biosynthesized. Isotope-labeling experiments in *Haminoea orbignyana* demonstrated that the alkaloid haminol-2 is formed via polyketide biosynthetic pathway (Cutignano et al. 2003, 2004). Similar studies allowed also establishing that polyketides, such as triophamine, and terpe-noids, such as terpenoic acid glycerides and tanolide B (Graziani et al. 1996), and also nanaimoal (Graziani and Andersen 1996), are biosynthesized de novo in sev-eral dorid nudibranchs (Graziani and Andersen 1996, Kubanek and Andersen 1997, 1999). Another example is that of the dendronotacean nudibranch *Melibe*, able to biosynthesize defensive terpenoids typical from octocorals through the mevalonate pathway (Barsby et al. 2002). This allowed a shift in their diet into a wide range of food items (amphipods, copepods, etc.) favoring the colonization of different habi-tats and extending its geographical distribution. Sesquiterpenoids are biosynthesized by *Dendrodoris* nudibranchs (Avila et al. 1991, Fontana et al. 1999, 2000), with the precursor, 7-deacetoxyolepupuane, being the only compound found in the juveniles (<1 mm long) (Avila 1993), and thus supporting the biosynthetic pathway proposed by Fontana et al. (1999). The diterpenoid glycerides of *D. verrucosa* are biosynthesized de novo as demonstrated by the incorporation of labeled precursors into the verru-cosins (Fontana et al. 2003). However, verrucosins do not seem to be present during the early larval stages, while xylosil-MTA is present only during the first days of development (Avila 2006). In both cases, *Dendrodoris* and *Doris*, the absence of the defensive metabolites in the young slugs fits well with the GDBH model (see above). Other examples of diterpenoid biosynthesis are the glycerides from the Antarctic *Doris kerguelenensis*, which were suggested to be de novo biosynthesized because of their absence in their sponge preys and their location in the mantle (Iken et al. 2002, Cutignano et al. 2011, Moles et al. 2017a). Interestingly, Wilson and co-authors (2013) linked the biosynthetic compounds of *D. kerguelenensis* to their genetic variability, suggesting that this species is actually a complex of cryptic species.

The dorid nudibranch *Doriopsilla areolata* is able to biosynthesize compounds that are also present in sponges (Gavagnin et al. 2001, Fontana et al. 2003). Explanations for this imply that similar enzymes could be present in prey and predator and act on similar substrates. Likewise, the enzymes are obtained through lateral gene transfer or by changes in the genes coding for the substrates, or by retrobiosynthesis (Cimino and Ghiselin 1999, Fontana et al. 2003). The retrosynthetic evolution hypothesis means that the final steps of biosynthesis evolved first, and earlier steps were added subsequently. For example, ancestral dendrodorids perhaps used pallescensin-A (or *ent*-pallescensin-A) from food as a defensive compound. The capacity to modify this compound into its derivatives would evolve next, and then they would become able to synthesize it from successively earlier precursors. This scenario has the advantage of proceeding one step at a time through modification of enzymes that were already present (Cimino and Ghiselin 2009). This also may indicate that the number of innovations needed along evolution to acquire the biosynthetic ability could be fewer than expected. It seems just too hard to believe that this is the result of a random evolutionary convergence. Interestingly, the dorid *Cadlina luteomarginata* seems to be able to regulate the de novo biosynthesis of its defensive compounds on the basis of its needs (Kubanek et al. 1997, 2000). Further studies in examples like these mentioned here may open new perspectives in the field of chemical ecology of sea slugs.

Biosynthesis seems to have evolved independently several times along the evolution of Heterobranchia (see below). It is an energetically costly mechanism, and thus it is not surprising that biosynthesized compounds could be used for several purposes (Avila 1995). Examples of these include sesquiterpenes in dendrodorids (for defense and reproduction), prostaglandins of *Tethys fimbria* (for defense, regeneration, and reproduction), as well as sacoglossan compounds (for defense and regeneration; see above). Concentrations of biosynthetic defensive compounds are usually much lower than those reported in a previous section for dietary compounds. Examples include 0.05%–0.15% of chemical defenses in total dry mass in the mantle of the Antarctic *Bathydoris hodgsoni*; 0.06%–0.09% in the mantle of the Antarctic *Doris kerguelenensis*; 0.25% in the mantle of the Mediterranean *Doris verrucosa*; and 10.2% in the mantle rim of the Mediterranean *Dendrodoris limbata* (C. Avila, unpublished results). This may be an energy saving strategy since the compounds are effective at very low concentrations. Curiously, most biosynthetic sea slug species are cryptic, and this deserves further investigations.

The bacterial origin of marine natural products in sea slugs has been so far proven only recently in very few cases (see below) (Cimino and Ghiselin 2009). Among these, some studies suggest that bacterial-derived products exist in sacoglossans, such as the kahalalide polypeptides in *Elysia* (Davis et al. 2014). Fontana and collaborators (2001) proposed that jorumycin from *Jorunna funebris* was also produced by symbiotic microorganisms due to the presence of related compounds in sponges, mollusks, and tunicates, but this remains to be proven. Wägele and Johnsen (2001) reviewed symbiotic chloroplasts and zooxanthellae in sea slugs, widening the previous knowledge in this field. Some nudibrachs, like the aeolidids *Piseinotecus*, *Phyllodesmium*, the dendronotid *Melibe*, and the arminid *Dermatobanchus*, establish symbiotic partnerships with unicellular algae (i.e., zooxanthellae) (Wägele and Johnsen 2001, Wägele 2004,

Burghardt et al. 2008). Finally, lipids and fatty acids in sea slugs showed a wide diversity (Avila et al. 2004, Zhukova 2014) and this has recently been proposed to be, at least in part, due to the presence of symbiotic bacteria (Zhukova 2014).

Heavily defended cone snails also occasionally contain abundant secondary metabolites, γ-pyrones known as nocapyrones, and these pyrones are synthesized by bacterial symbionts (Lin et al. 2013). The natural roles of nocapyrones are not known, but they are active in neurological assays at low concentrations, revealing again that shelled mollusks are an overlooked source of natural products. In this case, in *Conus rolani* and *C. tribblei*, actinomycetes synthesize polyketide pyrones. As mentioned previously, pyrones and related polyketides have been isolated from cephalaspideans, sacoglossans, and siphonarids from diverse habitats, where they are produced de novo (Sleeper and Fenical 1977, Manker et al. 1988, Di Marzo et al. 1991a, Di Marzo et al. 1993, Gavagnin et al. 1994, Ireland and Scheuer 1979, Davies-Coleman and Garson 1998, Darias et al. 2006). They are involved in different roles, such as defense, regeneration, and communication. However, now we know that the source could be related to symbionts too. The pyrone C-methyl groups in sacoglossans arise from propionate, which might indicate a bacterial source (Di Marzo et al. 1991a, Cutignano et al. 2012). This is due to the fact that α-methylated polyketides often arise from incorporation of the polyketide precursor methylmalonate (propionate) in bacteria, whereas in animals and other eukaryotes α-methylated polyketides usually originate from later methylation of polyketide intermediates (Lin et al. 2013). Pulmonates are sister taxa to sacoglossans, which is consistent with the prevalent biosynthesis of polyproprionates (Garson 1993, 2001, Cimino and Ghiselin 2001, Fontana 2006, Moore 2006) across a range of Heterobranchia, including the Siphonaridae (basommatophoran pulmonates) (Darias et al. 2006), *Placobranchus* sp. (saccoglossan) (Manzo et al. 2005, Fu et al. 2000), haminoeids (Cutignano et al. 2003, 2004), and *Pleurobranchus* sp. (pleurobranchomorph) (Ciavatta et al. 1993).

Recent studies described a renewable, dividing population of symbiotic bacteria in the epidermis of *Dendrodoris nigra* (Zhukova and Eliseikina 2012). No bacteria were found in the digestive system or the gonads. The only previous record reporting symbiotic bacteria in nudibranchs was in the gonad and eggs of the same species (Klussmann-Kolb and Brodie 1999). The odd-chain fatty acids in *D. nigra* have been suggested to be produced by these bacteria (Zhukova and Eliseikina 2012), in a similar way than that reported for the bone eating, siboglinid worms, *Osedax* (Goffredi et al. 2005). Symbiotic bacteria would be facultative and obtain nutrition from the nudibranch. Perhaps this could as well be the case in the pteropod *Clione limacina* (Kattner et al. 1998).

Finally, one of the hot topics in marine natural products research in recent years is investigating metagenomic biodiversity to map natural products diversity and identify producing organisms. Metagenomic pre-screening, together with eco-targeting chemical screening, would increase the hit rate significantly and speed up the refill of the pipelines of new compounds in biotechnological applications. However, not so many studies have involved genomics of marine organisms and the relationship with marine natural products (Lane and Moore 2011). Most of these studies nowadays focus on microorganisms, and only a tiny fraction deal with macroorganisms (Piel 2009, Lane and Moore 2011, Blunt et al. 2015), while some studies have also looked at genomics of chemically-prolific marine actinomycetes (Bull and Stach 2007, Nett

2009). Examples include symbionts of more than 20 sponges (Piel et al. 2004, Flatt et al. 2005, Fieseler et al. 2007, He et al. 2012, Hentschel et al. 2012), one bryozoan (Hildebrand et al. 2004, Sudek et al. 2007), two tunicates (Long et al. 2005, Schmidt et al. 2005, Milne et al. 2006, Donia et al. 2011), and the two aforementioned mollusks (Lin et al. 2013). Surprisingly, none of these studies has been carried out with sea slugs yet. Therefore, there is a huge potential for the study of sea slugs' marine natural compounds and their potential applications when combining it with genome mining. Furthermore, knowing the complete genome allows determining the biosynthetic gene clusters of the organisms, whether they will be micro- or macroorganisms, thus increasing our chances to find (and produce) novel bioactive marine compounds.

3.5 BIOGEOGRAPHICAL CONSIDERATIONS

Sea slugs are distributed all over the world oceans, from the tropics to the poles. It was predicted that animals in higher latitudes would have low levels of chemical defenses (Bakus 1974, 1981, Bakus and Green 1974). However, this has been proven to be false, with many evidences coming from the studies of polar (mostly Antarctic) chemical ecology (Lebar et al. 2007, Avila et al. 2008, Núñez-Pons and Avila 2015, von Salm et al. 2018). For sea slugs in particular, many examples of natural products from Antarctica have been reported in the previous sections for the different groups (Table 3.1). Ironically, Davies-Coleman (2006) pointed out that it seems to be more data from Antarctic slugs than for South America or South Africa. In general, the Arctic has been poorly studied also, like some areas of Antarctica (e.g., Amundsen, Davis, and Bellingshausen seas), as well as the deep-sea and some areas in the central ocean basins. This is due to the challenges in reaching those areas, although this is gradually improving over the years.

The extreme specialization of many sea slugs, as well as their large reliance on particular hosts, have both characterized their capabilities of dispersal around the planet. Some groups that are highly dependent on algae for food and chemical protection, such as sacoglossans and anaspideans, are only reported from low-to-mid latitudes (except for a few sacoglossans found in the Malvinas [Falkland] Islands [Linse 1999]). The absence in polar regions of these groups is likely related to temperature and marked seasonality of light irradiation. Consequently, an inconsistent algal cover during the year possibly limits survival of these herbivore taxa. Global climate change, therefore, is a clear threat, with increasing temperatures favoring some of these species to expand their geographical range, since algae could reach higher latitudes and then open the possibility for some of these species to expand their geographical range (Hughes et al. 2015). These changes may produce unpredictable consequences in the existing benthic communities. Contrastingly, some genera within Nudibranchia (*Bathydoris, Doris, Cuthona, Doto, Tritonia*), Pleurobranchomorpha (*Bathyberthella*), Cephalaspidea (*Cylichna, Diaphana, Philine, Scaphander, Toledonia*), and Pteropoda (*Clione, Spongiobranchaea*), are able to colonize habitats from the tropics to the Poles. Amazingly, the chemical defensive strategies of these sea slugs are not so different in the distinct geographic areas, even when the chemicals are used against very different kinds of predators. Effective protection from potential predators, thus, is achieved by similar patterns of chemical defensive strategies in very different ecosystems. Examples of this include sesquiterpenes from Mediterranean and temperate

Dendrodoris species and hodgsonal from the Antarctic *Bathydoris hodgsoni*, or the wide distributed *Doris* species and similar compounds from *Doris kerguelenensis* from Antarctica (Avila 2006) (Figure 3.2). Perhaps not surprisingly, most Antarctic species present biosynthetic defensive compounds (with the exception of *Tritoniella belli*), and this is probably related to the fact that no herbivore sea slugs have been reported from the poles, and thus no algal chemicals can be obtained. Also, Antarctic sponge, bryozoan, and cnidarian feeders seem to rely on biosynthesized compounds for defense. The evolutionary implications of this may be related to the absence of algae in winter, on the one side, and to the patchy, scarce distribution of other prey items (e.g., sponges) in Antarctic ecosystems, on the other. Therefore, in these ecosystems it became more effective to rely on biosynthesis, rather than on diet, for defense. The wide variety of sponge species described in the diet of *Doris kerguelenensis* (McDonald and Nybakken 1997) further supports the fact that patchiness and habitat heterogeneity may influence diet availability. All these data fit well with both RAM and EST models (see above). However, our data on Antarctic sea slugs are still scarce, and therefore general assumptions should be regarded with caution.

Some authors argued that powerful chemical defense combined with aposematic coloration tends to be most pronounced in sea slugs that live in communities where the climate is stable and biodiversity levels are high (Mollo et al. 2008). Biodiversity, of course, is a key factor in all sort of ecological interactions (Margalef 1982). The description of a stable community with high biodiversity levels fits well with Antarctic benthic communities (Dayton et al. 1974, 1994, Arntz et al. 1994, 1997). However, most Antarctic chemically defended slugs seem to be cryptic. This fact is probably related to the different putative predators present in these communities. Typical generalist predators in tropical and temperate waters are fish, which are visually oriented. In Antarctic ecosystems though, predation is mainly ruled by sea stars, which are chemically oriented. These facts indicate that predators and food availability are actually the key factors in the evolution of defensive trends in sea slugs, rather than climate stability. As far as we know, and making a simplistic assumption, we could say that in polar zones the typical sea slugs would be cryptic and use biosynthetic defenses, while tropical sea slugs would be mostly aposematic and contain diet-derived defensive chemicals. However, exceptions may also occur and more data are needed to further develop this and other assumptions.

3.6 CHEMICAL DEFENSE AND EVOLUTION

Several reviews have dealt with chemical defense and evolution in sea slugs during recent years, either for specific groups or as a whole (Marín and Ros 2004, Wägele 2004, Wägele and Klussmann-Kolb 2005, Cimino and Ghiselin 2009). While the driving force of sea slugs' evolution had been traditionally considered to be the dietary specialization (Thompson 1976, Mikkelsen 2002, Göbbeler and Klussmann-Kolb 2011), other authors claimed that the true driving force is chemical defense (Cimino and Ghiselin 2009). Since evolution is a complex, global process, probably both factors, along with predation, competition, and others, played a role synergistically. Actually, there is no reason to assume that only one factor is responsible for the evolution and diversification of heterobranchs. It may well be that in the different clades and at different times, these diverse factors had different impacts along the process.

Defensive mechanisms work also synergistically to protect sea slugs from generalist predators. Many sea slugs combine behavioral, morphological, and chemical mechanisms to obtain protection from predators. Examples include chemistry combined with autotomy (e.g., sacoglossans, dorids, dendronotids), crypsis by homomorphy and homochromy (e.g., aeolidids, dorids, tylodinoideans), camouflage within ink secretions (e.g., aplysids), swimming (e.g., dorids, anaspidea), acidic secretions (e.g., dorids, pleurobranchids, philinids), mantle spicules (e.g., dorids), and others. The combination of defense mechanisms with natural products may in fact target different sorts of predators or competitors. A fascinating example is that of nudibranchs that feed on cnidarians (e.g., aeolidids; see above); which may camouflage on them (homochromy, homomorphy), steal their nematocysts (cleptocnidae), and use their chemical defenses (e.g., *Phyllodesmium*), while protecting their skin and stomach epithelium from nematocysts by special multivacuolated epithelium, containing chitinous inclusions called spindles (Martin et al. 2007). Actually, the protection of sea slugs against their own toxins and that of their prey has received scarce attention in the literature (except for the aforementioned detoxification mechanisms). The dendronotid *Tritonia hamnerorum* is one of the few cases studied, where inmunochemical protection against the toxins from its gorgonian prey was described (Whalen et al. 2010). Cost-benefit trade-offs of consuming toxic prey presumably exist in sea slugs and do affect their coevolution. While costs may include low nutritional content, intoxication, or physiological resistance, benefits may include the use of their prey as substrata and chemical protected habitat, exploitation of food sources unavailable for other organisms, and the use of chemicals for its own defense (Sotka et al. 2009, Williams et al. 2012). Only when the physiological mechanisms of resistance evolved, the sequestration and further use of bioactive compounds was possible, switching from passive to active selection. Another interesting point is that, while specialist predators of sea slugs (i.e., other sea slugs and pycnogonids) may have to deal with the unpleasant taste of their food, generalist predators, like those mentioned previously, may use the strategy of eating small numbers of different preys, accumulating just small amounts of different noxious chemicals, thus avoiding the effect of a single, very potent compound. This has been described in insects and was proposed also for marine systems in the past (Hay et al. 1987, 1989, Hay 1991).

A controversial defense case has been that of the aeolidid *Cratena peregrina*. An experimental study tried to evaluate chemical or nematocyst-based defenses against fish, using colors and artificial models (Aguado and Marin 2007, Edmunds 2009, Marín 2009, Penney 2009). The problem with the experiments is that they combined too many factors, chemistry, nematocysts, and colors, and these were not clearly separated methodologically. Penney (2009) argued reasonably that, even if the study clearly showed that *C. peregrina* is deterrent to fish predators and that fish can associate shape and color with deterrent factors, several flaws make it impossible to distinguish whether the reason behind this was the nematocysts or the chemistry. This has been a recurrent topic in the literature, as pointed out by Edmunds (2009). Further studies should thus revisit this topic clearly separating the different putative defensive traits, if possible. This should be done also keeping in mind, as mentioned for other defense mechanisms, that some species of slugs may use mainly chemistry (e.g., *Phyllodesmium*, since they do not possess functional nematocysts), other slugs mainly nematocysts (Putz et al. 2010), or perhaps both simultaneously, and may be even synergistic. Edmunds (2009) suggested some interesting points worth to consider when planning those experiments. For example, some results indicate that *Flabellina verrucosa* uses both nematocysts and secondary chemicals to deter fish, but the role of the nematocysts did not seem to be crucial in that case (Penney et al. 2010).

One of the most common combinations of defensive systems is warning coloration and deterrent chemicals. Predators can learn to avoid distasteful prey because they may associate color and noxiousness (Edmunds 1974, 1987, Endler 1988, Giménez-Casalduero et al. 1999), but they may also learn to avoid distasteful food independently of colors (Long and Hay 2006). In fact, learning to avoid unpleasant food by generalist consumers is common in marine environments. Very few examples exist, however, proving that color is really aposematic in sea slugs (Tullrot and Sundberg 1991, Tullrot 1994, Cortesi and Cheney 2010). These authors further suggested that unpalatability and warning colorations in sea slugs initially evolved by individual, and not by kin selection, on the basis of their low abundances and their ability to survive predators' attacks. Some results also support that in *C. peregrina* defenses could arise via individual selection (Aguado and Marin 2007), as noted previously (Penney 2004) (also see below). The variable abundances, however, and the fact that some sea slugs are usually found living in groups (e.g., phillidids, chromodorids) may question whether this is a general rule, or it could in fact vary with the sea slug group studied.

The case of *Hypselodoris* species provides a good example of what a Müllerian mimicry group represents, with all studied species possessing deterrent chemicals, even if some of them do not possess MFDs, and all sharing similar color patterns (Avila 1993, 1995, Wägele et al. 2006, Haber et al. 2010). Müllerian mimicry occurs, thus, not also in color groups, but also in chemical groups, since they share the same arsenal of natural products. The same happens for *Chromodoris* species (Rudman 1984, 1987, 1991, Gosliner 2001) and phyllidids (Brunckhorst 1991, Brunckhorst 1993, Nuzzo et al. 2012). Specimens from other taxa, such as cephalaspideans, or even turbellarians and juvenile holothurians (e.g., *Pearsonothuria graeffei*) may mimic the color patterns of some of these slugs (Putz et al. 2010). Batesian mimicry, however, is very difficult to prove, since species, initially thought to be chemically unprotected and just using the same color pattern as the protected ones, may have not been well studied yet, as happened with viceroy butterflies. The case of these butterflies, the once classical example of Batesian mimicry, was further proven to be a Müllerian mimicry example (Ritland and Brower 1991, Vane-Wright 1991). In addition, Valdés and co-authors (2013) recently warned about variation in color pattern in different populations of a single species of the cephalaspidean *Philinopsis pusa*, thus adding more complexity to the topic of the evolution of coloration in sea slugs. In this case, however, it may well be that color was not selected because it was not crucial for the survival of the species, due to the presence of alternative defensive mechanisms (burrowing; homochromy with different substrata). Color polymorphism, instead, may be more effective in this case, avoiding learning in predators and forcing them to keep a broad search image (Avila 1995). Furthermore, coloration and color groups in chromodorids might not be the result of convergent evolution, but of the phylogeny of the group (Wilson and Lee 2005). This could be true for other groups too. The result, however, is the same in the sense that these color groups seem to have provided selective advantage to them in their respective localities, and thus have been conserved and radiated quite widely (e.g., *Chromodoris*, *Hypselodoris*, phyllidids). The characteristics of the natural products and their origin in chromodorids correlate well with current phylogenies for the group, with *Cadlina* species being most primitive and *Chromodoris* and *Hypselodoris* more derived forms (Wilson and Lee 2005, Johnson and Gosliner 2012). Biosynthetic species are thus at the base of the phylogenetic trees in all cases, differently to what happens in sacoglossans (see above). Johnson and Gosliner (2012) provided a new classification of chromodorids that indicates that some of the names used in this chapter should be updated; further specific studies in the group, however, are needed to clarify the relationship between their chemistry and the new classification.

Along evolution, cryptic models are considered to be the most ancestral condition, and aposematic colorations would be derived (Cortesi and Cheney 2010). This, again, needs to be further evaluated with more detailed reviews relating coloration with the rest of characters, since it may also be dependent of the environment, their predators, as well as be different in the different clades and at different times. Moreover, since warning colorations may be energetically expensive, it has been suggested that only species that are strongly protected may afford them, and thus, the most conspicuous sea slugs would be the most toxic (Cortesi and Cheney 2010). Further studies are needed to demonstrate that, as well as to ascertain the difference between toxicity and unpalatability in chemical defenses from sea slugs, since usually both terms are used inconsistently in the literature even if they mean different things.

It is worth mentioning that metabolite accumulation from the prey is a mechanism to store toxic chemicals for further elimination and not necessarily to be used as chemical defenses by the consumer in the first place (Pennings and Paul 1993). The use as chemical defenses of dietary compounds probably evolved later. Similarly, nematocysts removal from the digestive system may have been the original function of the cnidosacs (Martin et al. 2009). All these mechanisms could have been selected during evolution to perform a defensive role, because they offered an advantage, as a Darwinian selected character. Lamarck believed that evolution happens according to a predetermined plan, where an organism changes during life to adapt to its environment, and these changes are passed on to its descendants. This is not, however, accepted anymore, for centuries, and as Darwin proposed, since all specimens are different, those possessing traits that help them to survive in their environments do in fact survive, and have more offspring (Kardong 2008). Accordingly, cnidosacs, MDFs, and other structures or metabolic pathways, were kept during evolution because the animals possessing them survived and reproduced. Obviously, the most favorable characters were selected for sea slugs' survival, contrarily to certain interpretations in the literature on sea slugs, which are somehow anthropogenically biased. Selection of the most active chemicals (see above), like that of the most potent nematocysts (Thompson 1976), was driven by natural selection.

Traditionally, the most primitive chemical defense in sea slugs was proposed to be the acquisition of chemicals from the diet, by incorporation from protected food items, and then bioaccumulation for their own protection (cleptochemistry). Afterwards, when the slugs developed the ability to transform the obtained bioactive compounds into modified products, their capability to diversify their niches and diet was enhanced. Evolutionarily, the last step is usually considered to be the adoption of biosynthetic pathways to construct from simple molecules their own secondary metabolites for defense. This seems to have happened nicely in sacoglossans, for instance (see above). However, other possibilities may exist in other groups, with different evolutionary trends, like in the previously mentioned chromodorids, as well as in other dorids (e.g., dendrodorids, bathydorids). Shell has been lost independently several times during evolution in "opisthobranchs" (Wägele and Klussmann-Kolb 2005). Herbivory is probably the ancestral feeding strategy, while carnivory evolved independently several times in different clades (Göbbeler and Klussmann-Kolb 2011). Similarly, biosynthesis probably appeared independently several times during evolution, in different clades, and may have been lost again in some groups. Chromodoridids, from *Cadlina* to *Hypselodoris*, could be a good example of this (see above).

Currently, there is increased awareness that certain natural products, previously thought to be exclusive of a characteristic taxon, are more and more frequently found in phylogenetically and ecologically distant organisms. One explanation for this is evolutionary convergence, as Margalef (1977, 1982) pointed out: "The number of models in nature is limited." These, for sea slugs, may be applied both for colors and for chemicals (Avila 1995). Just to mention one example, polygodial, was first found in the plant *Polygonum hydropiper* (Barnes and Loder 1962), and it is used as a spice in Southern Asia. Polygodial is also found in dendrodorids, where it has been proven to be de novo biosynthesized. Another reason for this is the presence of symbiotic microorganisms. In sea slugs, although the involvement of symbiotic microorganisms cannot be completely ruled out (see above), it clearly seems an example of evolutionary convergence. In the aforementioned example, polygodial is used against predators no matter what, which are insects in the case of the plant, and fishes or crabs for the slugs. Similarly, longifolin (Figure 3.2) is found in plants, in sponges, and in their slug predators, and dendrolasin is found in plants, in ants, and in sponges, as well as in their slug predators (Avila 1995). Contrarily, *Aplysia dactylomela* was recently shown to possess dactylomelatriol, derived from an omphalane skeleton previously described only in terrestrial fungi (Díaz-Marrero et al. 2012). The alkaloid bursatellin found in *Bursatella leachii* is strictly related to chloramphenicol, derived from terrestrial bacteria (Gopichand and Schmitz 1980). Homarine is also a good example of a defensive compound found in largely separated clades, either due to evolutionary convergence or symbiosis. It is found in mollusks, hydroids, gorgonians, soft corals, and crustaceans. Homarine has been described to possess a broad spectrum of bioactivities, including antimicrobial against bacteria and diatoms, antifeedant, and morphogenesis inductor (Paul et al. 1997, McClintock and Baker 2001, Affeld et al. 2006, Paul et al. 2006).

Many kinds of predators have been reported to eat, or try to eat, sea slugs, including other opisthobranchs, nemerteans, flatworms, crabs, pycnogonids, starfishes, fishes, birds, turtles, and even cone snails, although very few specific predators have been shown to really eat them (Todd 1981, Vermeij 1987, Piel 1991, Avila 1995, Rogers et al. 2000, Rudman 2000, Valdés et al. 2013). Usually, sea slugs are rejected after only a bite. However, different sorts of possible predators may have diverse foraging trends, and different perceptual and learning abilities. It is not the same to be attacked by a crab, a sea star, a fish or by bacteria, and thus colors may be useful against some enemies but useless against others. Some predators may be visually oriented while others may be chemically oriented. And even within visually oriented predators, some may see a pattern as aposematic while others may not see it at all (crypsis) (Endler 1988). Also, the chemical perception of visual predators, such as fish, for a particular compound may be different for different species (Long and Hay 2006). How all these mechanisms work together to efficiently protect sea slugs against predators is complex and still poorly understood. A recent review analyzes how marine invertebrates use chemical cues to track and select food, including some heterobranchs (Kamio and Derby 2017). Furthermore, as mentioned previously, chemical defense in sea slugs is not only used against potential predators, but also against all sort of enemies and competitors, including microorganisms, as antibacterials, antifouling, antifungals, and so on (e.g., landulaurencianols A–C from the anaspidean *Aplysia punctata* and julianin-S from *A. juliana*). Chemical defenses may also affect the small-sized organisms living nearby (or larvae, or sperm) by toxicity or cytotoxicity (e.g., polyunsaturated fatty acids from Arctic specimens of the cephalaspidean *Scaphander lignarius*, aplysianin A peptide from the egg masses and albumen gland of *Aplysia kurodai*, or dolastatins from *Dolabella auricularia*), or they may act as pheromones (e.g., alarm pheromones in cephalaspideans; sexual pheromones in anaspideans; see aforementioned references). In fact, since allelochemicals carry ecological information, sometimes the natural products are called "infochemicals." (Ianora et al. 2006). After all, chemical defense is all about communication. In the case of chemical communication, this means a long range, a slow transmission, a relatively low cost, and a variable localization (Alcock 1989, Krebs and Davies 1993). If sea slugs should be placed in the classical triangle of communication channels among organisms (Wilson 1980), they would be placed near butterflies (Figure 3.11). Aposematic slugs would be next to the diurnal butterflies, and the cryptic ones next to the nocturnal butterflies, as predicted by Thompson (1976), quoted at the beginning of this chapter.

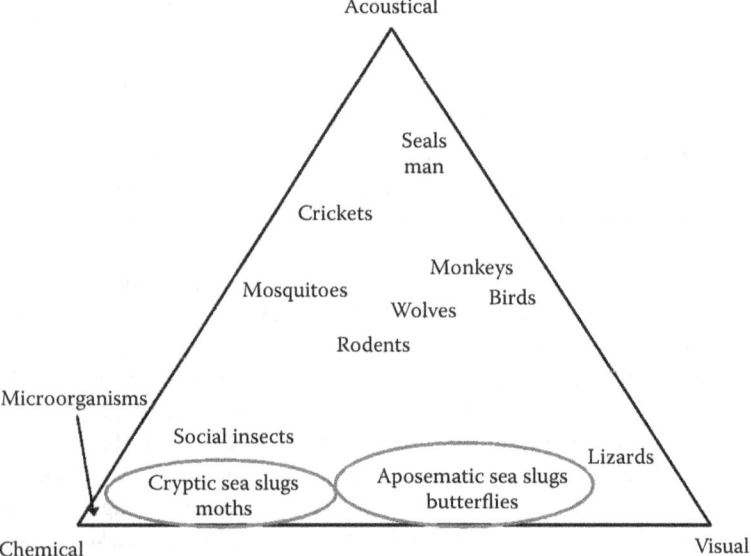

FIGURE 3.11 Relative importance of the communication channels in selected organisms, considering only chemical, visual, and acoustic channels. (Modified from Wilson, E.O., *Sociobiology*, Harvard University Press, Cambridge, 1980.)

3.7 CONCLUDING REMARKS

Overall, less than 1% of the known mollusk species have been investigated regarding their natural products (Benkendorff 2010, Blunt 2015). Sea slugs offer a serendipitous world of chemical defensive strategies, in which all sorts of mechanisms have appeared to favor their survival. Often their defenses include the use of chemicals obtained from their prey, the transformation of these compounds, or even de novo biosynthesis of bioactive defensive compounds. These are usually combined with other defensive systems, such as cryptic appearances or aposematic colors, particular behavior, or autotomy. Over the last 25 years, a great deal of advances from intense research efforts was addressed to understand the chemical defense strategies of sea slugs from different latitudes, from the tropics to the poles. Amazingly, the defensive strategies of marine slugs and the compounds used are not as variable in the distinct geographic areas, even if the kinds of predators, competitors, and threats may be remarkably different. Effective protection from potential enemies thus, is achieved by similar patterns of chemical defensive strategies in very different ecosystems.

Modern chemical techniques now allow for detecting and identifying organic compounds in very small quantities. However, there are still many limitations due to the available amounts of compounds, instability, and lability, as well as natural variability. As seen in this chapter, much has been done in recent years thanks to the fruitful collaboration of biologists, ecologists, and chemists. Nevertheless, much effort is still needed for understanding the role of natural products in structuring marine benthic communities, as well as to know how they may directly affect sea slugs' fitness and survival. Further field experiments against sympatric predators are urgently needed, where natural concentrations should always be determined. Moreover, the mechanisms allowing for the detection of natural products by consumers, and how do these learn to avoid them, is a topic that has been almost neglected so far. Symbiont analyses as well as metagenomics, are currently emerging exciting topics in sea slugs. To address all these issues, novel and ecologically relevant methods are needed to test for deterrence, antifouling, antimicrobial, and other possible roles of natural products, as well as their origin. As Valiela (2001) wrote: "Scientists, unlike most other people, tend to be interested in what they do not know: this is what drives their work forward." In the field of sea slug chemical ecology, much remains to be known, and this should make us move forward. More than ever, we must move from a descriptive chemical ecology to an experimental chemical ecology of sea slugs. The fascinating world of sea slugs still has a lot to offer and we should be able to deep into all these open aspects in the current frame of a changing environment.

ACKNOWLEDGMENTS

We thank the editors for their kind invitation to write this chapter. It is in fact based on the original ideas presented at a conference given by C. Avila in 2001 in Vienna (Avila 2001) From the Tropics to Antarctica: A Comparison of Chemical Defensive Strategies in Opisthobranch Mollusks. Abstracts World Congress of Malacology 2001, Vienna, Austria. Unitas Malacologica, 2001 p. 14, L Salvini-Plawen, J Voltzow, H Sattmann, and G Steiner, eds.). Many researchers contributed during years to our studies and we are thankful for this to G Cimino, A Fontana, M Gavagnin, A Cutignano, G Villani, K Iken, W Arntz, H Wägele, VJ Paul, R Sardà, S Taboada, C Angulo-Preckler, J Vázquez, B Figuerola, FJ Cristobo, A Riesgo, M Bas, M Nappo, M Ballesteros, and S Agustí. Thanks are also due to L Sardà-Avila and J Sardà-Avila for helping with the reference list, and to J Giménez, L Orlando, and P Pons for support. Financial support from our government is also acknowledged through grants: REN2003-00545/ANT, CGL2004-03356/ANT, CGL2007-65453/ANT, CTM2010-17415/ANT, and CTM2013-42667/ANT (ECOQUIM, ECOQUIM-2; ACTIQUIM-I and -II, DISTANTCOM) to C. Avila. Fellowships to L. Núñez-Pons (PhD – FPU Spanish government; Postdoctoral – Fundación Ramón Areces and Postdoctoral Program Beatriu de Pinós Marie Curie CO-FUND) and J. Moles (PhD – FPI Spanish government; Postdoctoral – Fundación Ramón Areces) are also acknowledged.

REFERENCES

Abramson, S. N., Z. Radic, D. Manker, D. J. Faulkner, and P. Taylor. 1989. Onchidal: A naturally occurring irreversible inhibitor of acetylcholinesterase with a novel mechanism of action. *Mol. Pharmacol.* 36:349–354.

Affeld, S., S. Kehraus, H. Wägele, and G. M. König. 2009. Dietary derived sesquiterpenes from *Phyllodesmium lizardensis*. *J. Nat. Prod.* 72:298–300.

Affeld, S., H. Wägele, C. Avila, S. Kehraus, and G. König. 2006. Distribution of homarine in some Opisthobranchia (Gastropoda: Mollusca). *Bonn Zoologische Beiträge* 3/4:181–190.

Agena, M., C. Tanaka, N. Hanif, M. Yasumoto-Hirose, and J. Tanaka. 2009. New cytotoxic spongian diterpenes from the sponge *Dysidea* cf. *arenaria*. *Tetrahedron* 65:1495–1499.

Aguado, F., and A. Marin. 2007. Warning coloration associated with nematocyst-based defences in aeolidiodean nudibranchs. *J. Molluscan Stud.* 73:23–28.

Alcock, J. 1989. *Animal Behavior. An Evolutionary Approach.* Sunderland, MA: Sinauer Associates.

Alexander, J., and A. Valdés. 2013. The ring doesn't mean a thing: Molecular data suggest a new taxonomy for two Pacific species of sea hares (Mollusca: Opisthobranchia, Aplysiidae). *Pac. Sci.* 67:283–294.

Alkon, D. L., H. Epstein, A. Kuzirian, M. C. Bennett, and T. J. Nelson. 2005. Protein synthesis required for long-term memory is induced by PKC activation on days before associative learning. *Proc. Nat. Acad. Sci.* 102:16432–16437.

Alvi, K. A., S. A. Abbas, T. Sultana, and M. Shameel. 1991. A diterpenoid lactone from *Aplysia juliana*. *J. Nat. Prod.* 54:886–888.

Amsler, C. D., and V. A. Fairhead. 2005. Defensive and sensory chemical ecology of brown algae. *Adv. Bot. Res.* 43:1–91.

Andersen, R. J., K. Desjardine, and K. Woods. 2006. Skin chemistry of nudibranchs from the West Coast of North America. In *Molluscs: From Chemo-ecological Study to Biotechnological Applications*, G. Cimino and M. Gavagnin (Eds.), pp. 277–301. Berlin, Germany: Springer.

Andersen, R. J., and D. J. Faulkner. 1972. Antibiotics from marine organisms of the Gulf of California. *Abstracts from a Conference on Food and Drugs from the Sea*, pp. 111–115. Kingston, RI.

Appleton, D. R., M. A. Sewell, M. V. Berridge, and B. R. Copp. 2002. A new biologically active malyngamide from a New Zealand collection of the sea hare *Bursatella leachii*. *J. Nat. Prod.* 65:630–631.

Arntz, W. E., T. Brey, and A. Gallardo. 1994. Antarctic zoobenthos. *Oceanogr. Mar. Biol.* 32:241–304.

Arntz, W. E., J. Gutt, and M. Klages. 1997. Antarctic marine biodiversity: An overview. In *Antarctic Communities: Species, Structure, and Survival*, B. Battaglia, J. Valencia and D. W. H. Walton (Eds.), pp. 3–14. Cambridge, MA: Cambridge University Press.

Ashour, M., R. A. Edrada, R. Ebel, V. Wray, W. W. K. Padmakumar, W. E. G. Müller, W. H. Lin, and P. Proksch. 2006. Kahalalide derivatives from the Indian sacoglossan mollusk *Elysia grandifolia*. *J. Nat. Prod.* 69:1547–1553.

Avila, C. 1992. A preliminary catalogue of natural substances of opisthobranch molluscs from Western Mediterranean and near Atlantic. *Sci. Mar.* 56:373–382.

Avila, C. 1993. Substancias naturales de moluscos Opistobranquios: Estudio de su estructura, origen y función en ecosistemas bentónicos. PhD dissertation. Universitat de Barcelona, Barcelona, Spain.

Avila, C. 1995. Natural products of opisthobranch molluscs: A biological review. *Oceanogr. Mar. Biol. Annu. Rev.* 33:487–559.

Avila, C. 1996. The growth of *Peltodoris atromaculata* Bergh, 1880 (Gastropoda: Nudibranchia) in the laboratory. *J. Molluscan Stud.* 62:151–157.

Avila, C. 1998. Chemotaxis in the nudibranch *Hermissenda crassicornis*: Does ingestive conditioning influence its behaviour in a Y-maze? *J. Molluscan Stud.* 64:215–222.

Avila, C. 2006. Molluscan natural products as biological models: Chemical ecology, histology, and laboratory culture. In *Molluscs: From Chemo-ecological Study to Biotechnological Applications*, G. Cimino and M. Gavagnin (Eds.), pp. 1–23. Berlin, Germany: Springer.

Avila, C., G. Cimino, A. Crispino, and A. Spinella. 1991. Drimane sesquiterpenoids in Mediterranean *Dendrodoris* nudibranchs: Anatomical distribution and biological role. *Experientia* 47:306–310.

Avila, C., and M. Durfort. 1996. Histology of epithelia and mantle glands of selected species of doridacean mollusks with chemical defensive strategies. *The Veliger* 39:148–163.

Avila, C., A. Fontana, M. Esposito, M. L. Ciavatta, and G. Cimino. 2004. Fatty acids of Antarctic gastropods: Distribution and comparison with Mediterranean species. *Iberus* 22:33–44.

Avila, C., K. Iken, A. Fontana, and G. Cimino. 2000. Chemical ecology of the Antarctic nudibranch *Bathydoris hodgsoni* Eliot, 1907: Defensive role and origin of its natural products. *J. Exp. Mar. Biol. Ecol.* 252:27–44.

Avila, C., and V. J. Paul. 1997. Chemical ecology of the nudibranch *Glossodoris pallida*: Is the location of diet-derived metabolites important for defense? *Mar. Ecol. Progr. Ser.* 150:171–180.

Avila, C., S. Taboada, and L. Núñez-Pons. 2008. Antarctic marine chemical ecology: What is next? *Mar. Ecol.* 29:1–71.

Awaad, A., and A. Y. Moustafa. 2016. Immunotoxicity of acid secretion produced by the sea slug *Berthellina citrina* in mice spleen: Histological and immunohistochemical study. *Acta Histochem.* 6:596–605.

Ayer, S. W., and R. J. Andersen. 1983. Degraded monoterpenes from the opisthobranch mollusc *Melibe leonina*. *Experientia* 39:255–256.

Baker, B., and P. Scheuer. 1994. The punaglandins: 10-chloroprostanoids from the octocoral *Telesto riisei*. *J. Nat. Prod.* 57:1346–1353.

Bakus, G. and G. Green. 1974. Toxicity in sponges and holothurians: A geographic pattern. *Science* 185:951–953.

Bakus, G. J. 1974. Toxicity in holothurians: A geographical pattern. *Biotropica* 6:229–236.

Bakus, G. J. 1981. Chemical defense mechanisms on the Great Barrier Reef, Australia. *Science* 211:497–499.

Bandaranayake, W. M. 2006. The nature and role of pigments of marine invertebrates. *Nat. Prod. Rep.* 23:223–255.

Barnes, C. S., and J. W. Loder. 1962. Structure of polygodial: A new sesquiterpene dialdehyde from *Polygonium hydropiper*. *Aust. J. Chem.* 15:322–327.

Barone, R., C. De Santi, F. Palma Esposito, P. Tedesco, F. Galati, M. Visone, A. Di Scala, and D. De Pascale. 2014. Marine metagenomics, a valuable tool for enzymes and bioactive compounds discovery. *Front. Mar. Sci.* 1:1–6.

Barrero, A. F., E. A. Manzaneda, J. Altarejos, S. Salido, J. M. Ramos, M. S. J. Simmonds, and W. M. Blaney. 1995. Synthesis of biologically active drimanes and homodrimanes from (–)-sclareol. *Tetrahedron* 51:7435–7450.

Barsby, T., R. G. Linington, and R. J. Andersen. 2002. De novo terpenoid biosynthesis by the dendronotid nudibranch *Melibe leonina*. *Chemoecology* 12:199–202.

Becerro, M. A., G. Goetz, V. J. Paul, and P. J. Scheuer. 2001. Chemical defenses of the sacoglossan mollusk *Elysia rufescens* and its host alga *Bryopsis* sp. *J. Chem. Ecol.* 27:2287–2299.

Becerro, M. A., J. A. Starmer, and V. J. Paul. 2006. Chemical defenses of cryptic and aposematic gastropterid molluscs feeding on their host sponge *Dysidea granulosa*. *J. Chem. Ecol.* 32:1491–1500.

Becerro, M. A., X. Turon, M. J. Uriz, and J. Templado. 2003. Can a sponge feeder be a herbivore? *Tylodina perversa* (Gastropoda) feeding on *Aplysina aerophoba*. *Biol. J. Linnean Soc.* 78:429–438.

Begon, M., C. R. Townsend, and J. L. Harper. 2006. Ecology: From individuals to ecosystems. Blackwell Publishing Ltd. Victoria, Australia. *Bioorg. Chem.* 7:125–131.

Benkendorff, K., A. R. Davis, C. N. Rogers, and J. B. Bremner. 2005. Free fatty acids and sterols in the benthic spawn of aquatic molluscs, and their associated antimicrobial properties. *J. Exp. Mar. Biol. Ecol.* 316:29–44.

Benkendorff, K. 2010. Molluscan biological and chemical diversity: Secondary metabolites and medicinal resources produced by marine molluscs. *Biol. Rev.* 85:757–775.

Berlinck, R. G. S., E. Hajdu, R. M. Rocha et al. 2004. Challenges and rewards of research in marine natural products chemistry in Brazil. *J. Nat. Prod.* 67:510–522.

Beukes, D. R., and M. T. Davies-Coleman. 1999. Novel polypropionates from the South African marine mollusc *Siphonaria capensis*. *Tetrahedron* 55:4051–4056.

Bian, W.-T., Z.-J. You, C.-Y. Wang, and C.-L. Shao. 2014. Brominated pimare diterpenoids from the sea hare *Aplysia pulmonica* from the South China Sea. *Chem. Nat. Comp.* 50:557–559.

Biskupiak, J. E., and C. M. Ireland. 1983. Pectinatone, a new antibiotic from the mollusc *Siphonaria pectinata*. *Tetrahedron Lett.* 24:3055–3058.

Biskupiak, J. E., and C. M. Ireland. 1985. Cytotoxic metabolites from the mollusc *Peronia peronii*. *Tetrahedron Lett.* 26:4307–4310.

Blunt, J. W., B. R. Copp, R. A. Keyzers, M. H. G. Munro, and M. R. Prinsep. 2015. Marine natural products. *Nat. Prod. Rep.* 32:116–211.

Bogdanov, A., C. Hertzer, S. Kehraus, S. Nietzer, S. Rohde, P. J. Schupp, H. Wägele, and G. M. König. 2016. Defensive diterpene from the Aeolidoidean *Phyllodesmium longicirrum*. *J. Nat. Prod.* 79:611–615.

Bogdanov, A., C. Hertzer, S. Kehraus, S. Nietzer, S. Rohde, P. J. Schupp, H. Wägele, and G. M. König. 2017. Secondary metabolome and its defensive role in the aeolidoidean *Phyllodesmium longicirrum*, (Gastropoda, Heterobranchia, Nudibranchia). *Beilstein J. Org. Chem.* 13:502–519.

Bogdanov, A., S. Kehraus, S. Bleidissel, G. Preisfeld, D. Schillo, J. Piel, A. O. Brachmann, H. Wägele, and G. M. König. 2014. Defense in the Aeolidoidean genus *Phyllodesmium* (Gastropoda). *J. Chem. Ecol.* 40:1013–1024.

Brecknell, D. J., L. A. Collett, M. T. Davies-Coleman, M. J. Garson, and D. D. Jones. 2000. New non-contiguous polypropionates from marine molluscs: A comment on their natural product status. *Tetrahedron* 56:2497–2502.

Brunckhorst, D. J. 1991. Do phyllidiid nudibranchs demonstrate behaviour consistent with their apparent warning coloration? Some field observations. *J Molluscan Stud.* 57:481–483.

Brunckhorst, D. J. 1993. The systematics and phylogeny of phyllidiid nudibranchs (Doridoidea). *Rec. Aust. Mus., Supplement* 16:1–107.

Bryan, P. J., J. B. McClintock, and B. J. Baker. 1998. Population biology and antipredator defenses of the shallow-water Antarctic nudibranch *Tritoniella belli. Mar. Biol.* 132:259–265.

Bryan, P. J., W. Y. Yoshida, J. B. McClintock, and B. J. Baker. 1995. Ecological role for pteroenone, a novel antifeedant from the conspicuous antarctic pteropod *Clione antarctica* (Gymnosomata: Gastropoda). *Mar. Biol.* 122:271–277.

Bull, A. T., and J. E. Stach. 2007. Marine actinobacteria: New opportunities for natural product search and discovery. *Trends Microbiol.* 15:491–499.

Burghardt, I., K. Stemmer, and H. Wägele. 2008. Symbiosis between *Symbiodinium* (Dinophyceae) and various taxa of Nudibranchia (Mollusca: Gastropoda), with analyses of long-term retention. *Org. Divers. Evol.* 8:66–76.

Caccamese, S., R. M. Toscano, S. Cerrini, and E. Gavuzzo. 1982. Laurencianol, a new halogenated diterpenoid from the marine alga *Laurencia obtusa. Tetrahedron Lett.* 23:114–116.

Cambie, R. C., P. A. Craw, M. J. Stone, and P. R. Bergquist. 1988. Chemistry of sponges, IV. Spongian diterpenes from *Hyatella intestinalis. J. Nat. Prod.* 51:293–297.

Capper, A., E. Cruz-Rivera, V. J. Paul, and I. R. Tibbetts. 2006. Chemical deterrence of a marine cyanobacterium against sympatric and non-sympatric consumers. *Hydrobiologia* 553:319–326.

Capper, A., I. R. Tibbetts, J. M. O'Neil, and G. R. Shaw. 2005. The fate of *Lyngbya majuscula* toxins in three potential consumers. *J. Chem. Ecol.* 31:1595–1606.

Carballeira, N. M., E. Anastacio, J. Salvá, and M. J. Ortega. 1992. Identification of the new 10, 15-eicosadienoic acid and related acids in the opisthobranch *Haminaea templadoi. J. Nat. Prod.* 55:1783–1786.

Carbone, M. 2007. I molluschi opistobranchi: Uno straordinario modello per selezionare molecole biologicamente attive. PhD dissertation. Salerno, Italy: Università degli Studi di, 217 p.

Carbone, M., C. Irace, F. Costagliola, F. Castelluccio, G. Villani, G. Calado, V. Padula, G. Cimino, J. L. Cervera, R. Santamaria, and M. Gavagnin. 2010. A new cytotoxic tambjamine alkaloid from the Azorean nudibranch *Tambja ceutae. Bioorg. Med. Chem. Lett.* 20:2668–2670.

Carbone, M., M. L. Ciavatta, V. Mathieu, A. Ingels, R. Kiss, P. Pascale, E. Mollo, N. Ungur, Y.-W. Guo, and M. Gavagnin. 2017. Marine terpenoid diacylguanidines: Structure, synthesis, and biological evaluation of naturally occurring actinofide and synthetic analogues. *J. Nat. Prod.* 80:1339–1346.

Carbone, M., M. L. Ciavatta, G. De Rinaldis, F. Castelluccio, E. Mollo, and M. Gavagnin. 2014. Identification of thuridillin-related aldehydes from Mediterranean sacoglossan mollusk *Thuridilla hopei. Tetrahedron* 70:3770–3773.

Carbone, M., M. L. Ciavatta, J. R. Wang, I. Cirillo, V. Mathieu, R. Kiss, E. Mollo, Y-W. Guo, and M. Gavagnin. 2013. Extending the record of bis-γ-pyrone polypropionates from marine pulmonate mollusks. *J. Nat. Prod.* 76:2065–2073.

Carbone, M., M. Gavagnin, M. Haber et al. 2013. Packaging and delivery of chemical weapons: A defensive Trojan horse stratagem in chromodorid nudibranchs. *PLoS One* 8:e62075.

Carbone, M., M. Gavagnin, E. Mollo, M. Bidello, V. Roussis, and G. Cimino. 2008. Further syphonosides from the sea hare *Syphonota geographica* and the sea-grass *Halophila stipulacea. Tetrahedron* 64:191–196.

Carbone, M., M. Gavagnin, C. A. Mattia, C. Lotti, F. Castelluccio, B. Pagano, E. Mollo, Y. W. Guo, and G. Cimino. 2009. Structure of onchidione, a bis-γ-pyrone polypropionate from a marine pulmonate mollusk. *Tetrahedron* 65:4404–4409.

Carbone, M., C. Muniain, F. Castelluccio, O. Iannicelli, and M. Gavagnin. 2013. First chemical study of the sacoglossan *Elysia patagonica*: Isolation of a g-pyrone propionate hydroperoxide. *Biochem. Syst. Ecol.* 49:172–175.

Carbone, M., Y. Li, C. Irace, E. Mollo, F. Castelluccio, A. Di Pascale, G. Cimino, R. Santamaria, Y.-W. Guo, and M. Gavagnin. 2011. Structure and cytotoxicity of phidianidines A and B: First finding of 1,2, 4-oxadiazole system in a marine natural product. *Org. Lett.* 13:2516–2519.

Carefoot, T. H. 1987. *Aplysia*: Its biology and ecology. *Oceanogr. Mar. Biol. Ann. Rev.* 25:139–568.

Carte, B. K. 1996. Biomedical potential of marine natural products. *BioScience* 46:271–287.

Carte, B., and D. J. Faulkner. 1983. Defensive metabolites from three nembrothid nudibranchs. *J. Org. Chem.* 48:2314–2318.

Cattaneo-Vietti, R., B. Burlando, and L. Senes. 1993. Life history and diet of *Pleurobranchaea meckelii* (Opisthobranchia: Notaspidea). *J. Molluscan Stud.* 59:309–313.

Cavalcanti, B. C., H. V. Júnior, M. H. Seleghim, R. G. Berlinck, G. M. Cunha, M. O. Moraes, and C. Pessoa. 2008. Cytotoxic and genotoxic effects of tambjamine D, an alkaloid isolated from the nudibranch *Tambja eliora*, on Chinese hamster lung fibroblasts. *Chem. -Biol. Inter.* 174:155–162.

Chang, E. S. 2014. Possible anti-predation properties of the egg masses of the marine gastropods *Dialula sandiegensis*, *Doris montereyensis* and *Haminoea virescens* (Mollusca, Gastropoda). *Friday Harbor Laboratories Student Research Papers* 528, http://hdl.handle.net/1773/34623.

Charupant, K., K. Suwanborirux, S. Amnuoypol, E. Saito, A. Kubo, and N. Saito. 2007. Jorunnamycins A–C, new stabilized renieramycin-type bistetrahydroisoquinolines isolated from the Thai nudibranch *Jorunna funebris*. *Chem. Pharm. Bull.* 55:81–86.

Chau, R., J. A. Kalaitzis, and B. A. Neilan. 2011. On the origins and biosynthesis of tetrodotoxin. *Aquat. Toxicol.* 104:61–72.

Cheney, K. L., F. Cortesi, M. J. How, N. G. Wilson, S. P. Blomberg, A. E. Winters, S. Umanzör, and N. J. Marshall. 2014. Conspicuous visual signals do not coevolve with increased body size in marine sea slugs. *J. Evol. Biol.* 27:676–687.

Christa, G., V. Zimorski, C. Woehle, A. G. Tielens, H. Wägele, W. F. Martin, and S. B. Gould. 2014. Plastid-bearing sea slugs fix CO_2 in the light but do not require photosynthesis to survive. *Proc. R. Soc. B: Biol. Sci.* 281:20132493.

Ciavatta, M. L., E. Manzo, E. Mollo, C. A. Mattia, C. Tedesco, C. Irace, Y.-W. Guo, X.-B. Li, G. Cimino, and M. Gavagnin. 2011. Tritoniopsins A–D, cladiellane-based diterpenes from the South China Sea nudibranch *Tritoniopsis elegans* and its prey *Cladiella krempfi*. *J. Nat. Prod.* 74:1902–1907.

Ciavatta, M. L., E. Trivellone, G. Villani, and G. Cimino. 1993. Membrenones: New polypropionates from the skin of the Mediterranean mollusc *Pleurobranchus membranaceus*. *Tetrahedron Lett.* 34(42):6791–6794.

Ciavatta, M. L., E. Trivellone, G. Villani, and G. Cimino. 1996. Prenylphenols from the skin of the aeolid mollusc *Cratena peregrina*. *Gazz. Chim. Ital.* 126:707–710.

Ciavatta, M. L., G. Nuzzo, K. Takada, V. Mathieu, R. Kiss, G. Villani, and M. Gavagnin. 2014. Sequestered fulvinol-related polyacetylenes in *Peltodoris atromaculata*. *J. Nat. Prod.* 77:1678–1684.

Ciavatta, M. L., G. Villani, E. Trivellone, and G. Cimino. 1995. Two new labdane aldehydes from the skin of the notaspidean *Pleurobranchaea meckelii*. *Tetrahedron Lett.* 36:8673–8676.

Ciavatta, M. L., E. Manzo, G. Nuzzo, G. Villani, G. Cimino, J. L. Cervera, M. A. E. Malaquias, and M. Gavagnin. 2009. Aplysiopsenes: An additional example of marine polyketides with a mixed acetate/propionate pathway. *Tetrahedron Lett.* 50:527–529.

Ciavatta, M. L., M. Gavagnin, R. Puliti, G. Cimino, E. Martinez, J. Ortea, and C. A. Mattia. 1996. Dolabriferol: A new polypropionate from the skin of the anaspidean mollusc *Dolabrifera dolabrifera*. *Tetrahedron* 52:12831–12838.

Ciavatta, M. L., M. Gavagnin, R. Puliti, G. Cimino, E. Martínez, J. Ortea, and C. A. Mattia. 1997. Dactylallene: A novel dietary C_{15} bromoallene from the Atlantic anaspidean mollusc *Aplysia dactylomela*. *Tetrahedron* 53:17343–17350.

Ciavatta, M. L., M. P. López Gresa, M. Gavagnin, E. Manzo, E. Mollo, L. D'Souza, and G. Cimino. 2006. New caulerpenyne-derived metabolites of an *Elysia* sacoglossan from the south Indian coast. *Molecules* 11:808–816.

Cimino, G., A. Crispino, A. Spinella, and G. Sodano. 1988. Two ichthyotoxic diacylglycerols from the opisthobranch mollusc *Umbraculum mediterraneum*. *Tetrahedron Lett.* 29:3613–3616.

Cimino, G., A. Crispino, V. Di Marzo, M. Gavagnin, and J. Ros. 1990. Oxytoxins, bioactive molecules produced by the marine opisthobranch mollusc *Oxynoeolivacea* from a diet-derived precursor. *Experientia* 46:767–770.

Cimino, G., A. Fontana, A. Cutignano, and M. Gavagnin. 2004. Biosynthesis in opisthobranch molluscs: General outline in the light of recent use of stable isotopes. *Phytochem. Rev.* 3:285–307.

Cimino, G., A. Fontana, and M. Gavagnin. 1999. Marine opisthobranch molluscs: Chemistry and ecology in sacoglossan and dorids. *Curr. Org. Chem.* 3 (1999):327–372.

Cimino, G., A. Passeggio, G. Sodano, A. Spinella, and G. Villani. 1991. Alarm pheromones from the Mediterranean opisthobranch *Haminoea navicula*. *Experientia* 47:61–63.

Cimino, G., A. Spinella, A. Scopa, and G. Sodano. 1989. Umbraculumin-B, an unusual 3-hydroxybutyric acid ester from the opisthobranch mollusc *Umbraculum mediterraneum*. *Tetrahedron Lett.* 30:1147–1148.

Cimino, G., and G. Sodano. 1993. Biosynthesis of secondary metabolites in marine mollusks. *Top. Curr. Chem.* 167:77–115.

Cimino, G., and G. Sodano. 1994. Transfer of sponge secondary metabolites to predators. In *Sponges in Time and Space: Biology, Chemistry, Paleontology*, R. W. M. van Soest, T. M. G. van Kempen, and J.-C. Braekman (Eds.), pp. 459–472. Rotterdam, the Netherlands: AA Balkema.

Cimino, G., and M. T. Ghiselin. 1998. Chemical defense and evolution in the Sacoglossa (Mollusca: Gastropoda: Opisthobranchia). *Chemoecology* 8:51–60.

Cimino, G., and M. T. Ghiselin. 1999. Chemical defense and evolutionary trends in biosynthetic capacity among dorid nudibranchs (Mollusca: Gastropoda: Opisthobranchia). *Chemoecology* 9:187–207.

Cimino, G., and M. T. Ghiselin. 2001. Marine natural products chemistry as an evolutionary narrative. In *Marine Chemical Ecology*, J. B. McClintock and B. J. Baker (Eds.), pp. 115–154. Boca Raton, FL: CRC Press LLC.

Cimino, G., and M. T. Ghiselin. 2006. Molluscs: From chemo-ecological study to biotechnological applications. In *Progress in Molecular and Subcellular Biology: Marine Molecular Biotechnology*, Vol. 43, W. E. G. Müller (Ed.). Berlin, Germany: Springer.

Cimino, G., and M. T. Ghiselin. 2009. Chemical defense and the evolution of opisthobranch gastropods. *Proc. Calif. Acad. Sci.* 60:175–422.

Cimino, G., G. Sodano, A. Spinella, and E. Trivellone. 1985. Aglajne-1, a polypropionate metabolite from the opisthobranch mollusk *Aglaja depicta*. *Tetrahedron Lett.* 26:3389–3392.

Cimino, G., G. Sodano, and A. Spinella. 1987. New propionate-derived metabolites from *Aglaja depicta* and from its prey *Bulla striata* (opisthobranch mollusks). *J. Org. Chem.* 52:5326–5331.

Cimino, G., M. Gavagnin, G. Sodano, A. Spinella, and G. Strazzullo. 1987. Revised structure of Bursatellin. *J. Org. Chem.* 52:2303–2306.

Cimino, G., M. Gavagnin, G. Sodano, R. Puliti, C. A. Mattia, and L. Mazzarella. 1988. Verrucosin-A and -B, ichthyotoxic diterpenoic acid glycerides with a new carbon skeleton from the dorid nudibranch *Doris verrucosa*. *Tetrahedron* 44:2301–2310.

Cimino, G., M. L. Ciavatta, A. Fontana, and M. Gavagnin. 2001. Metabolites of marine opisthobranchs: Chemistry and biological activity. In *Bioactive Compounds from Natural Sources—Isolation, Characterization and Biological Properties*, C. Tringali (Ed.), pp. 579–637. London, UK: Taylor & Francis Group.

Cimino, G., S. De Rosa, S. De Stefano, A. Spinella, and G. Sodano. 1984. The zoochrome of the sponge *Verongia aerophoba* ("Uranidine"). *Tetrahedron Lett.* 25:2925–2928.

Cimino, G., S. De Rosa, S. De Stefano, and G. Sodano. 1980. Cholest-4-en-4,16β,18,22R-tetrol-3-one 16,18-diacetate a novel polyhydroxilated steroid from the hydroid *Eudendrium* sp. *Tetrahedron Lett.* 21:3303–3304.

Cimino, G., S. De Rosa, S. De Stefano, and G. Sodano. 1985. Observations on the toxicity and metabolic relationships of polygodial, the chemical defense of the nudibranch *Dendrodoris limbata*. *Experientia* 41:1335–1336.

Cimino, G., S. De Rosa, S. De Stefano, and G. Sodano. 1986. Marine natural products: New results from Mediterranean invertebrates. *Pure Appl. Chem.* 58:375–386.

Clardy, J., and C. Walsh. 2004. Lessons from natural molecules. *Nature* 432:829–837.

Clark, K. B., and M. Busacca. 1978. Feeding specificity and chloroplast retention in four tropical Ascoglossa, with a discussion of the extent of chloroplast symbiosis and the evolution of the order. *J. Molluscan Stud.* 44:272–282.

Coley, P. D., J. P. Bryant, and F. S. Chapin. 1985. Resource availability and plant herbivore defense. *Science* 230:895–899.

Coll, J., B. Bowden, D. Tapiolas, R. Willis, P. Djura, M. Streamer, and L. Trott. 1985. Studies of Australian soft corals—XXXV: The terpenoid chemistry of soft corals and its implications. *Tetrahedron* 41:1085–1092.

Cortesi, F., and K. L. Cheney. 2010. Conspicuousness is correlated with toxicity in marine opisthobranchs. *J Evol. Biol.* 23:1509–1518.

Coval, S. J., and P. J. Scheuer. 1985. An intriguing C_{16}-alkadienone-substituted 2-pyridine from a marine mollusk. *J. Org. Chem.* 50:3024–3025.

Coval, S. J., G. R. Schulte, G. K. Matsumoto, D. M. Roll, and P. J. Scheuer. 1985. Two polypropionate metabolites from the cephalaspidean mollusk *Philinopsis speciosa*. *Tetrahedron Lett.* 26:5359–5362.

Cragg, G. M., P. G. Grothaus, and D. J. Newman. 2014. New horizons for old drugs and drug leads. *J. Nat. Prod.* 77:703–723.

Cronin, G. 2001. Resource allocation in seaweeds and marine invertebrates: Chemical defense patterns in relation to defense theories. In *Marine Chemical Ecology*, J. B. McClintock and B. J. Baker (Eds.), pp. 325–353. Boca Raton, FL: CRC Press.

Cronin, G., and M. E. Hay. 1996. Susceptibility to herbivores depends on recent history of both the plant and animal. *Ecology* 77:1531–1543.

Cronin, G., M. Hay, W. Fenical, and N. Lindquist. 1995. Distribution, density, and sequestration of host chemical defenses by the specialist nudibranch *Tritonia hamnerorum* found at high densities on the sea fan *Gorgonia ventalina*. *Mar. Ecol. Progr. Ser.* 119:177–189.

Cruz-Rivera, E. 2011. Evidence for chemical defence in the cephalaspidean *Nakamigawaia spiralis* Kuroda and Habe, 1961. *J. Molluscan Stud.* 77:95–97.

Cueto, M., L. D'Croz, J. L. Maté, A. San-Martín, and J. Darias. 2005. Elysiapyrones from *Elysia diomedea*. Do such metabolites evidence an enzymatically assisted electrocyclization cascade for the biosynthesis of their bicyclo[4.2.0]octane core? *Org. Lett.* 7:415–418.

Cutignano, A., A. Tramice, S. De Caro, G. Villani, G. Cimino, and A. Fontana. 2003. Biogenesis of 3-alkylpyridine alkaloids in the marine mollusc *Haminoea orbignyana*. *Angew. Chem. Int. Ed.* 42:2633–2636

Cutignano, A., A. Fontana, V. Renzulli, and G. Cimino. 2003. Placidenes C–F, novel alpha-pyrone propionates from the Mediterranean sacoglossan *Placida dendritica*. *J. Nat. Prod.* 66:1399–1401.

Cutignano, A., C. Avila, A. Domenech-Coll, G. d'Ippolito, G. Cimino, and A. Fontana. 2008. First biosynthetic evidence on the phenyl-containing polyketides of the marine mollusc *Scaphander lignarius*. *Org. Lett.* 10:2963–2966.

Cutignano, A., C. Avila, A. Rosica, G. Romano, B. Laratta, A. Domenech-Coll, G. Cimino, E. Mollo, and A. Fontana. 2012. Biosynthesis and cellular localization of functional polyketides in the gastropod mollusc *Scaphander lignarius*. *Chembiochem* 13:1759–1766.

Cutignano, A., D. Blihoghe, A. Fontana, G. Villani, G. d'Ippolito, and G. Cimino. 2007. Fusaripyrones, novel polypropionates from the Mediterranean mollusc *Haminoea fusari*. *Tetrahedron* 63:12935–12939.

Cutignano, A., G. Calado, H. Gaspar, G. Cimino, and A. Fontana. 2011. Polypropionates from *Bulla occidentalis*: Chemical markers and trophic relationships in cephalaspidean molluscs. *Tetrahedron Lett.* 52:4595–4597.

Cutignano, A., G. Cimino, A. Giordano, G. d'Ippolito, and A. Fontana. 2004. Polyketide origin of 3-alkylpyridines in the marine mollusc *Haminoea orbignyana*. *Tetrahedron Lett.* 45:2627–2629.

Cutignano, A., G. Cimino, G. Villani, and A. Fontana. 2009. Origin of the C_3-unit in placidenes: Further insights into taxa divergence of polypropionate biosynthesis in marine molluscs and fungi. *Tetrahedron* 65:8161–8164.

Cutignano, A., G. Cimino, G. Villani, and A. Fontana. 2009. Shaping the polypropionate biosynthesis in the solar-powered mollusc *Elysia viridis*. *Chem. Bio. Chem.* 10:315–322.

Cutignano, A., G. Villani, and A. Fontana. 2012. One metabolite, two pathways: Convergence of polypropionate biosynthesis in fungi and marine molluscs. *Org. Lett.* 14:992–995.

Cutignano, A., J. Moles, C. Avila, and A. Fontana. 2015. Granuloside, a unique linear homosesterterpene from the Antarctic nudibranch *Charcotia granulosa*. *J. Nat. Prod.* 78:1761–1764.

Cutignano, A., V. Notti, G. d'Ippolito, A. D. Coll, G. Cimino, and A. Fontana. 2004. Lipase-mediated production of defensive toxins in the marine mollusc *Oxynoe olivacea*. *Org. Biomol. Chem.* 2:3167–3171.

Cutignano, A., W. Zhang, C. Avila, G. Cimino, and A. Fontana. 2011. Intrapopulation variability in the terpene metabolism of the Antarctic opisthobranch mollusc *Austrodoris kerguelenensis*. *Europ. J. Org. Chem.* 2011:5383–5389.

Czeczuga, B. 1984. Investigations of carotenoids in some animals of the Adriatic Sea—VI. Representatives of sponges, annelids, molluscs and echinodermates. *Comp. Biochem. Physiol.—Part B: Comp. Biochem.* 78:259–264.

Da Cruz, J. F., H. Gaspar, and G. Calado. 2012. Turning the game around: Toxicity in a nudibranch–sponge predator–prey association. *Chemoecology* 22:47–53.

Dalisay, D. S., E. W. Rogers, A. S. Edison, and T. F. Molinski. 2009. Structure elucidation at the nanomole scale. 1. Trisoxazole macrolides and thiazole-containing cyclic peptides from the nudibranch *Hexabranchus sanguineus*. *J. Nat. Prod.* 72:732–738.

Daoust, J., A. Fontana, C. E. Merchant, N. J. De Voogd, B. O. Patrick, T. J. Kieffer, and R. J. Andersen. 2010. Ansellone A, a sesterterpenoid isolated from the nudibranch *Cadlina luteromarginata* and the sponge *Phorbas* sp., activates the cAMP signaling pathway. *Org. Lett.* 12:3208–3211.

Darias, J., M. Cueto, and A. R. Díaz-Marrero. 2006. The chemistry of marine pulmonate gastropods. In *Molluscs: From Chemo-ecological Study to Biotechnological Applications*, G. Cimino and M. Gavagnin (Eds.), pp. 105–131. Berlin, Germany: Springer.

Davidson, S. K. 1999. The biology of bryostatins in the bryozoan *Bugula neritina*. PhD dissertation. San Diego, CA: University of California San Diego.

Davies-Coleman, M. T. 2006. Secondary metabolites from the marine gastropod molluscs of Antarctica, Southern Africa and South America. In *Molluscs: From Chemo-ecological Study to Biotechnological Applications*, G. Cimino and M. Gavagnin (Eds.), pp. 133–157. Berlin, Germany: Springer.

Davies-Coleman, M. T., and D. J. Faulkner. 1991. New diterpenoic acid glycerides from the Antarctic nudibranch *Austrodoris kerguelensis*. *Tetrahedron* 47:9743–9750.

Davies-Coleman, M. T., and M. J. Garson. 1998. Marine polypropionates. *Nat. Prod. Rep.* 15:477–493.

Davis, J., W. F. Fricke, M. T. Hamann, E. Esquenazi, P. C. Dorrestein, and R. T. Hill. 2014. Characterization of the bacterial community of the chemically defended Hawaiian sacoglossan *Elysia rufescens*. *Appl. Environ. Microbiol.* 79:7073–7081.

Dawe, R. D., and J. L. C. Wright. 1986. The major polypropionate metabolites from the sacoglossan mollusc *Elysia chlorotica*. *Tetrahedron Lett.* 27:2559–2562.

Dayrat, B., and S. Tillier. 2002. Evolutionary relationships of euthyneuran gastropods (Mollusca): A cladistic re-evaluation of morphological characters. *Zool. J. Linn. Soc.* 135:403–470.

Dayton, P. K., B. J. Mordida, and F. Bacon. 1994. Polar marine communities. *Am. Zool.* 34:90–99.

Dayton, P. K., G. A. Robilliard, R. T. Paine, and L. B. Dayton. 1974. Biological accommodation in the benthic community at McMurdo Sound, Antarctica. *Ecol. Monogr.* 44:105–128.

De Silva, E. D., and P. J. Scheuer. 1982. Furanoditerpenoids from the dorid nudibranch *Casella atromarginata*. *Heterocycles* 17:167–170.

De Silva, E. D., R. E. Schwartz, and P. J. Scheuer. 1983. Srilankyene, a new metabolite from the sea hare *Aplysia oculifera*. *J. Org. Chem.* 48:395–396.

Della Sala, G., A. Cutignano, A. Fontana, A. Spinella, G. Calabrese, A. Domènech, G. D'Ippolito, C. D. Monica, and G. Cimino. 2007. Towards the biosynthesis of the aromatic products of the Mediterranean mollusc *Scaphander lignarius*: Isolation and synthesis of analogues of lignarenones. *Tetrahedron* 63:7256–7263.

Derby, C. D. 2007. Escape by inking and secreting: Marine molluscs avoid predators through a rich array of chemicals and mechanisms. *Biol. Bull.* 213:274–289.

Di Marzo, V., Cimino, G., Crispino, A., Minardi, C., Sodano, G., and Spinella, A. 1991b. A novel multifunctional metabolic pathway in a marine mollusc leads to unprecedented prostaglandin derivatives (prostaglandin 1, 15-lactones). *Biochem. J.* 273:593–600.

Di Marzo, V., A. Marín, R. R. Vardaro, L. De Petrocellis, G. Villani, and G. Cimino. 1993. Histological and biochemical bases of defense mechanisms in four species of Polybranchoidea ascoglossan molluscs. *Mar. Biol.* 117:367–380.

Di Marzo, V., R. R. Vardaro, L. De Petrocellis, G. Villani, R. Minei, and G. Cimino. 1991a. Cyercenes, novel pyrones from the ascoglossan mollusc *Cyerce cristallina*. Tissue distribution, biosynthesis and possible involvement in defense and regenerative processes. *Experientia* 47:1221–1227.

Di Prisco, C. L., F. Dessì-Fulgheri, and M. Tomasucci. 1973. Identification and biosynthesis of steroids in the marine mollusc *Aplysia depilans*. *Comp. Biochem. Physiol.—Part B: Comp. Biochem.* 45:303–310.

Díaz-Marrero, A. R., M. Cueto, L. D'Croz, and J. Darias. 2008. Validating and endoperoxide as a key intermediate in the biosynthesis of elysiapyrones. *Org. Lett.* 10:3057–3060.

Díaz-Marrero, A. R., J. M. De La Rosa, I. Brito, J. Darias, and M. Cueto. 2012. Dactylomelatriol, a biogenetically intriguing omphalane-derived marine sesquiterpene. *J. Nat. Prod.* 75:115–118.

Díaz-Marrero, A. R., E. Dorta, M. Cueto, J. Rovirosa, A. San-Martín, A. Loyola, and J. Darias. 2003. Labdane diterpenes with a new oxidation pattern from the marine pulmonate *Trimusculus peruvianus*. *Tetrahedron* 59:4805–4809.

Dieter, R. K., R. Kinnel, J. Meinwald, and T. Eisner. 1979. Brasudol and isobrasudol: Two bromosesquiterpenes from a sea hare (*Aplysia brasiliana*). *Tetrahedron Lett.* 19:1645–1648.

Dinapoli, A., and A. Klussmann-Kolb. 2010. The long way to diversity—Phylogeny and evolution of the Heterobranchia (Mollusca: Gastropoda). *Mol. Phylogenetics Evol.* 55:60–76.

Diyabalanage, T., K. B. Iken, J. B. McClintock, C. D. Amsler, and B. J. Baker. 2010. Palmadorins A—C, diterpene glycerides from the Antarctic nudibranch *Austrodoris kerguelenensis. J. Nat. Prod.* 73:416–421.

Domènech, A., C. Avila, and M. Ballesteros. 2006. Opisthobranch molluscs from the subtidal trawling grounds off Blanes (Girona, north-east Spain). *J. Mar. Biol. Assoc. U.K.* 86:383–389.

Donia, M. S., D. E. Ruffner, S. Cao, and E. W. Schmidt. 2011. Accessing the hidden majority of marine natural products through metagenomics. *Chem. Bio. Chem.* 12:1230–1236.

Doronina, S. O., B. E. Toki, M. Y. Torgov et al. 2003. Development of potent monoclonal antibody auristatin conjugates for cancer therapy. *Nat. Biotechnol.* 21:778–784.

Doty, M. S., and G. Aguilar-Santos. 1970. Transfer of toxic algal substances in marine food chains. *Science* 24:351–355.

Dumdei, E. J., A. E. Flowers, M. J. Garson, and C. J. Moore. 1997. The biosynthesis of sesquiterpene isocyanides and isothiocyanates in the marine sponge *Acanthella cavernosa* (Dendy); evidence for dietary transfer to the dorid nudibranch *Phyllidiella pustulosa. Comp. Biochem. Physiol.Part A: Physiol.* 118:1385–1392.

Dumdei, E. J., J. Kubanek, J. E. Coleman, J. Pika, R. J. Andersen, J. R. Steiner, and J. Clardy. 1997. New terpenoid metabolites from the skin extracts, an egg mass, and dietary sponges of the Northeastern Pacific dorid nudibranch *Cadlina luteomarginata. Canad. J. Chem.* 75:773–789.

Ebel, R., A. Marin, and P. Proksch. 1999. Organ-specific distribution of dietary alkaloids in the marine opisthobranch *Tylodina perversa. Biochem. Syst. Ecol.* 27:769–777.

Edlinger, K. 1982. Colour adaption in *Haminoea navicula* (Da Costa) (Mollusca—Opisthobranchia). *Malacologia* 22:593–600.

Edmunds, M. 1974. *Defence in Animals. A Survey of Antipredator Defences.* Harlow, UK: Longman.

Edmunds, M. 1987. Color in opisthobranchs. *Am. Malacol. Bull.* 5:185–196.

Edmunds, M. 2009. Do nematocysts sequestered by aeolid nudibranchs deter predators?— A background to the debate. *J. Molluscan Stud.* 75:203–205.

Eilertsen, M. H., and M. A. E. Malaquias. 2015. Speciation in the dark: Diversification and biogeography of the deep-sea gastropod genus *Scaphander* in the Atlantic Ocean. *J. Biogeogr.* 42:843–855.

Ellingson, R. A., and P. J. Krug. 2006. Evolution of poecilogony from planktotrophy: Cryptic speciation, phylogeography, and larval development in the gastropod genus *Alderia. Evolution (NY)* 60:2293–2310.

Endler, J. A. 1988. Frequency-dependent predation, crypsis and aposematic coloration. *Phil. Trans. Royal Soc. London B* 319:505–522.

Enge, S., G. M. Nylund, T. Harder, and H. Pavia. 2012. An exotic chemical weapon explains low herbivore damage in an invasive alga. *Ecology* 93:2736–2745.

Engene, N., E. C. Rottacker, J. Kaštovský, T. Byrum, H. Choi, M. H. Ellisman, J. Komárek, and W. H. Gerwick. 2012. *Moorea producens* gen. nov., sp. nov. and *Moorea bouillonii* comb. nov., tropical marine cyanobacteria rich in bioactive secondary metabolites. *Int. J. Syst. Evol. Microbiol.* 62:1171–1178.

Fahey, S. J., and A. R. Carroll. 2007. Natural products isolated from species of *Halgerda* Bergh, 1880 (Mollusca: Nudibranchia) and their ecological and evolutionary implications. *J. Chem. Ecol.* 33:1226–1234.

Fahey, S. J., and M. J. Garson. 2002. Geographic variation of natural products of tropical nudibranch *Asteronotus cespitosus. J. Chem. Ecol.* 28:1773–1785.

Faircloth, G., and C. Cuevas. 2006. Kahalalide F and ES285: Potent anticancer agents from marine molluscs. In *Molluscs: From Chemo-ecological Study to Biotechnological Applications*, G. Cimino and M. Gavagnin (Eds.), pp. 363–379. Berlin, Germany: Springer.

Faulkner, D. J., and M. T. Ghiselin. 1983. Chemical defense and the evolutionary ecology of dorid nudibranchs and some other opisthobranch gastropods. *Mar. Ecol. Prog. Ser.* 13:295–301.

Faulkner, D. J., T. F. Molinski, R. J. Andersen, E. J. Dumdei, and E. D. De Silva. 1990. Geographical variation in defensive chemicals from Pacific Coast dorid nudibranchs and some related marine molluscs. *Comp. Biochem. Physiol.* 97C:233–240.

Faulkner, D. J., and M. O. Stallard. 1973. 7-chloro-3,7-dimethyl-1,4,6-trobromo-1-octen-3-ol, a novel monoterpene alcohol from *Aplysia californica*. *Tetrahedron Lett.* 14:1171–1174.

Faulkner, D. J., M. O. Stallard, J. Fayos, and J. Clardy. 1973. (3R,4S,7S)-trans,trans-3,7-dimethyl-1,8,8-tribromo-3,4,7-trichloro-1,5-octadiene, a novel monoterpene from the sea hare, *Aplysia californica*. *J. Am. Chem. Soc.* 95:3413–3414.

Feeny, P. 1976. Plant apparency and chemical defenses. In *Recent Advances in Phytochemistry*, J. W. Wallace and R. L. Mansell (Eds.), pp. 1–40. New York: Plenum Press.

Fenical, W., H. L. Sleeper, V. J. Paul, M. O. Stallard, and H. H. Sun. 1979. Defensive chemistry of *Navanax* and related opisthobranch molluscs. *Pure Appl. Chem.* 51:1865–1874.

Fernández, R., J. Rodríguez, E. Quiñoá, R. Riguera, L. Muñoz, M. Fernández-Suárez, and C. Debitus. 1996. Onchidin B: A new cyclodepsipeptide from the mollusc *Onchidium* sp. *J. Am. Chem. Soc.* 118:11635–11643.

Fieseler, L., U. Hentschel, L. Grozdanov, A. Schirmer, G. Wen, M. Platzer, S. Hrvatin, D. Butzke, K. Zimmermann, and J. Piel. 2007. Widespread occurrence and genomic context of unusually small polyketide synthase genes in microbial consortia associated with marine sponges. *Appl. Environ. Microbiol.* 73:2144–2155.

Findlay, J. A., and G. Li. 2002. Novel terpenoids from the sea hare *Aplysia punctata*. *Can. J. Chem.* 80:1697–1707.

Finer, J., J. Clardy, W. Fenical, L. Minale, R. Riccio, J. Battaile, M. Kirkup, and R. E. Moore. 1979. Structures of dictyodial and dictyolactone, unusual marine diterpenoids. *J. Org. Chem.* 44:2044–2047.

Flatt, P. M., J. T. Gautschi, R. W. Thacker, M. Musafija-Girt, P. Crews, and W. H. Gerwick. 2005. Identification of the cellular site of polychlorinated peptide biosynthesis in the marine sponge *Dysidea (Lamellodysidea) herbacea* and symbiotic cyanobacterium *Oscillatoria spongeliae* by CARD-FISH analysis. *Mar. Biol.* 147:761–774.

Fontana, A. 2006. Biogenetic proposals and biosynthetic studies on secondary metabolites of opisthobranch molluscs. In *Molluscs*, Cimino, G. and Gavagnin, M. (Eds.), pp. 303–332. Berlin, Germany: Springer.

Fontana, A., P. Cavaliere, N. Ungur, L. D'Souza, P. S. Parameswaram, and G. Cimino. 1999. New scalaranes from the nudibranch *Glossodoris atromarginata* and its sponge prey. *J. Nat. Prod.* 62:1367–1370.

Fontana, A., P. Cavaliere, S. Wahidulla, C. G. Naik, and G. Cimino. 2000. A new antitumor isoquinoline alkaloid from the marine nudibranch *Jorunna funebris*. *Tetrahedron* 56:7305–7308.

Fontana, A., M. L. Ciavatta, L. D'Souza, E. Mollo, C. G. Naik, P. S. Parameswaran, S. Wahidulla, and G. Cimino. 2001. Selected chemo-ecological studies of marine opisthobranchs from Indian coasts. *J. Ind. Ins. Sci.* 81:403–415.

Fontana, A., M. L. Ciavatta, T. Miyamoto, A. Spinella, and G. Cimino. 1999. Biosynthesis of drimane terpenoids in dorid molluscs: Pivotal role of 7-deacetoxyolepupuane in two species of *Dendrodoris* nudibranchs. *Tetrahedron* 55:5937–5946.

Fontana, A., M. L. Ciavatta, E. Mollo, C. D. Naik, S. Wahidulla, L. D'Sousa, and G. Cimino. 1999. Volvatellin, cauerpenyne-related product from the sacoglossan *Volvatella* sp. *J. Nat. Prod.* 62:931–933.

Fontana, A., A. Cutignano, A. Giordano, A. D. Coll, and G. Cimino. 2004. Biosynthesis of aglajnes, polypropionate allomones of the opisthobranch mollusc *Bulla striata*. *Tetrahedron Lett.* 45:6847–6850.

Fontana, A., M. Gavagnin, E. Mollo, E. Trivellone, J. Ortea, and G. Cimino. 1995. Chemical studies of *Cadlina* molluscs from the Cantabrian Sea (Atlantic Ocean). *Comp. Biochem. Physiol.—Part B: Biochem.* 111:283–290.

Fontana, A., F. Giménez, A. Marín, E. Mollo, and G. Cimino. 1994. Transfer of secondary metabolites from the sponges *Dysidea fragilis* and *Pleraplysilla spinifera* to the mantle dermal formations (MDFs) of the nudibranch *Hypselodoris webbi*. *Experientia* 50:510–516.

Fontana, A., E. Mollo, D. Ricciardi, I. Fakhr, and G. Cimino. 1997. Chemical studies of Egyptian opisthobranchs: Spongian diterpenoids from *Glossodoris atromarginata*. *J. Nat. Prod.* 60:444–448.

Fontana, A., E. Mollo, J. Ortea, M. Gavagnin, and G. Cimino. 2000. Scalarane and homoscalarane compounds from the nudibranchs *Glossodoris sedna* and *Glossodoris dalli*: Chemical and biological properties. *J. Nat. Prod.* 63:527–530.

Fontana, A., C. Muniaín, and G. Cimino. 1998. First chemical study of patagonian nudibranchs: A new seco-11, 12-spongiane, tyrinnal, from the defensive organs of *Tyrinna nobilis*. *J. Nat. Prod.* 61:1027–1029.

Fontana, A., A. Tramice, A. Cutignano, G. d'Ippolito, L. Renzulli, and G. Cimino. 2003. Studies of the biogenesis of verrucosins, toxic diterpenoid glycerides of the Mediterranean mollusc *Doris verrucosa*. *Europ. J. Org. Chem.* 2003:3104–3108.

Fontana, A., A. Tramice, A. Cutignano, G. d'Ippolito, M. Gavagnin, and G. Cimino. 2003. Terpene biosynthesis in the nudibranch *Doriopsilla areolata*. *J. Org. Chem.* 68:2405–2409.

Fontana, A., G. Villani, and G. Cimino. 2000. Terpene biosynthesis in marine molluscs: Incorporation of glucose in drimane esters of *Dendrodoris* nudibranchs via classical mevalonate pathway. *Tetrahedron Lett.* 41:2429–2433.

Forster, L. C., G. K. Pierens, A. M. White, K. L. Cheney, P. Dewapriya, R. J. Capon, and M. J. Garson. 2017. Cytotoxic spiroepoxide lactone and its putative biosynthetic precursor from *Goniobranchus splendidus*. *ACS Omega* 2:2672–2677.

Fu, X., A. J. Palomar, E. P. Hong, F. J. Schmitz, and F. A. Valeriote. 2004. Cytotoxic lissoclimide-type diterpenes from the molluscs *Pleurobranchus albiguttatus* and *Pleurobranchus forskalii*. *J. Nat. Prod.* 67:1415–1418.

Fu, X., E. P. Hong, and F. J. Schmitz. 2000. New polypropionate pyrones from the Philippine sacoglossan mollusc *Placobranchus ocellatus*. *Tetrahedron* 56:8989–8993.

Fusetani, N., H. Hirota, T. Okino, Y. Tomono, and E. Yoshimura. 1996. Antifouling activity of isocyanoterpenoids and related compounds isolated from a marine sponge and nudibranchs. *J. Nat. Tox.* 5:249–259.

Fusetani, N., H. J. Wolstenholme, S. Matsunaga, and H. Hirota. 1991. Two new sesquiterpene isonitriles from the nudibranch *Phyllidia pustulosa*. *Tetrahedron Lett.* 32:7291–7294.

Gallimore, W. A., and P. J. Scheuer. 2000. Malyngamides O and P from the sea hare *Stylocheilus longicauda*. *J. Nat. Prod.* 63:1422–1424.

Gallimore, W. A., D. L. Galario, C. Lacy, Y. Zhu, and P. J. Scheuer. 2000. Two complex proline esters from the sea hare *Stylocheilus longicauda*. *J. Nat. Prod.* 63:1022–1026.

Gamberale, G. 1998. Aposematism and gregariousness: The combined effect of group size and coloration on signal repellence. *Proc. Royal Soc. London. Series B, Biol. Sci.* 265:889–894.

Garson, M. 2010. Marine natural products as antifeedants. In *Comprehensive Natural Products II. Chemistry and Biology*, L. Mander and H. W. Liu (Eds.), pp. 503–537. Oxford, UK: Elsevier Science.

Garson, M. J. 2006. Marine mollusks from Australia and New Zealand: Chemical and ecological studies. In *Molluscs: From Chemo-ecological Study to Biotechnological Applications*, G. Cimino and M. Gavagnin (Eds.), pp. 159–174. Berlin, Germany: Springer.

Garson, M., and J. S. Simpson. 2004. Marine isocyanides and related natural products—Structure, biosynthesis and ecology. *Nat. Prod. Rep.* 21:164–179.

Garson, M. J. 1993. The biosynthesis of marine natural products. *Chem. Rev.* 93:1699–1733.

Garson, M. J. 2001. Ecological perspectives on marine natural product biosynthesis. In *Marine Chemical Ecology*, J. B. McClintock and B. Baker (Eds.), pp. 71–114. Boca Raton, FL: CRC Press LLC.

Gaspar, H., A. Cutignano, T. Ferreira, G. Calado, G. Cimino, and A. Fontana. 2008. Biosynthetic evidence supporting the generation of terpene chemodiversity in marine mollusks of the genus *Doriopsilla*. *J. Nat. Prod.* 71:2053–2056.

Gaspar, H., M. Gavagnin, G. Calado, F. Castelluccio, E. Mollo, and G. Cimino. 2005. Pelseneeriol-1 and-2: New furanosesquiterpene alcohols from porostome nudibranch *Doriopsilla pelseneeri*. *Tetrahedron* 61:11032–11037.

Gavagnin, M., A. De Napoli, F. Castelluccio, and G. Cimino. 1999a. Austrodorin-A and -B: First tricyclic diterpenoid 2'-monoglyceryl esters from an Antarctic nudibranch. *Tetrahedron Lett.* 40:8471–8475.

Gavagnin, M., A. De Napoli, G. Cimino, K. Iken, C. Avila, and F. J. Garcia. 1999b. Absolute configuration of diterpenoid diacylglycerols from the Antarctic nudibranch *Austrodoris kerguelenensis*. *Tetrah.: Asym.* 10:2647–2650.

Gavagnin, M., A. Marín, E. Mollo, A. Crispino, G. Villani, and G. Cimino. 1994. Secondary metabolites from Mediterranean Elysioidea: Origin and biological role. *Comp. Biochem. Physiol. Part B: Comp. Biochem.* 108:107–115.

Gavagnin, M., A. Marín, F. Castelluccio, G. Villani, and G. Cimino. 1994. Defensive relationships between *Caulerpa prolifera* and its shelled sacoglossan predators. *J. Exp. Mar. Biol. Ecol.* 175:197–210.

Gavagnin, M., A. Spinella, A. Crispino, R. D. A. Epifanio, A. Marn, and G. Cimino. 1993. Chemical-components of the Mediterranean ascoglossan *Thuridilla hopei*. *Gazzetta Chim. Ital.* 123:205–208.

Gavagnin, M., and A. Fontana. 2000. Diterpenes from marine opisthobranch molluscs. *Curr. Org. Chem.* 4:1201–1248.

Gavagnin, M., E. Mollo, D. Montanaro, J. Ortea, and G. Cimino. 2000. Chemical studies of Caribbean sacoglossans: Dietary relationships with green algae and ecological implications. *J. Chem. Ecol.* 26:1563–1578.

Gavagnin, M., E. Mollo, F. Castelluccio, D. Montanaro, J. Ortea, and G. Cimino. 1997. A novel dietary sesquiterpene from the marine sacoglossan *Tridachia crispata*. *Nat. Prod. Lett.* 10:151–156.

Gavagnin, M., E. Mollo, F. Castelluccio, M. T. Ghiselin, G. Calado, and G. Cimino. 2001. Can molluscs biosynthesize typical sponge metabolites? The case of the nudibranch *Doriopsilla areolata*. *Tetrahedron* 57:8913–8916.

Gavagnin, M., E. Mollo, G. Calado, S. Fahey, M. T. Ghiselin, J. Ortea, and G. Cimino. 2001. Chemical studies of porostome nudibranchs: Comparative and ecological aspects. *Chemoecology* 11:131–136.

Gavagnin, M., E. Mollo, G. Cimino, and J. Ortea. 1996. A new γ-dihydropyrone-propionate from the Caribbean ascoglossan *Tridachia crispata*. *Tetrahedron Lett.* 37:4259–4262.

Gavagnin, M., E. Mollo, T. Docimo, Y.-W. Guo, and G. Cimino. 2004. Scalarane metabolites of the nudibranch *Glossodoris rufomarginata* and its dietary sponge from the South China Sea. *J. Nat. Prod.* 67:2104–2107.

Gavagnin, M., E. Trivellone, F. Castelluccio, G. Cimino, and R. Cattaneo-Vietti. 1995. Glyceryl ester of a new halimane diterpenoic acid from the skin of the Antarctic nudibranch *Austrodoris kerguelenensis*. *Tetrah. Lett.* 36:7319–7322.

Gavagnin, M., M. Carbone, E. Mollo, and G. Cimino. 2003a. Austrodoral and austrodoric acid: Nor-sesquiterpenes with a new carbon skeleton from the Antarctic nudibranch *Austrodoris kerguelenensis*. *Tetrah. Lett.* 44:1495–1498.

Gavagnin, M., M. Carbone, E. Mollo, and G. Cimino. 2003b. Further chemical studies on the Antarctic nudibranch *Austrodoris kerguelenensis*: New terpenoid acylglycerols and revision of the previous stereochemistry. *Tetrahedron* 59:5579–5583.

Gavagnin, M., M. Carbone, M. Nappo, E. Mollo, V. Roussis, and G. Cimino. 2005. First chemical study of anaspidean *Syphonota geographica*: Structure of degraded sterols aplykurodinone-1 and -2. *Tetrahedron* 61:617–621.

Gavagnin, M., N. Ungur, E. Mollo, J. Templado, and G. Cimino. 2002. Structure and synthesis of a progesterone homologue from the skin of the dorid nudibranch *Aldisa smaragdina*. *Europ. J. Org. Chem.* 2002:1500–1504.

Gavagnin, M., N. Ungur, F. Castelluccio, C. Muniain, and G. Cimino. 1999. New minor diterpenoid diacylglycerols from the skin of the nudibranch *Anisodoris fontaini*. *J. Nat. Prod.* 62:269–274.

Gavagnin, M., Spinella, A., Cimino, G., and Sodano, G. 1990. Stereochemistry of ichthyotoxic diacylglycerols from opisthobranch molluscs. *Tetrahedron Lett* 31:6093–6094.

Gavagnin, M., Ungur, N., Castelluccio, F., and Cimino, G. 1997. Novel verrucosins from the skin of the Mediterranean nudibranch *Doris verrucosa*. *Tetrahedron* 53:1491–1504.

Gemballa, S., and F. Schermutzki. 2004. Cytotoxic haplosclerid sponges preferred: A field study on the diet of the dotted sea slug *Peltodoris atromaculata* (Doridoidea: Nudibranchia). *Mar. Biol.* 144:1213–1222.

Gerwick, W. H., W. Fenical, N. Fritsch, and J. Clardy. 1979. Stypotriol and stypoldione; ichthyotoxins of mixed biogenesis from the marine alga *Stypopodium zonale*. *Tetrahedron Lett.* 2:145–148.

Gillette, R., M. Saeki, and R.-C. Huang. 1991. Defensive mechanisms in notaspid snails: Acid humor and evasiveness. *J. Exp. Biol.* 156:335–347.

Giménez-Casalduero, F., R. W. Thacker, and V. J. Paul. 1999. Association of color and feeding deterrence by tropical reef fishes. *Chemoecology* 9:33–39.

Ginsburg, D. W., and V. J. Paul. 2001. Chemical defenses in the sea hare *Aplysia parvula*: Importance of diet and sequestration of algal secondary metabolites. *Mar. Ecol. Progr. Ser.* 215:261–274.

Giordano, A., G. Andreotti, E. Mollo, and A. Trincone. 2004. Transglycosylation reactions performed by glycosyl hydrolases from the marine anaspidean mollusc *Aplysia fasciata*. *J. Mol. Catal. B: Enzym.* 30:51–59.

Göbbeler, K., and A. Klussmann-Kolb. 2011. Molecular phylogeny of the Euthyneura (Mollusca, Gastropoda) with special focus on Opisthobranchia as a framework for reconstruction of evolution of diet. *Thalassas* 27:121–153.

Goetz, G., Y. Nakao, and P. J. Scheuer. 1997. Two acyclic kahalalides from the sacoglossan mollusk *Elysia rufescens*. *J. Nat. Prod.* 60:562–567.

Goffredi, S. K., V. J. Orphan, G. W. Rouse, L. Jahnke, T. Embaye, K. Turk, R. Lee, and R. C. Vrijenhoek. 2005. Evolutionary innovation: A bone-eating marine symbiosis. *Environ. Microbiol.* 7: 1369–1378.

Gopichand, Y., and F. J. Schmitz. 1980. Bursatellin: A new diol dinitrile from the sea hare *Bursatella leachii pleii*. *J. Org. Chem.* 45:5383–5385.

Gopichand, Y., F. J. Schmitz, J. Shelly, A. Rahman, and D. van der Helm. 1981. Marine natural products: Halogenated acetylenic ethers from the sea hare *Aplysia dactylomela*. *J. Org. Chem.* 46:5192–5197.

Gosliner, T. M. 1995. The genus *Thuridilla* (Opisthobranchia: Elysiidae) from the tropical Indo-Pacific, with a revision of the phylogeny and systematics of the Elysiidae. *Proc. Calif. Acad. Sci.*, Ser. 4, 49:1–54.

Gosliner, T. M. 2001. Aposematic coloration and mimicry in opisthobranch mollusks: New phylogenetic and experimental data. *Boll. Malacol.* 37:163–170.

Gotsbacher, M. P., and P. Karuso. 2015. New antimicrobial bromotyrosine analogues from the sponge *Pseudoceratina purpurea* and its predator *Tylodina corticalis*. *Mar. Drugs* 13:1389–1409.

Granato, A. C., R. G. S. Berlinck, A. Magalhaes, A. B. Schefer, A. G. Ferreira, B. De Sanctis, J. C. De Freitas, E. Hajdu, and A. E. Migotto. 2000. Natural products from the marine sponges *Aaptos* sp. and *Hymeniacidon* aff. *heliophila*, and from the nudibranch *Doris* aff. *verrucosa*. *Quimica Nova* 23:594–599.

Gray, C. A., M. T. Davies-Coleman, and C. McQuaid. 1998. Labdane diterpenes from the South African marine pulmonate *Trimusculus costatus*. *Nat. Prod. Lett.* 12:47–53.

Graziani, E. I., and R. J. Andersen. 1998. Limaciamine, a new diacylguanidine isolated from the North Sea nudibranch *Limacia clavigera*. *J. Nat. Prod.* 61:285–286.

Graziani, E. I., R. J. Andersen, P. J. Krug, and D. J. Faulkner. 1996. Stable isotope incorporation evidence for the de novo biosynthesis of terpenoic acid glycerides by dorid nudibranchs. *Tetrahedron* 52:6869–6878.

Graziani, E. I., T. M. Allen, and R. J. Andersen. 1995. Lovenone, a cytotoxic degraded triterpenoid isolated from skin extracts of the North Sea dorid nudibranch *Adalaria loveni*. *Tetrahedron Lett.* 36:1763–1766.

Graziani, E. I., and R. J. Andersen. 1996. Investigations of sesquiterpenoid biosynthesis by the dorid nudibranch *Acanthodoris nanaimoensis*. *J. Am. Chem. Soc.* 118:4701–4702.

Graziani, E. I., and R. J. Andersen. 1996. Stable isotope incorporation evidence for a polyacetate origin of the acyl residues in triophamine, a diacylguanidine metabolite obtained from the dorid nudibranch *Triopha catalinae*. *Chem. Commun.* 20:2377–2378.

Grkovic, T., D. R. Appleton, and B. R. Copp. 2005. Chemistry and chemical ecology of some of the common opisthobranch molluscs found on the shores of NE New Zealand. *Chem. New Zealand* 69:12–15.

Guerriero, A., M. D'Ambrosio, and F. Pietra. 1987. Verecynarmin A, a novel briarane diterpenoid isolated from both the Mediterranean nudibranch mollusc *Armina maculata* and its prey, the pennatulacean octocoral *Veretillum cynomorium*. *Helv. Chim. Acta* 70:984–991.

Guerriero, A., M. D'Ambrosio, and F. Pietra. 1988. Slowly interconverting conformers of the briarane diterpenoids verecynarmin B, C, and D, isolated from the nudibranch mollusc *Armina maculata* and the octocoral *Veretillum cynomorium* of the East Pyrenean waters. *Helv. Chim. Acta* 71:472–485.

Guerriero, A., M. D'Ambrosio, and F. Pietra. 1990. Isolation of the cembranoid preverecynarmin alongside some briaranes, the verecynarmins, from both the nudibranch mollusc *Armina maculata* and the octocoral *Veretillum cynomorium* of the East Pyrenean Mediterranean Sea. *Helv. Chim. Acta* 73:277–283.

Guilford, T., and L. Cuthill. 1991. Evolution of aposematism in marine gastropods. *Evolution* 45:449–451.

Gulavita, N. K., E. D. De Silva, M. R. Hagadone, P. Karuso, P. J. Scheuer, G. D. van Duyne, and J. Clardy. 1986. Nitrogenous bisabolene sesquiterpenes from marine invertebrates. *J. Org. Chem.* 51:5136–5139.

Guo, Y.-W. 1997. Chemical studies of the novel bioactive secondary metabolites from the benthic invertebrates: Isolation and structure characterization. Ph. D. thesis, Naples, Italy: University of Naples.

Guo, Y.-W., Gavagnin, M., Mollo, E., Trivellone, E., and Cimino, G. 1998. Structure of the pigment of the Red Sea nudibranch *Hexabranchus sanguineus*. *Tetrahedron Lett.* 39:2635–2638.

Guo, Y.-W., M. Gavagnin, M. Carbone, E. Mollo, and G. Cimino. 2012. Recent Sino-Italian collaborative studies on marine organisms from the South China Sea. *Pure Appl. Chem.* 84:1391–1405.

Haber, M., S. Cerfeda, M. Carbone, G. Calado, H. Gaspar, R. Neves, V. Maharajan, G. Cimino, M. Gavagnin, M. T. Ghiselin, and E. Mollo, . 2010. Coloration and defense in the nudibranch gastropod *Hypselodoris fontandraui*. *Biol. Bull.* 218:181–188.

Haefner, B. 2003. Drugs from the deep: Marine natural products as drug candidates. *Drug Disc. Today* 8:536–544.

Hamann, M. T., and P. J. Scheuer. 1993. Kahalalide F: A bioactive depsipeptide from the sacoglossan mollusk *Elysia rufescens* and the green alga *Bryopsis* sp. *J. Am. Chem. Soc.* 115:5825–5826.

Hamann, M. T., C. S. Otto, P. J. Scheuer, and D. C. Dunbar. 1996. Kahalalides: Bioactive peptides from a marine mollusk *Elysia rufescens* and its algal diet *Bryopsis sp. J. Org. Chem.* 61:6594–6600.

Händeler, K., Y. P. Grzymbowski, P. J. Krug, and H. Wägele. 2009. Functional chloroplasts in metazoan cells—A unique evolutionary strategy in animal life. *Front. Zool.* 6:28.

Härlin, C., and M. Härlin. 2003. Towards a historization of aposematism. *Evolut. Ecol.* 17:197–212.

Harvell, C. D. 1984. Why nudibranchs are partial predators: Intracolonial variation in bryozoan palatability. *Ecology* 65:716–724.

Hay, M. E. 1991. Marine terrestrial contrasts in the ecology of plant-chemical defenses against herbivores. *Trends Ecol. Evol.* 6:362–365.

Hay, M. E. 1996. Marine chemical ecology: What's known and what's next? *J. Exp. Mar. Ecol. Biol.* 200:103–134.

Hay, M. E., and P. D. Steinberg. 1992. The chemical ecology of plant–herbivore interactions in marine versus terrestrial communities. In *Herbivores:Their Interactions with Plant Secondary Metabolites*, G. A. Rosenthal and M. R. Berenbaum (Eds.), pp. 371–413. New York: Academic Press.

Hay, M. E., and W. Fenical. 1988. Marine plant–herbivore interactions: The ecology of chemical defense. *Ann. Rev. Ecol. Syst.* 19:111–145.

Hay, M. E., J. E. Duffy, and C. A. Pfister. 1987. Chemical defense against different marine herbivores—Are amphipods insect equivalents? *Ecology* 68:1567–1580.

Hay, M. E., J. E. Duffy, V. J. Paul, P. E. Renaud, and W. Fenical. 1990. Specialist herbivores reduce their susceptibility to predation by feeding on the chemically defended seaweed *Avrainvillea longicaulis*. *Limnol. Oceanogr.* 35:1734–1747.

Hay, M. E., J. R. Pawlik, J. E. Duffy, and W. Fenical. 1989. Seaweed-herbivore-predator interactions: Host–plant specialization reduces predation on small herbivores. *Oecologia* 81:418–427.

He, H., and D. J. Faulkner. 1989. Renieramycins E and F from the sponge *Reniera* sp. Reassignment of the stereochemistry of the renieramycins. *J. Org. Chem.* 54:5822–5824.

He, R., T. Wakimoto, Y. Takeshige, Y. Egami, H. Kenmoku, T. Ito, B. Wang, Y. Asakawa, and I. Abe. 2012. Porphyrins from a metagenomic library of the marine sponge *Discodermia calyx. Mol. BioSys.* 8:2334–2338.

He, W. F., Y. Li, M. T. Feng, M. Gavagnin, E. Mollo, S. C. Mao, and Y. W. Guo. 2014. New isoquinolinequinone alkaloids from the South China Sea nudibranch *Jorunna funebris* and its possible sponge-prey *Xestospongia* sp. *Fitoterapia* 96:109–114.

Heck, K. L., and M. P. Weinstein. 1978. Mimetic relationships between tropical burrfishes and opisthobranchs. *Biotropica* 10:78–79.

Hellou, J., R. J. Andersen, and J. E. Thompson. 1982. Terpenoids from the dorid nudibranch *Cadlina luteomarginata*. *Tetrahedron* 38:1875–1879.

Hentschel, U., J. Piel, S. M. Degnan, and M. W. Taylor. 2012. Genomic insights into the marine sponge microbiome. *Nat. Rev. Microbiol.* 10:641–654.

Herms, D. A., and W. J. Mattson. 1992. The dilemma of plants: To grow or defend. *Quart. Rev. Biol.* 67:283–335.

Hertzberg, S., S. Liaaen-Jensen, C. R. Enzell, and G. W. Francis. 1969. Animal carotenoids. *Acta Chem. Scandinavica* 23:3290–3312.

Hildebrand, M., L. E. Waggoner, G. E. Lim, K. H. Sharp, C. P. Ridley, and M. G. Haygood. 2004. Approaches to identify, clone, and express symbiont bioactive metabolite genes. *Nat. Prod. Rep.* 21:122–142.

Hinde, R., and D. C. Smith. 1974. Chloroplast symbiosis and the extent to which it occurs in Sacoglossa (Gastropoda: Mollusca). *Biol. J. Linn. Soc.* 6:349–356.

Hirota, H., T. Okino, E. Yoshimura, and N. Fusetani. 1998. Five new antifouling sesquiterpenes from two marine sponges of the genus *Axinyssa* and the nudibranch *Phyllidia pustulosa*. *Tetrahedron* 54:13971–13980.

Hochlowski, J. E., and D. J. Faulkner. 1980. A diterpene related to cladiellin from a Pacific soft coral. *Tetrahedron Lett.* 21:4055–4056.

Hochlowski, J., J. Coll, D. J. Faulkner, and J. Clardy. 1984. Novel metabolites of four *Siphonaria* species. *J. Am. Chem. Soc.* 106:6748–6750.

Hochlowski, J. E., and D. J. Faulkner. 1983. Antibiotics from the marine pulmonate *Siphonaria diemenensis*. *Tetrahedron Lett.* 24:1917–1920.

Hochlowski, J. E., and D. J. Faulkner. 1984. Metabolites of the marine pulmonate *Siphonaria australis*. *J. Org. Chem.* 49:3838–3840.

Hochlowski, J. E., D. J. Faulkner, G. K. Matsumoto, and J. Clardy. 1983. The denticulatins, two propionate metabolites from the pulmonate *Siphonaria denticulata*. *J. Am. Chem. Soc.* 105:7413–7415.

Hochlowski, J. E., J. C. Coll, D. J. Faulkner, J. E. Biskupiak, C. M. Ireland, Z. Qi-Tai, H. Cun-Heng, and J. Clardy. 1983. Novel metabolites of four *Siphonaria* species. *J. Am. Chem. Soc.* 106:6748–6750.

Högberg, H.-E., R. H. Thompson, and T. J. King. 1976. The cymopols, a group of prenylated bromohydroquinones from the green calcareous alga *Cymopolia barbata*. *J. Chem. Soc., Perkin Trans., Ser.* 1:1696–1701.

Horgen, F. D., D. B. de los Santos, G. Goetz, B. Sakamoto, Y. Kan, H. Nagai, and P. J. Scheuer. 2000. A new depsipeptide from the sacoglossan mollusk *Elysia ornata* and the green alga *Bryopsis* species. *J. Nat. Prod.* 63:152–154.

Hu, G. P., J. Yuan, L. Sun, Z. G. She, J. H. Wu, X. J. Lan, X. Zhu, Y.-C. Lin, and S. P. Chen. 2011. Statistical research on marine natural products based on data obtained between 1985 and 2008. *Mar. Drugs* 9:514–525.

Huang, R. Y., W. T. Chen, T. Kurtán, A. Mándi, J. Ding, J. Li, and Y. W. Guo. 2016. Bioactive isoquinolinequinone alkaloids from the South China Sea nudibranch *Jorunna funebris* and its sponge-prey *Xestospongia* sp. *Fut. Med. Chem.* 8:17–27.

Hughes, K. A., L. R. Pertierra, M. A. Molina-Montenegro, and P. Convey. 2015. Biological invasions in terrestrial Antarctica: What is the current status and can we respond? *Biodivers. Conserv.* 24:1031–1055.

Hunt, B. P. V., E. A. Pakhomov, G. W. Hosie, V. Siegel, P. Ward, and K. Bernard. 2008. Pteropods in Southern Ocean ecosystems. *Prog. Oceanogr.* 78:193–221.

Ianora, A., M. Boersma, R. Casotti, A. Fontana, J. Harder, F. Hoffmann, H. Pavia, P. Potin, S. A. Poulet, and G. Toth. 2006. New trends in marine chemical ecology. *Estuaries and Coasts* 29:531–551.

Ichiba, T., and T. Higa. 1986. New cuparene-derived sesquiterpenes with unprecedented oxygenation patterns from the sea hare *Aplysia dactylomela*. *J. Org. Chem.* 51:3364–3366.

Iijima, R., J. Kisugi, and M. Yamazaki. 1995. Antifungal activity of aplysianin E, a cytotoxic protein of sea hare (*Aplysia kurodai*) eggs. *Develop Comp Immunol* 19:13–19.

Iijima, R., J. Kisugi, and M. Yamazaki. 2003. A novel antimicrobial peptide from the sea hare *Dolabella auricularia*. *Develop Comp Immunol* 27:305–311.

Iken, K., C. Avila, A. Fontana, and M. Gavagnin. 2002. Chemical ecology and origin of defensive compounds in the Antarctic nudibranch *Austrodoris kerguelenensis* (Opisthobranchia: Gastropoda). *Mar. Biol.* 141:101–109.

Iken, K., C. Avila, M. L. Ciavatta, A. Fontana, and G. Cimino. 1998. Hodgsonal, a new drimane sesquiterpene from the mantle of the Antarctic nudibranch *Bathydoris hodgsoni.* *Tetrahedron Lett.* 39:5635–5638.

Imperato, F., L. Minale, and R. Riccio. 1977. Constituents of the digestive gland of molluscs of the genus *Aplysia.* II. Halogenated monoterpenes from *Aplysia limacina.* *Experientia* 33:1273–1274.

Imre, S., S. Islimyeli, A. Öztunç, R. H. Thomson. 1981. Obtusenol, a sesquiterpene from *Laurencia obtusa.* *Phytochemistry* 20:833–834.

Ioannou, E., M. Nappo, C. Avila, C. Vagias, and V. Roussis. 2009. Metabolites from the sea hare *Aplysia fasciata. J. Nat. Prod.* 72:1716–1719.

Ireland, C., and D. J. Faulkner. 1977. Diterpenes from *Dolabella californica. J. Org. Chem.* 42:3157–3162.

Ireland, C., and D. J. Faulkner. 1978. The defensive secretion of the opisthobranch mollusc *Onchidella binneyi. Bioorg. Chem.* 7:125–131.

Ireland, C., and D. J. Faulkner. 1981. The metabolites of the marine molluscs *Tridachiella diomedea* and *Tridachia crispata. Tetrahedron* 37:233–240.

Ireland, C., and P. J. Scheuer. 1979. Photosynthetic marine molluscs: In vivo ^{14}C incorporation into metabolites of the sacoglossan *Placobranchus ocellatus. Science* 205:922–923.

Ireland, C., D. J. Faulkner, J. Finer, and J. Clardy. 1976. A novel diterpene from *Dollabella californica. J. Am. Chem. Soc.* 98:4664–4665.

Ireland, C., M. O. Stallard, and D. J. Faulkner. 1976. Some chemical constituents of the digestive gland of the sea hare *Aplysia californica. J. Org. Chem.* 41:2461–2465.

Ireland, C. D., D. J. Faulkner, B. A. Solheim, and J. Clardy. 1978. Tridachione, a propionate-derived metabolite of the opisthobranch mollusc *Tridachiella diomedea. J. Am. Chem. Soc.* 100:1002–1003.

Ireland, C. M., J. E. Biskupiak, G. J. Hite, M. Rapposch, P. J. Scheuer, and J. R. Ruble. 1984. Ilikonapyrone esters, likely defense allomones of the mollusc *Onchidium verruculatum. J. Org. Chem.* 49:559–561.

Ishibashi, M., Y. Yamaguchi, and Y. J. Hirano. 2006. Bioactive natural products from nudibranchs. In *Biomaterials from Aquatic and Terrestrial Organisms,* M. Fingerman and R. Nagabhushanam (Eds.), pp. 513–535. Enfield, New Hampshire: Science Publishers.

Iwashima, M., I. Terada, K. Iguchi, and T. Yamori. 2002. New biologically active marine sesquiterpenoid and steroid from the Okinawan sponge of the genus *Axinyssa. Chem. Pharm. Bull.* 50:1286–1289.

Jacobs, R. S., M. A. Bober, I. Pinto, A. B. Williams, P. B. Jacobson, and M. S. de Carvalho. 1993. Pharmacological studies of marine novel marine metabolites. In *Advances in Marine Biotechnology,* Vol 1, D. H. Attaway and O. R. Zaborsky (Eds.), pp. 77–99. New York: Plenum.

Jaisamut, S., S. Prabpai, C. Tancharoen, S. Yuenyongsawad, S. Hannongbua, P. Kongsaeree, and A. Plubrukarn. 2013. Bridged tricyclic sesquiterpenes from the tubercle nudibranch *Phyllidia coelestis* Bergh. *J. Nat. Prod.* 76:2158–2161.

Jensen, K. R. 1980. A review of sacoglossan diets, with comparative notes on radular and buccal anatomy. *Malacol. Rev.* 13:55–77.

Jensen, K. R. 1984. Defensive behavior and toxicity of the ascoglossan opisthobranch *Mourgona germaineae* Marcus. *J. Chem. Ecol.* 10:475–486.

Jensen, K. R. 1992. Anatomy of some Indo-Pacific Elysiidae (Opisthobranchia, Sacoglossa (=Ascoglossa)), with a discussion of the generic division and phylogeny. *J. Molluscan Stud.* 58:257–296.

Jensen, K. R. 1993. Morphological adaptations and plasticity of radular teeth of the Sacoglossa (=Ascoglossa) (Mollusca, Opisthobranchia) in relation to their food plants. *Biol. J. Linn. Soc.* 48:135–155.

Jensen, K. R. 1996. Phylogenetic systematics and classification of the Sacoglossa (Mollusca, Gastropoda, Opisthobranchia). *Philos. Trans. R. Soc. Lond.*, ser. B, 351:91–122.

Jensen, K. R. 1997. Evolution of the Sacoglossa (Mollusca, Opisthobranchia) and the ecological associations with their food plants. *Evol. Ecol.* 11:301–335.

Jiménez, C., E. Quiñoá, L. Castedo, and R. Riguera. 1986. Epidioxy sterols from the tunicates *Dendrodoa grossularia* and *Ascidiella aspersa* and the Gastropoda *Aplysia depilans* and *Aplysia punctata. J. Nat. Prod.* 49:905–909.

Jiménez-Romero, C., A. M. S. Mayer, and A. D. Rodríguez. 2014. Dactyloditerpenol acetate, a new prenylbisabolane-type diterpene from *Aplysia dactylomela* with significant in vitro anti-neuroinflammatory activity. *Bioorg. Med. Chem. Lett.* 24:334–348.

Jiménez-Romero, C., K. González, and A. D. Rodríguez. 2012. Dolabriferols B and C, non-contiguous polypropionate esters from the tropical sea hare *Dolabrifera dolabrifera. Tetrahedron Lett.* 53:6641–6645.

Jing, J., and Gillette, R. 1999. Central pattern generator for escape swimming in the notaspid sea slug *Pleurobranchaea californica. J. Neurophysiol.* 81:654–667.

Johnson, P. M., C. E. Kicklighter, M. Schmidt, M. Kamio, H. Yang, D. Elkin, W. C. Michel, P. C. Tai, and C. D. Derby. 2006. Packaging of chemicals in the defensive secretory glands of the sea hare *Aplysia californica. J. Exp. Biol.* 209:78–88.

Johnson, R. F., and T. M. Gosliner. 2012. Traditional taxonomic groupings mask evolutionary history: A molecular phylogeny and new classification of the chromodorid nudibranchs. *PLoS One* 7:e33479.

Jörger, K. M., J. L. Norenburg, N. G. Wilson, and M. Schrödl. 2012. Barcoding against a paradox? Combined molecular species delineations reveal multiple cryptic lineages in elusive meiofaunal sea slugs. *BMC Evol. Biol.* 12:245.

Jung, V., and G. Pohnert. 2001. Rapid wound-activated transformation of the green algal defensive metabolite caulerpenyne. *Tetrahedron* 57:7169–717.

Kaiser, C. R., L. F. Pitombo, and A. C. Pinto. 1998. C-13 and H-1 NMR assignments of the chamigrenes prepacifenol and dehydroxyprepacifenol epoxioes. *Spectroscopy Lett.* 31:573–585.

Kamio, M., and C. D. Derby. 2017. Finding food: How marine invertebrates use chemical cues to track and select food. *Nat. Prod. Rep.* 34:514–528.

Kamio, M., C. E. Kicklighter, L. Nguyen, M. W. Germann, and C. D. Derby. 2011. Isolation and structural elucidation of novel mycosporine-like amino acids as alarm cues in the defensive ink secretion of the sea hare *Aplysia californica. Helv. Chim. Acta* 94:1012–1018.

Kamio, M., K. C. Ko, S. Zheng, B. Wang, S. L. Collins, G. Gadda, P. C. Tai, and C. D. Derby. 2009. The chemistry of escapin: Identification and quantification of the components in the complex mixture generated by an L-amino acid oxidase in the defensive secretion of the sea snail *Aplysia californica. Chem.-A Europ. J.* 15:1597–1603.

Kamio, M., L. Nguyen, S. Yaldiz, and C. D. Derby. 2010b. How to produce a chemical defense: Structural elucidation and anatomical distribution of aplysioviolin and phycoerythrobilin in the sea hare *Aplysia californica. Chem. Biodivers.* 7:1183–1197.

Kamio, M., T. V. Grimes, M. H. Hutchins, R. van Dam, and C. D. Derby. 2010a. The purple pigment aplysioviolin in sea hare ink deters predatory blue crabs through their chemical senses. *Animal Behav.* 80:89–100.

Kamiya, H., K. Muramoto, R. Goto, M. Sakai, Y. Endo, and M. Yamazaki. 1989. Purification and characterization of an antibacterial and antineoplastic protein secretion of a sea hare, *Aplysia juliana. Toxicon* 27:1269–1277.

Kamiya, H., R. Sakai, and M. Jimbo. 2006. Bioactive molecules from sea hares. In *Molluscs: From Chemo-ecological Study to Biotechnological Applications*, G. Cimino and M. Gavagnin (Eds.), pp. 215–239. Berlin, Germany: Springer.

Kandel, E. R. 1979. *Behavioral Biology of Aplysia: A Contribution to the Comparative Study of Opisthobranch Mollusks*. San Francisco, CA: Freemann.

Kandyuk, R. P. 2006. Sterols and their functional role in Mollusks (a review). *Hydrobiol. J.* 42:56–66.

Kardong, K. V. 2008. *An Introduction to Biological Evolution*. New York: McGraw-Hill.

Karuso, P., and P. J. Scheuer. 2002. Natural products from three nudibranchs: *Nembrotha kubaryana*, *Hypselodoris infucata* and *Chromodoris petechialis*. *Molecules* 7:1–6.

Kashman, Y., A. Croweiss, and U. Shmueli. 1980. Latrunculin, a new 2-thiazolidinone macrolide from the marine sponge *Latrunculia magnifica*. *Tetrahedron Lett.* 21:3629–3632.

Kashman, Y., A. Croweiss, R. Kidor, D. Blasberger, and S. Carmely. 1985. Latrunculins: NMR study, two new toxins and a synthetic approach. *Tetrahedron* 41:1905–1914.

Katavic, P. L., P. Jumaryatno, J. N. Hooper, J. T. Blanchfield, and M. J. Garson. 2012. Oxygenated terpenoids from the Australian sponges *Coscinoderma matthewsi* and *Dysidea* sp., and the nudibranch *Chromodoris albopunctata*. *Aust. J. Chem.* 65:531–538.

Katayama, A., K. Ina, H. Nozaki, and M. Nakayama. 1982. Structural elucidation of kurodainol, a novel halogenated monoterpene from sea hare (*Aplysia kurodai*). *Agr. Biol. Chem.* 46:859–860.

Kato, Y., and P. J. Scheuer. 1974. Aplysiatoxin and debromoaplysiatoxin, constituents of the marine mollusk *Stylocheilus longicauda* (Quoy and Gaimard, 1824). *J. Am. Chem. Soc.* 96:2245–2246.

Kato, Y., and P. J. Scheuer. 1975. The aplysiatoxins. *Pure Appl. Chem.* 41:1–14.

Kattner, G., W. Hagen, M. Graeve, and C. Albers. 1998. Exceptional lipids and fatty acids in the pteropod *Clione limacina* (Gastropoda) from both polar oceans. *Mar. Chem.* 61:219–228.

Kaviarasan, T., S. R. Siva, and A. Yogamoorthi. 2012. Antimicrobial secondary metabolites from marine gastropod egg capsules and egg masses. *Asian Pac J Trop Biomed* 2:916–922.

Kicklighter, C. E., and C. D. Derby. 2006. Multiple components in ink of the sea hare *Aplysia californica* are aversive to the sea anemone *Anthopleura sola*. *J. Exp. Mar. Biol. Ecol.* 334:256–268.

Kicklighter, C. E., S. Shabani, P. M. Johnson, and C. D. Derby. 2005. Sea hares use novel antipredatory chemical defenses. *Curr. Biol.* 15:549–554.

Kigoshi, H., and M. Kita. 2015. Antitumor effects of sea hare-derived compounds in cancer. In *Handbook of Anticancer Drugs from Marine Origin*, S.-K. Kim (Ed.), pp. 701–739. Cham, Switzerland: Springer International Publishing.

Kigoshi, H., Y. Imamura, K. Yoshikawa, and K. Yamada. 1990. Three new cytotoxic alkaloids, aplaminone, neoaplaminoe and neoaplaminone sulgate from the marine mollusc *Aplysia kurodai*. *Tetrahedron Lett.* 31:4911–4914.

Kinnel, R. B., R. K. Dieter, J. Meinwald, D. van Engen, J. Clardy, T. Eisner, M. O. Stallard, and W. Fenical. 1979. Brasilenyne and cis-dihydrorhodophytin: Antifeedant medium-ring haloethers from a sea hare (*Aplysia brasiliana*). *Proc. Nat. Acad. Sci. U.S. Am.* 76:3576–3579.

Kinnel, R., A. J. Duggan, T. Eisner, J. Meinwald, and I. Miura. 1977. Panacene: An aromatic bromoallene from a sea hare (*Aplysia brasiliana*). *Tetrahedron Lett.* 44:3913–3916.

Kisugi, J., H. Ohye, H. Kamiya, and M. Yamazaki. 1992. Biopolymers from marine invertebrates. XIII. Characterization of an antibacterial protein, dolabellanin A, from the albumen gland of the sea hare, *Dolabella auricularia*. *Chem. Pharm. Bull.* 40:1537–1539.

Kitano, Y., T. Ito, T. Suzuki, Y. Nogata, K. Shinshima, E. Yoshimura, K. Chiba, M. Tada, and I. Sakaguchi. 2002. Synthesis and antifouling activity of 3-isocyanotheonellin and its analogues. *J. Chem. Soc., Perkin Trans.* 1:2251–2255.

Kladi, M., D. Ntountaniotis, M. Zervou, C. Vagias, E. Ioannou, and V. Roussis. 2014. Glandulaurencianols A-C, brominated diterpenes from the red alga, *Laurencia glandulifera* and the sea hare, *Aplysia punctata*. *Tetrahedron Lett.* 55:2835–2837.

Klussmann-Kolb, A., and A. Dinapoli. 2006. Systematic position of the pelagic Thecosomata and Gymnosomata within Opisthobranchia (Mollusca, Gastropoda)—Revival of the Pteropoda. *J. Zool. Syst. Evol. Res.* 44:118–129.

Klussmann-Kolb, A., and G. D. Brodie. 1999. Internal storage and production of symbiotic bacteria in the reproductive system of a tropical marine gastropod. *Mar. Biol.* 133:443–447.

Kobayashi, M., F. Kanda, and H. Kamiya. 1991. Occurrence of pyropheophorbides a and b in the viscera of the sea hare *Aplysia juliana*. *Nippon Suisan Gakkaishi* 57:1991.

Koehler, E. 1999. 'Trailing' behaviour in *Risbecia tryoni*. *Sea Slug Forum*. Australian Museum, Sydney. http://www.seaslugforum.net/find/760 accessed April 28, 2015.

Kohmoto, S., O. J. McConnell, A. Wrigth, and S. Cross. 1987. Isospongiadiol, a cytotoxic and antiviral diterpene from a Caribbean deep water marine sponge, *Spongia* sp. *Chem. Lett.* 16:1687–1690.

Krebs, J. R., and N. B. Davies. 1993. *An Introduction to Behavioural Ecology*. Oxford, UK: Blackwell Science.

Krug, P. J., K. G. Boyd, and D. J. Faulkner. 1995. Isolation and synthesis of tanyolides A and B, metabolites of the nudibranch *Sclerodoris tanya*. *Tetrahedron* 51:11063–11074.

Ksebati, M. B., and F. J. Schmitz. 1985. Tridachiapyrones: Propionate-derived metabolites from the sacoglossan mollusc *Tridachia crispata*. *J. Org. Chem.* 50:5637–5642.

Kubanek, J., and R. J. Andersen. 1999. Evidence for de novo biosynthesis of the polyketide fragment of diaulusterol A by the Northeastern Pacific dorid nudibranch *Diaulula sandiegensis*. *J. Nat. Prod.* 62:777–779.

Kubanek, J., and R. J. Andersen. 1997. Evidence for the incorporation of intact butyrate units in the biosynthesis of triophamine. *Tetrahedron Lett.* 38:6327–6330.

Kubanek, J., D. J. Faulkner, and R. J. Andersen. 2000. Geographic variation and tissue distribution of endogenous terpenoids in the Northeastern Pacific dorid nudibranch *Cadlina marginata*: Implications for the regulation of de novo biosynthesis. *J. Chem. Ecol.* 26:377–389.

Kubanek, J., E. I. Graziani, and R. J. Andersen. 1997. Investigations of terpenoid biosynthesis by the dorid nudibranch *Cadlina luteomarginata*. *J. Org. Chem.* 62:7239–7246.

Kuniyoshi, M., K. Yamada, and T. Higa. 1985. A biologically active diphenyl ether from the green alga *Cladophora fascicularis*. *Experientia* 41:523–524.

Kuroda, T., and H. Kigoshi. 2008. Aplaminal: A novel cytotoxic aminal isolated from the sea hare *Aplysia kurodai*. *Org. Lett.* 10:489–491.

Kusumi, T., H. Uchida, Y. Inouye, M. Ishitsukam, H. Yamamoto, H. Kakisawa. 1987. Novel cytotoxic monoterpenes having a halogenated tetrahydropyran from *Aplysia kurodai*. *J. Org. Chem.* 52:4597–4600.

Laetz, E. M. J., V. C. Moris, L. Moritz, A. N. Haubrich, and H. Wägele. 2017. Photosynthate accumulation in solarpowered sea slugs—Starving slugs survive due to accumulated starch reserves. *Front. Zool.* 14:4.

Lane, A. L., and B. S. Moore. A sea of biosynthesis: Marine natural products meet the molecular age. *Nat. Prod. Rep.* 28 (2011): 411–428.

Lebar, M. D., J. L. Heimbegner, and B. J. Baker. 2007. Cold-water marine natural products. *Nat. Prod. Rep.* 24:774–797.

Lin, Z., J. P. Torres, M. A. Ammon et al. 2013. A bacterial source for mollusk pyrone polyketides. *Chem. Biol.* 20:73–81.

Linse, K. 1999. Mollusca of the Magellan region. A checklist of the species and their distribution. *Sci. Mar.* 63:399–407.

Long, J. D., and M. E. Hay. 2006. Fishes learn aversions to a nudibranch's chemical defense. *Mar. Ecol. Progr. Ser.* 307:199–208.

Long, P. F., W. C. Dunlap, C. N. Battershill, and M. Jaspars. 2005. Shotgun cloning and heterologous expression of the patellamide gene cluster as a strategy to achieving sustained metabolite production. *Chem. Bio. Chem.* 6:1760–1765.

López-Vera, E., E. P. Heimer de la Cotera, M. Maillo, J. R. Riesgo-Escovar, B. M. Olivera, and A. M. Aguilar. 2004. A novel structural class of toxins: The methionine-rich peptides from the venoms of turrid marine snails (Mollusca, Conoidea). *Toxicon* 43:365–374.

Luesch, H., W. Y. Yoshida, R. E. Moore, and V. J. Paul. 1999. Lyngbyastatin 2 and norlyngbyastatin 2, analogues of dolastatin G and nordolastatin G from the marine cyanobacterium *Lyngbya majuscula. J. Nat. Prod.* 62:1702–1706.

Lyakhova, E. G., S. A. Kolesnikova, A. I. Kalinovskii, and V. A. Stonik. 2010. Secondary metabolites of the Vietnamese nudibranch mollusc *Phyllidiella pustulosa. Chem. Nat. Comp.* 46:534–538.

Malaquias, M. A. E., J. Mackenzie-Dodds, P. Bouchet, T. Gosliner, and D. G. Reid. 2009. A molecular phylogeny of the Cephalaspidea *sensu lato* (Gastropoda: Euthyneura): Architectibranchia redefined and Runcinacea reinstated. *Zool. Scripta* 38:23–41.

Manker, D. C., and D. J. Faulkner. 1989. Vallartanones A and B, polypropionate metabolites of *Siphonaria maura* from Mexico. *J. Org. Chem.* 54:5374–5377.

Manker, D. C., and D. J. Faulkner. 1996. Investigation of the role of diterpenes produced by marine pulmonates *Trimusculus reticulatus* and *T. conica. J. Chem. Ecol.* 22:23–35.

Manker, D. C., D. J. Faulkner, T. J. Stout, and J. Clardy. 1989. The baconipyrones. Novel polypropionates from the pulmonate *Siphonaria baconi. J. Org. Chem.* 54:5371–5374.

Manker, D. C., M. J. Garson, and D. J. Faulkner. 1988. De novo biosynthesis of polypropionate metabolites in the marine pulmonate *Siphonaria denticulata. J. Chem. Soc., Chem. Commun.* 16:1061–1062.

Manzo, E., M. Carbone, E. Mollo, C. Irace, A. Di Pascale, Y. Li, M. L. Ciavatta, G. Cimino, Y.-W. Guo, and M. Gavagnin. 2011. Structure and synthesis of a unique isonitrile lipid isolated from the marine mollusk *Actinocyclus papillatus. Org. Lett.* 13:1897–1899.

Manzo, E., M. Gavagnin, G. Bifulco, P. Cimino, S. Di Micco, M. L. Ciavatta, Y.-W. Guo, G. Cimino. 2007. Aplysiols A and B, squalene-derived polyethers from the mantle of the sea hare *Aplysia dactylomela. Tetrahedron* 63:9970–9978.

Manzo, E., M. Gavagnin, M. J. Somerville, S.-C. Mao, M. L. Ciavatta, E. Mollo, P. J. Schupp, M. J. Garson, V. Guo, and G. Cimino. 2007. Chemistry of *Glossodoris* nudibranchs: Specific occurrence of 12-keto scalaranes. *J. Chem. Ecol.* 33:2325–2336.

Manzo, E., M. L. Ciavatta, M. Gavagnin, E. Mollo, Y.-W. Guo, and G. Cimino. 2004. Isocyanide terpene metabolites of *Phyllidiella pustulosa*, a nudibranch from the South China Sea. *J. Nat. Prod.* 67:1701–1704.

Manzo, E., M. L. Ciavatta, M. Gavagnin, E. Mollo, S. Wahidulla, and G. Cimino. 2005. New γ-pyrone propionates from the Indian Ocean sacoglossan *Placobranchus ocellatus. Tetrahedron Lett.* 46:465–468.

Manzo, E., M. L. Ciavatta, M. Gavagnin, R. Puliti, E. Mollo, Y.-W. Guo, C. A. Mattia, L. Mazzarella, and G. Cimino. 2005. Structure and absolute stereochemistry of novel C_{15}-halogenated acetogenins from the anaspidean mollusc *Aplysia dactylomela. Tetrahedron* 61:7456–7460.

Mao, S. C., M. Gavagnin, E. Mollo, and Y.-W. Guo. 2011. A new rare asteriscane sesquiterpene and other related derivatives from the Hainan aeolid nudibranch *Phyllodesmium magnum. Biochem. Syst. Ecol.* 39:408–411.

Margalef, R. 1977. Comunicación y engaño. Aspectos e implicaciones de la cripsis, advertencia y mimetismo. *Graellsia* 31:341–356.

Margalef, R. 1982. *Ecologia*, pp. 1–951. Barcelona, Spain: Omega.

Marín, A. 2009. Chemical or nematocyst-based defence in the nudibranch *Cratena peregrina*?—A reply to B. K. Penney. *J. Molluscan Stud.* 75:201–202.

Marín, A., and J. Ros. 2004. Chemical defenses in sacoglossan opisthobranchs: Taxonomic trends and evolutive implications. *Sci. Mar.* 68:227–241.

Marín, A., L. A. Álvarez, G. Cimino, and A. Spinella. 1999. Chemical defence in cephalaspidean gastropods: Origin, anatomical location and ecological roles. *J. Molluscan Stud.* 65:121–131.

Marín, A., M. D. López Belluga, G. Scognamiglio, and G. Cimino. 1997. Morphological and chemical camouflage of the Mediterranean nudibranch *Discodoris indecora* on the sponges *Ircinia variabilis* and *Ircinia fasciculata*. *J. Molluscan Stud.* 63:431–439.

Márquez, B. L., K. S. Watts, A. Yokochi, M. A. Roberts, P. Verdier-Pinard, J. I. Jiménez, E. Hamel, P. J. Scheuer, and W. H. Gerwick. 2002. Structure and absolute stereochemistry of hectochlorin, a potent stimulator of actin assembly. *J. Nat. Prod.* 65:866–871.

Martin, R., K.-H. Tomaschko, and P. Walther. 2007. Protective skin structures in shell-less gastropods. *Mar. Biol.* 150:807–817.

Martin, R., M. Hess, M. Schrödl, and K.-H. Tomaschko. 2009. Cnidosac morphology in dendronotacean and aeolidacean nudibranch molluscs: From expulsion of nematocysts to use in defense? *Mar. Biol.* 156:261–268.

Maschek, J. A., E. Mevers, T. Diyabalanage, L. Chen, Y. Ren, J. B. McClintock, C. D. Amsler, J. Wu, and B. J. Baker. 2012. Palmadorin chemodiversity from the Antarctic nudibranch *Austrodoris kerguelenensis* and inhibition of Jak2/STAT5-dependent HEL leukemia cells. *Tetrahedron* 68:9095–9104.

Matsunaga, S., N. Fusetani, K. Hashimoto, K. Koseki, M. Noma, H. Noguchi, and U. Sankawa. 1989. Bioactive marine metabolites. 25. Further kabiramides and halichondramides, cytotoxic macrolides embracing trisoxazole, from the *Hexabranchus* egg masses. *J. Org. Chem.* 54:1360–1363.

Matsuura, B. S., P. Kölle, D. Trauner, R. de Vivie-Riedle, and R. Meier. 2016. *Unravelling photochemical relationships among natural products from Aplysia dactylomela. ACS Central Sci.* 3:39–46.

McClintock, J. B., and B. J. Baker. 1997. Palatability and chemical defense in the eggs, embryos and larvae of shallow-water Antarctic marine invertebrates. *Mar. Ecol. Progr. Ser.* 154:121–0131.

McClintock, J. B., and B. J. Baker. 2001. *Marine Chemical Ecology.* Boca Raton, FL: CRC Marine Science Series Press.

McClintock, J. B., and J. Janssen. 1990. Pteropod abduction as a chemical defence in a pelagic antarctic amphipod. *Nature* 346:462–464.

McClintock, J. B., B. J. Baker, M. Slattery, J. N. Heine, P. J. Bryan, W. Yoshida, M. T. Davies-Coleman, and D. J. Faulkner. 1994b. Chemical defense of common antarctic shallow-water nudibranch *Tritoniella belli* Eliot (Mollusca: Tritonidae) and its prey, *Clavularia frankliniana* Rouel (Cnidaria: Octocorallia). *J. Chem. Ecol.* 20:3361–3372.

McClintock, J. B., C. D. Amsler, and B. J. Baker. 2010. Overview of the chemical ecology of benthic marine invertebrates along the Western Antarctic Peninsula. *Integr. Comp. Biol.* 50:967–980.

McClintock, J. B., P. J. Bryan, M. Slattery, B. J. Baker, W. Y. Yoshida, M. Hamann, and J. N. Heine. 1994a. Chemical ecology of three antarctic gastropods. *Antarct. J. US* 29:151–154.

McDonald, F. J., D. C. Campbell, D. J. Vanderah, F. J. Schmitz, D. M. Washecheck, J. E. Burks, and D. van der Helm. 1975. Marine natural products. Dactylyne, an acetylenic dibromochloro ether from the sea hare *Aplysia dactylomela. J Org. Chem.* 40:665–666.

McDonald, G. R., and J. W. Nybakken. 1997. A list of the worldwide food habits of nudibranchs. I. Introduction and the suborder Arminacea. *Veliger* 40:1–425.

McKey, D. 1974. Adaptive patterns in alkaloid physiology. *Am. Nat.* 108:305–320.

McNabb, P., A. I. Selwood, R. Munday, S. A. Wood, D. I. Taylor, L. A. MacKenzie, and R. van Ginkel. 2010. Detection of tetrodotoxin from the grey side-gilled sea slug—*Pleurobranchaea maculata*, and associated dog neurotoxicosis on beaches adjacent to the Hauraki Gulf, Auckland, New Zealand. *Toxicon* 56:466–473.

McPhail, K. L., and M. T. Davies-Coleman. 1997. New spongiane diterpenes from the East African nudibranch *Chromodoris hamiltoni*. *Tetrahedron* 53:4655–4660.

McPhail, K. L., and M. T. Davies-Coleman. 2005. (3Z)-Bromofucin from a South African sea hare. *Nat. Prod. Res.* 19:449–452.

McPhail, K. L., M. T. Davies-Coleman, and J. Starmer. 2001. Sequestered chemistry of the arminacean nudibranch *Leminda millecra* in Algoa Bay, South Africa. *J. Nat. Prod.* 64:1183–1190.

McPhail, K. L., M. T. Davies-Coleman, and P. Coetzee. 1998. A new furanosesterterpene from the South African nudibranch *Hypselodoris capensis* and a Dictyoceratida sponge. *J. Nat. Prod.* 61:961–964.

McPhail, K. L., M. T. Davies-Coleman, R. C. B. Copley, and D. S. Eggleston. 1999. New halogenated sesquiterpenes from South African specimens of the circumtropical sea hare *Aplysia dactylomela*. *J. Nat. Prod.* 62:1618–1623.

Mebs, D. 1985. Chemical defense of a dorid nudibranch, *Glossodoris quadricolor*, from the Red Sea. *J. Chem. Ecol.* 11:713–716.

Medina, M., S. Lal, Y. Vallès, T. L. Takaoka, B. A. Dayrat, J. L. Boore, and T. Gosliner. 2011. Crawling through time: Transition of snails to slugs dating back to the Paleozoic, based on mitochondrial phylogenomics. *Mar. Genomics* 4:51–59.

Medina, M., T. Collins, and P. J. Walsh. 2005. Phylogeny of the sea hares in the *Aplysia* clade based on mitochondrial DNA sequence data. *Bull. Mar. Sci.* 76:691–698.

Midland, S. L., R. M. Wing, and J. J. Sims. 1983. New crenulides from the sea hare *Aplysia vaccaria*. *J. Org. Chem.* 48:1906–1909.

Mikkelsen, P. M. 2002. Shelled Opisthobranchs. *Adv. Mar. Biol.* 42:67–136.

Miller, M. D. 1999. Mimicry: Flatworm and *C. geometrica*. *Sea Slug Forum*. Australian Museum, Sydney. http://www.seaslugforum.net/find/472 accessed April 28, 2015.

Milne, B. F., P. F. Long, A. Starcevic, D. Hranueli, and M. Jaspars. 2006. Spontaneity in the patellamide biosynthetic pathway. *Org. Biomol. Chem.* 4:631–638.

Minale, L., and R. Riccio. 1976, Constituents of the digestive gland of the molluscs of the genus *Aplysia*—I. Novel diterpenes from *Aplysia depilans*. *Tetrahedron Lett.* 31:2711–2714.

Miyamoto, T. 2006. Selected bioactive compounds from Japanese anaspideans and nudibranchs. In *Molluscs: From Chemo-ecological Study to Biotechnological Applications*, G. Cimino and M. Gavagnin (Eds.), pp. 199–214. Berlin, Germany: Springer.

Miyamoto, T., K. Sakamoto, H. Amano, Y. Arakawa, Y. Nagarekawa, T. Komori, R. Higuchi, and T. Sasaki. 1999. New cytotoxic sesterterpenoids from the nudibranch *Chromodoris inornata*. *Tetrahedron* 55:9133–9142.

Miyamoto, T., K. Sakamoto, K. Arao, T. Komori, R. Higuchi, and T. Sasaki. 1996. Dorisenones, cytotoxic spongian diterpenoids, from the nudibranch *Chromodoris obsoleta*. *Tetrahedron* 52:8187–8198.

Miyamoto, T., R. Higuchi, and T. Komori. 1986. Isolation and structures of aplykurodins A and B, two new isoprenoids from the marine mollusk *Aplysia kurodai*. *Tetrahedron Lett.* 27:1153–1156.

Miyamoto, T., R. Higuchi, N. Marubayashi, and T. Komori. 1988. Two new polyhalogenated monoterpenes from the sea hare *Aplysia kurodai*. *Liebigs Ann. Chem.* 12:1191–1193.

Miyamoto, T., Y. Ebisawa, and R. Higuchi. 1995. Aplyparvunin, a bioactive acetogenin from the sea hare *Aplysia parvula*. *Tetrahedron Lett.* 36:6073–6074.

Mohammed, K. A., C. F. Hossain, L. Zhang, R. K. Bruick, D. Zhou, and D. G. Nagle. 2004. Laurenditerpenol, a new diterpene from the tropical marine alga *Laurencia intricata* potently inhibits HIF-1 mediated hypoxic signaling in breast tumor cells. *J. Nat. Prod.* 67:2002–2007.

Moles, J., G. Mas, I. Figueroa, R. Fernández-Vilert, X. Salvador, and J. Giménez. 2017b. As fast as a hare: Colonization of the heterobranch *Aplysia dactylomela* (Mollusca: Gastropoda: Anaspidea) into the western Mediterranean Sea. *Cah. de Biol. Mar.* 58:341–345.

Moles, J., H. Wägele, A. Cutignano, A. Fontana, and C. Avila. 2016. Distribution of granu-loside in the Antarctic nudibranch *Charcotia granulosa* (Gastropoda: Heterobranchia: Charcotiidae). *Mar. Biol.* 163:54.

Moles, J., H. Wägele, A. Cutignano, A. Fontana, M. Ballesteros, and C. Avila. 2017a. Giant embryos and hatchlings of Antarctic nudibranchs (Mollusca: Gastropoda: Heterobranchia). *Mar. Biol.* 164:114.

Moles, J., L. Núñez-Pons, S. Taboada, B. Figuerola, J. Cristobo, and C. Avila. 2015. Anti-predatory chemical defences in Antarctic benthic fauna. *Mar. Biol.* 162:1813–1821.

Molinski, T. F., D. S. Dalisay, S. L. Lievens, and J. P. Saludes. 2009. Drug development from marine natural products. *Nat. Rev. Drug Disc.* 8:69–85.

Mollo, E., M. Gavagnin, M. Carbone, F. Castelluccio, F. Pozone, V. Roussis, J. Templado, M. T. Ghiselin, and G. Cimino. 2008. Factors promoting marine invasions: A chemoeco-logical approach. *Proc. Nat. Acad. Sci. US Am.* 105:4582–4586.

Mollo, E., M. Gavagnin, M. Carbone, Y.-W. Guo, and G. Cimino. 2005. Chemical studies on Indopacific *Ceratosoma* nudibranchs illuminate the protective role of their dorsal horn. *Chemoecology* 15:31–36.

Moore, B. S. 2006. Biosynthesis of marine natural products: Macroorganisms (Part B). *Nat. Prod. Rep.* 23:615–629.

Mudianta, I. W., V. L. Challinor, A. E. Winters, K. L. Cheney, J. J. De Voss, and M. J. Garson. 2013. Synthesis and determination of the absolute configuration of (−)-(5R, 6Z)-dendrolasin-5-acetate from the nudibranch *Hypselodoris jacksoni. Beilstein J. Org. Chem.* 9:2925–2933.

Muscatine, L., and R. W. Greene. 1973. Chloroplasts and algae as symbionts in molluscs. *Int. Rev. Cytol.* 36:137–169.

Mutou, T., T. Kondo, M. Ojika, and K. Yamada. 1996. Isolation and stereostructures of dol-astatin G and nordolastatin G, cytotoxic 35-membered cyclodepsipeptides from the Japanese sea hare *Dolabella auricularia. J. Org. Chem.* 61:6340–6345.

Nakano, T., J. Villamizar, and M. A. Maillo. 1995. A new entry to (+)-albicanol and (+)-bicyclofarnesol. *J. Chem. Res. Synopses* 8:330–331.

Nakao, Y., W. Y. Yoshida, and P. J. Scheuer. 1996. Pupukeamide, a linear tetrapeptide from a cephalaspidean mollusk *Philinopsis speciosa. Tetrahedron Lett.* 37:8993–8996.

Nakao, Y., W. Y. Yoshida, C. M. Szabo, B. J. Baker, and P. J. Scheuer. 1998. More peptides and other diverse constituents of the marine mollusk *Philinopsis speciosa. J. Org. Chem.* 63:3272–3280.

Napolitano, J. G., M. L. Souto, J. J. Fernández, and M. Norte. 2008. Micromelones A and B, noncontiguous polypropionates from *Micromelo undata. J. Nat. Prod.* 71:281–284.

Nett, M., H. Ikeda, and B. S. Moore. 2009. Genomic basis for natural product biosynthetic diversity in the actinomycetes. *Nat. Prod. Rep.* 26:1362–1384.

Neves, R., H. Gaspar, and G. Calado. 2009. Does a shell matter for defence? Chemical deterrence in two cephalaspidean gastropods with calcified shells. *J. Molluscan Stud.* 75:127–131.

Newman, D. J., and G. M. Cragg. 2004. Marine natural products and related compounds in clinical and advanced preclinical trials. *J. Nat. Prod.* 67:1216–1238.

Newman, D. J., and G. M. Cragg. 2014. Marine-sourced anti-cancer and cancer pain control agents in clinical and late preclinical development. *Mar. Drugs* 12:255–278.

Norte, M., F. Cataldo, A. G. González, M. L. Rodríguez, and C. Ruiz-Pérez. 1990. New metab-olites from the marine mollusc *Siphonaria grisea. Tetrahedron* 46:1669–1678.

Norte, M., J. J. Fernández, and A. Padilla. 1994. Isolation and synthesis of siphonarienal a new polypropionate from *Siphonaria grisea. Tetrahedron Lett.* 35:3413–3416.

Núñez-Pons, L., and C. Avila. 2015. Natural products mediating ecological interactions in Antarctic benthic communities: A mini-review of the known molecules. *Nat. Prod. Rep.* 32:1114–1130.

Nusnbaum, M., and C. D. Derby. 2010. Effects of sea hare ink secretion and its escapin-generated components on a variety of predatory fishes. *Biol. Bull.* 218:282–292.

Nuzzo, G., A. Cutignano, J. Moles, C. Avila, and A. Fontana. 2015. Exiguapyrone and exiguaone, new polypropionates from the Mediterranean cephalaspidean mollusc *Haminoea exigua. Tetrahedron Lett.* 57:71–74.

Nuzzo, G., M. L. Ciavatta, R. Kiss et al. 2012. Chemistry of the nudibranch *Aldisa andersoni*: Structure and biological activity of phorbazole metabolites. *Mar. Drugs* 10:1799–1811.

OBIS. 2014. Global biodiversity indices from the Ocean Biogeographic Information System. Intergovernmental Oceanographic Commission of UNESCO. http://www.iobis.org Accessed September 4.

Ojika, M., H. Kigoshi, K. Suenaga, Y. Imamura, K. Yoshikawa, T. Ishigaki, A. Sakakura, T. Mutou, and K. Yamada. 2012. Aplyronines D-H from the sea hare *Aplysia kurodai*: Isolation, structures, and cytotoxicity. *Tetrahedron* 68:982–987.

Ojika, M., H. Kigoshi, K. Yoshikawa, Y. Nakayama, and K. Yamada. 1992. A new bromo diterpene, epi-aplysin-20, and ent-isoconcinndiol from the marine mollusc *Aplysia kurodai. Bull. Chem. Soc. Jpn.* 65:2300–2302.

Ojika, M., T. Nagoya, and K. Yamada. 1995. Dolabelides A and B, cytotoxic 22-membered macrolides isolated from the sea hare *Dolabella auricularia. Tetrahedron Lett.* 36:7491–7494.

Ojika, M., T. Nemoto, and K. Yamada, K. 1993. Doliculols A and B, the non-halogenated C_{15} acetogenins with cyclic ether from the sea hare *Dolabella auricularia. Tetrahedron Lett.* 34:3461–3462.

Ojika, M., T. Yoshida, K. Yamada. 1993. Aplysepine, a novel 1,4-benzodiazepine alkaloid from the sea hare *Aplysia kurodai. Tetrahedron Lett.* 34:5307–5308.

Ojika, M., Y. Yoshida, M. Okumura, S. Ieda, and K. Yamada. 1990. Aplysiadiol, a new brominated diterpene from the marine mollusc *Aplysia kurodai. J. Nat. Prod.* 53:1619–1622.

Ojika, M., Y. Yoshida, Y. Nakayama, and K. Yamada. 1990. Aplydilactone, a novel fatty acid metabolite from the marine mollusc *Aplysia kurodai. Tetrahedron Lett.* 31:4907–4910.

Okino, T., E. Yoshimura, H. Hirota, and N. Fusetani. 1996. New antifouling sesquiterpenes from four nudibranchs of the family Phyllidiidae. *Tetrahedron* 52:9447–9454.

Okuda, R. K., D. Klein, R. B. Kinnel, M. Li, and P. J. Scheuer. 1982. Marine natural products. The past twenty years and beyond. *Pure Appl. Chem.* 54:1907–1914.

Oliveira, A. P., A. Lobo-da-Cunha, M. Taveira, M. Ferreira, P. Valentão, and P. B. Andrade. 2015. Digestive gland from *Aplysia depilans* Gmelin: Leads for inflammation treatment. *Molecules* 20:15766–15780.

Olivera, B. M. 2002. *Conus* venom peptides: Reflections from the biology of clades and species. *Ann. Rev. Ecol., Evol. Syst.* 33:25–47.

Opitz, S. E. W., and C. Muller. 2009. Plant chemistry and insect sequestration. *Chemoecology* 19:117–154.

Ortega, M. J., E. Zubía, and J. Salvá. 1997. 3-epi-aplykurodinone B, a new degraded sterol from *Aplysia fasciata. J. Nat. Prod.* 60:488–489.

Ortega, M. J., E. Zubía, and J. Salvá. 1997. New polyhalogenated monoterpenes from the sea hare *Aplysia punctata. J. Nat. Prod.* 60:482–484.

Oskars, T. R., P. Bouchet, and M. A. E. Malaquias. 2015. A new phylogeny of the Cephalaspidea (Gastropoda: Heterobranchia) based on expanded taxon sampling and gene markers. *Mol. Phylogenet. Evol.* 89:130–150.

Palaniveloo, K., and C. S. Vairappan. 2013. Chemical relationship between red algae genus *Laurencia* and sea hare (*Aplysia dactylomela* Rang) in the North Borneo Island. *J. Appl. Phycol.* 26:1199–1205.

Paul, M. C., E. Zubía, M. J. Ortega, and J. Salvá. 1997. New polypropionates from *Siphonaria pectinata. Tetrahedron* 53:2303–2308.

Paul, V. J. 1992. *Ecological Roles of Marine Natural Products*. New York: Comstock Publishications Association.

Paul, V. J., K. E. Arthur, R. Ritson-Williams, C. Ross, and K. Sharp. 2007. Chemical defenses: From compounds to communities. *Biol. Bull.* 213:226–251.

Paul, V. J., and K. L. van Alstyne. 1988. Use of ingested algal diterpenoids by *Elysia halimedae* Macnae (Opisthobranchia: Ascoglossa) as antipredator defenses. *J. Exp. Mar. Biol. Ecol.* 119:15–29.

Paul, V. J., and K. L. van Alstyne. 1992. Activation of chemical defenses in the tropical green algae *Halimeda* spp. *J. Exp. Mar. Biol. Ecol.* 160:191–203.

Paul, V. J., and M. P. Puglisi. 2004. Chemical mediation of interactions among marine organisms. *Nat. Prod. Rep.* 21:189–209.

Paul, V. J., and S. C. Pennings. 1991. Diet-derived chemical defenses in the sea hare *Stylocheilus longicauda* (Quoy et Gaimard 1824). *J. Exp. Mar. Biol. Ecol.* 151:227–243.

Paul, V. J., and W. Fenical. 1983. Isolation of Halimedatrial: Chemical defense adaptation in the calcareous reef-building alga *Halimeda*. *Science* 221:747–749.

Paul, V. J., and W. Fenical. 1984. Bioactive terpenoids from Caribbean marine algae of the genera *Penicillus* and *Udotea* (Chlorophyta). *Tetrahedron* 40:2913–2918.

Paul, V. J., E. Cruz-Rivera, and R. W. Thacker. 2001. Chemical mediation of macroalgal-herbivore interactions: Ecological and evolutionary perspectives. In *Marine Chemical Ecology*, J. B. McClintock and B. J. Baker (Eds.), pp. 227–266. Boca Raton, FL: CRC Press.

Paul, V. J., H. H. Sun, and W. Fenical. 1982. Udoteal, a linear diterpenoid feeding deterrent from the tropical green alga *Udotea flabellum*. *Phytochemistry* 21:468–469.

Paul, V. J., M. P. Puglisi, and R. Ritson-Williams. 2006. Marine chemical ecology. *Nat. Prod. Rep.* 23:153–180.

Pavia, H., and G. B. Toth. 2000. Inducible chemical resistance to herbivory in the brown seaweed *Ascophyllum nodosum*. *Ecology* 81:3212–3225.

Pawlik, J. R., K. F. Albizati, and D. J. Faulkner. 1986. Evidence of a defensive role for limatulone, a novel triterpene from the limpet *Collisella limatula*. *Mar. Ecol. Progr. Ser.* 30:251–260.

Payo, D. A., J. Colo, H. Calumpong, and O. de Clerck. 2011. Variability of non-polar secondary metabolites in the red alga *Portieria*. *Mar. Drugs* 9:2438–2468.

Penney, B. K. 2004. Individual selection and the evolution of chemical defence in nudibranchs: Experiments with whole *Cadlina luteomarginata* (Nudibranchia: Doridina). *J. Molluscan Stud.* 70:399–400.

Penney, B. K. 2009. A comment on F. Aguado & A. Marín: Warning coloration associated with nematocyst-based defences in aeolidioidean nudibranchs'. *J. Molluscan Stud.* 75:199–200.

Penney, B. K. 2013. How specialized are the diets of Northeastern Pacific sponge-eating dorid nudibranchs? *J. Molluscan Stud.* 79:64–73.

Penney, B. K., L. H. LaPlante, J. R. Friedman, and M. O. Torres. 2010. A noninvasive method to remove kleptocnidae for testing their role in defence. *J. Molluscan Stud.* 76:296–298.

Pennings, S. C. 1990. Multiple factors promoting narrow host range in the sea hare, *Aplysia californica*. *Oecologia* 82:192–200.

Pennings, S. C., and V. J. Paul. 1993. Sequestration of dietary secondary metabolites by three species of sea hares: Location, specificity and dynamics. *Mar. Biol.* 117:535–546.

Pennings, S. C., V. J. Paul, D. C. Dunbar, M. T. Hamann, W. A. Lumbang, B. Novack, and R. S. Jacobs. 1999. Unpalatable compounds in the marine gastropod *Dolabella auricularia*: Distribution and effect of diet. *J. Chem. Ecol.* 25:735–755.

Pereira, F. R., M. F. C. Santos, D. E. Williams, R. J. Andersen, V. Padula, A. G. Ferreira, and R. G. S. Berlinck. 2014. Rodriguesic acids, modified diketopiperazines from the gastropod mollusc *Pleurobranchus areolatus*. *J. Braz. Chem. Soc.* 25:788–794.

Pereira, F. R., R. G. S. Berlinck, E. Rodrigues Filho, K. Veloso, A. G. Ferreira, and V. Padula. 2012. Metabólitos Secundários dos nudibrânquios *Tambja stegosauriformis*, *Hypselodoris lajensis* e *Okenia zoobotryon* e dos briozoários *Zoobotryon verticillatum* e *Bugula dentata* da costa do Brasil. *Quimica Nova* 35:2194–2201.

Pereira, R. B., P. B. Andrade, and P. Valentão. 2016. Chemical diversity and biological properties of secondary metabolites from sea hares of *Aplysia* genus. *Mar. Drugs* 14:39.

Perrone, A. S. 1989. Duplicità funcionale nell'aposematismo distruttivo in *Peltodoris atromaculata* Bergh, 1880 (Opisthobranchia: Nudibranchia). *Boll. Malacol.* 24: 187–188.

Petraki, A., E. Ioannou, P. Papazafiri, and V. Roussis. 2015 Dactylomelane diterpenes from the sea hare *Aplysia depilans*. *J. Nat. Prod.* 78:462–467.

Pettit, G. R., C. L. Herald, M. S. Allen, R. B. von Dreele, L. D. Vanell, J. P. Y. Kao, and W. Blake. 1977. The isolation and structure of aplysistatin. *J. Am. Chem. Soc.* 99:262–263.

Pettit, G. R., C. L. Herald, R. H. Ode, P. Brown, D. J. Gust, and C. Michel. 1980. The isolation of loliolide from an Indian Ocean opisthobranch mollusc. *J. Nat. Prod.* 43:752–755.

Pettit, G. R., R. H. Ode, C. L. Herald, R. B. von Dreele, and C. Michel. 1976. The isolation and structure of dolatriol. *J. Am. Chem. Soc.* 98:4677–4678.

Pettit, G. R., Y. Kamano, C. L. Herald et al. 1993. Isolation of dolastatins 10–15 from the marine mollusc *Dolabella auricularia*. *Tetrahedron* 49:9151–9170.

Pettit, G. R., Y. Kamano, P. Brown, D. Gust, M. Inoue, and C. L. Herald. 1982. Structure of the cyclic peptide dolastatin 3 from *Dolabella auricularia*. *J. Am. Chem. Soc.* 104:905–907.

Piel, J., D. Hui, G. Wen, D. Butzke, M. Platzer, N. Fusetani, and S. Matsunaga. 2004. Antitumor polyketide biosynthesis by an uncultivated bacterial symbiont of the marine sponge *Theonella swinhoei*. *Proc. Natl. Acad. Sci. USA* 101:16222–16227.

Piel, W. H. 1991. Pycnogonid predation on nudibranchs and ceratal autotomy. *Veliger* 34:366–367.

Piel. J. 2009. Metabolites from symbiotic bacteria. *Nat. Prod. Rep.* 26:338–362.

Pika, J., and D. J. Faulkner. 1994. Four sesquiterpenes from the South African nudibranch *Leminda millecra*. *Tetrahedron* 50:3065–3070.

Pika, J., and D. J. Faulkner. 1995. Unusual chlorinated homo-diterpenes from the South African nudibranch *Chromodoris hamiltoni*. *Tetrahedron* 51:8189–8198.

Poiner, A., V. J. Paul, and P. J. Scheuer. 1989. Kumepaloxane, a rearranged trisnor sesquiterpene from the bubble shell *Haminoea cymbalum*. *Tetrahedron* 45:617–622.

Ponder, W. F., and D. R. Lindberg. 1997. Towards a phylogeny of gastropod molluscs: An analysis using morphological characters. *Zool. J. Linnean Soc.* 119:83–265.

Potin, P., K. Bouarab, J. P. Salaun, G. Pohnert, and B. Kloareg. 2002. Biotic interaction of marine algae. *Curr. Opin. Plant Biol.* 5:1–10.

Powell, K. J., J. L. Richens, J. P. Bramble, L.-C. Han, P. Sharma, P. O'Shea, and J. E. Moses. 2017. Photochemical activity of membrane-localised polyketide derived marine natural products. *Tetrahedron* doi:10.1016/j.tet.2017.10.056.

Puglisi, M. P., J. M. Sneed, K. H. Sharp, R. Ritson-Williams, and V. J. Paul. 2014. Marine chemical ecology in benthic environments. *Nat. Prod. Rep.* 31:1510–1553.

Putz, A., G. M. König, and H. Wägele. 2010. Defensive strategies of Cladobranchia (Gastropoda, Opisthobranchia). *Nat. Prod. Rep.* 27:1386–1402.

Putz, A., S. Kehraus, G. Díaz-Agras, H. Wägele, and G. M. König. 2011. Dotofide, a guanidine-interrupted terpenoid from the marine slug *Doto pinnatifida* (Gastropoda, Nudibranchia). *Europ. J. Org. Chem.* 2011:3733–3737.

Quiñoá, E., L. Castedo, and R. Riguera. 1989. The halogenated monoterpenes of *Aplysia punctata*. A comparative study. *Comp. Biochem. Physiol.—Part B: Comp. Biochem.* 92:99–101.

Radjasa, O. K., Y. M. Vaske, G. Navarro, H. C. Vervoort, K. Tenney, R. G. Linington, and
P. Crews. 2011. Highlights of marine invertebrate-derived biosynthetic products: Their
biomedical potential and possible production by microbial associants. *Bioorg. Med.
Chem.* 19:6658–6674.

Ramya, M. S., K. Sivasubramanian, S. Ravichandran, and R. Anbuchezhian. 2014. Screening
of antimicrobial compound from the sea slug *Armina babai. Bangladesh J Pharmacol*
9:268–274.

Rapson, T. D. 2004. Bioactive 4-methoxypyrrolic natural products from two South African
marine invertebrates. MSc thesis. Rhodes, South Africa: Rhodes University.

Reese, M. T., N. K. Gulavita, Y. Nakao, M. T. Hamann, W. Y. Yoshida, S. J. Coval, and
P. J. Scheuer. 1996. Kulolide: A cytotoxic depsipeptide from a cephalaspidean mollusk,
Philinopsis speciosa. J. Am. Chem. Soc. 118:11081–11084.

Rhoades, D. F. 1979. Evolution of plant chemical defense against herbivores. In *Herbivores:
Their Interaction with Plant Secondary Metabolites* G. A. Rosenthal and D. H. Janzen
(eds.), pp. 1–55. New York: Academic Press.

Rhoades, D. F., and R. G. Gates. 1976. Toward a general theory of plant antiherbivore chemistry. *Rec. Adv. Phytochem.* 10:168–213.

Ritland, D. B., and L. P. Brower. 1991. The viceroy butterfly is not a batesian mimic. *Nature*
350:497–498.

Ritson-Williams, R., and V. J. Paul. 2007. Marine benthic invertebrates use multimodal cues
for defense against reef fish. *Mar. Ecol. Progr. Ser.* 340:29–39.

Rodríguez, J., R. Fernández, E. Quiñoá, R. Riguera, C. Debitus, and P. Bouchet. 1994. Onchidin:
A cytotoxic depsipeptide with C2 symmetry from a marine mollusc. *Tetrahedron Lett.*
35:9239–9242.

Rodríguez, J., R. Riguera, and C. Debitus. 1992. The natural polypropionate-derived esters of
the mollusc *Onchidium* sp. *J. Org. Chem.* 57:4624–4632.

Rogers, C. N., R. de Nys, and P. D. Steinberg. 2000. Predation on juvenile *Aplysia parvula*
and other small Anaspidean, Ascoglossan and Nudibranch Gastropods by Pycnogonids.
Veliger 43(4):330–337.

Rogers, S. D., and V. J. Paul. 1991. Chemical defenses of three *Glossodoris* nudibranchs and
their dietary *Hyrtios* sponges. *Mar. Ecol. Progr. Ser.* 77:221–232.

Roll, D. M., J. E. Biskupiak, C. L. Mayne, and C. M. Ireland. 1986. Muamvatin, a novel
tricyclic spiro ketal from the Fijian mollusk *Siphonaria normalis. J. Am. Chem. Soc.*
108:6680–6682.

Ros, J. 1976. Sistemas de defensa en los opistobranquios. *Oecol. Aquatica* 2:41–77.

Ros, J., and A. Marín. 1991. Adaptative advantages of the "symbiosis" between algal chloroplasts and sacoglossan molluscs. *Oecol. Aquat.* 10:271–298.

Rose, A. F., P. J. Scheuer, J. P. Springer, and J. Clardy. 1978. Stylocheilamide, an unusual
constituent of the sea hare *Stylocheilus longicauda. J. Am. Chem. Soc.* 100:7665–7670.

Roussis, V., J. R. Pawlik, M. E. Hay, and W. Fenical. 1990. Secondary metabolites of the
chemically rich ascoglossan *Cyerce nigricans. Experientia* 49:327–329.

Rovirosa, J., E. Quezada, and A. San-Martín. 1992. New diterpene from the mollusc
Trimusculus peruvianus. Boletín de la Sociedad Chilena de Química 37:143–145.

Rudman, W. B., and R. C. Willan. 1998. Opisthobranchia introduction. In *Mollusca:
The Southern Synthesis. Fauna of Australia, Part B,* Vol. 5, P. L. Beesley, G. J. B. Ross,
and A. Wells (Eds.), pp. 915–942. Melbourne, Australia: CSIRO Publishing.

Rudman, W. B. 1984. The Chromodorididae (Opisthobranchia: Mollusca) of the Indo-West
Pacific: A review of the genera. *Zool. J. Linn.* Soc. 81:115–273.

Rudman, W. B. 1987. The chromodorididae (Opisthobranchia: Mollusca) of the Indo-West
Pacific: *Chromodoris epicuria, C. aureopurpurea, C. annulata, C. coi* and *Risbecia
tryoni* colour groups. *Zool. J. Linn. Soc.* 90:305–407.

Rudman, W. B. 1991. Purpose in pattern: The evolution of colour in chromodorid nudibranchs. *J. Molluscan Stud.* 57:5–21.

Rudman, W. B. What eats sea slugs? Sea Slug Forum. 2000. Available from http://www.seaslugforum.net/factsheet/predrecord accessed April 16, 2015.

Rudman, W. B., and P. R. Bergquist. 2007. A review of feeding specificity in the sponge-feeding Chromodorididae (Nudibranchia: Mollusca). *Molluscan Res.* 27:60–88.

Rumpho, M. E., E. J. Summer, and J. R. Manhart. 2000. Solar-powered sea slugs. Mollusc/algal chloroplast symbiosis. *Plant Physiol.* 123:29–38.

Sakio, Y., Y. J. Hirano, M. Hayashi, K. Komiyama, and M. Ishibashi. 2001. Dendocarbins A–N, new drimane sesquiterpenes from the nudibranch *Dendrodoris carbunculosa*. *J. Nat. Prod.* 64:726–731.

Salvitti, L. R., S. A. Wood, L. Winsor, and S. C. Cary. 2015. Intracellular Immunohistochemical Detection of Tetrodotoxin in *Pleurobranchaea maculata* (Gastropoda) and *Stylochoplana* sp. (Turbellaria). *Mar. Drugs* 13:756–769.

Salvitti, L. R., S. Wood, R. Fairweather, and S. C. Cary. 2016. In situ accumulation of tetrodotoxin in non-toxic *Pleurobranchaea maculata* (Opisthobranchia). *Aquat. Sci.* 79:1–10.

Schmidt, E. W., J. T. Nelson, D. A. Rasko, S. Sudek, J. A. Eisen, M. G. Haygood, and J. Ravel. 2005. Patellamide A and C biosynthesis by a microcin-like pathway in *Prochloron didemni*, the cyanobacterial symbiont of *Lissoclinum patella*. *Proc. Nat. Acad. Sci. USA* 102:7315–7320.

Schmitz, F. J., and F. J. McDonald. 1974. Marine natural products: Dactyloxene-B, a sesquiterpene ether from the sea hare, *Aplysia dactylomela*. *Tetrahedron Lett.* 29:2541–2544.

Schmitz, F. J., D. P. Michaud, and K. H. Hollenbeak. 1980. Marine natural products: Dihydroxydeodactol monoacetate, a halogenated sesquiterpene ether from the sea hare *Aplysia dactylomela*. *J. Org. Chem.* 45:1525–1528.

Schmitz, F. J., D. P. Michaud, and P. G. Schmidt. 1982. Marine natural products: Parguerol, deoxyparguerol, and isoparguerol. New brominated diterpenes with modified pimarane skeletons from the sea hare *Aplysia dactylomela*. *J. Am. Chem. Soc.* 104:6415–6423.

Schmitz, F. J., F. J. McDonald, and D. J. Vanderah. 1978. Marine natural products: Sesquiterpenes alcohols and ethers from the sea hare *Aplysia dactylomela*. *J. Org. Chem.* 43:4220–4225.

Schrödl, M., K. M. Jörger, A. Klussmann-Kolb, and N. G. Wilson. 2011. Bye bye 'Opisthobranchia'! A review on the contribution of mesopsammic sea slugs to euthyneuran systematics. *Thalassas* 27:101–112.

Schulte, G. R., M. C. H. Chung, and P. J. Scheuer. 1981. Two bicyclic C_{15} enynes from the sea hare *Aplysia oculifera*. *J. Org. Chem.* 46:3870–3873.

Sheybani, A., M. Nusnbaum, J. Caprio, C. D. Derby. 2009. Responses of the sea catfish *Ariopsis felis* to chemical defenses from the sea hare *Aplysia californica*. *J. Exp. Mar. Biol. Ecol.* 368:153–160.

Shimomura, M., H. Miyaoka, and Y. Yamada. 1999. Absolute configuration of marine diterpenoid kalihinol A. *Tetrahedron Lett.* 40:8015–8017.

Shin, J., and Y. Seo. 1995. Isolation of new ceramides from the gorgonian *Acabaria undulata*. *J. Nat. Prod.* 58:948–953

Slattery, M., C. Avila, J. Starmer, and V. J. Paul. 1998. A sequestered soft coral diterpene in the aeolid nudibranch *Phyllodesmium guamensis*. *J. Exp. Mar. Biol. Ecol.* 226:33–49.

Sleeper, H. L., and W. Fenical. 1977. Navenones A-C trail-breaking alarm pheromones from the marine opisthobranch *Navanax inermis*. *J. Am. Chem. Soc.* 99:2367–2368.

Sleeper, H. L., V. J. Paul, and W. Fenical. 1980. Alarm pheromones from the marine opisthobranch *Navanax inermis*. *J. Chem. Ecol.* 6:57–70.

Sodano, S., and A. Spinella. 1986. Janolusimide, a lipophilic tripeptide toxin from the nudibranch mollusc *Janolus cristatus*. *Tetrahedron Lett.* 27:2505–2508.

Somerville, M. J., E. Mollo, G. Cimino, W. Rungprom, and M. J. Garson. 2006. Spongian diterpenes from Australian nudibranchs: an anatomically guided chemical study of *Glossodoris atromarginata. J. Nat. Prod.* 69:1086–1088.

Somerville, M. J., P. L. Katavic, L. K. Lambert, G. K. Pierens, J. T. Blanchfield, G. Cimino, E. Mollo, M. Gavagnin, M. G. Banwell, and M. J. Garson. 2012. Isolation of thuridillins D-F, diterpene metabolites from the Australian sacoglossan mollusk *Thuridilla splendens*; relative configuration of the epoxylactone ring. *J. Nat. Prod.* 75:1618–1624.

Sone, H., H. Kigoshi, and K. Yamada. 1996. Aurisides A and B, cytotoxic macrolide glycosides from the Japanese sea hare *Dolabella auricularia. J. Org. Chem.* 61:8956–8960.

Sone, H., T. Kondo, M. Kiryu, H. Ishiwata, M. Ojika, and K. Yamada. 1995. Dolabellin, a cytotoxic bisthiazole metabolite from the sea hare *Dolabella auricularia*: Structural determination and synthesis. *J. Org. Chem.* 60:4774–4781.

Sone, H., T. Nemoto, H. Ishiwata, M. Ojika, and K. Yamada. 1993. Isolation, structure, and synthesis of dolastatin D, a cytotoxic cyclic depsipeptide from the sea hare *Dolabella auricularia. Tetrahedron Lett.* 34:8449–8452.

Sone, H., T. Nemoto, M. Ojika, and K. Yamada. 1993. Isolation, structure, and synthesis of dolastatin C, a cytotoxic cyclic depsipeptide from the sea hare *Dolabella auricularia. Tetrahedron Lett.* 34:8445–8448.

Sone, H., T. Shibata, T. Fujita, M. Ojika, and K. Yamada. 1996. Dolastatin H and isodolastatin H, potent cytotoxic peptides from the sea hare *Dolabella auricularia*: Isolation, stereo-structures, and synthesis. *J. Am. Chem. Soc.* 118:1874–1880.

Sotka, E. E., J. Forbey, M. Horn, A. G. B. Poore, D. Raubenheimer, and K. E. Whalen. 2009. The emerging role of pharmacology in understanding consumer-prey interactions in marine and freshwater systems. *Integr. Comp. Biol.* 49:291–313.

Spinella, A., E. Mollo, E. Trivellone, and G. Cimino. 1997. Testudinariol A and B, two unusual triterpenoids from the skin and the mucus of the marine mollusc *Pleurobranchus testudinarius. Tetrahedron* 53:16891–16896.

Spinella, A., E. Zubía, E. Martínez, J. Ortea, and G. Cimino. 1997. Structure and stereochemistry of Aplyolides A-E, lactonized dihydroxy fatty acids from the skin of the marine mollusk *Aplysia depilans. J. Org. Chem.* 62:5471–5475.

Spinella, A., L. A. Álvarez, and G. Cimino. 1993. Predator-prey relationship between *Navanax inermis* and *Bulla gouldiana*: A chemical approach. *Tetrahedron* 49:3203–3210.

Spinella, A., L. A. Álvarez, C. Avila, and G. Cimino. 1994. New acetoxy-ent-pallescensin-A sesquiterpenoids from the skin of the porostome nudibranch *Doriopsilla areolata. Tetrahedron Lett.* 35:8665–8668.

Spinella, A., L. A. Álvarez, and G. Cimino. 1998. Alkylphenols from the cephalaspidean mollusc *Haminoea callidegenita. Tetrahedron Lett.* 39:2005–2008.

Spinella, A., M. Gavagnin, A. Crispino, and G. Cimino. 1992. 4-acetylaplykurodin B and aplykurodinone B, two ichthyotoxic degraded sterols from the Mediterranean mollusk *Aplysia fasciata. J. Nat. Prod.* 5:989–993.

Stachowicz, J. J., and N. Lindquist. 2000. Hydroid defenses against predators: The importance of secondary metabolites versus nematocysts. *Oecologia* 124:280–288.

Stallard, M. O., and D. J. Faulkner. 1974. Chemical constituents of the digestive gland of the sea hare *Aplysia californica*-II. Chemical transformations. *Comp. Biochem. Physiol.— Part B: Biochem. Mol. Biol.* 49:37–41.

Stallard, M. O., and W. Fenical. 1978. The brasilenols, rearranged sesquiterpene alcohols isolated from the marine opisthobranch *Aplysia brasiliana. Tetrahedron* 34:2077–2081.

Stasek, C. R. 1967. Autotomy in the Mollusca. *Occ. Papers Calif. Acad Sci* 61:1–44.

Storvick, J. M., E. Ankoudinova, B. R. King, H. Van Epps, and G. W. O'Neil. 2011. Total synthesis of haminol A: An analysis of vinylpyridine metathesis reactivity. *Tetrahedron Lett.* 52:5858–5861.

Suciati, L., K. Lynette, and M. J. Garson. 2011. Structures and anatomical distribution of oxygenated diterpenes in the Australian nudibranch *Chromodoris reticulata*. *Austral. J. Chem.* 646:757–765.

Sudek, S., N. B. Lopanik, L. E. Waggoner, M. Hildebrand, C. Anderson, H. Liu, A. Patel, D. H. Sherman, and M. G. Haygood. 2007. Identification of the putative bryostatin polyketide synthase gene cluster from "*Candidatus* Endobugula sertula", the uncultivated microbial symbiont of the marine bryozoan *Bugula neritina*. *J. Nat. Prod.* 70:67–74.

Suenaga, K., H. Kigoshi, and K. Yamada. 1996. Auripyrones A and B, cytotoxic polypropionates from the sea hare *Dolabella auricularia*: Isolation and structures. *Tetrahedron Lett.* 37:5151–5154.

Suenaga, K., T. Mutou, T. Shibata et al. 2004. Aurilide, a cytotoxic depsipeptide from the sea hare *Dolabella auricularia*: Isolation, structure determination, synthesis, and biological activity. *Tetrahedron* 60:8509–8527.

Suenaga, K., T. Mutou, T. Shibata, T. Itoh, H. Kigoshi, and K. Yamada. 1996. Isolation and stereostructure of aurilide, a novel cyclodepsipeptide from the Japanese sea hare *Dolabella auricularia*. *Tetrahedron Lett.* 37:6771–6774.

Suenaga, K., T. Nagoya, T. Shibata, H. Kigoshi, and K. Yamada. 1997. Dolabelides C and D, cytotoxic macrolides isolated from the sea hare *Dolabella auricularia*. *J. Nat. Prod.* 60:155–157.

Suenaga, K., T. Shibata, N. Takada, H. Kigoshi, and K. Yamada. 1998. Aurilol, a cytotoxic bromotriterpene isolated from the sea hare *Dolabella auricularia*. *J. Nat. Prod.* 61:515–518.

Sun, H. H., F. J. McEnroe, and W. Fenical. 1983. Acetoxycrenulide, a new bicyclic cyclopropane-containing diterpenoid from the brown seaweed *Dictyota crenulata*. *J. Org. Chem.* 48:1903–1906.

Suntornchashwej, S., N. Chaichit, M. Isobe, and K. Suwanborirux. 2005. Hectochlorin and morpholine derivatives from the Thai sea hare, *Bursatella leachii*. *J. Nat. Prod.* 68:951–955.

Szabo, C. M., Y. Nakao, W. Y. Yoshida, and P. J. Scheuer. 1996. Two diverse constituents of the cephalaspidean mollusk *Smaragdinella calyculata*. *Tetrahedron* 52:9681–9686.

Taboada, S., L. Núñez-Pons, and C. Avila. 2013. Feeding repellence of Antarctic and sub-Antarctic benthic invertebrates against the omnivorous sea star *Odontaster validus* Koehler, 1906. *Polar Biol.* 36:13–25.

Tan, K. C., T. Wakimoto, K. Takada, T. Ohtsuki, N. Uchiyama, G. Goda, and I. Abe. 2013. Cycloforskamide, a cytotoxic macrocyclic peptide from the sea slug *Pleurobranchus forskalii*. *J. Nat. Prod.* 76:1388–1391.

Tanaka, J., and T. Higa. 1988. The absolute configuration of kurospongin a new furanoterpene from a marine sponge, *Spongia* sp. *Tetrahedron* 44:2805–2810.

Tanaka, J., and T. Higa. 1999. Two new cytotoxic carbonimidic dichlorides from the nudibranch *Reticulidia fungia*. *J. Nat. Prod.* 62:1339–1340.

Teeyapant, R., P. Kreis, V. Wray, L. Witte, and P. Proksch. 1993. Brominated secondary compounds from the marine sponge *Verongia aerophoba* and the sponge feeding gastropod *Tylodina perversa*. *Zeitschrift für Naturforschung* 48C:630–644.

Terlau, H., and B. M. Olivera. 2004. *Conus* venoms, a rich source of novel ion channel-targeted peptides. *Physiol. Rev.* 84:41–68.

Thompson, J. E., R. P. Walker, S. J. Wratten, and D. J. Faulkner. 1982. A chemical defense mechanism for the nudibranch *Cadlina luteomarginata*. *Tetrahedron Lett.* 23:1865–1873.

Thompson, T. E. 1976. *Biology of Opisthobranch Molluscs 1*. London, UK: Ray Society.

Thompson, T. E. 1986. Investigation of the acidic allomone of the gastropod mollusc *Philine aperta* by means of ion chromatography and histochemical localisation of sulphate and chloride ions. *J. Molluscan Stud.* 52:38–44.

Thompson, T. E., and D. J. Slinn. 1959. On the biology of the opisthobranch *Pleurobranchus membranaceus*. *J. Mar. Biol. Assoc. U.K.* 38:507–524.

Thompson, T. E., and D. R. Seaward. 1989. Ecology and taxonomic status of the aplysiomorph *Akera bullata* in the British Isles. *J. Molluscan Stud.* 55:489–496.

Thompson, T. E. 1960. Defensive acid-secretion in marine gastropods. *J. Mar. Biol. Assoc. U.K.* 39:115–122.

Thompson, T. E. 1988. Acidic allomones in marine organisms. *J. Mar. Biol. Assoc. U.K.* 68:499–517.

Thoms, C., M. Wolff, K. Padmakumar, R. Ebel, and P. Proksch. 2004. Chemical defense of Mediterranean sponges *Aplysina cavernicola* and *Aplysina aerophoba*. *Zeitschrift für Naturforschung*, ser. C, 59:113–122.

Tilvi, S., and C. G. Naik. 2007. Tandem mass spectrometry of kahalalides: Identification of two new cyclic depsipeptides, kahalalide R and S from *Elysia grandifolia*. *J. Mass Spectrom.* 42:70–80.

Todd, C. 1981. The ecology of nudibranch molluscs. *Oceanography Mar. Biol. Ann. Rev.* 19:141–234.

Todd, J. S., and W. H. Gerwick. 1995. Malyngamide I from the tropical marine cyanobacterium *Lyngbya majuscula* and the probable structure revision of stylocheilamide. *Tetrahedron Lett.* 36:7837–7840.

Toth, G. B., O. Langhamer, and H. Pavia. 2005. Inducible and constitutive defenses of valuable seaweed tissues: Consequences for herbivore fitness. *Ecology* 86:612–618.

Tringali, C., A. Parisi, M. Piattelli, and G. M. di San Lio. 1993. Phomenins A and B, bioactive polypropionate pyrones from culture fluids of *Phoma tracheiphila*. *Nat. Prod. Lett.* 3:101–106.

Tullrot, A. 1994. The Evolution of unpalatability and warning coloration in soft-bodied marine invertebrates. *Evolution* 48:925–928.

Tullrot, A., and P. Sundberg. 1991. The conspicuous nudibranch *Polycera quadrilineata*: Aposematic coloration and individual selection. *Animal Behav.* 41:175–176.

Uddin, M. H., M. Otsuka, T. Muroi, A. Ono, N. Hanif, S. Matsuda, T. Higa, and J. Tanaka. 2009. Deoxymanoalides from the nudibranch *Chromodoris willani*. *Chem. Pharm. Bull.* 57:885–887.

Valdés, A., and P. Lozouet. 2000. Opisthobranch molluscs from the Tertiary of the Aquitaine Basin (south-western France), with descriptions of seven new species and a new genus. *Palaeontology* 43:457–479.

Valdés, A., E. Ornelas-Gatdula, and A. Dupont. 2013. Color pattern variation in a shallow-water species of opisthobranch mollusc. *Biol. Bull.* 224:35–46.

Valdés, A., L. Blanchard, and W. Marti. 2013. Caught naked: First report a nudibranch sea slug attacked by a cone snail. *Am. Malacol. Bull.* 31:337–338.

Valiela, I. 2001. *Doing Science*. Oxford, UK: Oxford University Press.

Van Wyk, A. W. W., C. A. Gray, C. E. Whibley, O. Osoniyi, D. T. Hendricks, R. Caira Mino, and M. T. Davies-Coleman. 2008. Bioactive metabolites from the South African marine mollusk *Trimusculus costatus*. *J. Nat. Prod.* 71:420–425.

Vanderah, D. J., and F. J. Schmitz. 1976. Marine natural products: Isodactylyne, a halogenated acetylenic ether from the sea hare *Aplysia dactylomela*. *J. Org. Chem.* 41:3480–3481.

Vane-Wright, R. I. 1991. A case of self-deception. *Nature* 350:460–461.

Vardaro, R. R., V. Di Marzo, A. Crispino, and G. Cimino. 1991. Cyercenes, novel polypropionate pyrones from the autotomizing Mediterranean mollusc *Cyerce cristallina*. *Tetrahedron* 47:5569–5576.

Vardaro, R. R., V. Di Marzo, A. Marín, and G. Cimino. 1992. A- and g-Pyrone-polypropionates from the Mediterranean ascoglossan mollusc *Ercolania funerea*. *Tetrahedron* 48:9561–9566.

Vardaro, R. R., V. Di Marzo, and G. Cimino. 1992. Placidenes: Cyercene-like polypropionate γ-pyrones from the Mediterranean ascoglossan mollusc *Placidia dendritica*. *Tetrahedron Lett.* 33:2875–2878.

Vasskog, T., J. H. Andersen, E. Hansen, and J. Svenson. 2012. Characterization and cytotoxicity studies of the rare 21:4 *n*-7 acid and other polyunsaturated fatty acids from the marine opisthobranch *Scaphander lignarius*, isolated using bioassay guided fractionation. *Mar. Drugs* 10:2676–2690.

Verdier-Pinard, P., J. A. Kepler, G. R. Pettit, and E. Hamel. 2000. Sustained intracellular retention of dolastatin 10 causes its potent antimitotic activity. *Mol. Pharmacol.* 57:180–187.

Vermeij, G. J. 1987. *Evolution and Escalation: An Ecological History of Life*. Princeton, NJ: Princeton University Press.

Vermeij, G. J. 1994. The evolutionary interaction among species: Selection, escalation, and coevolution. *Annu. Rev. Ecol., Evol., Syst.* 25:219–236.

Vincent, E., J. Saxton, C. Baker-Glenn, I. Moal, J. D. Hirst, G. Pattenden, and P. E. Shaw. 2007. Effects of ulapualide A and synthetic macrolide analogues on actin dynamics and gene regulation. *Cell. Mol. Life Sci.* 64:487–497.

von Salm et al. 2018. The status of marine chemical ecology in Antarctica: Form and function of unique high-latitude chemistry. In *Chemical Ecology: The Ecological Impacts of Marine Natural Products*.

Wägele, H. 2004. Potential key characters in Opisthobranchia (Gastropoda, Mollusca) enhancing adaptive radiation. *Anim. Divers. Evol.* 4:175–188.

Wägele, H., A. Klussmann-Kolb, E. Verbeek, and M. Schrödl. 2014. Flashback and foreshadowing—A review of the taxon Opisthobranchia. *Org. Divers. Evol.* 14:133–149.

Wägele, H., and A. Klussmann-Kolb. 2005. Opisthobranchia (Mollusca, Gastropoda)—More than just slimy slugs. Shell reduction and its implications on defence and foraging. *Front. Zool.* 2:1–18.

Wägele, H., and G. Johnsen. 2001. Observations on the histology and photosynthetic performance of "solar-powered" opisthobranchs (Mollusca, Gastropoda, Opisthobranchia) containing symbiotic chloroplasts or zooxanthellae. *Org., Divers. Evol.* 1:193–210.

Wägele, H., Ballesteros, M., and Avila, C. 2006. Defensive grandular structures in opistobranch molluscs—From histology to ecology. *Oceanography Mar. Biol.—An Annu. Rev.* 44:197–276.

Wägele, H., K. Knezevic, and A. Y. Moustafa. 2017. Distribution and morphology of defensive acid-secreting glands in Nudipleura (Gastropoda: Heterobranchia), with an emphasis on Pleurobranchomorpha. *J. Molluscan Stud.* 83:422–433.

Wägele, H., K. Knezevic, and A. Y. Moustafa. 2017. Distribution and morphology of defensive acid-secreting glands in Nudipleura (Gastropoda: Heterobranchia), with an emphasis on Pleurobranchomorpha. *J. Molluscan Stud.* 83:422–433.

Wahidulla, S., Y. W. Guo, I. M. I. Fakhr, and E. Mollo. 2006. Chemical diversity in opisthobranch molluscs from scarcely investigated Indo-Pacific areas. In *Molluscs: From Chemo-ecological Study to Biotechnological Applications*, G. Cimino and M. Gavagnin (Eds.), pp. 175–198. Berlin, Germany: Springer.

Wakimoto, T., K. C. Tan, and I. Abe. 2013. Ergot alkaloid from the sea slug *Pleurobranchus forskalii*. *Toxicon* 72:1–4.

Wang, J. R., W. F. He, and Y. W. Guo. 2013. Chemistry, chemoecology, and bioactivity of the South China Sea opisthobranch molluscs and their dietary organisms. *J. Asian Nat. Prod. Res.* 15:185–197.

Wang, J. R., M. Carbone, M. Gavagnin, A. Mándi, S. Antus, L. G. Yao, G. Cimino, T. Kurtán, and Y. W. Guo. 2012. Assignment of absolute configuration of bis-γ-pyrone polypropionates from marine pulmonate molluscs. *Europ. J. Org. Chem.* 2012:1107–1111.

Wesson, K. J., and M. T. Hamann. 1996. Keenamide A, a bioactive cyclic peptide from the marine mollusk *Pleurobranchus forskalii*. *J. Nat. Prod.* 59:629–631.

Whalen, K. E., E. E. Sotka, J. V. Goldstone, and M. E. Hahn. 2010. The role of multixenobiotic trasnporters in predatory marine molluscs as counter-defense mechanisms against dietary allelochemicals. *Comp. Biochem. Physiol.* C152:288–300.

Whibley, C. E., K. L. McPhail, R. A. Keyzers, M. F. Maritz, V. D. Leaner, M. J. Birrer, M. T. Davies-Coleman, and D. T. Hendricks. 2007. Reactive oxygen species mediated apoptosis of esophageal cancer cells induced by marine triprenyl toluquinones and toluhydroquinones. *Mol. Cancer Ther.* 6:2535–2543.

Whitehead, K., D. Karentz, and J. Hedges. 2001. Mycosporine-like amino acids (MAAs) in phytoplankton, a herbivorous pteropod (*Limacina helicina*), and its pteropod predator (*Clione antarctica*) in McMurdo Bay, Antarctica. *Mar. Biol.* 139:1013–1019.

Willan, R. C. 1984. A review of the diets in the Notaspidea (Mollusca: Opisthobranchia). *J. Malacol. Soc. Aust.* 6:125–142.

Williams, B. L., C. T. Hanifin, E. D. Brodie, and E. D. Brodie. III. 2012. Predators usurp prey defenses? Toxicokinetics of tetrodotoxin in common garter snakes after consumption of rough-skinned newts. *Chemoecology* 22:179–185.

Williams, D. E., and R. J. Andersen. 1987. Terpenoid metabolites from skin extracts of the dendronotid nudibranch *Tochuina tetraquetra*. *Can. J. Chem.* 65:2244–2247.

Wilson, E. O. 1980. *Sociobiology*. Cambridge, UK: Harvard University Press.

Wilson, N. G., J. A. Maschek, and B. J. Baker. 2013. A species flock driven by predation? Secondary metabolites support diversification of slugs in Antarctica. *PLoS One* 8:e80277.

Wilson, N. G., and M. S. Y. Lee. 2005. Molecular phylogeny of *Chromodoris* (Mollusca, Nudibranchia) and the identification of a planar spawning clade. *Mol. Phylogenet. Evol.* 36:722–727.

Wright, A. D. 2003. GC-MS and NMR analysis of *Phyllidiella pustulosa* and one of its dietary sources, the sponge *Phakellia carduus*. *Comp. Biochem. Physiol.* 134A:307–313.

Wyeth, R. C., and A. O. D. Willows. 2006. Odours detected by rhinophores mediate orientation to flow in the nudibranch mollusc *Tritonia diomedea*. *J. Exp. Biol.* 209:1441–1453.

Yamada, K., and H. Kigoshi. 1997. Bioactive compounds from the sea hares of two genera: *Aplysia* and *Dolabella*. *Bull. Chem. Soc. Jpn.* 70:1479–1489.

Yamada, K., M. Ojika, H. Kigoshi, and K. Suenaga. 2010. Cytotoxic substances from two species of Japanese sea hares: Chemistry and bioactivity. *Proc. Jpn. Acad. Ser. B, Phys. Biol. Sci.* 86:176–189.

Yamada, K., M. Ojika, T. Ishigaki, Y. Yoshida, H. Ekimoto, and M. Arakawa. 1993. Aplyronine A, a potent antitumor substance, and the congeners aplyronines B and C isolated from the sea hare *Aplysia kurodai*. *J. Am. Chem. Soc.* 115:11020–11021.

Yamamura, S., and Y. Hirata. 1963. Structures of aplysin and aplysinol, naturally occurring bromo-compounds. *Tetrahedron* 19:1485–1496.

Yamamura, S., and Y. Hirata. 1971. A naturally-occurring bromo-compound, aplysin-20 from *Aplysia kurodai*. *Bull Chem Soc Jpn* 44:2560–2562.

Yamamura, S., and Y. Terada. 1977. Isoaplysin-20, a natural bromine-containing diterpene, from *Aplysia kurodai*. *Tetrahedron Lett.* 25:2171–2172.

Yasman, Y., R. A. Edrada, V. Wray, and P. Proksch. 2003. New 9-thiocyanatopupukeanane sesquiterpenes from the nudibranch *Phyllidia varicosa* and its sponge-prey *Axinyssa aculeata*. *J. Nat. Prod.* 66:1512–1514.

Yong, K. W., A. A. Salim, and M. J. Garson. 2008. New oxygenated diterpenes from an Australian nudibranch of the genus *Chromodoris*. *Tetrahedron* 64:6733–6738.

Yoshida, W. Y., P. J. Bryan, B. J. Baker, and J. B. McClintock. 1995. Pteroenone: A defensive metabolite of the abducted Antarctic pteropod *Clione antarctica*. *J. Org. Chem.* 60:780–782.

Young, C. M., P. G. Greenwood, and C. J. Powell. 1986. The ecological role of defensive secretions in the intertidal pulmonate *Onchidella borealis*. *Biol. Bull.* 171:391–404.

Young, R. M., and B. J. Baker. 2015. The defensive chemistry of the Irish nudibranch *Archidoris pseudoargus* (Gastropoda: Opisthobranchia). *Planta Medica* 81:PQ26.

Zapata, F., N. G. Wilson, M. Howison, S. C. Andrade, K. M. Jörger, M. Schrödl, F. E. Goetz, G. Giribet, and C. W. Dunn. 2014. Phylogenomic analyses of deep gastropod relationships reject Orthogastropoda. *Proc. Royal Soc. B: Biol. Sci.* 281:20141739.

Zhang, W., M. Gavagnin, Y.-W. Guo, E. Mollo, M. T. Ghiselin, and G. Cimino. 2007. Terpenoid metabolites of the nudibranch *Hexabranchus sanguineus* from the South China Sea. *Tetrahedron* 63:4725–4729.

Zhukova, N. V. 2014. Lipids and fatty acids of nudibranch mollusks: Potential sources of bioactive compounds. *Mar. Drugs* 12:4578–4592.

Zhukova, N. V., and M. G. Eliseikina. 2012. Symbiotic bacteria in the nudibranch mollusk *Dendrodoris nigra*: Fatty acid composition and ultrastructure analysis. *Mar. Biol.* 159:1783–1794.

4 The Chemical Ecology of Seagrasses

Kathryn L. Van Alstyne and Dianna K. Padilla

CONTENTS

4.1 INTRODUCTION

Seagrass communities are found in marine and estuarine environments around the world, from cold north and south temperate areas to the tropics (Hemminga and Duarte 2000; den Hartog and Kuo 2006; Larkum et al. 2006). They occur in a variety of habitat types, from soft sediment environments to hard rocky shores, and from quiet bays to very wave-swept areas. The average productivity of seagrass beds has been estimated at 2.7 g DW m^{-2} day^{-1}, making them among the most productive plant communities on Earth (Duarte and Chiscano 1999). These highly productive and complex habitats provide a variety of ecologically and economically important services. For example, they attenuate waves and protect coastal areas from erosion and storm surge, stabilize sediments, provide nursery grounds for commercially and ecologically important finfish and shellfish, and take up nutrients and pollutants (Orth et al. 2006; Koch et al. 2009; Barbier et al. 2011). Globally, seagrass beds are in decline (Hauxwell et al. 2001; Orth et al. 2006; Hughes et al. 2009; Waycott et al. 2009). Their loss is attributed to habitat loss as a result of anthropogenic activities, eutrophication, sedimentation, overgrowth by macroalgae, and disease, particularly "wasting disease" caused by the pathogenic fungus *Labyrinthula zosterae.*

The seagrass plants that form the foundation for these communities are angiosperms but are not related to terrestrial grasses. Their taxonomy is in flux, but it is clear that they represent a number of different lineages of plants that evolved on land but have invaded marine and estuarine habitats. Most seagrasses share a number of adaptations that allow them to live in saline waters (Hemminga and Duarte 2000; Larkum et al. 2006). They tend to have flexible, strap-shaped or oval leaves with relatively low lignin content; they lack stomata in their epidermal cells; they can uptake nutrients from the water as well as the sediments; they are anchored to the substratum by rhizomes; they can transport oxygen from their leaves to their rhizomes and detoxify sulfide, which allows them to inhabit anoxic sediments; and many have carbon concentrating mechanisms that increase the efficiency of carbon dioxide uptake (Raven et al. 2008). Like terrestrial plants, they can reproduce vegetatively, asexually, or sexually, but unlike terrestrial plants, pollination and seed production occurs underwater.

The distribution of seagrasses is affected by several abiotic environmental factors. Light can be important in limiting the depth to which seagrasses can grow. Typically, seagrasses are limited to depths with >10% of surface light levels (Duarte 1991). All but two seagrass genera, *Phyllospadix* and *Amphibolis*, are restricted to soft sediment habitats, which have limited wave motion (Williams and Heck 2001). Seagrasses tend to be nitrogen, rather than phosphorus, limited (Short 1987) and in many areas, sediment nutrients are more limiting for seagrasses than light (Hughes et al. 2009). However, nutrient over-enrichment is detrimental to seagrass communities; high concentrations of ammonia can be toxic to seagrasses and nutrients stimulate the growth of algae that compete with seagrasses, and epiphytes that foul their surfaces, shading leaves (Burkholder et al. 2007).

Although the diversity of seagrass species is low (~60 species) relative to terrestrial plants (den Hartog and Kuo 2006), these species form the foundation of complex and dynamic communities (Heck et al. 2008). Biological interactions including herbivory, predation, detritivory, competition, epiphytism, and mutualisms are all important forces structuring seagrass communities (Orth and Van Montfrans 1984; Jernakoff et al. 1996; Williams and Heck 2001; Heck and Orth 2006; Heck and Valentine 2006; Valentine and Duffy 2006; van der Heide 2012), but their relative importance varies among seagrass species, sites, and biogeographic areas.

Seagrasses provide both food and habitat for many other organisms, which often include commercially important finfish and shellfish (Beck et al. 2001). Epiphytic microbes including bacteria, fungi, and protists as well as macroalgae and invertebrates can inhabit seagrass leaves. As much as half of the biomass of primary producers in seagrass beds can consist of epiphytic algae (Borowitzka et al. 2006), and light limitation due to epiphytes can be an important factor leading to the decline of seagrass beds (Orth and Montfrans 1984). The herbivores that consume seagrasses range in size from small invertebrate mesograzers to megaherbivores such as water-fowl, turtles, manatees, and dugongs (Thayer et al. 1984; Valentine and Duffy 2006). Many of the herbivores in seagrass beds are also consumers of the epiphytes that grow on the seagrasses, and thus some grazers may provide a net benefit to the plants (Orth and Montfrans 1984). The importance of herbivores in the movement of carbon and nutrients through seagrass communities varies widely; estimates of the percentage of seagrass biomass moved into food webs by herbivores range from 3% to 100% (Heck and Valentine 2006). Thus, in some communities, more of the biomass generated by seagrasses is consumed by detritivores than herbivores.

Many of the ecological interactions between seagrasses and the organisms that occur in seagrass beds are mediated by the natural products produced by seagrasses (Sieg and Kubanek 2013), and the production of those compounds can be influenced by local abiotic and biotic conditions. Here, we review the natural products that are known to be produced by seagrasses, with an emphasis on phenolic compounds, which are the best studied of the seagrass secondary metabolites. We then examine the factors that regulate the amounts of these compounds in the plants and describe the role of these metabolites in interactions between the seagrasses producing these compounds and the organisms that feed on, compete with, grow on, and infect them.

4.2 BIOACTIVE NATURAL PRODUCTS ISOLATED FROM SEAGRASSES

4.2.1 PHENOLIC COMPOUNDS

The best-known allelochemicals associated with seagrasses are the phenolic compounds, which include simple phenols, such as phenolic acids, and more complex compounds such as flavones, condensed tannins, and lignins. Most seagrass genera, with the exception of *Halophila*, *Zostera*, and *Phyllospadix*, produce specialized "tannin cells" (Den Hartog and Kuo 2006) where tannins, and possibly other phenolic compounds, are thought to be produced or stored (McMillan 1984; Dumay et al. 2004). In terrestrial plants, phenolic compounds are produced via the shikimic acid and phenylpropenoid pathways from the amino acids phenylalanine, tyrosine, and tryptophan (reviewed by Arnold and Targett 2002). However, in seagrasses, the synthesis of these compounds and the regulation of their production have not been well studied.

Phenolic acids are metabolites that combine a hydroxylated aromatic ring with a carboxylic acid functional group. The phenolic compounds in seagrasses fall into two groups, the hydroxycinnamic acids (Figure 4.1) and the hydroxybenzoic acids (Figure 4.2). Some of the more commonly reported hydroxycinnamic acids reported from seagrasses include *para*-coumaric acid (**1**), caffeic acid (**2**), *trans*-ferulic acid (**3**), chicoric acid (**4**), and rosmarinic acid (**5**) (Zapata and McMillan 1979; Dumay et al. 2004; Haznedaroglu and Zeybek 2007; Achamlale et al. 2009b; Nuissier et al. 2010; Wang et al. 2012; Grignon-Dubois and Rezzonico 2013). Among the more commonly reported hydroxybenzoic acids reported from seagrasses are *para*-hydroxybenzoic acid (**6**), protocatechuic acid (**7**), gentisic acid (**8**), vanillic acid (**9**), and gallic acid (**10**) (Quackenbush et al. 1986; Agostini et al. 1998; Dumay et al. 2004). Sulfated phenolic *p*-coumaric acid (**1**)

Caffeic acid (**2**)

Ferulic acid (**3**)

Chicoric acid (**4**)

Rosmarinic acid (**5**)

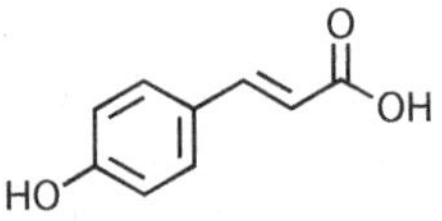

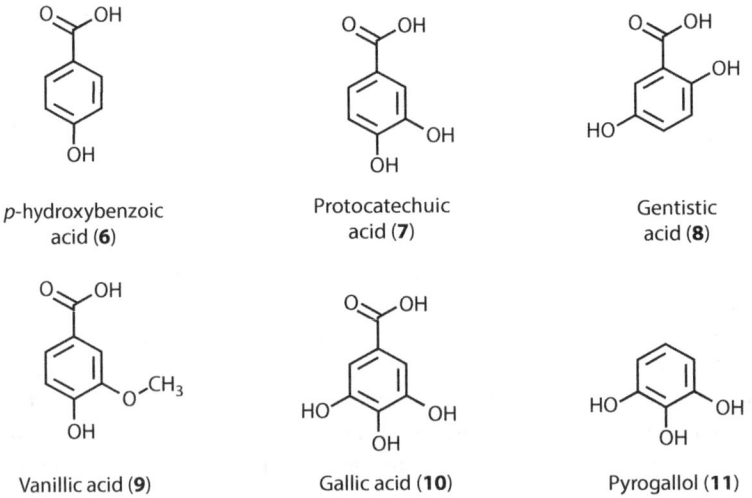

FIGURE 4.1 Hydroxycinnamic acids produced by seagrasses.

p-hydroxybenzoic acid (**6**)

Protocatechuic acid (**7**)

Gentistic acid (**8**)

Vanillic acid (**9**)

Gallic acid (**10**)

Pyrogallol (**11**)

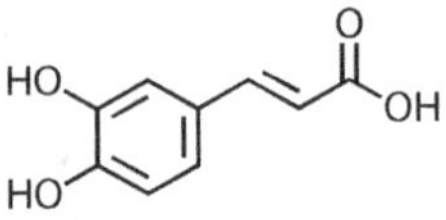

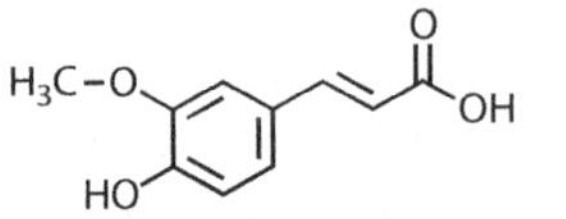

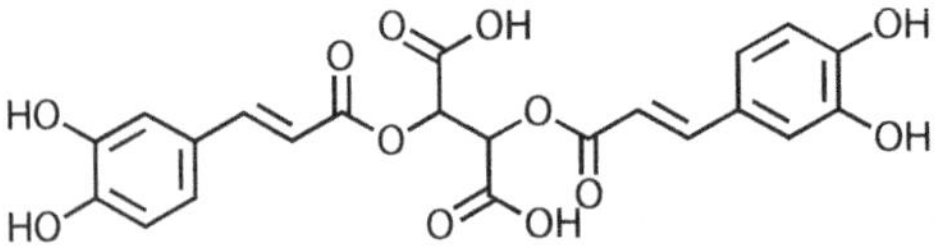

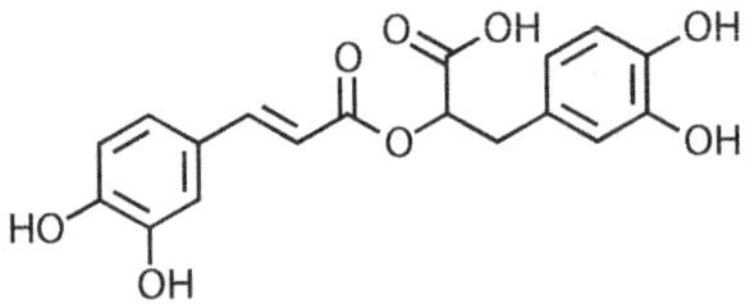

FIGURE 4.2 Hydroxybenzoic acids and related compounds produced by seagrasses.

acids (Figure 4.3) can be found in most seagrass genera (McMillan et al. 1980; Todd et al. 1993; Achamlale et al. 2009a). Phenolic acids tend to be biologically active. They can function as herbivore deterrents and inhibit the growth of many organisms including microbes, pathogens, and fouling organisms (Arnold and Targett 2002; Sieg and Kubanek 2013). These functions are described in more detail in Section 4.4. Lignins are polymers of phenolic acids that are often highly methoxylated (Hagerman and Butler 1991). In *Zostera marina*, lignins are more abundant in roots and rhizomes than in leaves (Klap et al. 2000). They are generally thought to serve a structural function but could affect herbivore food preferences by increasing the toughness of tissues (but see Arnold et al. 2014).

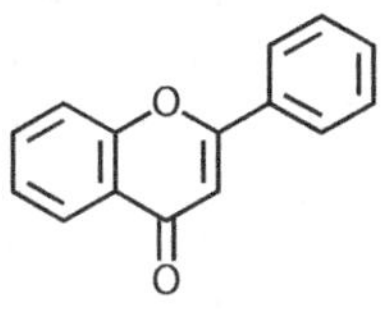

Flavone (**12**)

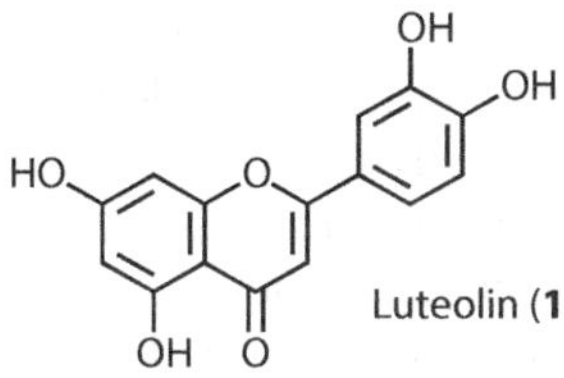

Luteolin (**13**)

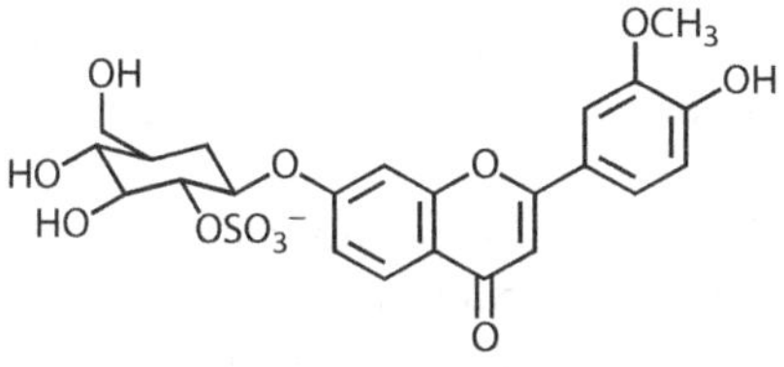

Thalassiolin B (**14**)

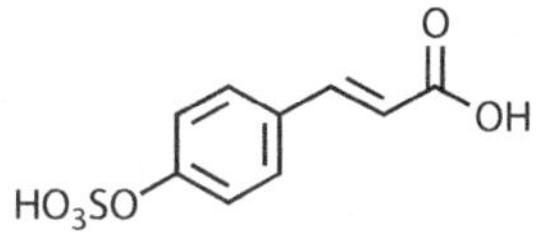

Zosteric acid (**15**)

FIGURE 4.3 Other representative phenolic compounds from seagrasses. Shown are flavone (**12**); a simple flavonoid, luteolin (**13**); a sulfated flavone glycoside, thalassiolin B (**14**); and a sulfated phenolic acid, zosteric acid (**15**).

Flavonoids (Figure 4.3) are pigments that are based on a flavone (2-phenyl-1,4 benzopyrone) backbone (**12**) (Harborne 1991). Seagrasses, especially *Halophila* spp., produce more than a dozen flavonoid compounds and several flavone glycosides (Meng et al. 2008; Bitam et al. 2010; Gavin and Durako 2011). Sulfated flavones have been reported from members of the genera *Zostera*, *Enhalus*, *Thalassia*, and *Halophila* (McMillan 1984, 1986; Rowley et al. 2002; Regalado et al. 2009; Garateix et al. 2011). Flavonoids can function as accessory pigments or provide protection from UV light, but they may also have secondary functions such as inhibiting the growth of microbes or protecting plants from herbivory (Harborne and Williams 2000). In seagrasses, flavonoids have been shown to have antioxidant (Gavin and Durako 2012) and antifungal (Jensen et al. 1998) properties. Condensed tannins, also known as proanthocyanidins, are oligomers and polymers of flavonoids (Hagerman and Butler 1991) that range in size from 400 to 3000 Da (McMillan 1984). In terrestrial plants, condensed tannins can be herbivore feeding deterrents (Hagerman and Butler 1991); however, their functions in seagrasses have not been clearly demonstrated.

4.2.2 Measurements of Phenolic Compounds

A variety of methods are used to quantify phenolic compounds in seagrasses, and each method produces different types of information about the composition of phenolic constituents or their activity. One of the simplest and least equipment-intensive involves the use of colorimetric assays based on the Folin–Denis method (Harrison 1977; Arnold et al. 2012; Sánchez-Rangel et al. 2013). Although these methods often report phenolic concentrations as amounts or percentages of phenolic compounds per mass of plant, they do not actually measure the amounts of compounds that the plants contain. Instead, they provide a rough measure of phenolic activity by measuring redox-based activity (Appel et al. 2001). In Folin assays, phosphomolybdic and phosphotungstic acid reagents react with hydroxylated aromatic compounds, which include phenolic compounds, as well as other metabolites such as ascorbic acid, phenylalanine, and tyrosine. The complexes formed by the reagent and reactive compounds produce a blue color under alkaline conditions and the intensity of the color, measured by absorbance with a spectrophotometer, is used as a proxy for phenolic activity. Other colorimetric methods are used to measure condensed tannins, which include the proanthocyanidin and vanillin methods. The proanthocyanidin method is often preferred because it tends to produce more repeatable results (Dudgeon et al. 1994).

There are several important caveats that should be considered when interpreting the results of Folin assays (Van Alstyne 1994; Stern et al. 1996; Appel et al. 2001). Some compounds that react with the reagents are not phenolic compounds; therefore, assays may provide an overestimate of phenolic activity. Reactions between reagents and compounds are also affected by the number of hydroxyl groups and their locations on the aromatic ring (Appel et al. 2001). As a result, the same amounts of two phenolic compounds can produce different absorbances, and different amounts of two compounds could produce the same absorbance. Thus, differences in absorbances in Folin assays can occur because the amounts of phenolic compounds differ or because the types of phenolic compounds differ. This creates a problem for selection of a standard for Folin assays because most seagrasses contain a mix of different phenolic compounds that react differently with the reagents. Many studies use a single phenolic acid as a standard, such as caffeic or ferulic acid, as these compounds are known to occur in seagrasses. However, a better, but much more methodologically difficult method is to use phenolic compounds that have been extracted from the population of plants being studied (Appel et al. 2001).

For measurements of individual compounds, such as phenolic acids and flavonoids, chromatographic methods are often used; however, these methods are more complex than colorimetric assays. A few studies have used gas chromatography to measure the concentrations of derivatized phenolic acids (Nomme and Harrison 1991; Citová et al. 2006), but more commonly underivatized phenolic acids are quantified directly by high performance liquid chromatography (Zapata and McMillan 1979; Posey 1988; Vergeer and Develi 1997; Citová et al. 2006; Arnold et al. 2012; Ferrat et al. 2012). Although these assays provide information about the distributions and concentrations of specific phenolic acids, they are more labor-intensive and time-consuming to perform, and they require more expensive equipment and reagents. They are also not well suited for measuring complex mixes of polymers that can be difficult to separate, such as condensed tannins.

4.2.3 OTHER NATURAL PRODUCTS

In addition to phenolic compounds, seagrasses are reported to produce a number of other natural products that have biological activity. These compounds include steroids

(Gillan et al. 1984; Kontiza et al. 2006), terpenes (DellaGreca et al. 2000; Carbone et al. 2008; Kontiza et al. 2008; Hammami et al. 2013), glycosides (Mohammed et al. 2014), sulfated polysaccharides (Aquino et al. 2004; Silva et al. 2012), pectins (Zaporozhets et al. 1991a, 1991b; Gloaguen et al. 2010), glycolipids (Milkova et al. 1994), triglycerols (Yuvaraj et al. 2012), fatty acids (Gillan et al. 1984), and numerous volatile compounds (Milkova et al. 1994; Kawasaki et al. 1998; Pino and Regalado 2010). However, Heglmeier and Zidorn (2010) point out that many studies that have reported metabolites from *Posidonia oceanica* did not use rigorous methods for identifying these compounds; this was particularly true if the studies were older. Therefore, some of these compounds may not have been properly identified or may be artifacts of the methods used to extract them. This is undoubtedly true for other seagrass species as well.

4.3 SPATIAL, TEMPORAL, AND ENVIRONMENTALLY-INDUCED VARIATION IN THE CONCENTRATIONS OF SEAGRASS PHENOLIC COMPOUNDS

The distribution of phenolic compounds in seagrasses varies greatly among tissues of a given plant, seasonally within species at given sites, and among sites, as well as among species.

4.3.1 WITHIN-PLANT VARIATION

Within individual plants, there can be differences in the types and amounts of phenolic compounds between leaves and rhizomes. In *Zostera marina*, the combined concentrations of caffeic acid and rosmarinic acid are usually higher in leaves than rhizomes, although these concentrations change seasonally. At some times of the year, concentrations in rhizomes are higher than those in leaves (Ravn et al. 1994). Likewise, concentrations of chicoric acid in the leaves of *Cymodocea nodosa* detritus are consistently greater than in rhizomes at five sites in Europe (Grignon-Dubois and Rezzonico 2013). Total phenolic concentrations do not differ among leaves, rhizomes, and shoots of *Thalassia testudinum* in the Virgin Islands; however, condensed tannin concentrations are higher in rhizomes than in older leaves, although not younger ones (Arnold et al. 2008). The higher concentrations of phenolic compounds in leaves relative to rhizomes may occur because leaves are accessible to herbivores. Rhizomes also contain more lignin than leaves (Klap et al. 2000) and thus may be better defended structurally.

In vascular plants, younger leaves are generally a carbon sink (net importer of carbohydrates), while older leaves tend to be carbon sources that export carbohydrates (Schultz et al. 2013). The availability of imported carbohydrates in younger leaves may allow these tissues to produce higher levels of phenolic compounds than older leaves. In seagrasses, phenolic concentrations tend to be higher in younger than older leaves. For example, total phenolic concentrations are higher in intermediate leaves than adult leaves in *Posidonia oceanica* from Europe (Dumay et al. 2004) and higher in younger leaves than older leaves of *Thalassia testudinum* from

Florida, USA (Steele and Valentine 2012). In *P. oceanica* from France, the ratios of phenolic acids in adults and intermediate leaves are similar to one another but differ from those found in the sheaths of the plant. Sheaths contain higher concentrations of 4-hydroxybenzoic acid as well as a mix of ferulic acid and an unidentified phenolic acid but have lower concentrations of caffeic acid than leaves (Dumay et al. 2004). In *P. oceanica* from the Aegean Sea, the distribution of phenolic acids differs between young and mature leaves (Haznedaroglu and Zeybek 2007). Mature leaves have high amounts of gentisic acid (394 µg/g dry weight) and low amounts (less than 40 µg/g dry weight) of other phenolic acids, whereas young leaves have less than half as much gentisic acid (147 µg/g dry weight), a similar amount of chicoric acid (139 µg/g dry weight), and lower amounts of other phenolic acids. Concentrations of condensed tannins also tend to be higher in younger than older leaves of *T. testudinum* and *Halodule wrightii* from the U.S. Virgin Islands (Arnold et al. 2008; Steele and Valentine 2012).

4.3.2 SEASONAL CHANGES

Concentrations of phenolic compounds in seagrass leaves change seasonally; however, there is little consistency in the patterns of seasonal variation among studies. For example, total phenolic compounds in *Z. marina* in British Columbia peak in early September and are lowest in the spring (Harrison and Durance 1989). However, the combined concentrations of rosmarinic and caffeic acid in the leaves of *Z. marina* from Denmark peak in the spring and are lowest in summer and fall, contrasting with concentrations of these compounds in the rhizomes, which do not have a strong seasonal pattern (Ravn et al. 1994). In *P. oceanica* from the Mediterranean, phenolic activity in leaves is highest in December and lower at other times of the year in intermediate leaves but is similar in concentration throughout the year in adult leaves and in sheaths (Boumaza et al. 2014).

4.3.3 AMONG-SITE AND ENVIRONMENTAL CORRELATES

Qualitative and quantitative differences in phenolic compounds have often been measured within and among sites to determine if phenolic compounds change across spatial scales or are correlated with differences in environmental factors among sites. Seawater pH appears to be one of the biggest determinants of the total amounts of phenolic compounds in seagrasses. Arnold and coworkers (2012, 2014) measured phenolic concentrations in seagrasses at two sites with natural gradients in pH, one near a groundwater discharge site (Queensland, Australia) and one near an undersea volcanic vent that was discharging CO_2 off the coast of Italy. At both sites, seawater pH was much lower close to the discharge area and decreased with distance from it; plants close to the discharge areas produced lower concentrations of phenolic compounds than plants that were farther away. In Australia, *Zostera muelleri* near the discharge site had phenolic concentrations over 87% lower than plants located approximately 30 m away. There were no differences in the carbon and nitrogen concentrations of the plants, but condensed tannins, caffeic acid, and rosmarinic acid were absent from the plants closer to the discharge. The gallic acid concentrations in plants close to the discharge were also a fraction of those found in plants that were more distant from it (Arnold et al. 2014). At the volcanic vent site, total phenolic compounds in *C. nodosa* were reduced by 14%, proanthocyanidins were reduced by 24%, and total phenolic acids were reduced by 49% in plants close to the vent relative to plants about 100 m away (Arnold et al. 2012). The concentrations of ferulic acid, coumaric acid, vanillin, acetovillione, and syringaldehyde and 4-hydroxybenzoic acid in the plants differed, but concentrations of gallic acid did not. Experimental decreases in the pH at an estuarine field site in Maryland, USA caused decreases in the concentrations of proanthocyanidins in the rhizomes, but not the shoots, of *Ruppia maritima*. However, no changes were seen in the concentrations of ferulic acid, coumaric acid, vanillic acid, acetovillone, or syringaldehyde plus 4-hydroxybenzoic acid. In contrast, concentrations of both total reactive phenolics and proanthocyanidins decrease dramatically in *Potamogeton perfoliatus* when pH is experimentally reduced (Arnold et al. 2012).

In some cases, the concentrations of nutrients in seawater correlate with phenolic concentrations in seagrasses, although both negative and positive correlations have been observed. For example, concentrations of proanthocyanidins and flavonols were significantly higher in *P. oceanica* adjacent to a fish farm than in plants 40 and 130 m further away (Cannac et al. 2006). Interstitial water near the farm has significantly higher concentrations of ammonia and phosphate. In contrast, Buchsbaum and coworkers (1990) measured total phenolic concentrations in *Z. marina* grown in two mesocosms, one containing mud and the other containing sand. *Zostera* grown in the tank with mud, which had higher concentrations of ammonia and phosphate, had about half the concentrations of phenolic compounds as plants grown in the tank with sand. They also found a strong negative correlation between plant tissue nitrogen concentrations and phenolic concentrations. However, not all studies have found similar correlations between seawater nutrients and phenolic concentrations in seagrasses. Total phenolic concentrations in the leaves of *P. oceanica* from an area in Algeria with high discharge of nutrient-rich urban effluent and a nearby site without discharge do not differ (Boumaza et al. 2014).

Other abiotic environmental factors have been shown to affect concentrations of phenolic compounds in seagrasses. Total phenolic activity in *Z. marina* is negatively correlated with temperature, positively correlated with light availability, but unaffected by salinity ranging from 10–30 (Vergeer et al. 1994). Likewise, McKone and Tanner (2009) found that the combined concentration of ferulic, gallic, and rosmarinic acid in *Z. marina* is not correlated with salinity unless the plants are infected by the pathogen *Labyrinthula zosterae*. In infected plants, the combined concentrations of the three phenolic acids increase with increasing salinity.

Interactions with other organisms, especially with grazers, may also affect concentrations of phenolic metabolites in seagrasses. However, how plant responses to grazers can depend on the grazer and time of year and can be different among seagrass species and even among different tissues within the same plant. For example, when *T. testudinum* was mechanically damaged in a way that simulated grazing by parrotfish, concentrations of condensed tannins increased significantly in older leaves, but not young leaves (Arnold et al. 2008). However, direct grazing by urchins (*Lytechinus variegatus*) did not result in an increase in total phenolic compounds or condensed tannins in *Thalassia* leaves. It did, however, cause an increase in condensed tannins, but not total phenolic activity, in plant rhizomes. Condensed tannins did increase in *T. testudinum* that was caged with either 0, 4, or 10 *L. variegatus* in a seagrass meadow off St. Joseph's Bay, Florida (Steele and Valentine 2012). Concentrations were higher in plants from cages with 10 urchins relative to plants from cages with 0–4 urchins (Steele and Valentine 2012). In these experiments, although total phenolic acid concentrations differed among the treatments, they were much more affected by season (summer versus fall) than by grazers. Grazers did not affect the overall concentrations of *p*-hydroxybenzoic acid in plants; however, in urchin-damaged plants, concentrations of the compound were higher at the site of damage and above the damaged area on the leaf than below it. The authors attributed this effect to the disruption of the plants' vascular tissues. They hypothesized that the damage prevented the movement of photosynthates from the shoots to the roots, and therefore plants then converted these carbohydrates to phenolic compounds. Steele and Valentine (2012) also measured concentrations of phenolic compounds from *H. wrightii* from the same cages. In the summer, condensed tannin concentrations were higher in plants exposed to more urchins, but the opposite effect was seen in the fall. Total phenolic concentrations and the concentrations of individual phenolic acids were unaffected by the presence of urchins in both summer and fall.

To date, there is no evidence that waterborne metabolites produced by grazing on conspecifics induce the production of phenolic compounds in neighboring plants. When *T. testudinum* was exposed to waterborne cues from conspecifics being eaten by urchins, there were no differences in condensed tannins or total phenolic compounds in the leaves of exposed and control plants (Arnold et al. 2008). There were also no changes in concentrations of condensed tannins or phenolic activity when *T. testudinum* was exposed to jasmonic acid (Arnold et al. 2008), a compound that is known to induce defense responses in many terrestrial plants (Ballaré 2011).

It is frequently assumed that the production of defensive compounds in plants incurs a metabolic cost that results in reduced growth or reproductive output. Evidence for such a cost for phenolic compounds in seagrasses is inconsistent. Arnold and coworkers (2008) reported a 1–2 mm/day reduction in the growth rate of *T. testudinum* leaves that contained increased condensed tannin concentrations as a result of simulated parrotfish grazing relative to growth rates of unwounded control plants. However, Steele and Valentine (2012) reported no differences in the growth rates of *T. testudinum* that were producing increased concentrations of condensed tannins in response to urchin grazing in caging experiments.

Competitors can induce the production of allelopathic compounds in terrestrial plants (Kegge and Pierik 2010; Meiners et al. 2012). However, this phenomenon has not been well studied in seagrasses. Most studies have relied on measuring differences in phenolic compounds at sites with and without potential competitors. Three studies found no difference in the concentration of phenolic compounds in plants from areas colonized by the invasive green alga *Caulerpa taxifolia* and areas where the alga was absent (Cuny et al. 1994; Agostini et al. 1998; Dumay et al. 2004). Comparisons have been made among sites where *Caulerpa* and *Posidonia* were present but there was limited contact between them, and sites where *Caulerpa* stolons and *Posidonia* rhizomes were intertwined (Dumay et al. 2004). Concentrations of an ester (methyl 12-aceto-oxyricinoleate) and a mixture of compounds known to contain ferulic acid were highest at the site where plants had the most contact and lowest at the sites where they had no contact (Dumay et al. 2004). The number of tannin cells in the leaves of these plants followed a similar trend, suggesting that the production of these compounds may be related to the interaction between *Posidonia* and *Caulerpa*, although further experimental work is needed to determine conclusively whether this is the case.

4.4 ECOLOGICAL INTERACTIONS INVOLVING NATURAL PRODUCTS

4.4.1 FEEDING DETERRENCE

Several studies have examined whether seagrasses contain allelochemicals deterrent to grazers by incorporating crude extracts of the plants into diets composed of agar and a palatable seaweed or by coating the extract dissolved in a carrier solvent onto a palatable seaweed and offering it to grazers. For example, Paul and coworkers (1990) coated pieces of the green alga *Enteromorpha clathrata* with extracts of *Enhalus acoroides* and *Halophila minor* at concentrations similar to those found in these plants and offered the coated algae along with pieces of solvent-coated algae (controls) to juvenile and adult rabbitfish (*Siganus argenteus*). *Enhalus* extracts deterred feeding by adult rabbitfish relative to controls, but not juvenile rabbitfish, and *Halophila* extracts were not deterrent to either adults or juveniles. Thus, the effects of these extracts as deterrents depended on the plant species and the age of the herbivore. Similarly, when *P. oceanica* extracts were incorporated into an *Ulva*-agar diet, they deterred an assemblage of fish in the field and three species of urchins, but not the gastropod *Cerithium vulgatum* (Vergés et al. 2007a).

Phenolic compounds can be potent anti-herbivore defenses. In a study examining the effects of pH on seagrasses near and far from a groundwater discharge site, Arnold and coworkers (2014) found that lower seawater pH caused a dramatic reduction in phenolic concentrations of about 87% (see previous section). Black rabbitfish (*Siganus fuscescens*) offered a choice of *Z. muelleri* from the two areas consumed significantly more of the low-phenolic, high-lignin plants growing close to the low pH site. These data suggest that the presence of phenolic compounds in plants growing further from the discharge site made them less palatable to the grazers, although it is possible that other changes in the plants that occurred as a result of the change in seawater chemistry might have affected the preferences of the fish. Although very large differences in phenolic concentrations over small spatial scales can affect herbivore preferences, smaller but still significant differences in concentrations over larger spatial scales may not impact herbivores. When sea urchins (*L. variegatus*) were offered *H. wrightii* and *T. testudinum* from two sites in Florida, USA where phenolic concentrations in these seagrasses differed, urchin preference for plants from the two sites did not differ (Steele and Valentine 2012).

Concentrations of phenolic compounds in seagrasses are also correlated with food preferences of herbivorous birds. The relatively low concentrations of phenolic compounds in *Z. marina* relative to concentrations in other marsh plants may make *Zostera* the preferred food of Canada geese (*Branta canadensis*) (Buchsbaum et al. 1984). Feeding by geese is significantly and negatively correlated with the phenolic concentrations in marsh plants. Geese are also deterred by the addition of several phenolic acids when coated onto a palatable food, including ferulic acid and *p*-coumaric acid, which are known to be present in some seagrasses (Zapata and McMillan 1979; Quackenbush et al. 1986).

Preferences for plant parts can be herbivore-specific, and the effects of grazer deterrents differ not only among plant species, but also among tissues of individual plants. For example, when extracts of young *Z. marina* leaves collected in British Columbia, Canada, were added to flasks containing amphipods and decaying *Z. marina* leaves, the amphipods reduced feeding relative to controls; however, the addition of extracts of old leaves had no effect on feeding (Harrison 1982). Sea urchins (*Paracentrotus lividus*) from the Spanish Mediterranean coast prefer older *P. oceanica* leaves over younger leaves, but fish (*Salpa salpa*) prefer young leaves over older ones (Vergés et al. 2011). The urchins also prefer inflorescences over leaves (Vergés et al. 2007b). The patterns of preferences among tissues in *P. oceanica* follow the predictions of optimal defense theory (Rhoades 1979), which states that more valuable tissues should be better protected than less valuable ones. Younger leaf tissue may be more valuable than older leaf tissue because the plants have invested resources in their production but have received little benefit from this investment. Inflorescences should be high-value tissues because of their direct effect on plant fitness. The concentrations of phenolic compounds within *P. oceanica* also follow this prediction. Young leaves have significantly higher concentrations of phenolic compounds than older leaves (Vergés et al. 2011) and, at two out of the three sites, inflorescences had significantly higher concentrations than leaves (Vergés et al. 2007b).

Plant tissues differ in morphology, which can also affect herbivore food choices. Vergés and coworkers (2011) incorporated freeze-dried, ground tissues of old and young *P. oceanica* leaves into agar-based foods, eliminating some of the morphology of the plant tissues. Urchins showed no preference when given a choice between the two diets. However, when crude extracts were incorporated into agar-based diets, urchins ate more of the diet containing extracts from old leaves, indicating that both plant chemistry and fine-scale differences in the morphology of the two leaf types affected their preferences. When similar experiments were conducted comparing preferences of urchins for leaves versus inflorescences, urchins consumed about twice as much freeze-dried, ground leaf tissue relative to inflorescences tissue; however, they consumed about 17 times more leaf tissue than inflorescence tissue when crude extracts were incorporated into agar-based diets. Thus, although plant chemistry was a significant determinant of food preference for urchins, structure was more important (Vergés et al. 2007b).

4.4.2 Signaling Herbivore Offenses

In addition to seagrasses producing induced chemical defenses in response to grazing, seagrass environments may trigger induced offenses in some grazers. *Lacuna* is a genus of small snails that are prominent members of marine temperate communities worldwide. In the macrophyte systems where they occur, they are typically the most abundant grazers (Miloslavich et al. 2013). *Lacuna* live for 6–12 months and are reproductive and abundant year-round. When in eelgrass (*Z. marina*) meadows, rather than feeding on eelgrass tissue, *Lacuna* feed primarily on diatom epiphytes (Padilla 1998, 2001), which cover the eelgrass and can reduce its productivity (Nelson 1994; Nelson and Waaland 1997; Peterson et al. 2007); however, in Japan, some species of *Lacuna* directly consume seagrasses (Kajihara et al. 2010). When on macroalgae, *Lacuna* feed on algal tissues and can cause extensive damage (Padilla 1998, 2001; Molis et al. 2010; Krumhansl and Scheibling 2011; Dubois and Iken 2012). Unusual among gastropods, *Lacuna* have phenotypically plastic tooth morphologies that are induced by the food or the environment where they are found. Tooth shapes that are produced are those that are most effective for consuming each kind of food (Padilla 1984, 1987, 1989, 1998); pointed teeth are produced when animals consume any fleshy macroalga (independent of type or species) and blunt-shaped teeth are produced in response to consuming epiphytes found on eelgrass (Padilla 1998, 2001). New teeth are constantly produced at the distal end of the radular ribbon at a constant rate (3 rows/day) while old teeth are shed anteriorly throughout the life of the snail (Padilla et al. 1996). Tooth morphology cannot be changed once a tooth is initially fabricated, and it takes just over three weeks for a newly produced tooth to be used to graze (Padilla et al. 1996), resulting in a lag time between induction and having a functional morphology, which could limit the adaptive value of this plasticity (Padilla and Adolph 1996; Grünbaum and Padilla 2014).

When *Lacuna variegata* is given macroalgae to consume but is held in water containing diatom covered eelgrass, they produce the eelgrass bed tooth morphology (Padilla 2001). When the food type and water-borne cues are reversed, snails still produce the eelgrass bed tooth morphology (Padilla 2001). Thus, for *L. variegata*, the change to blunt teeth appears to be triggered by a chemical signal associated with eelgrass or diatoms. This result precludes the possibility that morphology is use-induced or results from a direct mechanical feedback of feeding. For *Lacuna vincta*, the patterns are somewhat different; diet appears to play a stronger role in controlling morphology for most individuals (Padilla 2001). In addition, both species produce pointed teeth when fed macroalgae in agar and blunt-shaped teeth when fed diatom covered eelgrass in agar (Padilla, unpublished data), again indicating that this switch is not mechanically triggered. Taken together, these data suggest three possibilities: (1) there is a single chemical cue produced by eelgrass or the eelgrass environment that induces a change in tooth morphology for *L. variegata*, but it does not induce change in *L. vincta*, (2) there is a single chemical cue that is used by both snails, but *L. vincta* requires a higher concentration to induce the alternative phenotype, or (3) different chemical cues induce changes in tooth morphology in the two species. Preliminary experiments feeding both species of *Lacuna* a neutral diet (Romaine lettuce) suggest that these snails produce pointed teeth in the absence of natural chemical cues (Yee et al. 2014), and that there is a chemical cue associated with eelgrass or the eelgrass bed environment that triggers the production of blunt-shaped teeth. More experimentation is needed to determine the precise chemicals that are responsible for controlling this inducible offense.

4.4.3 INTERACTIONS WITH MARINE MICROBES AND FOULING ORGANISMS

Seagrasses live in a virtual soup of other organisms and can be subject to fouling or infection by bacteria, fungi, microalgae, macroalgae, invertebrates, and viruses (Engel et al. 2002). Because they live in nearshore environments where nutrient levels and concentrations of dissolved and particulate organic matter in the water tend to be high, there is abundant food for heterotrophic microorganisms, which in turn support the growth of large numbers of consumers. These organisms can live on and around seagrasses and their impacts can range from being beneficial, as in the case of mutualists, to neutral, to extremely detrimental, as in the case of pathogens (Engel et al. 2002). Although relatively little is known about the role of natural products produced by seagrasses in mediating interactions between the plants and neighboring microorganisms, it is assumed that they might be used to encourage the growth of beneficial species while inhibiting the growth of detrimental ones.

Extracts or specific compounds produced by seagrasses are frequently assayed for their ability to inhibit bacterial, viral, macroalgal, and fungal growth, and even the growth of cancer cells (Bernard and Pesando 1989; Premnathan et al. 1992, 1994; Ballesteros et al. 1992; Bhosale et al. 2002; Folmer et al. 2010; Laabir et al. 2013; Mohammed et al. 2014). These studies are typically used to screen extracts and compounds for the development of commercial products, not to determine the ecological importance of the measured activity. Often the amounts of compounds used in assays are not ecologically relevant, the methods do not mimic how the compounds would be deployed on a plant surface or in seawater, and the organisms used in the screening process are not found in the marine environment. Frequently, these studies examine the activity of compounds extracted from plants without determining whether the compounds would actually be released by the plants and the concentrations that target organisms would experience in the field are not determined (Henrikson and Pawlik 1994). Therefore, whether the compounds that are responsible for this activity have similar functions in nature is unknown.

With the exception of studies that have examined the role of natural products in inhibiting the growth of pathogenic fungi in the genus *Labyrinthula* (see the following), there are only a handful of studies that have assayed the effects of seagrass metabolites on marine microorganisms in an ecologically-relevant way. One of the first studies to look at the inhibition of the growth of microorganisms found that methanolic extracts of *Z. marina* that had recently been washed up on a beach following a storm inhibited the growth of several benthic diatoms species, marine flagellates, and a terrestrial bacterium (Harrison and Chan 1980). Similar assays with older beach wrack found that the inhibitory activity decreased as the eelgrass leaves decayed. Cinnamic acid isolated from *Z. marina* inhibits the growth of the fouling bacterium *Acinetobacter* sp. in a dose-dependent manner (Todd et al. 1993). Aging also plays a role in the ability of *R. maritima* to inhibit the growth of bacteria (Bushmann and Ailstock 2006). Methanolic extracts from field-collected plants from the Chesapeake Bay area inhibit the growth of 12 of 12 gram-positive bacterial species and three of 11 gram-negative species. However, this inhibitory effect is only present in plants collected in the spring; extracts from plants collected in the fall have little effect on bacterial growth.

Tropical and subtropical seagrasses also produce metabolites that have antimicrobial activity. Hydrophilic extracts of *E. acoriodes* from Guam inhibit the growth of the bacterium *Pseudoalteromonas bacteriolytica*, which causes red spot disease in kelps; however, the bacterium's growth is not inhibited by lipophilic or hydrophilic extracts of *Halophila minor* (Puglisi et al. 2007). Hydrophilic extracts of both seagrasses also inhibit the growth of the marine saprophyte *Schizochytrium aggregatum* and lipophilic extracts inhibit the growth of the saprophytic ascomycete *Dendryphiella salina*. Growth of the saprophytic oomycete *Halophytophthora spinosa* is inhibited by lipophilic extracts of *Enhalus* and hydrophilic extracts of *Halophila*. These results suggest that these two seagrasses produce several different bioactive metabolites that selectively inhibit the growth of several saprophytes, which might prevent saprophytes from infecting healthy seagrass tissues. Interestingly, neither the hydrophilic nor the lipophilic extracts of either seagrass is an effective inhibitor of the growth of the pathogenic ascomycete *Lindra thalassiae* (Puglisi et al. 2007); thus, the ability to resist the effects of the secondary metabolites that inhibit the growth of saprophytic organisms may contribute to *L. thalassiae*'s ability to be an effective pathogen.

Crude extracts from four of five species of seagrasses from the same area in Florida inhibit the growth of at least one species of local fungi (Ross et al. 2008). Only *Syringodium filiforme* extracts showed no inhibitory activity. However, unlike the Indo-Pacific seagrasses, there are no clear trends in terms of whether or not the inhibited fungi are saprophytic or pathogenic. An earlier study found that extracts from healthy *T. testudinum* from the Bahamas inhibits the growth of the thraustochytrid *Schizochytrium aggregatum*, whereas extracts from leaves of dead plants do not. A bioassay-guided fractionation localized the inhibitory activity to the flavone glycoside luteolin 7-O-β-D- glucopyranosyl-2"-sulfate, which is found in healthy leaves at concentrations of about 4 mg mL^{-1}, about 14 times higher than the amount of the compound needed to reduce *S. aggregatum*'s growth by 40% (Jensen et al. 1998).

Invertebrate animals can also foul seagrass leaves (Borowitzka et al. 2006). In most cases, colonization occurs when a larva recruits to the surface of the leaf and undergoes metamorphosis. Therefore, natural products that are toxic to larvae, inhibit settlement, or prevent attachment could ultimately reduce the biomass of invertebrate epibionts on seagrass leaves. Crude ethanol extracts from *E. acoroides* have little effect on the survival of larvae of the bryozoan *Bugula neretina* but do inhibit larval settlement in a dose-dependent manner at concentrations ranging from 0.4 to 40 mg of extract mL^{-1} (Qi et al. 2008). Zosteric acid has been shown to reduce attachment rates of freshwater mussels detached from a surface, but only at relatively high concentrations of 400–1000 ppm (Ram et al. 2012). This compound has no effect on detachment rates of mussels that have already settled.

4.4.4 Defenses against Disease—*Labyrinthula*

During the 1930s, eelgrass beds on the East Coast of North America were nearly decimated when ~90% of the *Z. marina* succumbed to a "wasting disease," which was later found to be caused by the by the pathogenic fungus *Labyrinthula zosterae* (Muehlstein et al. 1991). Populations of *Z. marina* in Europe were affected as well

(Brakel et al. 2014). Since then, it has been determined that *L. zosterae* and other species of *Labyrinthula* occur in many coastal areas and inhabit most seagrass genera (Vergeer and Den Hartog 1994; Garcias-Bonet et al. 2011; Bockelmann et al. 2013; Sullivan et al. 2013; Brakel et al. 2014); however, in many of these locations, the virulence of the fungus is limited and it can be found in apparently healthy plants (Bockelmann et al. 2013; Brakel et al. 2014) or is primarily associated with older, senescing leaves (Vergeer and Den Hartog 1994). The pathogenicity of the fungus is affected by environmental factors such as low salinities (Muehlstein et al. 1988; McKone and Tanner 2009) and, in some situations, by high light intensities (Vergeer et al. 1994).

Phenolic compounds have been suggested to inhibit the growth of *Labyrinthula*, and these compounds may serve as the first line of defense in protecting seagrasses from an initial fungal infection. Concentrations of caffeic acid similar that to those found in infected leaves of *Z. marina* have been shown to inhibit the growth of *L. zosterae* in culture experiments, but gallic acid, *p*-coumaric acid, and ferulic acid have no inhibitory activity (Vergeer and Develi 1997). Laboratory experiments have also found a correlation between total phenolic activity in *Z. marina* and mortality resulting from subsequent *Labyrinthula* infection (Buchsbaum et al. 1990).

Several studies have found differences in phenolic compounds in healthy and infected tissues from field collections and in laboratory experiments in which healthy seagrass tissues were infected with *Labyrinthula*, suggesting that plants alter the types and quantities of phenolic compounds following infection by fungi. Concentrations of gallic acid, caffeic acid, *p*-coumaric acid, and ferulic acid are substantially higher in infected relative to uninfected leaves of *Z. marina* collected in the spring in Roscoff, France (Vergeer and Develi 1997). Similarly, in Maryland, USA, McKone and Tanner (2009) found a strong positive correlation between the combined concentrations of gallic, rosmarinic, and ferulic acid and sizes of lesions caused by *Labyrinthula* infection in *Z. marina* leaves grown at a salinity of 24, but not at a salinity of 10 (caffeic acid was only present in trace amounts). Differences in the types of phenolic acids present in plants could be due to genetic differences between North American and European *Z. marina*. Alternatively, different strains of *Labyrinthula* may occur in the two areas, and the plant responses may be strain-specific.

Changes in production of phenolic compounds induced by *Labyrinthula* can cause differences in their distributions within single leaves. For example, total phenolic activity is greatest near lesions caused by *Labyrinthula* on *Z. marina* leaves, with the highest concentrations occurring within 2 cm of lesions and decreasing with increased distance from them (Vergeer et al. 1994). At 10 cm from lesions, phenolic activity levels are about 33% of those near lesions. However, condensed tannins are lowest at the site of *Labyrinthula* lesions in infected *T. testudinum* and increase significantly with distance in young leaves (Steele et al. 2004). In older leaves, condensed tannin concentrations are lower at the site of lesions than in tissues 2 cm away, but there is no difference in condensed tannin concentrations in tissues 2 cm or farther from lesions. The combined concentrations of 3,4-dihydroxybenzoic acid, *p*-hydroxybenzoic acid, *p*-coumaric acid, and vanillin are also lower at the sites of lesions caused by *Labyrinthula* infection than they are at either 2 cm below lesions or 2 cm above lesions. The concentrations of all four phenolic acids are highest 2 cm above lesions and significantly higher than at lesions for all compounds except *p*-hydroxybenzoic acid.

The production of phenolic compounds can be upregulated following infection by *Labyrinthula*, but whether or not this is an adaptive inducible defense has been questioned. When *Z. marina* from a European population was infected with *L. zosterae*, the expression of the CYP73A gene, which is known to be associated with the synthesis of phenolic compounds, increased 80-fold, whereas many other genes associated with the production of defense in other species of terrestrial plants were downregulated (Brakel et al. 2014). Steele et al. (2004) attributed lower concentrations of condensed tannins and phenolic acids in *Zostera* at the sites where lesions occurred to a disruption of vascular tissues by the fungus. Since carbohydrates, which are necessary for phenolic synthesis, could not be imported to the site of infection, the production of phenolic compounds was disrupted. Disruptions in the transport of carbohydrates at the sites of lesions could also account for an accumulation of carbohydrates in areas above lesions, leading to increases in the synthesis of phenolic compounds and higher phenolic levels at those locations. Thus, increases in phenolic compounds near lesions could be a storage mechanism for fixed carbon, which can no longer be transported to the rhizomes, or it could be an adaptive, induced defense to prevent further infection.

4.5   PRODUCTION OF ROS AND ANTIOXIDANTS

When plants are subjected to mechanical damage, fouling, or physiological stresses, one of the consequences is the release of reactive oxygen species (ROS). ROS are small oxygen-containing molecules such as singlet oxygen, superoxide (O_2^-), hydrogen peroxide (H_2O_2), and hydroxyl radicals (OH.) (reviewed in Halliwell and Gutteridge 2015; Apel and Hirt 2004; Murphy 2009). Singlet oxygen is generated in Photosystem II when excitation energy is transferred from chlorophyll to O_2 and O_2^- radicals are generated during respiration and in Photosystem I when an electron is transferred to O_2. O_2^- can be reduced enzymatically to form H_2O_2, which can subsequently form extremely reactive and damaging OH. ROS in plants can have physiological and ecological functions, such as serving as signaling molecules and anti-pathogen defenses (Beck 2004; Miller et al. 2008); however, if excess ROS are not scavenged by enzymatic and non-enzymatic antioxidants they can damage the plant's DNA, lipids, and proteins (Halliwell and Gutteridge 2015).

Because there is interest in using antioxidants for commercial purposes, seagrass extracts are often assayed for their antioxidant activity (Athiperumalsami et al. 2008; Kesraoui et al. 2011; Rengasamy et al. 2012). Phenolic compounds typically have antioxidant activity (Rice-Evans et al. 1996; Silva et al. 2000; Prior et al. 2004) and most seagrasses contain them. As a result, screening assays generally find that seagrass extracts display some measure of antioxidant activity, and this activity has been shown to correlate with measures of phenolic compounds in the plants (Rengasamy et al. 2012). However, other antioxidants such as vitamins A, B_3, C, and E, beta-carotene, and a sulfated polysaccharide have also been isolated from seagrasses (Athiperumalsami et al. 2010; Silva et al. 2012; Kannan et al. 2013).

In vitro studies of antioxidant activity conducted with crude extracts do not necessarily provide information that is useful in understanding the ecological and physiological roles of seagrass antioxidant compounds in the marine environment. There are only a handful of studies that have examined how antioxidants produced by seagrasses function in marine systems.

Like many terrestrial plants, seagrasses can produce "oxidative bursts," large releases of ROS into the extracellular environment that are part of the hypersensitive response activated by plants following infection by pathogens (Lamb and Dixon 1997; Potin 2008). Ross and coworkers (2008) documented releases of H_2O_2 by *S. filiforme, H. wrightii, T. testudinum, R. maritima, and Halophila decipens* in response to infection by a suite of 10 fungal species. Although there was some variation in the amounts of H_2O_2 released, all species of plants responded to all of the fungi. However, seagrass species identity was generally a stronger determinant of the amount of H_2O_2 released than fungal species. Within 24 h of exposure to *L. thalassiae* or *D. salina, T. testudinum* increased caspase activity, indicating the activation of programmed cell death. When plants were treated with diphenylene iodonium, an NADPH oxidase inhibitor that restricts the plant's ability to produce H_2O_2, infection with the fungal homogenates did not cause increased caspase activity. A follow-up study found that the pathogen-associated molecular pattern (PAMP) that leads to the oxidative burst response in *T. testudinum* can be initiated by a lipopolysaccharide (Loucks et al. 2013).

Similar responses may occur when seagrasses are epiphytized by macroalgae. *P. oceanica* collected in the Western Mediterranean where they were epiphytized by the invasive brown alga *Lophocladia lallemandii* (Sureda et al. 2008) produced higher amounts of H_2O_2 than plants collected from a site where they were not epiphytized. Plants from the site with the epiphytes also had higher levels of activity in the antioxidant enzymes catalase, glutathione peroxidase, and superoxide dismutase, as well as indications of oxidative damage to lipids and proteins. Sureda et al. (2008) suggested that the higher level of release of H_2O_2 may be acting as a defense against epiphytism. However, it is also possible that these plants were producing responses as a result of another environmental stress that was also weakening the plants and reducing their ability to respond to the epiphytes.

4.6   PRODUCTION OF ALLELOCHEMICALS BY ASSOCIATED ORGANISMS

In addition to producing allelochemicals themselves, seagrasses can harbor other organisms that synthesize and utilize bioactive compounds. These include fungi (Jones et al. 2008; Supaphon et al. 2013), bacteria (Ravikumar et al. 2012), and microalgae (Flewelling et al. 2004; Hitchcock et al. 2012), which can produce a variety of bioactive metabolites that have been shown to have antioxidant, antimicrobial, antifungal, and anticancer activities (Belofsky et al. 1999; Arunpanichlert et al. 2011; Ravikumar et al. 2012; Rukachaisirikul et al. 2013; Supaphon et al. 2013). However, whether these compounds have ecologically-important functions has not been explored.

One example of ecologically relevant bioactive metabolites produced by an organism associated with seagrasses is the production of brevetoxins by the red tide-forming dinoflagellate *Karenia brevis*. Following blooms of *Karenia* along the coast of Florida, USA, brevetoxins are found in the leaves and rhizomes of *T. testudinum* and in the surrounding sediments (Flewelling et al. 2004; Hitchcock et al. 2012). Consumption of seagrasses harboring epiphytes containing brevetoxins has been implicated in the deaths of manatees, whose stomachs were packed with seagrasses (Flewelling et al. 2004). Interestingly, seagrasses can also harbor organisms that have detrimental effects on harmful microalgae. For example, *Z. marina* can harbor bacteria that produce compounds that inhibit the growth of several species of toxic microalgae (Onishi et al. 2014).

4.7 FUTURE DIRECTIONS

Seagrasses are an ecologically and economically important group of plants that serve as foundation species for many nearshore communities. Many of their interactions with other organisms in these communities are mediated by a complex suite of natural products whose identities and functions are not well characterized. Knowing more about the role of natural products in mediating these interactions will lead to a better understanding of how these complex communities function and the impacts that human activities are having on them. Understanding how the production of these compounds is regulated and their impacts on marine communities will also provide information important for better conservation and restoration practices.

The increasing availability of complex analytical chemistry instrumentation is now making it easier to isolate, identify, and quantify individual metabolites. Earlier studies often used colorimetric assays, such as Folin analyses, as indicators of the amounts of chemical defenses in seagrasses. While there is still a role for these kinds of analyses in understanding the dynamics of seagrass communities, studies of seagrass chemical ecology will increasingly need to employ methods in which the activities of individual compounds are measured and assayed, rather than relying on general measurements of structurally-similar groups of compounds and correlating these with ecological functions.

Little is known about the production and release of seagrass secondary metabolites, the cellular and molecular mechanisms by which these processes are regulated, or the costs associated with producing these metabolites. A better understanding is needed of the synthetic pathways involved in producing these metabolites. In many cases, these may have already been discovered as a large number of seagrass secondary metabolites are also produced by terrestrial plants. We also need a better understanding of the genes involved in regulating their production, and the mechanisms by which they are absorbed, degraded, or released into the environment.

There is already evidence that many environmental factors affect the production of secondary metabolites in seagrasses. We need a better understanding of the role of the external environment in altering the production and use of seagrass natural products, especially anthropogenically-generated changes to the marine environment. These environmental factors include nutrient over-enrichment, air and seawater temperature changes, chemical alterations to seawater pH and carbonate chemistry due to ocean acidification, and changes in the salinities of coastal waters. Transcriptomic and metabolomic analyses of plants experiencing different environmental conditions would help elucidate how specific environmental changes affect not only the production of secondary metabolites, but a host of other ecologically-relevant, metabolic processes.

Much more work is needed to understand the effects of seagrass natural products on other organisms and the consequences of these effects on community dynamics. Presently, relatively little is known about the effects of specific metabolites on ecological functions. The use of bioassay-guided fractionation methods will be particularly important for identifying seagrass metabolites that have ecological activity. Many studies of seagrass ecology assume that phenolic concentrations correlate with ecological function, such as chemical defense or anti-pathogen activity; however, there is limited direct experimental evidence to support these conclusions. In many experiments, observed activity is likely due to one or several metabolites, rather than the group as a whole. It is also possible that other types of metabolites, whose concentrations are correlated with phenolic concentrations may be responsible for the biological activity observed. Bioassay-guided fractionation methods will enable researchers to isolate the individual compounds responsible for the ecological activity of interest and direct future research efforts towards understanding the roles of these individual compounds.

More work is also needed to determine the effects of seagrass natural products on interactions among species, within communities, and between trophic levels. Changes in the amounts of bioactive metabolites in seagrasses that result from changing environmental conditions are likely to provide both positive and negative feedbacks for seagrasses and the communities they form. Herbivores may move to alternate foods in response to changing plant chemistry, or there may be reductions or increases in herbivore growth and reproduction rates as a function of seagrass chemistry. The effects of pathogens can lead to alterations in the availability of habitat and food for organisms living in seagrass beds. Understanding the role of seagrass natural products in modulating the dynamics of these communities, especially as they are being impacted by anthropogenic changes, will be important to helping preserve and restore these important marine habitats.

REFERENCES

Achamlale, S., B. Rezzonico, and M. Grignon-Dubois. 2009a. Evaluation of *Zostera* detritus as a potential new source of zosteric acid. *J. Appl. Phycol.* 21:347–342.

Achamlale, S., B. Rezzonico, and M. Grignon-Dubois. 2009b. Rosmarinic acid from beach waste: Isolation and HPLC quantification in *Zostera* detritus from Arcachon lagoon. *Food Chem.* 113:878–883.

Agostini, S., J. M. Desjobert, and G. Pergent. 1998. Distribution of phenolic compounds in the seagrass *Posidonia oceanica*. *Phytochemistry* 48:611–617.

Apel, K., and H. Hirt. 2004. Reactive oxygen species: Metabolism, oxidative stress, and signal transduction. *Annu. Rev. Plant Biol.* 44:373–399.

Appel, H. M., H. L. Govenor, M. D'Ascenzo, E. Siska, and J. C. Schultz. 2001. Limitations of Folin assays of foliar phenolics in ecological studies. *J. Chem. Ecol.* 27:761–778.

Aquino, R. S., A. M. Landeira-Fernandez, A. P. Valente, L. R. Andrade, and P. A. Mourão. 2004. Occurrence of sulfated galactans in marine angiosperms: Evolutionary implications. *Glycobiology* 14:11–20.

Arnold, T. A., C. E. Tanner, M. Rothen, and J. Bullington. 2008. Wound-induced accumulations of condensed tannins in turtlegrass, *Thalassia testudinum*. *Aquat. Bot.* 89:27–33.

Arnold, T. M., and N. M. Targett. 2002. Marine tannins: The importance of a mechanistic framework for predicting ecological roles. *J. Chem. Ecol.* 28:1919–1934.

Arnold, T., C. Mealey, H. Leahey et al. 2012. Ocean acidification and the loss of phenolic substances in marine plants. *PLoS One* 7:10.

Arnold, T., G. Freundlich, T. Weilnau, A. Verdi, and I. R. Tibbetts. 2014. Impacts of groundwater discharge at Myora Springs (North Stradbroke Island, Australia) on the phenolic metabolism of eelgrass, *Zostera muelleri*, and grazing by the juvenile rabbitfish, *Siganus fuscescens*. *PLoS One* 9:e104738.

Arunpanichlert, J., V. Rukachaisirikul, Y. Sukpondma, S. Phongpaichit, O. Supaphon, and J. Sakayaroj. 2011. A β-resorcylic macrolide from the seagrass-derived fungus *Fusarium* sp. PSU-ES73. *Arch. Pharm. Res.* 34:1633–1637.

Athiperumalsami, T., V. D. Rajeswari, S. H. Poorna, V. Kumar, and L. L. Jesudass. 2010. Antioxidant activity of seagrasses and seaweeds. *Bot. Mar.* 43:241–247.

Athiperumalsami, T., V. Kumar, and L. L. Jesudass. 2008. Survey and phytochemical analysis of seagrasses in the Gulf of Mannar, southeast coast of India. *Bot. Mar.* 41:269–277.

Ballaré, C. L. 2011. Jasmonate-induced defenses: A tale of intelligence, collaborators and rascals. *Trends Plant Sci.* 16:249–247.

Ballesteros, E., D. Martin, and M. J. Uriz. 1992. Biological activity of extracts from some Mediterranean macrophytes. *Bot. Mar.* 34:481–484.

Barbier, E. B., S. D. Hacker, C. Kennedy, E. W. Koch, A. C. Stier, and B. R. Silliman. 2011. The value of estuarine and coastal ecosystem services. *Ecol. Monograph* 81:169–193.

Beck, M. W., K. L. Heck, Jr., K. W. Able et al. 2001. The identification, conservation, and management of estuarine and marine nurseries for fish and invertebrates: A better understanding of the habitats that serve as nurseries for marine species and the factors that create site-specific variability in nursery quality will improve conservation and management of these areas. *Bioscience* 41:633–641.

Beck, C. F. 2004. Signaling pathways from the chloroplast to the nucleus. *Planta* 222:743–746.

Belofsky, G. N., P. R. Jensen, and W. Fenical. 1999. Sansalvamide: A new cytotoxic cyclic depsipeptide produced by a marine fungus of the genus *Fusarium*. *Tetrahedron Lett.* 40:2913–2916.

Bernard, P., and D. Pesando. 1989. Antibacterial and antifungal activity of extracts from the rhizomes of the Mediterranean seagrass *Posidonia oceanica* (L) DeLile. *Bot. Mar.* 32:84–88.

Bhosale, S. H., V. L. Nagle, and T. G. Jagtap. 2002. Antifouling potential of some marine organisms from India against species of *Bacillus* and *Pseudomonas*. *Mar. Biotech.* 4:111–118.

Bitam, F., M. L. Ciavatta, M. Carbone, E. Manzo, E. Mollo, and M. Gavagnin. 2010. Chemical analysis of flavonoid constituents of the seagrass *Halophila stipulacea*: First finding of malonylated derivatives in marine phanerogams. *Biochem. Syst. Ecol.* 38:686–690.

Bockelmann, A.-C., V. Tams, J. Ploog, P. R. Schubert, and T. B. Reusch. 2013. Quantitative PCR reveals strong spatial and temporal variation of the wasting disease pathogen, *Labyrinthula zosterae* in northern European eelgrass (*Zostera marina*) beds. *PLoS One* 8:e62169.

Borowitzka, M., P. Lavery, and M. Van Keulen. 2006. Epiphytes of seagrasses. In *Seagrasses: Biology, Ecology and Conservation*, A. W. D. Larkum, R. J. Orth, and C. M. Duarte (Eds.), pp. 463–401, Dordrecht, the Netherlands: Springer.

Boumaza, S., N. Boudefoua, R. Boumaza, and R. Semroud. 2014. Effects of urban effluents on spatial structure, morphology and total phenols of *Posidonia oceanica*: Comparison with a reference site. *J. Exp. Mar. Biol. Ecol.* 447:113–119.

Brakel, J., F. J. Werner, V. Tams, T. B. Reusch, and A.-C. Bockelmann. 2014. Current European *Labyrinthula zosterae* are not virulent and modulate seagrass (*Zostera marina*) defense gene expression. *PLoS One* 9:e92448.

Buchsbaum, R. N., F. T. Short, and D. P. Cheney. 1990. Phenolic-nitrogen interactions in eelgrass, *Zostera marina* L.: Possible implications for disease resistance. *Aquat. Bot.* 37:291–297.

Buchsbaum, R., I. Valiela, and T. Swain. 1984. The role of phenolic compounds and other plant constituents in feeding by Canada geese in a coastal marsh. *Oecologia* 63:343–349.

Burkholder, J. M., D. A. Tomasko, and B. W. Touchette. 2007. Seagrasses and eutrophication. *J. Exp. Mar. Biol. Ecol.* 340:46–72.

Bushmann, P. J., and M. S. Ailstock. 2006. Antibacterial compounds in estuarine submersed aquatic plants. *J. Exp. Mar. Biol. Ecol.* 331:41–40.

Cannac, M., L. Ferrat, C. Pergent-Martini, G. Pergent, and V. Pasqualini. 2006. Effects of fish fanning on flavonoids in *Posidonia oceanica*. *Sci. Total Environ.* 370:91–98.

Carbone, M., M. Gavagnin, E. Mollo, M. Bidello, V. Roussis, and G. Cimino. 2008. Further syphonosides from the sea hare *Syphonota geographica* and the sea-grass *Halophila stipulacea*. *Tetrahedron* 64:191–196.

Citová, I., R. Sladkovský, and P. Solich. 2006. Analysis of phenolic acids as chloroformate derivatives using solid phase microextraction–gas chromatography. *Anal. Chim. Acta* 473:231–241.

Cuny, P., L. Serve, H. Jupin, and C. F. Boudouresque. 1994. Water soluble phenolic compounds of the marine phanerogam *Posidonia oceanica* in a Mediterranean area colonised by the introduced chlorophyte *Caulerpa taxifolia*. *Aquat. Bot.* 42:237–242.

DellaGreca, M., A. Fiorentino, M. Isidori, P. Monaco, and A. Zarrelli. 2000. Antialgal *ent*-labdane diterpenes from *Ruppia maritima*. *Phytochemistry* 44:909–913.

den Hartog, C., and J. Kuo. 2006. Taxonomy and biogeography of seagrasses. In *Seagrasses: Biology, Ecology and Conservation*, A. W. D. Larkum, R. J. Orth, and C. M. Duarte, (Eds.), pp. 1–23, Dordrecht, the Netherlands: Springer.

Duarte, C. M. 1991. Seagrass depth limits. *Aquat. Bot.* 40:363–377.

Duarte, C. M., and C. L. Chiscano. 1999. Seagrass biomass and production: A reassessment. *Aquat. Bot.* 64:149–174.

Dubois, A., and K. Iken. 2012. Seasonal variation in kelp phlorotannins in relation to grazer abundance and environmental variables in the Alaskan sublittoral zone. *Algae* 27:9–19.

Dudgeon, S. R., J. Kubler, R. Vadas, and I. R. Davison. 1994. Physiological responses to environmental variation in intertidal red algae: Does thallus morphology matter? *Mar. Ecol. Prog. Ser.* 117:193–206.

Dumay, O., J. Costa, J. M. Desjobert, and G. Pergent. 2004. Variations in the concentration of phenolic compounds in the seagrass *Posidonia oceanica* under conditions of competition. *Phytochemistry* 64:3211–3220.

Engel, S., P. R. Jensen, and W. Fenical. 2002. Chemical ecology of marine microbial defense. *J. Chem. Ecol.* 28:1971–1984.

Ferrat, L., S. Wylie-Echeverria, R. Cates et al. 2012. *Posidonia oceanica* and *Zostera marina* as potential biomarkers of heavy metal contamination in coastal systems. In *Ecological Water Quality-Water Treatment and Reuse,* K. Voudouris and D. Voutsa, (Eds.), pp. 978–943, Rijeka, Croatia: InTech.

Flewelling, L. J., J. P. Naar, J. P. Abbott et al. 2004. Brevetoxicosis: Red tides and marine mammal mortalities. *Nature* 434:744–746.

Folmer, F., M. Jaspars, M. Dicato, and M. Diederich. 2010. Photosynthetic marine organisms as a source of anticancer compounds. *Phytochem. Rev.* 9:447–479.

Garateix, A., E. Salceda, R. Menendez et al. 2011. Antinociception produced by *Thalassia testudinum* extract BM-21 is mediated by the inhibition of acid sensing ionic channels by the phenolic compound thalassiolin B. *Molec. Pain* 7:10.

Garcias-Bonet, N., T. D. Sherman, C. M. Duarte, and N. Marbà. 2011. Distribution and pathogenicity of the protist *Labyrinthula* sp. in western Mediterranean seagrass meadows. *Estuar. Coasts* 34:1161–1168.

Gavin, N. M., and M. J. Durako. 2011. Localization and antioxidant capacity of flavonoids from intertidal and subtidal *Halophila johnsonii* and *Halophila decipiens. Aquat. Bot.* 94:242–247.

Gavin, N. M., and M. J. Durako. 2012. Localization and antioxidant capacity of flavonoids in *Halophila johnsonii* in response to experimental light and salinity variation. *J. Exp. Mar. Biol. Ecol.* 416:32–40.

Gillan, F. T., R. W. Hogg, and E. A. Drew. 1984. The sterol and fatty-acid compositions of 7 tropical seagrasses from North Queensland, Australia. *Phytochemistry* 23:2817–2821.

Gloaguen, V., V. Brudieux, B. Closs et al. 2010. Structural characterization and cytotoxic properties of an apiose-rich pectic polysaccharide obtained from the cell wall of the marine phanerogam *Zostera marina. J. Nat. Prod.* 73:1087–1092.

Grignon-Dubois, M., and B. Rezzonico. 2013. The economic potential of beach-cast seagrass—*Cymodocea nodosa:* A promising renewable source of chicoric acid. *Bot. Mar.* 46:303–311.

Grünbaum, D., and D. K. Padilla. 2014. An integrated modeling approach to assessing linkages between environment, organism, and phenotypic plasticity. *Integr. Comp. Biol.* 54(2):323–335.

Hagerman, A., and L. G. Butler. 1991. Tannins and lignins. In *Herbivores: Their Interactions with Secondary Plant Metabolites,* G. Rosenthal and M. Berenbaum, B. Halliwell, and J. Gutteridge (Eds.); *Free Radicals in Biology and Medicine,* pp. 344–388, Oxford, UK: Oxford University Press.

Halliwell, B. and J. M. C. Gutteridge. 2015. *Free Radicals in Biology and Medicine.* Oxford, UK: Oxford University Press.

Hammami, S., A. Ben Salem, M. L. Ashour, J. Cheriaa, G. Graziano, and Z. Mighri. 2013. A novel methylated sesquiterpene from seagrass *Posidonia oceanica* (L.) Delile. *Nat. Prod. Res.* 27:1264–1270.

Harborne, J. 1991. Flavonoid pigments. In *Herbivores: Their Interactions with Secondary Plant Metabolites: The Chemical Participants,* G. Rosenthal, and M. Berenbaum (Eds.), San Diego, CA: Academic Press.

Harborne, J. B., and C. A. Williams. 2000. Advances in flavonoid research since 1992. *Phytochemistry* 44:481–404.

Harrison, P. G. 1977. *Zostera japonica* (Aschers. & Graebn.) in British Columbia, Canada. *Syesis* 9:349–360.

Harrison, P. G. 1982. Control of microbial growth and of amphipod grazing by water-soluble compounds from leaves of *Zostera marina. Mar. Biol.* 67:224–230.

Harrison, P. G., and A. T. Chan. 1980. Inhibition of the growth of micro-algae and bacteria by extracts of eelgrass (*Zostera marina*) leaves. *Mar. Biol.* 61:21–26.

Harrison, P. G., and C. Durance. 1989. Seasonal variation in phenolic content of eelgrass shoots. *Aquat. Bot.* 34:409–413.

Hauxwell, J., J. Cebrián, C. Furlong, and I. Valiela. 2001. Macroalgal canopies contribute to eelgrass (*Zostera marina*) decline in temperate estuarine ecosystems. *Ecology* 82:1007–1022.

Haznedaroglu, M. Z., and U. Zeybek. 2007. HPLC determination of chicoric acid in leaves of *Posidonia oceanica. Pharm. Biol.* 44:744–748.

Heck, K. L., T. J. B. Carruthers, C. M. Duarte et al. 2008. Trophic transfers from seagrass meadows subsidize diverse marine and terrestrial consumers. *Ecosystems* 11:1198–1210.

Heck, K. L., and J. F. Valentine. 2006. Plant–herbivore interactions in seagrass meadows. *J. Exp. Mar. Biol. Ecol.* 330:420–436.

Heck, K. L., Jr, and R. J. Orth. 2006. Predation in seagrass beds. In *Seagrasses: Biology, Ecology and Conservation,* A. W. D. Larkum, R. J. Orth, and C. M. Duarte, (Eds.), pp. 437–40, Dordrecht, the Netherlands: Springer.

Heglmeier, A., and C. Zidorn. 2010. Secondary metabolites of *Posidonia oceanica* (Posidoniaceae). *Biochem. Syst. Ecol.* 38:964–970.

Hemminga, M. A., and C. M. Duarte. 2000. *Seagrass Ecology.* Cambridge, UK: Cambridge University Press

Henrikson, A. A., and J. R. Pawlik. 1994. A new antifouling assay method: Results from field experiments using extracts of four marine organisms. *J. Exp. Mar. Biol. Ecol.* 194:147–164.

Hitchcock, G. L., J. W. Fourqurean, J. L. Drake, R. N. Mead, and C. A. Heil. 2012. Brevetoxin persistence in sediments and seagrass epiphytes of east Florida coastal waters. *Harmful Algae* 13:89–94.

Hughes, A. R., S. L. Williams, C. M. Duarte, K. L. Heck, Jr., and M. Waycott. 2009. Associations of concern: Declining seagrasses and threatened dependent species. *Front. Ecol. Environ.* 7:242–246.

Jensen, P. R., K. M. Jenkins, D. Porter, and W. Fenical. 1998. Evidence that a new antibiotic flavone glycoside chemically defends the sea grass *Thalassia testudinum* against zoosporic fungi. *Appl. Environ. Microbiol.* 64:1490–1496.

Jernakoff, P., A. Brearley, and J. Nielsen. 1996. Factors affecting grazer-epiphyte interactions in temperate seagrass meadows. *Oceanogr. Mar. Biol. Ann. Rev.* 34:109–162.

Jones, E. B. G., S. J. Stanley, and U. Pinruan. 2008. Marine endophyte sources of new chemical natural products: A review. *Bot. Mar.* 41:163–170.

Kajihara, R., T. Komorita, A. Hamada, S. Shibanuma, T. Yamada, and S. Montani. 2010. Possibility of direct utilization of seagrass and algae as main food resources by small gastropod, *Lacuna decorata*, in a subarctic lagoon, Hichirippu, eastern Hokkaido, Japan with stable isotope evidences of carbon and nitrogen. *Plankton Benth. Res.* 4:90–97.

Kannan, R. R. R., R. Arumugam, T. Thangaradjou, and P. Anantharaman. 2013. Phytochemical constituents, antioxidant properties and *p*-coumaric acid analysis in some seagrasses. *Food Res. Int.* 44:1229–1236.

Kawasaki, W., K. Matsui, Y. Akakabe, N. Itai, and T. Kajiwara. 1998. Volatiles from *Zostera marina. Phytochemistry* 47:27–29.

Kegge, W., and R. Pierik. 2010. Biogenic volatile organic compounds and plant competition. *Trends Plant Sci.* 14:126–132.

Kesraoui, O., M. N. Marzouki, T. Maugard, and F. Limam. 2011. In vitro evaluation of anti-oxidant activities of free and bound phenolic compounds from *Posidonia oceanica* (l.) Delile leaves. *African J. Biotech.* 10:3176–184.

Klap, V. A., M. A. Hemminga, and J. J. Boon. 2000. Retention of lignin in seagrasses: Angiosperms that returned to the sea. *Mar. Ecol. Prog. Ser.* 194:1–11.

Koch, E. W., E. B. Barbier, B. R. Silliman et al. 2009. Non-linearity in ecosystem services: Temporal and spatial variability in coastal protection. *Front. Ecol. Environ.* 7:29–37.

Kontiza, I., D. Abatis, K. Malakate, C. Vagias, and V. Roussis. 2006. 3-Keto steroids from the marine organisms *Dendrophyllia cornigera* and *Cymodocea nodosa. Steroids* 71.177–181.

Kontiza, L., M. Stavri, M. Zloh, C. Vagias, S. Gibbons, and V. Roussis. 2008. New metabolites with antibacterial activity from the marine angiosperm *Cymodocea nodosa. Tetrahedron* 64:1696–1702.

Krumhansl, K. A., and R. E. Scheibling. 2011. Spatial and temporal variation in grazing damage by the gastropod *Lacuna vincta* in Nova Scotian kelp beds. *Aquat. Biol.* 13:163–173.

Laabir, M., M. Grignon-Dubois, E. Masseret et al. 2013. Algicidal effects of *Zostera marina* L. and *Zostera noltii* Hornem. extracts on the neuro-toxic bloom-forming dinoflagellate *Alexandrium catenella. Aquat. Bot.* 111:16–24.

Lamb, C., and R. Dixon. 1997. The oxidative burst in plant disease resistance. *Annu. Rev. Plant Physiol.* 48:241–274.

Larkum A. W. D., R. J. Orth, and C. M. Duarte. 2006. *Seagrasses: Biology, Ecology and Conservation*. Dordrecht, the Netherlands: Springer.

Loucks, K., D. Waddell, and C. Ross. 2013. Lipopolysaccharides elicit an oxidative burst as a component of the innate immune system in the seagrass *Thalassia testudinum*. *Plant Physiol. Biochem.* 70:294–303.

McKone, K. L., and C. E. Tanner. 2009. Role of salinity in the susceptibility of eelgrass *Zostera marina* to the wasting disease pathogen *Labyrinthula zosterae*. *Mar. Ecol. Prog. Ser.* 377:123–130.

McMillan, C. 1984. The condensed tannins (proanthocyanidins) in seagrasses. *Aquat. Bot.* 20:341–347.

McMillan, C. 1986. Sulfated flavonoids and leaf morphology in the *Halophila ovalis*-h-minor complex (Hydrocharitaceae) of the Indo-Pacific Ocean. *Aquat. Bot.* 24:63–72.

McMillan, C., O. Zapata, and L. Escobar. 1980. Sulfated phenolic-compounds in seagrasses. *Aquat. Bot.* 8:267–278.

Meiners, S.J., C.-H. Kong, L.M. Ladwig, N.L. Pisula, and K.A. Lang. 2012. Developing an ecological context for allelopathy. *Plant Ecol.* 213:1221–1227.

Meng, Y. H., A. J. Krzysiak, M. J. Durako, J. I. Kunzelman, and J. L. C. Wright. 2008. Flavones and flavone glycosides from *Halophila johnsonii*. *Phytochemistry* 69:2603–2608.

Milkova, T., R. Petkova, R. Christov, S. Popov, and S. Dimitrovakonaklieva. 1994. Comparative study of the chemical composition of *Zostera marina* L and *Zostera nana* Roth from the Black Sea. *Bot. Mar.* 38:99–101.

Miller, G., V. Shulaev, and R. Mittler. 2008. Reactive oxygen signaling and abiotic stress. *Physiol. Plant.* 133:481–489.

Miloslavich, P., J. J. Cruz-Motta, E. Klein et al. 2013. Large-scale spatial distribution patterns of gastropod assemblages in rocky shores. *PLoS One* 8:e71396.

Mohammed, M. M. D., A. H. A. Hamdy, N. M. El-Fiky, W. S. A. Mettwally, A. A. El-Beih, and N. Kobayashi. 2014. Anti-influenza A virus activity of a new dihydrochalcone diglycoside isolated from the Egyptian seagrass *Thalassodendron ciliatum* (Forsk.) den Hartog. *Nat. Prod. Res.* 28:377–382.

Molis, M., A. Enge, and U. Karsten. 2010. Grazing impact of, and indirect interactions between mesograzers associated with kelp (*Laminaria digitata*). *J. Phycol.* 46:76–84.

Muehlstein, L. K., D. Porter, and F. T. Short. 1988. *Labyrinthula* sp., a marine slime mold producing the symptoms of wasting disease in eelgrass, *Zostera marina*. *Mar. Biol.* 99:464–472.

Muehlstein, L. K., D. Porter, and F. T. Short. 1991. *Labyrinthula zosterae* sp. nov., the causative agent of wasting disease of eelgrass, *Zostera marina*. *Mycologia* 83(2):180–191.

Murphy, M. P. 2009. How mitochondria produce reactive oxygen species. *Biochem. J.* 417:1–13.

Nelson, T. A. 1994. Interactions and dynamics of eelgrass (*Zostera marina* L.), epiphytes, and grazers in subtidal meadows of Puget Sound. PhD dissertation. Seattle, WA: University of Washington.

Nelson, T. A., and J. R. Waaland. 1997. Seasonality of eelgrass, epiphyte, and grazer biomass and productivity in subtidal eelgrass meadows subjected to moderate tidal amplitude. *Aquat. Bot.* 46:41–74.

Nomme, K. M., and P. G. Harrison. 1991. Evidence for interaction between the seagrasses *Zostera marina* and *Zostera japonica* on the Pacific coast of Canada. *Can. J. Bot.* 69:2004–2010.

Nuissier, G., B. Rezzonico, and M. Grignon-Dubois. 2010. Chicoric acid from *Syringodium filiforme*. *Food Chem.* 120:783–788.

Onishi, Y., Y. Mohri, A. Tuji, K. Ohgi, A. Yamaguchi, and I. Imai. 2014. The seagrass *Zostera marina* harbors growth-inhibiting bacteria against the toxic dinoflagellate *Alexandrium tamarense*. *Fish. Sci.* 80:343–362.

Orth R. J., and J. Van Montfrans. 1984. Epiphyte-seagrass relationships with an emphasis on the role of micrograzing: A review. *Aquat. Bot.* 18:43–69.

Orth, R. J., T. J. B. Carruthers, W. C. Dennison et al. 2006. A global crisis for seagrass ecosystems. *Bioscience* 46:987–996.

Padilla, D. K. 1984. Structural resistance of algae to herbivores: A biomechanical approach. *Mar. Biol.* 90:103–109.

Padilla, D. K. 1987. Relationships among plant calcification, plant form, and herbivore mode of feeding in marine plant–herbivore interactions. Ph. D. diss. Edmonton, AB: University of Alberta.

Padilla, D. K. 1989. Algal structure defenses: Form and calcification in resistance to tropical limpets. *Ecology* 70:834–842.

Padilla, D. K. 1998. Inducible phenotypic plasticity of the radula in *Lacuna* (Gastropoda: Littorinidae). *Veliger* 41:201–204.

Padilla, D. K. 2001. Food and environmental cues trigger an inducible offence. *Evol. Ecol. Res.* 3:14–24.

Padilla, D. K., and S. C. Adolph. 1996. Plastic inducible morphologies are not always adaptive: The importance of time delays in a stochastic environment. *Evol. Ecol.* 10:104–117.

Padilla, D. K., D. E. Dittman, J. Franz, and R. Sladek. 1996. Radular production rates in two species of *Lacuna turton* (Gastropoda: Littorinidae). *J. Molluscan. Stud.* 62:274–280.

Paul, V. J., S. G. Nelson, and H. R. Sanger. 1990. Feeding preferences of adult and juvenile rabbitfish *Siganus argenteus* in relation to chemical defenses of tropical seaweeds. *Mar. Ecol. Prog. Ser.* 60:23–34.

Peterson, B. J., T. A. Frankovich, J. C. Zieman. 2007. Response of seagrass epiphyte load to field manipulations of fertilization, gastropod grazing and leaf turnover rates. *J. Exp. Mar. Biol. Ecol.* 349:61–72.

Pino, J. A., and E. L. Regalado. 2010. Volatile constituents of *Thalassia testudinum* Banks ex König leaves. *J. Essential Oil Res.* 22:421–423.

Posey, M. H. 1988. Community changes associated with the spread of an introduced seagrass, *Zostera japonica*. *Ecology* 69:974–983.

Potin, P. 2008. Oxidative burst and related responses in biotic interactions of algae. In *Algal Chemical Ecology*, C. D. Amsler (Ed.), pp. 244–272, Berlin, Germany: Springer-Verlag.

Premnathan, M., K. Chandra, S. K. Bajpai, and K. Kathiresan. 1992. A survey of some Indian marine plants for antiviral activity. *Bot. Mar.* 34:321–324.

Premnathan, M., K. Kathiresan, and K. Chandra. 1994. Antiviral evaluation of some marine plants against Semliki Forest virus. *Int. J. Pharmacogn.* 33:74–77.

Prior, R. L., X. Wu, and K. Schaich. 2004. Standardized methods for the determination of antioxidant capacity and phenolics in foods and dietary supplements. *J. Agric. Food Chem.* 43:4290–4302.

Puglisi, M. P., S. Engel, P. R. Jensen, and W. Fenical. 2007. Antimicrobial activities of extracts from Indo-Pacific marine plants against marine pathogens and saprophytes. *Mar. Biol.* 140:431–440.

Qi, S. H., S. Zhang, P. Y. Qian, and B. G. Wang. 2008. Antifeedant, antibacterial, and anti-larval compounds from the South China Sea seagrass *Enhalus acoroides*. *Bot. Mar.* 41:441–447.

Quackenbush, R. C., D. Bunn, and W. Lingren. 1986. HPLC determination of phenolic-acids in the water-soluble extract of *Zostera marina* L (eelgrass). *Aquat. Bot.* 24:83–89.

Ram, J. L., S. Purohit, B. M. Z. Newby, and T. J. Cutright. 2012. Evaluation of the natural product antifoulant, zosteric acid, for preventing the attachment of quagga mussels—A preliminary study. *Nat. Pro. Res.* 26:480–484.

Raven J. A., C. S. Cockell, and C. L. De La Rocha. 2008. The evolution of inorganic carbon concentrating mechanisms in photosynthesis. *Phil. Trans. Royal Soc. B.* 363:2641–2640.

Ravikumar, S., M. Gnanadesigan, A. Saravanan, N. Monisha, V. Brindha, and S. Muthumari. 2012. Antagonistic properties of seagrass associated *Streptomyces* sp. RAUACT-1: A source for anthraquinone rich compound. *Asian Pac. J. Trop. Med.* 4:887–890.

Ravn, H., M. F. Pedersen, J. Borum et al. 1994. Seasonal variation and distribution of 2 phenolic compounds, rosmarinic acid and caffeic acid, in leaves and roots-rhizomes of eelgrass (*Zostera-marina* L). *Ophelia* 40:41–61.

Regalado, E. L., M. Rodriguez, R. Menendez et al. 2009. Repair of UVB-damaged skin by the antioxidant sulphated flavone glycoside thalassiolin b isolated from the marine plant *Thalassia testudinum* Banks ex Konig. *Mar. Biotech.* 11:74–80.

Rengasamy, R. R. K., A. Rajasekaran, G. D. Micheline, and A. Perumal. 2012. Antioxidant activity of seagrasses of the Mandapam coast, India. *Pharm. Biol.* 40:182–187.

Rhoades, D. F. 1979. Evolution of plant chemical defense against herbivores. In *Herbivores: Their Interaction with Secondary Plant Metabolites*, G. A. Rosenthal and D. H. Janzen, (Eds.), pp. 3–44, New York: Academic Press.

Rice-Evans, C. A., N. J. Miller, and G. Paganga. 1996. Structure-antioxidant activity relationships of flavonoids and phenolic acids. *Free Rad. Biol. Med.* 20:933–946.

Ross, C., M. P. Puglisi, and V. J. Paul. 2008. Antifungal defenses of seagrasses from the Indian River Lagoon, Florida. *Aquat. Bot.* 88:134–141.

Rowley, D. C., M. S. T. Hansen, D. Rhodes et al. 2002. Thalassiolins A-C: New marine-derived inhibitors of HIV cDNA integrase. *Bioorg. Med. Chem.* 10:3619–3624.

Rukachaisirikul, V., S. Kannai, S. Klaiklay, S. Phongpaichit, and J. Sakayaroj. 2013. Rare 2-phenylpyran-4-ones from the seagrass-derived fungi *Polyporales* PSU-ES44 and PSU-ES83. *Tetrahedron* 69:6981–6986.

Sánchez-Rangel, J. C., J. Benavides, J. B. Heredia, L. Cisneros-Zevallos, and D. A. Jacobo-Velázquez. 2013. The Folin–Ciocalteu assay revisited: Improvement of its specificity for total phenolic content determination. *Analyt. Meth.* 4:4990–4999.

Schultz, J. C., H. M. Appel, A. P. Ferrieri, and T. M. Arnold. 2013. Flexible resource allocation during plant defense responses. *Front. Plant Sci.* 4:1–11.

Short, F. T. 1987. Effects of sediment nutrients on seagrasses: Literature review and mesocosm experiment. *Aquat. Bot.* 27:41–47.

Sieg, R. D., and J. Kubanek. 2013. Chemical ecology of marine angiosperms: Opportunities at the interface of marine and terrestrial systems. *J. Chem. Ecol.* 39:687–711.

Silva, F. A., F. Borges, C. Guimarães, J. L. Lima, C. Matos, and S. Reis. 2000. Phenolic acids and derivatives: Studies on the relationship among structure, radical scavenging activity, and physicochemical parameters. *J. Agric. Food Chem.* 48:2122–2126.

Silva, J. M. C., N. Dantas-Santos, D. L. Gomes et al. 2012. Biological activities of the sulfated polysaccharide from the vascular plant *Halodule wrightii*. Brazil. *J. Pharmacogn.* 22:94–101.

Steele, L., and J. F. Valentine. 2012. Idiosyncratic responses of seagrass phenolic production following sea urchin grazing. *Mar. Ecol. Prog. Ser.* 466:81–92.

Steele, L., M. Caldwell, A. Boettcher, and T. Arnold. 2004. Seagrass-pathogen interactions: "Pseudo-induction" of turtlegrass phenolics near wasting disease lesions. *Mar. Ecol. Prog. Ser.* 303:123–131.

Stern, J. L., A. E. Hagerman, P. D. Steinberg, F. C. Winter, and J. A. Estes. 1996. A new assay for quantifying brown algal phlorotannins and comparisons to previous methods. *J. Chem. Ecol.* 22:1273–1293.

Sullivan, B. K., T. D. Sherman, V. S. Damare, O. Lilje, and F. H. Gleason. 2013. Potential roles of *Labyrinthula* spp. in global seagrass population declines. *Fungal Ecol.* 6:328–338.

Supaphon, P., S. Phongpaichit, V. Rukachaisirikul, and J. Sakayaroj. 2013. Antimicrobial potential of endophytic fungi derived from three seagrass species: *Cymodocea serrulata*, *Halophila ovalis* and *Thalassia hemprichii*. *PLoS One* 8:e72420.

Sureda, A., A. Box, J. Terrados, S. Deudero, and A. Pons. 2008. Antioxidant response of the seagrass *Posidonia oceanica* when epiphytized by the invasive macroalgae *Lophocladia lallemandii*. *Mar. Env. Res.* 66:349–363.

Thayer, G. W., K. A. Bjorndal, J. C. Ogden, S. L. Williams, and J. C. Zieman. 1984. Role of larger herbivores in seagrass communities. *Estuaries* 7:341–376.

Todd, J. S., R. C. Zimmerman, P. Crews, and R. S. Alberte. 1993. The antifouling activity of natural and synthetic phenolic acid sulfate esters. *Phytochemistry* 34:401–404.

Valentine J. F., and J. E. Duffy. 2006. The central role of grazing in seagrass ecology. In *Seagrasses: Biology, Ecology and Conservation*, A. W. D. Larkum, R. J. Orth, and C. M. Duarte (Eds.), pp. 463–401, Dordrecht, the Netherlands: Springer.

Van Alstyne, K. L. 1994. Comparison of three methods for quantifying brown algal polyphenolic compounds. *J. Chem. Ecol.* 21:44–48.

Van der Heide, T., L. L. Govers, J. de Fouw, H. Olff, M. vander Geest, M. M. Van Katwijk, T. Piersma, J. van de Koppel, B. R. Silliman, A. J. P. Smolders, and J. A. va Gils. 2012. A three-stage symbiosis from the foundation of seagrass ecosystems. *Science* 336:1432–1434.

Vergeer, L. H. T., and A. Develi. 1997. Phenolic acids in healthy and infected leaves of *Zostera marina* and their growth-limiting properties towards *Labyrinthula zosterae*. *Aquat. Bot.* 48:64–72.

Vergeer, L. H. T., T. L. Aarts, and J. D. Degroot. 1994. The wasting disease and the effect of abiotic factors (light intensity, temperature, salinity) and infection with *Labyrinthula zosterae* on the phenolic content of *Zostera marina* shoots. *Aquat. Bot.* 42:34–44.

Vergeer, L., and C. den Hartog. 1994. Omnipresence of Labyrinthulaceae in seagrasses. *Aquat. Bot.* 48:1–20.

Vergés, A., T. Alcoverro, and J. Romero. 2011. Plant defences and the role of epibiosis in mediating within-plant feeding choices of seagrass consumers. *Oecologia* 166:381–390.

Vergés, A., M. A. Becerro, T. Alcoverro, and J. Romero. 2007a. Experimental evidence of chemical deterrence against multiple herbivores in the seagrass *Posidonia oceanica*. *Mar. Ecol. Prog. Ser.* 343:107–114.

Vergés, A., M. A. Becerro, T. Alcoverro, and J. Romero. 2007b. Variation in multiple traits of vegetative and reproductive seagrass tissues influences plant–herbivore interactions. *Oecologia* 141:674–686.

Wang, J. Y., X. R. Pan, Y. Han, D. S. Guo, Q. Q. Guo, and R. G. Li. 2012. Rosmarinic acid from eelgrass shows nematicidal and antibacterial activities against pine wood nematode and its carrying bacteria. *Mar. Drugs* 10:2729–2740.

Waycott, M., C. M. Duarte, T. J. B. Carruthers et al. 2009. Accelerating loss of seagrasses across the globe threatens coastal ecosystems. *Proc Nat. Acad. Sci.* 106:12377–12381.

Williams, S. L., and K. L. Heck. 2001. Seagrass community ecology. In *Marine Community Ecology*, M.D. Bertness, S. D. Gaines, and M. E Hay (Eds.), pp. 317–337, Sunderland, MA: Sinauer Associates.

Yee, A. K., K. L. Van Alstyne, and D. K Padilla. 2014. Chemical signaling in an inducible offence. *Int. Comp. Biol.* http://www.sicb.org/meetings/2014/SICB2014AbstractBook.pdf.

Yuvaraj, N., P. Kanmani, R. Satishkumar, A. Paari, V. Pattukumar, and V. Arul. 2012. Seagrass as a potential source of natural antioxidant and anti-inflammatory agents. *Pharm. Biol.* 40:448–467.

Zapata, O., and C. McMillan. 1979. Phenolic acids in seagrasses. *Aquat. Bot.* 7:307–317.

Zaporozhets, T. S., N. N. Besednova, G. P. Lyamkin, Y. N. Loenko, and A. A. Popov. 1991a. Antibacterial and therapeutic efficacy of pectin from *Zostera*, a seagrass. *Antibiot. Khim.* 36:24–26.

Zaporozhets, T. S., N. N. Besednova, G. P. Lyamkin, Y. N. Loenko, and A. A. Popov. 1991b. Immunomodulating properties of pectin from *Zostera* seagrass. *Antibiot. Khim.* 36:31–34.

5 The Evolution of Marine Herbivores in Response to Algal Secondary Metabolites

Erik E. Sotka, Veijo Jormalainen,
and Alistair G. B. Poore

CONTENTS

5.1 INTRODUCTION

Interactions between macrophytes and marine herbivores play central roles in regulating and structuring nearshore communities, biodiversity, and their cycling of nutrients and materials, and can determine the success of human introductions and poleward expansion of seaweeds due to a warming ocean (Lubchenco and Gaines 1981; Schiel and Foster 1985; Hay and Fenical 1988; Duffy and Hay 2001; Steneck et al. 2002; Stachowicz et al. 2007; Ling et al. 2008; Poore et al. 2012; Verges et al. 2015). These impacts occur despite the fact that marine herbivores face profound challenges when feeding on seaweeds. Although seaweeds are generally of higher quality food than terrestrial plants they are among the low-quality producers in marine ecosystems (Cebrian and Lartique 2004). Seaweeds contain low levels of nitrogen relative to herbivore tissues, produce physical structures (e.g., crusts and calcification) that make the algae tough and more difficult to digest, and contain an arsenal of secondary metabolites that deter herbivores (Mattson 1980; Littler and Littler 1980; Paul 1992; Paul et al. 2001; Targett and Arnold 2001; Amsler and Fairhead 2006).

In response to seaweed defenses, herbivores employ multiple strategies (termed "herbivore offenses," Karban and Agrawal 2002) that allow them to tolerate or avoid poorer-quality foods. Offensive traits are determined from the herbivores' point of view and represent their evolutionary solutions to the challenges of feeding on structurally and chemically defended seaweeds (Sotka et al. 2008, 2009; Jormalainen 2015). In this chapter, we outline our current understanding of the evolution of herbivore offenses, with a focus principally on responses to algal chemical defenses. We briefly outline the spatial and temporal variation in seaweed defenses, highlight the range of herbivore responses to seaweed defenses, and review the evidence for their genetic basis and ecological constraints on herbivore responses. We end with a review of micro- and macroevolutionary patterns in herbivore offenses documented to date.

5.2 VARIATION IN SEAWEED DEFENSIVE TRAITS

Variation in defensive traits of aquatic macrophytes is a precondition for the evolution of feeding preferences and locally varying selection for herbivore tolerance. This variation occurs in different scales, from within plant variation among different tissues (Van Alstyne et al. 2001a; Jormalainen and Honkanen 2008; Pavia and Toth 2008) to variation among genotypes (e.g., Wright et al. 2004; Jormalainen et al. 2011; Tomas et al. 2011), populations (e.g., Hemmi and Jormalainen 2004b; Koivikko et al. 2008), and species (Paul et al. 2001; Rasher et al. 2013). In addition, chemical defenses are notoriously plastic and are known to vary across individuals in response to local abiotic environments (reviewed in Amsler and Fairhead 2006; Jormalainen and Honkanen 2008) and the induction of chemical defenses following damage (reviewed in Toth and Pavia 2007; Pavia and Toth 2008).

The net result of both within- and between-species variation is that macrophyte assemblages will consistently differ with respect to the frequency and strength of defenses they harbor. Assemblages that are exposed to intensive and persistent herbivory will likely have greater numbers of chemically defended species and genotypes, relative to assemblages with lower herbivory (e.g., Siska et al. 2002). Differences in the mean palatability of multi-seaweed assemblages have been tested only rarely. For example, Thornber and Stachowicz (2008) offered intertidal and subtidal pairs of closely related species to intertidal and subtidal consumers, testing the notion that subtidal species will have greater herbivory pressure than intertidal species. Their results did not find a general intertidal gradient in palatability or morphological or chemical defenses. Other gradients in mean palatability across macrophyte assemblages that have been tested are across biogeographic gradients (among hemi-spheres or between tropical and temperate latitudes), but these are discussed in more detail in the context of macroevolutionary patterns (Section 5.5).

5.3 HERBIVORE OFFENSIVE TRAITS

In response to spatial and temporal variation in seaweed chemical defenses, herbivores employ a range of behavioral, physiological, and morphological adaptations, collectively known as herbivore offenses (Karban and Agrawal 2002). These represent their evolu-tionary solutions to the challenges of feeding on structurally and chemically defended seaweeds (Sotka et al. 2009). Our focus here is principally on responses to chemical defenses (i.e., herbivore feeding behavior and physiology). Herbivore adaptations that evolved primarily to cope with seaweed structural defenses in crustaceans (Jormalainen 2015), mollusks (Steneck and Watling 1982; Padilla 1985), and herbivorous fish (Horn and Messer 1992; Horn and Ferry-Graham 2006) have been reviewed elsewhere.

5.3.1 HERBIVORE BEHAVIOR

Despite the prevalence of generalism among marine herbivores, choosiness among seaweed species (e.g., Duffy and Hay 1991; Pennings et al. 1993; Poore et al. 2000; Cruz-Rivera and Hay 2000; Jormalainen et al. 2001; Raubenheimer et al. 2005; Taylor and Brown 2006; Crawley and Hyndes 2007) and among tissues within indi-viduals (e.g., Honkanen et al. 2002; Pavia et al. 2002; Taylor et al. 2002) is ubiquitous and well documented. These feeding preferences can be mediated by nutritive qual-ity and secondary metabolites and by non-nutritional factors such as habitat choice (especially for mesograzers). It is, however, worth noting that evolution of feeding specialization through selection for food utilization ability or through selection for habitat are not mutually exclusive but rather act together: the associations of small mesograzers with their seaweed hosts may well be initially driven by predation avoidance but the association as such will then select for better host use ability (Hay et al. 1989, 1990b; Duffy and Hay 1991).

Several kinds of evidence imply that secondary chemicals play a major role in feed-ing preferences. First, induced resistance of macroalgae, which is typically measured by feeding bio-assays, often covaries with the increasing contents of secondary metabolites. Second, bioassay-guided fractionation of herbivore-deterrent extracts has provided ample evidence for the role of a wide variety of both polar and lipid soluble macroalgal second-ary metabolites on feeding preferences (Pennings et al. 1999; Becerro et al. 2001; Taylor et al. 2003; Kubanek et al. 2004; Van Alstyne et al. 2006; Enge et al. 2012; reviewed in Paul et al. 2001 and Van Alstyne et al. 2001). Third, field observations have demonstrated that macrophyte susceptibility to grazing negatively covaries with the contents of their secondary metabolites (Steinberg 1984; Jormalainen and Ramsay 2009).

Although most marine herbivores display feeding choices among alternate diets, analyses of their gut contents commonly indicate a diverse resource base. Some spe-cies actively maintain a mixed diet, with evidence that individuals feed at greater rates on macrophytes that were not encountered previously (e.g., the amphipod *Peramphithoe parmerong*, Poore and Hill 2006; the sea urchin *Strongylocentrotus droebachiensis*, Lyons and Scheibling 2007; and the opisthobranch *Dolabella auricularia,* Pennings et al. 1993). In many mesograzers, diet changes with ontogeny thus giving rise to life-history dietary generalism (e.g., Hemmi and Jormalainen 2004a; Hultgren and Stachowicz 2010; Williamson and Steinberg 2012).

The benefit of mixing diets is that it broadens the available resource base and provides the herbivore equal if not better fitness than when reared on a single species diet (e.g., Pennings et al. 1993; Cruz-Rivera and Hay 2000, 2001; Hemmi and Jormalainen 2004a; Vesakoski et al. 2008; Aquilino et al. 2012; see review in Stachowicz et al. 2007 and Lefcheck et al. 2013). A prominent mechanistic hypothesis is that this dietary generalism dilutes plant toxins by adding less toxic species or species containing toxins that are dealt with a different kind of absorbance limitation or detoxification pathway (e.g., Barreiro et al. 2007; Sotka and Gantz 2013). Mixing diets is also thought to allow herbivores to balance their nutrient intake. In practice, both nutrients and secondary metabolites occur simultaneously within macrophyte tissue and it can be difficult to disentangle their independent and interactive effects on feeding behavior and herbivore fitness (Forbey et al. 2013).

Decorating behavior by majoidean crabs (Hultgren and Stachowicz 2011) represents another offensive behavior, as decoration can co-opt plant chemical defenses for the herbivore's own protection from larger consumers (Stachowicz and Hay 2000; Cruz-Rivera 2001; Rorandelli et al. 2007; Vasconcelos et al. 2009). As an example, the chemically rich *Dictyota* minimizes predation rates by omnivorous pinfishes on juveniles of the crab *Libinia dubia* because fishes avoid consuming the diterpene compounds of *Dictyota* (Stachowicz and Hay 1999). In most cases, the chemically defended alga used as protection décor is a low preference food (see also Amsler et al. 1999 for an urchin example).

5.3.2 PHYSIOLOGICAL ADAPTATIONS

After choosing to consume a chemically rich food, herbivore responses will largely be mediated by their physiological traits, including the length and conditions of the digestive tract (Horn and Messer 1992; Choat and Clements 1998; Targett and Arnold 2001), regulation of absorption and detoxification of the dietary toxins (McLean and Duncan 2006), and sequestration of secondary metabolites from the diet. The detailed physiological mechanisms of how herbivores deal with plant defensive compounds are still relatively poorly known (Sotka and Whalen 2008; Sotka et al. 2009; Forbey et al. 2013). We briefly outline herbivore traits that respond to seaweed chemical defenses and that may serve as targets of natural selection.

Gut conditions: Herbivore gut conditions can modify the potentially harmful effects of brown algal phlorotannins and red algal phenolics. Phlorotannins are not easily absorbed due to their polymeric nature and large size. Their harmful action has conventionally been attributed to their ability to form insoluble precipitates with dietary and endogenous proteins in the gut, thereby preventing nutrient absorption (Stern et al. 1996). On the other hand, as phenolic compounds have high oxidative capacity they may also act as auto-oxidants in the gut causing oxidative damage and cytotoxic effects on gut epithelium

(Barbehenn et al. 2005). The protein precipitation reactions require acidic to neutral conditions while oxidation occurs particularly under high pH (Salminen and Karonen 2011). The slightly alkaline nature of the sea water, when reflected in gut environment, may increase the importance of oxidation activity but this still remains to be shown in marine herbivores. Other chemical traits of the gut also mediate phlorotannin effectiveness (surfactants, microbial activity, redox potential; reviewed in Tugwell and Branch 1992; Targett and Arnold 2001) and have the potential to create herbivore offence traits.

Efflux transporters and detoxification: The absorption of secondary metabolites through the gut epithelium is mainly based on passive diffusion down a concentration gradient and depends on lipid solubility of the compound, with the more lipophilic compounds permeating membranes more easily (McLean and Duncan 2006). Absorbed compounds can, however, be actively transported back to the gut and excreted into feces (Sorensen and Dearing 2006). Such excretion is based on so-called efflux transporters, permeability glycoproteins, and other multidrug-resistance-associated proteins, which pump foreign molecules out of the body into the gut. These are aided by cytochromes P-450 and glutathione S-transferases (GST), which together limit or prevent absorption of secondary metabolites through several, non-exclusive mechanisms (Sorensen and Dearing 2006). While secondary metabolites act as substrates for efflux transporters, their presence or the presence of other diet derived metabolites may either facilitate or inhibit excretion activity of efflux transporters (McLean and Duncan 2006; Forbey et al. 2013).

The occurrence of xenobiotic transporter mechanisms has been frequently demonstrated in aquatic animals (clams, mussels, snails, crabs, shrimps, fish), particularly in the context of organism response to pollutants (Bard 2000). Furthermore, xenobiotic transporter mechanisms in marine consumers can be induced as a response to diet-derived toxin substrates (See Kuhajek and Schlenk 2003 for red algal metabolite effects on a chiton; Ame et al. 2009 for cyanotoxin effect on a fish; Whalen et al. 2010 for efflux transporters in predatory gastropods; Huang et al. 2014 for dinoflagellate toxin effects on a mussel). In addition, the activities of biotransformation enzymes have been found to be particularly high in herbivorous fish (P-450s; Stegeman et al. 1997) and sea turtles (GSTs; Richardson et al. 2009), as compared to carnivorous ones, suggesting a role for these enzymes in metabolizing plant-derived secondary compounds.

Sequestration of secondary metabolites: Sequestration of secondary metabolites from the diet and distributing them among tissues is an offensive trait used by many species of opisthobranch mollusks. Such sequestration of secondary metabolites from food algae is understood as a diet-derived chemical defense mechanism (reviewed in Avila 1995; Ginsburg and Paul 2001; Rogers et al. 2002; Marin and Ros 2004; Baumgartner et al. 2009). The compounds sequestered from algae include mono-, di-, tri- and sesquiterpenoids, steroids, halogenated furanones, nitrogenated compounds, and phenolics (Avila 1995). Known algal hosts include mainly red and green algae, but also some brown algae and cyanobacteria (Avila 1995).

The detailed physiological mechanisms how the sequestered chemicals are chosen and transported are poorly known, but the alga-feeding opisthobranches store defensive metabolites in various epidermal, subepithelial, and specific glandular structures (Wägele et al. 2006). In addition, algal metabolites have been found in mucous and opaline secretions, egg masses, and the defensive ink sprayed towards attacking predator (Avila 1995; Rogers et al. 2000; Johnson et al. 2006). Absorbed algal metabolites are either transported and used as such or biotransformed to novel defensive compounds. For example, certain sacoglossan species transform caulerpenyne from green algae into more toxic oxytoxins (Marin and Ros 2004). The occurrence of algal metabolites in the tegument, mucous secretions, and ink highlight their antipredator function although they may have other functions such as acting as antimicrobial compounds. Accumulation of compounds in glands may have initially been a way to excrete and avoid autotoxicity of dietary chemicals, which have then evolved to a defensive mechanism (Wägele et al. 2006). There is recent evidence to suggest that other marine herbivores may gain protection from fish predation after consuming chemically defended algae (the herbivorous amphipod *Paradexamine fissicauda* consuming the red alga *Plocamium cartilagineum*, Amsler et al. 2013). It is not yet known whether this protection derives from active sequestration of metabolites or passive accumulation of dietary metabolites.

5.4 POTENTIAL FOR HERBIVORE ADAPTATION

In order for herbivore behavior and physiology to evolve in response to plant secondary metabolites, there must be fitness consequences of consuming metabolites, and a heritable basis to variation in herbivore traits must exist. Here, we review the evidence for each.

5.4.1 FITNESS CONSEQUENCES

Fitness consequences of consuming macrophyte secondary metabolites are relatively poorly known, given the large number of studies that effects of metabolites on herbivore feeding behavior (Section 5.2). Most fitness studies using marine herbivores isolate individuals with whole tissue from macrophyte species (e.g., Arrontes 1990; Duffy and Hay 1991; Poore and Steinberg 1999; Jormalainen et al. 2001; Sotka and Hay 2002; Taylor and Brown 2006). Relatively few studies have established that variation in fitness across macrophytes is due to variation in the identity and concentration of secondary metabolites. Similarly, there are few studies that have considered the fitness consequences to herbivores of intraspecific variation in secondary metabolites (e.g., Taylor et al. 2003; Toth et al. 2005, Haavisto et al. 2010) despite the fact that such variation is common (Section 5.1).

The fitness consequences of consuming brown algal phlorotannins has mixed effects on herbivorous fish, urchins, gastropods, and crustaceans (reviewed by Targett and Arnold 2001; Amsler and Fairhead 2005). Assimilation efficiency and growth decline with increasing phlorotannin concentrations for some herbivores (e.g., the isopod *Idotea baltica*, Jormalainen et al. 2005; the fish *Xiphister mucosus*, Boettcher and Targett 1993), while other herbivores are unaffected (e.g., urchins and gastropods of south-eastern Australia, Steinberg and van Altena 1992) or display enhanced performance on diets rich in phlorotannins (the amphipod *Ampithoe valida*, Kubanek et al. 2004).

In most studies using non-polar metabolites, herbivore fitness generally declines with increasing concentrations of metabolites. For non-polar metabolites, Schnitzler et al. (2001) demonstrated that survival of the amphipod *Ampithoe longimana* was reduced by the inclusion of dithiepanone and 9-oxo acid from the brown alga *Dictyopteris membranacea*. Diterpene alcohols from the brown alga *Dictyota menstrualis* reduced growth, survival, and fecundity of four species of amphipod, but enhanced survival of the isopod *Paracerceis caudata*, and effects were modified by the nutritional quality of the diets in some cases (Cruz-Rivera and Hay 2003).

Thus, while the body of evidence remains relatively weak, the strong impacts of metabolites on herbivore behavior and at least in some cases, on fitness, indicates that metabolites can be a selective agent on herbivore response.

5.4.2 Genetic Variation

The potential for selection by secondary metabolites to engender evolutionary change depends on heritability of herbivore traits. The feeding behaviors and traits closely associated with fitness of several mesograzers are known to be heritable. The growth and feeding preferences of the amphipod *Amphithoe longimana* when feeding on chemically rich *Dictyota* vary significantly among full-sib families (Duffy and Hay 1991; Sotka et al. 2003), while preference for the palatable red *Hypnea musciformis* did not vary. Hybrids of *A. longimana* populations with high versus low preference for *Dictyota* had intermediate levels of preference, indicating that *Dictyota* preference was likely an autogenic and quantitative trait (Sotka 2003). In the isopod *Idotea baltica*, growth rate was highly heritable when raised on diets that vary in breadth (the brown alga *Fucus vesiculosus* alone vs *F. vesiculosus* and its epiphyte *Pilayella littoralis*) and nutrient enrichment of those diets (Hemmi and Jormalainen 2004a). For the amphipod *Peramphithoe parmerong*, survival on a poor-quality host, the brown alga *Padina crassa*, varies significantly among full-sib families, while there was no evidence of heritable variation for performance on the high-quality *Sargassum linearifolium* (Poore and Steinberg 2001).

Specific genetic markers have been associated with feeding preference for seaweeds in the amphipod *Gammarus palustris*. The frequency of different isozymes of the amylase varies with feeding preferences for two species of green algae (Borowsky et al. 1985). The isozymes differ in their efficiency in metabolizing the principal starches within each food, and amphipods preferred to consume the alga for which they had the most appropriate isozyme.

Genetic variation of feeding preference can manifest itself as individual specialization (Araújo et al. 2011). Within populations of two generalist marine herbivores, relatively specialized individuals are known to differ in their feeding behavior. Individuals of the sacoglossan *Elysia viridis* preferred green algal hosts similar to the one from which they collected, but subsequent growth was independent of the original host (Baumgartner et al. 2014). In a more extreme example of individual specialization, individuals of the sacoglossan *Placida dendritica* do not survive when switched from their original green algal host to species consumed by other individuals of this herbivore (Trowbridge 1991). Other tests for individual specialization have found that feeding preferences were strongly influenced by recent diet and a rather weak effect of feeding canalization by individuals (Poore and Hill 2006). The role, if any, of seaweed chemical defenses in mediating the individual specialization of these examples was not reported.

While the studies are relatively few in number, especially in contrast to the extensive research with terrestrial insect herbivores (e.g., Futuyma and Peterson 1985), they do indicate considerable heritable variation in feeding behavior or performance within populations of marine herbivores. Further testing of the genetic basis to macrophyte use, however, is required to better predict evolutionary responses to secondary metabolites. In particular, we have no knowledge of possible trade-offs in performance among diets (an important concept underlying models of dietary specialization (Dethier 1954; Via 1984), or trade-offs between diet choice and other traits relating to fitness. Testing for negative genetic correlations among suites of traits is required to establish whether such trade-offs may constrain responses to directional selection on tolerance to metabolites (Section 5.5.1).

5.4.3 CONSTRAINTS ON EVOLUTION

Herbivore behavior and physiology may respond strongly to macrophyte secondary metabolites, but their interaction does not occur in a vacuum. The dynamics of herbivore populations depend on other ecological factors that vary spatially and temporally and that can profoundly alter an herbivore's interaction with macrophytes. We cover three factors that have garnered much attention: competition with other herbivores, predation risk, and interactions with gut microbiota.

One long-standing hypothesis for the evolution of feeding preferences is that herbivores will specialize or evolve greater feeding preferences for plants that minimize competition with other herbivores (Strong et al. 1984). Such competition can play a role in determining marine herbivore density and host use. For example, Best et al. (2013) showed that the number of coexisting species in a community of grazing amphipods increased with the diversity of their feeding traits, although it seems likely that these mesograzers are competing for sheltered habitats, rather than for food (Best and Stachowicz 2014). A meta-analysis of the strength of competition over ecosystems and trophic levels found statistically significant intermediate-sized effects of competition in marine herbivorous mollusks and echinoderms (Gurevitch et al. 1992). Competition for food may occur periodically during mass occurrences of herbivores, for example echinoderms (Estes et al. 2004), amphipods (Tegner and Dayton 1991), and isopods (Engkvist et al. 2000), during which the host availability has been seriously depleted. However, under most conditions, marine grazers are rarely food-limited because seaweeds typically have high primary production, and herbivores tend to be very generalized in their feeding preferences and opportunistically feed on animal tissue or dead, decomposed, or allochthonous material (e.g., Pennings et al. 2000; Norderhaug et al. 2003; Lastra et al. 2010).

Another, perhaps more common, modifier of the ecology of marine herbivores is predator risk. Herbivores lie between "devil and the great blue sea" (i.e., predators and producers) and their fitness is determined not just by their food utilization ability but also by their ability to avoid predators. Avoiding predation may often be more important than competition for food for their population dynamics, and, consequently, traits related to predation avoidance may be strongly selected for. These involve various behavioral strategies and camouflage (Hacker and Madin 1991; Jormalainen et al. 1995; Merilaita 2001; Hultgren and Stachowicz 2008), but also chemical mediation where algal chemical defenses are used as the herbivore's antipredator traits: For example, several amphipods use chemically deterrent host algae to build domiciles (Hay et al. 1990a; Sotka et al. 1999; Gutow et al. 2011), and, as discussed previously, decorator crabs and opisthobranches gain protection against predators by using algal metabolites.

Hence, predation may modify the evolutionary outcome of the plant–herbivore interaction: with decreasing grazing pressure, the fitness effects of herbivores on producers diminish (Haavisto and Jormalainen 2014). This implies weaker selection for producer traits while selection for the herbivore traits may remain intense. Variation in predation may therefore contribute not just to the magnitude of grazing but also the evolutionary outcomes of plant–herbivore interactions.

Finally, many herbivores possess obligate or facultative microbial gut symbionts, the effects of which may vary from pathogenic to mutualistic (Dillon and Dillon 2004). In particular, herbivores that feed on low-quality diets may rely on microbial symbionts that can contribute to digestion enzymatically and/or provide essential nutrition lacking from the diet. Microbial symbionts may also contribute to the detoxification of diet-derived secondary metabolites either in the gut or in the digestive glands (Engel and Moran 2013; Forbey et al. 2013). Although our knowledge on the gut microbiota and its contribution to digestion in marine herbivores is limited, gut bacteria with digestive activity have been found in isopods, amphipods, decapods, mollusks, and echinoderms (reviewed in Harris 1993; Pinn et al. 1997; Klussmann-Kolb and Brodie 1999; Zimmer and Bartholme 2003; Mattila et al. 2014) as well as in vertebrate macrograzers such as fish (reviewed in Clements et al. 2014), marine turtles (Arthur et al. 2014), and dugongs (Nelson et al. 2013). Symbiotic bacteria have also been suggested to play a role also in zooplankton herbivory (Tang et al. 2010). It is, thus, possible that environmental availability of bacteria, or, the diet composition itself, may modify the gut microbiota and its contribution to the host utilization ability. For example, the isopod *Idotea balthica* harbors bacterial symbionts that differ in community composition between populations living in different host assemblages (Mattila et al. 2014). Furthermore, as the digestive contributions of gut microbiota can be seen as an herbivore offense trait, characteristics of gut or digestive glands may evolve to foster certain kinds of gut microbiota. Thus, mutualistic interactions with microbiota can mediate host-use making them a potentially important factor in host–herbivore coevolution.

5.5 MICROEVOLUTION OF HERBIVORE OFFENSES

5.5.1 FIELD-COLLECTED POPULATIONS AND RECIPROCAL FEEDING EXPERIMENTS IN COMMON-GARDEN

Marine mesograzers are analogous to terrestrial herbivorous insects in their small size, their intimate interactions with their host plants (for food and shelter), and their critical roles in their community as both prey and proficient consumers (Hay et al. 1987; Duffy et al. 2001; Jaschinski and Sommer 2008; Duffy et al. 2015). Terrestrial insects readily evolve population-level differences in host use traits (Thompson 2005), suggesting that marine mesograzers should similarly local adapt. Surprisingly, however, we have relatively few such examples among marine mesograzers (Sotka 2005) despite their potential to evolve (Section 5.3) and the spatial variation in the distribution and abundance of seaweeds and their chemical defenses.

To our knowledge, the first test of local adaptation in host preference or performance in a marine consumer was with the amphipod *Peramphithoe parmerong* (Poore and Steinberg 2001). Mothers were collected from either *Sargassum lineari-folium* or *Padina crassa* individuals within a single bay (meters apart), and from bays in which *P. crassa* did not occur, in southeastern Australia. Despite strong heritable variation in choosing and performing on at least one of the hosts (*P. crassa*), there was no evidence of local adaptation to this host on either spatial scale tested. It is likely that movement among these host plants is too frequent (Poore and Hill 2006) to allow for any differential selection to generate locally adapted demes on these plants.

The first demonstration of geographic variation in feeding preferences comes from *Ampithoe longimana*, an abundant amphipod of estuarine communities along the U.S. Atlantic shoreline. In warm-temperate waters of North Carolina, *A. longimana* preferentially lives and feeds on chemically defended brown seaweeds in the genus *Dictyota*, in large part because diterpene alcohols produced by the genus provide a refuge from consumption by large omnivorous fishes (Duffy and Hay 1991, 1994). *A. longimana* populations sympatric with *Dictyota* more readily consume two species of *Dictyota* and its lipophilic metabolites and had greater growth and fecundity when reared on *Dictyota*, than did allopatric populations within the colder waters of New England where *Dictyota* does not occur (Sotka and Hay 2002; Sotka et al. 2003). It is likely that natural selection, driven largely by omnivorous fish predation pressure, has favored greater feeding preferences for *Dictyota* in warm-water areas where amphipods co-occur with *Dictyota*.

A subsequent study (McCarty and Sotka 2012) tested whether this *A. longimana*-*Dictyota* interaction represents an exceptional case of local evolution in feeding preference, or alternatively, whether *A. longimana* will locally adapt to any seaweeds that occur at locally high densities. The study found significant regional variation or population-level variation in feeding preferences to seaweed genera that had known chemical defenses (*Dictyota, Caulerpa, Padina, Chondrus,* and *Fucus*). In contrast, feeding preference for seaweeds that generally do not produce anti-herbivore metabolites did not differ among herbivore populations (*Acanthophora, Ectocarpus, Gracilaria,* and *Hincksia/Feldmannia*). These patterns suggest that while evolution for greater feeding preference of locally important hosts can occur, such evolutionary responses are not uniform across seaweed hosts, and instead seem to depend more on whether the seaweed produces secondary metabolites.

One limitation of the *A. longimana* studies is that populations were compared across large spatial scales (100s of kilometers) that have been separated for 10s or 100s of thousands of generations (Sotka et al. 2003). Thus, it is not clear the extent to which selection or genetic drift may explain geographic variation in feeding patterns. In contrast, the isopod *Idotea balthica* shows geographic variation in juvenile fitness for alternative hosts between populations collected only a few kilometers apart. In the northern Baltic Sea, *I. balthica* occurs in various habitats but prefers *Fucus vesiculosus* as a host plant, a plant that can produce phlorotannins in quantities high enough (e.g., ~15%) to lower isopod growth. As with *A. longimana* on *Dictyota*, the isopod's preference for *F. vesiculosus* is related to its value as a shelter from visually hunting predatory fish (Jormalainen et al. 2001; Vesakoski et al. 2008). Vesakoski et al. (2009) reared populations, both mothers collected in areas dominated by either *Fucus* or the seagrass *Zostera marina* and their laboratory-born offspring and showed that while all populations performed best on *Fucus*, isopods from both assemblages performed better with their sympatric dominant host species than did isopods allopatric to this host.

Extending this pattern, Bell and Sotka (2012) compared New England and Virginia populations (100s of kilometers apart) of *I. balthica* that occur either on *Fucus* versus *Zostera,* respectively. *Fucus vesiculosus* was more readily consumed by northern populations sympatric with *Fucus* than by southern populations allopatric with *Fucus*. This geographic variation in feeding behavior was mediated by both water-soluble and lipophilic secondary metabolites. Similarly, southern *I. balthica* populations occur in habitats that are dominated by seagrasses and *U. linza*, and both hosts were consumed significantly more by southern than northern animals.

A final example is in the urchin *Arbacia punctulata* (Craft et al. 2013). When offered lipophilic extracts from nine subtropical seaweeds, subtropical populations of *Arbacia* consumed significantly more extract from *Dictyota ciliolata, D. pulchella,* and *S. zonale* than did cold-temperate populations that are naïve to these subtropical seaweeds. Unlike the previous examples, these animals were collected from the field, and thus, the study cannot rule out the possibility that previous experience affected feeding preferences.

All three species for which geographic variation in herbivore offensive traits have been detected are generalists. One consequence is that there must be geographic variation in the preference ranking of foods, as with some terrestrial generalist insects like grasshoppers and some beetles and butterflies (Thompson 2005; Singer 2008). This suggests that polyphagous herbivores are typically composed of populations diverged in their preferences rather than cryptic specialists at the individual or population level as commonly accepted for polyphagous insects (Fox and Morrow 1981; Bolnick et al. 2003; Sword et al. 2005). We propose that microevolutionary shifts in preference ranks are common because they allow generalist herbivores to minimize any fitness or ecological trade-offs that are incurred (Jormalainen et al. 2001; Sotka and Reynolds 2011), while maintaining the benefits of marine polyphagy (Hay and Steinberg 1992; Stachowicz et al. 2007; see Section 5.6.1).

5.5.2 SELECTION EXPERIMENTS

Selection experiments are a powerful approach to understanding evolutionary potential and constraints (Fry 2003). To our knowledge, Sotka and Reynolds
(2011) represent the only study that uses a controlled natural-selection experiment
to test responses to alternative diets in a marine herbivore. Replicate lines of the
polyphagous herbivore *Ampithoe longimana* isolated with only *Dictyota* displayed
greater feeding tolerance for *Dictyota* and its secondary metabolites compared to
lines isolated on alternative hosts (*Sargassum, Hypnea,* and *Ulva*). In addition,
Dictyota-line females become reproductive more quickly when on *Dictyota* than
did mixed-seaweed-line females. However, such adaptation appeared to come at a
fitness cost when on non-*Dictyota* seaweeds, as mixed-seaweed-lines grew more
quickly and produced more eggs when isolated on *Sargassum* and *Ulva* than did
Dictyota-line individuals. The observed cost does not appear to be symmetrical; that
is, a greater cost is incurred when *Dictyota*-adapted individuals are isolated on non-
Dictyota species, compared to when mixed-seaweed-lines are isolated on *Dictyota*.
Fitness costs when evolving greater preference for *Dictyota* may be strong enough to
drive the evolution of specialist feeding preferences for *Dictyota*. That *A. longimana*
does not specialize on *Dictyota* may be explained by the seasonal disappearance of
the tropically distributed genus during winter months of North Carolina.

5.5.3 NOVEL ASSOCIATIONS

Novel associations between herbivores and seaweeds arise when one is introduced
accidentally or deliberately. For example, Scottish populations of the herbivorous
sea slug *Elysia viridis* can be found abundantly on two green seaweeds; the native
Cladophora ruprestis and the recently (<50 years) introduced *Codium fragile* spp.
tomentosoides (hereafter *C. fragile*). In an admirable series of logistically difficult
assays, Trowbridge and Todd (2001) were able to demonstrate that *E. viridis* have
undergone a "host-switch" onto the introduced *C. fragile* within the last 50 years.
Adult sea slugs from *Cladophora* and from *C. fragile* will strongly prefer their
recent host relative to the alternative. The authors also collected batches of eggs
that were attached to the introduced *C. fragile* fronds and raised the emerged larvae for a month in the laboratory. Although these sea slug larvae metamorphosed
on both native and introduced hosts, neither postlarvae nor adults were able to
grow on the native *Cladophora* even though they were able to feed and grow on
the introduced *C. fragile*.

Unfortunately, as pointed by the authors, "it is not yet known whether the constraints (on the diet change) are based on genetic differences or on irreversible host-
induced changes triggered by parental diet" (pg. 235 in Trowbridge and Todd 2001).
Though it seems likely that genetic differences do account for the host-switch, phenotypic plasticity operating during an early life stage is also consistent with their
results. The ecological differences among individuals are particularly striking for an
animal with such tremendous potential for broad dispersal (i.e., *E. viridis* larvae are
planktonic for over 30 days). It is also unclear the extent to which chemical defenses
may underlie these patterns.

It is likely that other examples are present, given that at least 200 seaweeds have been introduced worldwide (Williams and Smith 2007), and chemical defenses have cited in facilitating success of several species (e.g., *Gracilaria vermiculophylla*, Hammann et al. 2013; *Fucus evanescens*, Wikström et al. 2006; *Bonnemaisonia hamifera*, Enge et al. 2012; *Caulerpa cylindracea*, Bulleri et al. 2015). It would be interesting to explore the frequency by which local herbivores add invasive seaweeds to their diet with or without evolutionary change.

5.6 MACROEVOLUTIONARY HYPOTHESES OF HERBIVORE OFFENSES

5.6.1 GENERALISM IN MARINE HERBIVORES

The host ranges of most small marine herbivore species remain undescribed. Previous reviews (Brawley 1992; Hay and Steinberg 1992) have argued that marine herbivores are largely generalists, feeding from more than 10 families from all divisions of seaweeds. These impressions have been mostly gleaned from studies that included large fish and urchin herbivores, animals that are as likely to display restricted host range as would large terrestrial mammals like bison or deer. The more appropriate herbivores to contrast with the well-described patterns of chemically mediated host specialization among terrestrial insects are among mesograzers such as slugs, isopods, amphipods, and small crabs because these herbivores are small relative to their host plants, and adults tend to be less mobile than those of fishes and urchins (Hay et al. 1987; Taylor and Steinberg 2005). Thus, mesograzers will be more sensitive to the selective environment imposed by any one seaweed species (Section 5.4.1).

The two groups of mesograzers for which host use has been well described are amphipods in the family Ampithoidae and the ascoglossan slugs. The two contrast profoundly. Ampithoid amphipods utilize a wide variety of taxonomically unrelated hosts from 20 orders across all three divisions of macroalgae and from 10 genera of seagrasses. This diversity is reflected within many individual species, with almost 60% of the amphipod species recorded on at least two or more host genera (Poore et al. 2008). While there are well-studied examples of relatively specialized ampithoid species (e.g., *Pseudamphithoides incurvaria* Hay et al. 1990a; *Peramphithoe tea* Sotka 2007), most of the species found on a single host genus were recorded from only one study, and thus the apparent specialization is likely to reflect insufficient data on these species. If amphipods are "insect-like" as has been previously proposed (Hay et al. 1987), then their rate of host specificity (~10% on one genus) is more in line with that of tropical root-feeders (Novotny and Basset 2005)

In contrast to the ampithoids, ascoglossans are far more restricted in their host range with virtually all ascoglossan slugs feeding on chlorophyte algae, and all shelled species feeding from the single genus *Caulerpa* (Jensen 1997; Marin and Ros 2004). The high specificity of ascoglossan slugs is analogous to that of lepidopteran caterpillars or the combined grouping of leaf-chewing and sap-sucking insects. In each of these groups, the majority of species utilize only one plant family (Novotny and Basset 2005). Unfortunately, we lack similar analyses of host use from most other groups of herbivores (isopods, crabs).

It is counter-intuitive that amphipods are generalists while sea slugs have restricted host use. Ampithoid amphipods and isopods are brooders that produce crawl-away offspring that disperse very locally (except when able to occasionally raft on floating seaweeds; e.g., Sotka et al. 2003) while ascoglossans have planktonically dispersed larvae that potentially disperse 10s of kilometers. Ampithoid mothers can essentially choose the host plant of their offspring in a manner analogous to butterflies (Poore and Steinberg 1999) while ascoglossan larvae must choose their host plants upon recruitment from the water column to the benthos. These traits predict that the amphipods, not the ascoglossans, should have restricted host use, while the opposite pattern is found (Poore et al. 2008).

There has been some recent development of hypotheses surrounding generalism in marine herbivores. First, while there is evidence for fitness tradeoffs in using alternative hosts, these tradeoffs are not strong enough to drive the evolution of specialized feeding preferences (sensu Fry 2003). As mentioned in Section 5.4.2, the isopod *Idotea balthica* and amphipod *Ampithoe longimana* both displayed tradeoffs that were likely mediated by chemical defenses. Similarly, molluscan herbivores (snails and slugs) produce radular types that differ in their effectiveness when grazing particular seaweeds (Steneck and Watling 1982), and tradeoffs arise when a mismatch between radular type and the available seaweed occurs (Padilla 1985; Trowbridge 1991).

At first glance, it is a paradox that tradeoffs exist within polyphagous herbivores given that tradeoffs should favor the evolution of specialization. One resolution of this conflict is that the benefit of using multiple hosts (i.e., broadening the resource base) outweighs its fitness cost (Poore et al. 2008). In *A. longimana*, polyphagy appears to be favored because it broadens resource use across seasons, and this benefit outweighs the fitness-based tradeoffs that favor the evolution of specialism on *Dictyota* (Sotka and Reynolds 2011). Similarly, the isopod *Idotea balthica* feeds on a variety of macrophytes but prefers the brown seaweed genus *Fucus* (Jormalainen et al. 2001) despite the fact that *Fucus* is a relatively poor-quality food for *Idotea* relative to alternative hosts. However, *Fucus* can be found year-round while those alternative foods largely disappear during colder winter months. For these herbivores, the temporal variability of host plants and host breadth are positively related, a hypothesis that has broad theoretical and empirical support (Sotka and Reynolds 2011 for discussion).

Second, including chemically rich seaweeds as part of a broad diet generates an even more consistent resource base. Poore et al. 2008 gathered diet information for ampithoid amphipods, ascoglossan slugs, and herbivorous fishes (Sotka and Whalen 2009) and assigned genera as producing (or not producing) classes of lipophilic chemistry that are known to generally deter many marine herbivores (e.g., terpenes, acetogenins). Including chemically rich macroalgae in the diet was associated with a dramatically increased diet breadth for herbivorous fishes and ampithoid amphipods. In the latter case, the advantage was proportionately greater than one would expect by chance (Poore et al. 2008). Interestingly, the host ranges of ascoglossan slugs, do not differ when chemically rich and chemically depauperate seaweeds are included. This suggests that their evolution may be driven by factors other than the presence of lipophilic chemistry.

A related point is that there arises an interesting contrast between terrestrial and marine systems in the relationship between specialization and chemically rich plants. Among terrestrial insects and mammals, specialists seem to be associated with chemically "difficult" plants (Berenbaum et al. 1996; Dearing et al. 2005), while in marine systems, generalists tend to avoid chemically rich plants (e.g., amphipods and fishes; Sotka et al. 2009; Sotka and Whalen 2009).

The final hypothesis surrounding generalism centers on differences between large and small herbivores. Nearly 30 years ago, Hay et al. (1987) predicted that small herbivores preferentially consume and inhabit seaweeds that are chemically defended against larger, more mobile omnivores such as fishes and urchins. Support for this hypothesis can be found in several contexts (Hay and Steinberg 1992; Paul et al. 2001; Taylor and Steinberg 2005 for reviews). The opposite pattern has also been suggested (Hay et al. 1998): relative to larger grazers, small mesograzers may be deterred more strongly by secondary metabolites produced by tissues vulnerable to these small grazers (e.g., algal eggs or zygotes).

5.6.2 PHYLOGENETIC INFLUENCE ON HERBIVORE OFFENSE

Most studies on phylogenetic constraint on herbivore preferences focus on terrestrial insects with relatively narrow host ranges or groups of insects with both specialist and generalist lineages (Winkler and Mitter 2008; Futuyma and Agrawal 2009). Phylogenetic constraints on the feeding preferences of marine herbivores have emerged from a relatively small number of studies. Species of ascoglossan slugs have a host specificity to seaweed genera (Jensen 1997). Genera of ampithoid amphipods show clear differences in tolerating seaweeds with lipophilic chemical defenses (Poore et al. 2008), with species in the genus *Peramphithoe* being the only genus that does not regularly utilize the chemically rich dictyotalean seaweeds (Poore et al. 2008). The feeding patterns of highly polyphagous urchins show surprisingly strong signals of phylogeny. Craft et al. (2013) offered lipophilic extracts from nine subtropical seaweeds at two concentrations to sea urchins and quantified urchin feeding resistance. Patterns of feeding resistance toward these lipophilic defenses were more similar within genera than across genera of urchins, indicating a substantial role for phylogenetic history. Species of fishes clearly differ in willingness to consume chemically rich seaweeds (e.g., Burkepile and Hay 2011; Rasher et al. 2013), but these patterns have not been explored in a phylogenetic context. A strong phylogenetic signal emerges from fish responses to mineral defenses (i.e., calcium carbonate), as parrotfishes are more likely to ingest calcified seaweeds than are surgeonfishes (acanthurids), because calcium carbonate neutralizes the acidic gut of the latter and thus lowers their feeding efficiency (Horn and Messer 1992).

5.6.3 DIFFUSE COEVOLUTION–AUSTRALASIA

Because the fitness of both herbivores and seaweeds are affected by each other's presence, it is possible, if not likely, that guilds of seaweeds and herbivores are reciprocally co-evolving. Here, we argue that such diffuse coevolution is common among marine seaweed–herbivore interactions. We should expect that because of the lack of intimate dependence between species, not all traits of either seaweeds nor herbivores are expected to be sculpted by their interaction. Rather the coevolution should be trait-specific (sensu Strauss et al. 2005) and largely centered on traits that mediate seaweed defenses and herbivore offenses.

One of the clearest examples comes from a comparison of Australasia with North American plant–herbivore interactions. Brown seaweeds of Australasia produce greater levels of water-soluble phlorotannins than do browns of the North-American Pacific coastline, and sea urchin and gastropod herbivores from Australia more readily consume phlorotannin-rich seaweeds than do North American herbivores (Estes and Steinberg 1988; Steinberg and van Altena 1992; Steinberg et al. 1995). The responses were largely consistent across seaweeds of phylogenetically distant origin, suggesting (although not proving) that entire guilds of seaweeds evolved greater levels of chemical defenses. The most parsimonious explanation for weaker seaweed defenses and herbivore offenses in North America is the presence of urchin predators (sea otters), which reduce herbivory by keeping sea urchin populations low (Estes and Steinberg 1988; Estes et al. 2004). This suggests that the predator trophic level may have a cascading effect on chemical defenses of producers. Similarly, urchins and snails from Australia had greater tolerance for phlorotannin-rich seaweeds than did American herbivores.

5.6.4 Diffuse Coevolution–Tropical Habitats

There is emerging evidence that latitudinal gradients in seaweed chemical defenses occur that have arisen in part from coevolutionary dynamics. That is, temperate and tropical seaweeds can be defended against local herbivores but the type of defense (water-soluble vs. lipophilic) that is employed changes with latitude. Similarly, there is some evidence that herbivores are responding to these latitudinal changes in plant defenses in a coevolutionary manner. Here is the preliminary evidence to date:

> *Tropical seaweeds have greater levels of lipophilic defenses than do temperate seaweeds*: Qualitative literature surveys, experimental evidence, and field survey data consistently document that tropical herbivores more frequently encounter terpenes and other lipophilic metabolites than do temperate herbivores. For example, qualitative literature reviews (Hay and Fenical 1988; Van Alstyne et al. 2001; Paul et al. 2001) and limited experimental evidence (Bolser and Hay 1996) indicate that the concentration and diversity of lipophilic secondary metabolites—including terpenes, prenylated quinones and hydroquinones, polyketides, and fatty acids—in tropical seaweeds are higher than those in temperate seaweeds. These studies are preliminary, however, in that no study has attempted to separate the co-varying effects of latitudinal origin and phylogenetic lineage on secondary metabolite production.
>
> *Tropical herbivores have greater feeding tolerance for lipophilic defenses than do temperate herbivores*: To our knowledge, only two studies have explored latitudinal decline in herbivore tolerance of lipophilic defenses, and both confirmed its existence. Cronin et al. (1997) found that three diterpenoid

metabolites from the tropical Pacific seaweed *Dictyota acutiloba* deterred North Carolina fishes and urchins at a lower concentration than the concentrations required to deter a fish and urchin species from Guam. A more robust test was recently published by Craft et al. (2013), who quantified the feeding tolerances of four tropical (*Arbacia punctulata, Diadema antillarum, Echinometra lucunter,* and *E. viridis*) and three temperate sea urchin populations (*Arbacia punctulata, Strongylocentrotus droebachiensis, and S. purpuratus*) when offered foods laden with lipophilic seaweed metabolites. Craft et al. (2013) collected nine seaweed species that are common to Caribbean reefs and known to produce lipophilic compounds (including terpenes) that strongly deter feeding by herbivores. Tropical urchins tended to consume more foods coated with lipophilic extract than did temperate populations. The latitudinal decline was apparent for seven of nine seaweeds, and the two exceptions (*Spatoglossum schroederi* and *Palisada poiteaui*) were among the most readily consumed (and by inference, the least deterrent) of the nine seaweeds tested. Thus, the combination of increased levels of lipophilic chemical defenses of tropical seaweeds and elevated feeding tolerance of these defenses by tropical herbivores is consistent with a coevolutionary arms race in tropical habitats that is mediated by lipophilic metabolites.

Latitudinal gradients mediated by water-soluble compounds? In contrast with the clear latitudinal decrease in lipophilic deterrence, there are conflicting lines of evidence on whether water-soluble chemistry, particularly polyphenolics, display a similar geographic pattern. Targett et al. (1992) found that tropical seaweeds in the Caribbean tended to have higher levels of water-soluble polyphenolics relative to more temperate seaweeds. However, Bolser and Hay (1996) found equivalent levels of deterrence among water-soluble compounds of temperate and tropical Atlantic seaweeds. Further, in the southwestern Pacific Ocean, temperate seaweeds have several fold greater levels of polyphenolics than tropical seaweeds (e.g., Steinberg 1986; Van Alstyne and Paul 1990), a strong contrast with the situation in the Atlantic Ocean. Interestingly, Antarctic seaweeds have phlorotannin levels that are relatively high (Amsler and Fairhead 2005), and some size classes of phlorotannins deterred some local consumers (Iken et al. 2009). Another widely cited class of water-soluble metabolites is dimethylsulfoniopropionate (DMSP), but there is little information on how this may vary with latitude (Van Alstyne 2008).

Similarly, the evidence for whether tropical herbivores have greater levels of feeding tolerance of water-soluble metabolites is relatively weak, although to our knowledge, no study has compared the feeding tolerance of both temperate vs. tropical consumers toward these compounds. When temperate, phenolic-rich seaweeds are offered to tropical consumers (fishes), the results were equivocal. Fishes sometimes avoid consuming the seaweeds (Van Alstyne and Paul 1990) and in other cases the fishes consumed the phenolic-rich seaweeds readily (Steinberg et al. 1991). There are hints that the physiologies of some tropical herbivores better tolerate greater levels of polyphenolics, relative to temperate herbivores. Specifically, Targett et al. (1995) and Boettcher and Targett (1993) report that high concentrations of polyphenolics

(~10% of tissue dry weight) lowered the assimilation efficiency of the temperate fish *Xiphaster mucosus* but did not affect two tropical fishes (*Sparisoma radians* and *S. chrysopterum*) and a tropical crab (*Mithrax sculptus*). Thus, latitudinal patterns of water-soluble defenses and herbivore offenses toward those defenses are equivocal, and, it remains unclear whether a coevolutionary arms race occurs on all temperate reefs (although it clearly occurs on some: Steinberg et al. 1995).

5.7 CONCLUSIONS AND OUTLOOK

Nearly 30 years of study on marine systems have provided several insights into herbivore responses to macrophyte secondary metabolites. First, it is now clear that macrophyte secondary metabolites alter the foraging behavior of consumers, their fitness, or both. The overwhelming pattern that emerges is one of variation: marine herbivores profoundly differ in feeding responses across individuals, populations, and species. Second, we have gained some clarity on why most (but not all) marine herbivores have a generalist diet. Fishes and urchins have a broad diet but tend to avoid chemically rich macrophytes. Mesograzers have a narrower diet than do fishes and urchins, but their diets are more generalized than those of terrestrial insects, which are their ecological analogue. Mesograzers that can include chemically rich macrophytes in their diet gain a broader and more consistent resource base, despite the fitness costs. Third, we now have strong evidence that some generalist herbivores (urchins and crustaceans) show among-population differentiation in their preference for and performance on chemically rich diets. Some of these are likely to represent local adaptations. These microevolutionary shifts in the preference ranks minimize fitness or ecological trade-offs that are incurred by a generalist diet. Finally, latitudinal and cross-basin comparisons of seaweed chemical defenses and herbivore offenses suggest diffuse coevolutionary arms races are common in the sea.

Our review has also pointed out some glaring gaps in the literature on marine herbivore responses to macrophyte chemical defenses. First, our understanding of how consumers may control the concentration of metabolites in the body and the mechanisms by which metabolites exert their effects (site of action and response) remains in its infancy. Second, with increasing recognition of the role of microbes in the tolerance of terrestrial herbivores to plant metabolites (Hammer and Bowers 2015) and the many new molecular tools to quantify their composition and function, we would expect that exciting new advances in this area will be made in the coming years. Third, preconditions and constraints for adaptive evolution of host ranges and feeding specializations, such as genetic variation and trade-offs in tolerance to specific macrophyte metabolites, still remain sparsely studied. Fourth, most studies of the macroevolution of marine herbivores lack an explicit phylogenetic component, either to test the effect of phylogenetic constraints, or to separate the effect of phylogeny and biogeography on the evolution of tolerance traits. The ease with which molecular phylogenies can be generated will facilitate such discoveries. Finally, the unprecedented rise in global ocean temperatures as a result of increased anthropogenic activity has facilitated the movement of tropical marine herbivores and seaweeds into temperate systems with dramatic consequences to ecosystem dynamics, function, and overall productivity (Verges et al. 2015 for review). Because emerging evidence suggests the diffuse coevolution of seaweed–herbivore interactions is intensified within tropical habitats, we predict that ongoing movement of marine herbivores and seaweeds into temperate habitats may both increase the relative success of herbivores with higher tolerance for chemical defenses and seaweeds with lower palatability. Similarly, the strength of diffuse coevolution of herbivore offenses and macrophyte defenses can help to predict the impacts of invasions of both seaweeds and herbivores.

REFERENCES

Ame, M. V., M. V. Baroni, L. N. Galanti, J. L. Bocco, and D. A. Wunderlin. 2009. Effects of microcystin-LR on the expression of P-glycoprotein in *Jenynsia multidentata*. *Chemosphere* 74:1179–1186.

Amsler, C. D., and V. A. Fairhead. 2005. Defensive and sensory chemical ecology of brown algae. *Adv. Bot. Res.* 43:1–91.

Amsler, C. D., J. B. McClintock, and B. J. Baker. 1999. An antarctic feeding triangle: Defensive interactions between macroalgae, sea urchins, and sea anemones. *Mar. Ecol. Prog. Ser.* 183:105–114.

Amsler, M., C. D. Amsler, J. von Salm, C. Aumack, J. McClintock, R. Young, and B. Baker. 2013. Tolerance and sequestration of macroalgal chemical defenses by an Antarctic amphipod: A "cheater" among mutualists. *Mar. Ecol. Prog. Ser.* 490:79–90.

Aquilino, K. M., M. E. Coulbourne, and J. J. Stachowicz. 2012. Mixed species diets enhance the growth of two rocky intertidal herbivores. *Marine Ecology Progress Series* 468:179–189.

Araújo, M. S., D. I. Bolnick, C. A. Layman. 2011. The ecological causes of individual specialisation. *Ecol. Lett.* 14(9):948–958.

Arrontes, J. 1999. On the evolution of interactions between marine mesoherbivores and algae. *Bot. Mar.* 42:137–155.

Arthur, K. E., S. Kelez, T. Larsen, C. A. Choy, and B. N. Popp. 2014. Tracing the biosynthetic source of essential amino acids in marine turtles using delta C-13 fingerprints. *Ecology* 95:1285–1293.

Avila, C. 1995. Natural products of opisthobranch molluscs: A biological review. *Oceanogr. Mar. Biol.* 33:487–559.

Barbehenn, R., S. Cheek, A. Gasperut, E. Lister, and R. Maben. 2005. Phenolic compounds in red oak and sugar maple leaves have prooxidant activities in the midgut fluids of *Malacosoma disstria* and *Orgyia leucostigma* caterpillars. *J. Chem. Ecol.* 31:969–988.

Bard, S. M. 2000. Multixenobiotic resistance as a cellular defense mechanism in aquatic organisms. *Aquat. Toxicol.* 48:357–389.

Barreiro, A., C. Guisande, I. Maneiro, A. R. Vergara, I. Riveiro, and P. Iglesias. 2007. Zooplankton interactions with toxic phytoplankton: Some implications for food web studies and algal defence strategies of feeding selectivity behaviour, toxin dilution and phytoplankton population diversity. *Acta Oecol.* 32:279–290.

Baumgartner, F. A., C.A. Motti, R. de Nys, and N. A. Paul. 2009. Feeding preferences and host associations of specialist marine herbivores align with quantitative variation in seaweed secondary metabolites. *Mar. Ecol. Prog. Ser.* 396:1–12.

Baumgartner, F. A., H. Pavia, and G. B. Toth. 2014. Individual specialization to non-optimal hosts in a polyphagous marine invertebrate herbivore. *PLoS ONE* 9.

Becerro, M. A., G. Goetz, V. J. Paul, and P. J. Scheuer. 2001. Chemical defenses of the sacoglossan mollusk *Elysia rufescens* and its host alga *Bryopsis* sp. *J. Chem. Ecol.* 27:2287–2299.

Bell, T. M., and E. E. Sotka. 2012. Local adaptation in adult feeding preference and juvenile performance in the generalist herbivore *Idotea balthica*. *Oecologia* 170:383–393

Berenbaum, M. R., C. Favret, and M. A. Schuler. 1996. On defining "key innovations" in an adaptive radiation: Cytochrome P450s and Papilionidae. *Am. Nat.* 148:S139–S155.

Best, R. J., and J. J. Stachowicz. 2014. Phenotypic and phylogenetic evidence for the role of food and habitat in the assembly of communities of marine amphipods. *Ecology* 95:775–786.

Best, R. J., N. C. Caulk, and J. J. Stachowicz. 2013. Trait vs. phylogenetic diversity as predictors of competition and community composition in herbivorous marine amphipods. *Ecol. Lett.* 16:72–80.

Boettcher, A. A., and N. M. Targett. 1993. Role of polyphenolic molecular size in reduction of assimilation efficiency in *Xiphister mucosus*. *Ecology* 74:891–903.

Bolnick, D. I., R. Svanbäck, J. A. Fordyce, L. H. Yang, J. M. Davis, C. D. Hulsey, and M. L. Forister. 2003. The ecology of individuals: Incidence and implications of individual specialization. *Am. Nat.* 161:1–28.

Bolser, R., and M. E. Hay. 1996. Are tropical plants better defended? Palatability and defenses of temperate vs. tropical seaweeds. *Ecology* 77:2269–2286.

Borowsky, R., B. Borowsky, H. Milani, and P. Greenberg. 1985. Amylase variation in the salt marsh amphipod, *Gammarus palustris*. *Genetics* 111:311–323.

Brawley, S. H. 1992. Mesoherbivores. In *Plant-animal Interactions in the Marine Benthos*, D. M. John, S. J. Hawkins, and J. H. Price (Eds.), pp. 235–263. Oxford, UK: Clarendon Press.

Bulleri, F., and F. Malquori. 2015. High tolerance to simulated herbivory in the clonal seaweed, *Caulerpa cylindracea*. *Mar. Environ. Res.* 107:61–65.

Burkepile, D. E., and M. E. Hay. 2011. Feeding complementarity versus redundancy among herbivorous fishes on a Caribbean reef. *Coral Reefs* 30:351–362.

Cebrian, J., and J. Lartigue. 2004. Patterns of herbivory and decomposition in aquatic and terrestrial ecosystems. *Ecol. Monogr.* 74:237–259.

Choat, J. H., and K. D. Clements. 1998. Vertebrate herbivores in marine and terrestrial environments: A nutritional ecology perspective. *Ann. Rev. Ecol. Syst.* 29:375–403.

Clements, K. D., E. R. Angert, W. L. Montgomery, and J. H. Choat. 2014. Intestinal microbiota in fishes: What's known and what's not. *Mol. Ecol.* 23:1891–1898.

Craft, J. D., V. J. Paul, and E. E. Sotka. 2013. Biogeographic and phylogenetic effects on feeding resistance of generalist herbivores toward plant chemical defenses. *Ecology* 94:18–24.

Crawley, K. R., and G. A. Hyndes. 2007. The role of different types of detached macrophytes in the food and habitat choice of a surf-zone inhabiting amphipod. *Mar. Biol.* 151:1433–1443.

Cronin, G., V. J. Paul, M. E. Hay, and W. Fenical. 1997. Are tropical herbivores more resistant than temperate herbivores to seaweed chemical defenses? Diterpenoid metabolites from *Dictyota acutiloba* as feeding deterrents for tropical versus temperate fishes and urchins. *J. Chem. Ecol.* 23:289–302.

Cruz-Rivera, E. 2001. Generality and specificity in the feeding and decoration preferences of three Mediterranean crabs. *J. Exp. Mar. Biol. Ecol.* 266:17–31.

Cruz-Rivera, E., and M. E. Hay. 2000. Can quantity replace quality? Food choice, compensatory feeding, and fitness of marine mesograzers. *Ecology* 81:201–219.

Cruz-Rivera, E., and M. E. Hay. 2001. Macroalgal traits and the feeding and fitness of an herbivorous amphipod: The roles of selectivity, mixing, and compensation. *Mar. Ecol. Prog. Ser.* 218:249–266.

Cruz-Rivera, E., and M. E. Hay. 2003. Prey nutritional quality interacts with chemical defenses to affect consumer feeding and fitness. *Ecol. Monograph.* 73(3):483–506.

Dearing, M. D., W. J. Foley, and S. McLean. 2005. The influence of plant secondary metabolites on the nutritional ecology of herbivorous terrestrial vertebrates. *Annu. Rev. Ecol. Evol. Syst.* 36:169–189.

Dethier, V. G. 1954. Evolution of feeding preferences in phytophagous insects. *Evolution* 8:33–54.

Dillon, R. J., and V. M. Dillon. 2004. The gut bacteria of insects: Nonpathogenic interactions. *Ann. Rev. Ent.* 49:71–92.

Duffy, J. E., and M. E. Hay. 1991. Food and shelter as determinants of food choice by an herbivorous marine amphipod. *Ecology* 72:1286–1298.

Duffy, J. E., and M. E. Hay. 1994. Herbivore resistance to seaweed chemical defense: The roles of mobility and predation risk. *Ecology* 75:1304–1319.

Duffy, J. E., and M. E. Hay. 2001. The ecology and evolution of marine consumer-prey interactions. In *Marine Community Ecology*, M. D. Bertness, S. D. Gaines, and M. E. Hay (Eds.), pp. 131–157. Sunderland, MA: Sinauer Publishers.

Duffy, J. E., P. L. Reynolds, C. Boström, J. Coyer, M. Cusson, S. Donadi, J. Douglass et al. 2015. Biodiversity mediates top-down control in eelgrass ecosystems: A global comparative-experimental approach. *Ecol. Lett.* 18:696–705.

Enge, S., G. M. Nylund, T. Harder, and H. Pavia. 2012. An exotic chemical weapon explains low herbivore damage in an invasive alga. *Ecology* 93:2736–2745.

Engel, P., and N. A. Moran. 2013. The gut microbiota of insects—Diversity in structure and function. *FEMS Microbiol. Rev.* 37:699–735.

Engkvist, R., T. Malm, and S. Tobiasson. 2000. Density dependent grazing effects of the isopod *Idotea baltica* Pallas on *Fucus vesiculosus* L in the Baltic Sea. *Aquat. Ecol.* 34:253–260.

Estes, J. A., and P. D. Steinberg. 1988. Predation, herbivory, and kelp evolution. *Paleobiology* 14:19–36.

Estes, J., E. Danner, D. Doak, B. Konar, A. Springer, P. D. Steinberg, M. Tinker, and T. Williams. 2004. Complex trophic interactions in kelp forest ecosystems. *Bull. Mar. Sci.* 74:621–638.

Forbey, J. S., M. D. Dearing, E. M. Gross, C. M. Orians, E. E. Sotka, and W. J. Foley. 2013. A pharm-ecological perspective of terrestrial and aquatic plant–herbivore interactions. *J. Chem. Ecol.* 39:465–480.

Fox, L., and P. Morrow. 1981. Specialization: Species property or local phenomenon? *Science* 211:887–893.

Fry, J. D. 1996. The evolution of host specialization: Are trade-offs overrated? *American Naturalist* 148:S84–S107.

Fry, J. D. 2003. Detecting ecological trade-offs using selection experiments. *Ecology* 84:1672–1678.

Futuyma, D. J., and S. C. Peterson. 1985. Genetic variation in the use of resources by insects. *Ann. Rev. Entomol.* 30:217–238.

Futuyma, D. J., and A. A. Agrawal. 2009. Macroevolution and the biological diversity of plants and herbivores. *Proc. Natl. Acad. Sci. U. S. A.* 106:18054–18061.

Ginsburg, D. W., and V. J. Paul. 2001. Chemical defenses in the sea hare *Aplysia parvula*: Importance of diet and sequestration of algal secondary metabolites. *Mar. Ecol. Prog. Ser.* 215:261–274.

Gurevitch, J., L. L. Morrow, A. Wallace, and J. S. Walsh. 1992. A metaanalysis of competition in field experiments. *Am. Nat.* 140:539–572.

Gutow, L., J. Long, O. Cerda, I. Hinojosa, E. Rothäusler, F. Tala, and M. Thiel. 2011. Herbivorous amphipods inhabit protective microhabitats within thalli of giant kelp *Macrocystis pyrifera*. *Mar. Biol.* 159(1):141–149.

Haavisto, F., and V. Jormalainen. 2014. Seasonality elicits herbivores' escape from trophic control and favors induced resistance in a temperate macroalga. *Ecology* 95:3035–3045.

Haavisto, F., T. Välikangas, and V. Jormalainen. 2010. Induced resistance in a brown alga: Phlorotannins, genotypic variation and fitness costs for the crustacean herbivore. *Oecologia* 162:685–695.

Hacker, S. D., and L. P. Madin. 1991. Why habitat architecture and color are important to shrimps living in pelagic *Sargassum*: Use of camouflage and plant-part mimicry. *Mar. Ecol. Prog. Ser.* 70:143–155.

Hammann, M., G. Wang, E. Rickert, S. Boo, and F. Weinberger. 2013. Invasion success of the seaweed *Gracilaria vermiculophylla* correlates with low palatibility. *Mar. Ecol. Prog. Ser.* 486:93–103.

Hammer, T. J., and M. D. Bowers. 2015. Gut microbes may facilitate insect herbivory of chemically defended plants. *Oecologia* 179(1):1–14.

Harris, J. M. 1993. The presence, nature, and role of gut microflora in aquatic invertebrates: A synthesis. *Microb. Ecol.* 25:195–231.

Hay, M. E., W. Boland, and I. Schnitzler. 1998. Seaweed sex pheromones and their degradation products frequently suppress amphipod feeding but rarely suppress sea urchin feeding. *Chemoecology* 8:91–98.

Hay, M. E., J. E. Duffy, and W. Fenical. 1990a. Host-plant specialization decreases predation on a marine amphipod: An herbivore in plant's clothing. *Ecology* 71:733–743.

Hay, M. E., J. E. Duffy, V. J. Paul, P. E. Renaud, and W. Fenical. 1990b. Specialist herbivores reduce their susceptibility to predation by feeding on the chemically defended seaweed *Avrainvillea longicaulis. Limnol. Oceanogr.* 35:1734–1743.

Hay, M. E., J. E. Duffy, C. A. Pfister, and W. Fenical. 1987. Chemical defense against different marine herbivores: Are amphipods insect equivalents? *Ecology* 68:1567–1580.

Hay, M. E., and W. Fenical. 1988. Marine plant–herbivore interactions: The ecology of chemical defense. *Annu. Rev. Ecol. Syst.* 19:111–146.

Hay, M. E., J. R. Pawlik, J. E. Duffy, and W. Fenical. 1989. Seaweed-herbivore-predator interactions: Host plant specialization reduces predation on small herbivores. *Oecologia* 81:418–427.

Hay, M. E., and P. D. Steinberg. 1992. The chemical ecology of plant–herbivore interactions in marine versus terrestrial communities. In: *Herbivores: Their Interactions with Secondary Plant Metabolites: Ecological and Evolutionary Processes*, G. Rosenthal, and M. Berenbaum (Eds.), pp. 371–413. San Diego, CA: Academic Press.

Hemmi, A., and V. Jormalainen. 2004a. Genetic and environmental variation in performance of a marine isopod: Effects of eutrophication. *Oecologia* 140:302–311.

Hemmi, A., and V. Jormalainen. 2004b. Geographic covariation of chemical quality of the host alga *Fucus vesiculosus* with fitness of the herbivorous isopod *Idotea baltica. Mar. Biol.* 145:759–768.

Honkanen, T., V. Jormalainen, A. Hemmi, A. Mäkinen, and N. Heikkilä. 2002. Feeding and growth of the isopod *Idotea baltica* on the brown alga *Fucus vesiculosus*: Roles of interpopulation and within-plant variation in plant quality. *Ecoscience* 9:332–338.

Horn, M. H., and L. A. Ferry-Graham. 2006. Feeding mechanisms and trophic interactions. In: *The Ecology of Marine Fishes: California and Adjacent Waters*, L. G. Allen, D. J. Pondella, and M. H. Horn (Eds.), pp. 387–410. Berkeley, CA: University of California Press.

Horn, M. H., and K. S. Messer. 1992. Fish guts as chemical reactors: A model of the alimentary canals of marine herbivorous fishes. *Mar. Biol.* 113:527–535.

Huang, L., J. Wang, W. C. Chen, H. Y. Li, J. S. Liu, T. Jiang, and W. D. Yang. 2014. P-glycoprotein expression in *Perna viridis* after exposure to *Prorocentrum lima*, a dinoflagellate producing DSP toxins. *Fish Shellfish Immunol.* 39:254–262.

Hultgren, K. M., and J. J. Stachowicz. 2008. Alternative camouflage strategies mediate predation risk among closely related co-occurring kelp crabs. *Oecologia* 155:519–528.

Hultgren, K. M., and J. J. Stachowicz. 2010. Size-related habitat shifts facilitated by positive preference induction in a marine kelp crab. *Behav. Ecol.* 21:329–336.

Hultgren, K. M., and J. J. Stachowicz. 2011. Camouflage in decorator crabs: Integrating ecological, behavioural and evolutionary approaches. In: *Animal Camouflage*, M. Stevens, and S. Merilaita (Eds.), pp. 214–229. Cambridge, UK: Cambridge University Press.

Iken, K., C. D. Amsler, M. O. Amsler, J. B. McClintock, and B. J. Baker. 2009. Field studies on deterrent roles of phlorotannins in Antarctic brown algae. *Bot. Mar.* 52:547–557.

Jaschinski, S., and U. Sommer. 2008. Functional diversity of mesograzers in an eelgrass-epiphyte system. *Mar. Biol.* 154:475–482.

Jensen, K. R. 1997. Evolution of the Sacoglossa (Mollusca, Opisthobranchia) and the ecological associations with their food plants. *Evol. Ecol.* 11:301–335.

Johnson, P. M., C. E. Kicklighter, M. Schmidt, M. Kamio, H. C. Yang, D. Elkin, W. C. Michel, P. C. Tai, and C. D. Derby. 2006. Packaging of chemicals in the defensive secretory glands of the sea hare *Aplysia californica. J. Exp. Biol.* 209:78–88.

Jormalainen, V. 2015. Grazers of macroalgae and higher plants. In: *Life Styles and Feeding Biology*, L. Watling, and M. Thiel (Eds.), pp. 502–534. New York: Oxford University Press.

Jormalainen, V., and T. Honkanen. 2008. Macroalgal chemical defenses and their roles in structuring temperate marine communities. In: *Algal Chemical Ecology*, C. D. Amsler (Ed.), pp. 57–89. Berlin, Germany: Springer.

Jormalainen, V. and T. Ramsay. 2009. Resistance of the brown alga *Fucus vesiculosus* to herbivory. Oikos 118:713–722.

Jormalainen, V., T. Honkanen, and N. Heikkilä. 2001. Feeding preferences and performance of a marine isopod on seaweed hosts: Cost of habitat specialization. *Mar. Ecol. Prog. Ser.* 220:219–230.

Jormalainen, V., T. Honkanen, O. Vesakoski, and R. Koivikko. 2005. Polar extracts of the brown alga *Fucus vesiculosus* (L.) reduce assimilation efficiency but do not deter the herbivorous isopod *Idotea baltica* (Pallas). *J. Exp. Mar. Biol. Ecol.* 317:143–157.

Jormalainen, V., R. Koivikko, V. Ossipov, and M. Lindqvist. 2011. Quantifying variation and chemical correlates of bladderwrack quality—Herbivore population makes a difference. *Funct. Ecol.* 25:900–909.

Jormalainen, V., S. Merilaita, and J. Tuomi. 1995. Differential predation on sexes affects colour polymorphism of the isopod *Idotea baltica* (Pallas). *Biol. J. Linn. Soc.* 55:45–68.

Karban, R., and A. A. Agrawal. 2002. Herbivore offense. *Annu. Rev. Ecol. Syst.* 33:641–664.

Klussmann-Kolb, A., and G. D. Brodie. 1999. Internal storage and production of symbiotic bacteria in the reproductive system of a tropical marine gastropod. *Mar. Biol.* 133:443–447.

Koivikko, R., J. K. Eränen, J. Loponen, and V. Jormalainen. 2008. Variation of phlorotannins among three populations of *Fucus vesiculosus* as revealed by HPLC and colorimetric quantification. *J. Chem. Ecol.* 34:57–64.

Kubanek, J., S. E. Lester, W. Fenical, and M. E. Hay. 2004. Ambiguous role of phlorotannins as chemical defenses in the brown alga *Fucus vesiculosus. Mar. Ecol. Prog. Ser.* 277:79–93

Kuhajek, J. M., and D. Schlenk. 2003. Effects of the brominated phenol, lanosol, on cytochrome P-450 and glutathione transferase activities in *Haliotis rufescens* and *Katharina tunicata. Comp. Biochem. Physiol. C Toxicol. Pharmacol.* 134:473–479.

Lastra, M., T. A. Schlacher, and C. Olabarria. 2010. Niche segregation in sandy beach animals: An analysis with surface-active peracarid crustaceans on the Atlantic coast of Spain. *Mar. Biol.* 157:613–625.

Lefcheck, J. S., M. A. Whalen, T. M. Davenport, J. P. Stone, and J. E. Duffy. 2013. Physiological effects of diet mixing on consumer fitness: A meta-analysis. *Ecology* 94:565–572.

Ling, S. D., C. R. Johnson, S. D. Frusher, and K. R. Ridgway. 2009. Overfishing reduces resilience of kelp beds to climate-driven catastrophic phase shift. *Proc. Natl. Acad. Sci. U. S. A.* 106:22341–22345.

Littler, M. M., and D. S. Littler. 1980. The evolution of thallus form and survival strategies in benthic marine macroalgae: Field and laboratory tests of a functional form model. *Am. Nat.* 116:25–44.

Lubchenco, J., and S. D. Gaines. 1981. A unified approach to marine plant–herbivore interactions. I. Populations and communities. *Annu. Rev. Ecol. Syst.* 12:405–437.

Lyons, D. A., and R. E. Scheibling. 2007. Effect of dietary history and algal traits on feeding rate and food preference in the green sea urchin *Strongylocentrotus droebachiensis*. *J. Exp. Mar. Biol. Ecol.* 349:194–204.

Marin, A., and J. Ros. 2004. Chemical defenses in *Sacoglossan* Opisthobranchs: Taxonomic trends and evolutive implications. *Sci. Mar.* 68:227–241.

Mattila, J. M., M. Zimmer, O. Vesakoski, and V. Jormalainen. 2014. Habitat-specific gut microbiota of the marine herbivore *Idotea balthica* (Isopoda). *J. Exp. Mar. Biol. Ecol.* 455:22–28.

Mattson, W. J. 1980. Herbivory in relation to plant nitrogen content. *Annu. Rev. Ecol. Syst.* 11:119–161.

McCarty, A. T., and E. E. Sotka. 2013. Geographic variation in feeding preference of a generalist herbivore: The importance of seaweed chemical defenses. *Oecologia* 172:1071–1083.

McLean, S., and A. J. Duncan. 2006. Pharmacological perspectives on the detoxification of plant secondary metabolites: Implications for ingestive behavior of herbivores. *J. Chem. Ecol.* 32:1213–1228.

Merilaita, S. 2001. Habitat heterogeneity, predation and gene flow: Colour polymorphism in the isopod, *Idotea baltica*. *Evol. Ecol.* 15:103–116.

Nelson, T. M., T. L. Rogers, and M. V. Brown. 2013. The gut bacterial community of mammals from marine and terrestrial habitats. *PLoS One* 8(12):e83655.

Norderhaug, K. M., S. Fredriksen, and K. Nygaard. 2003. Trophic importance of *Laminaria hyperborea* to kelp forest consumers and the importance of bacterial degradation to food quality. *Mar. Ecol. Prog. Ser.* 255:135–144.

Novotny, V., and Y. Basset. 2005. Host specificity of insect herbivores in tropical forests. *Proc. Biol. Soc. Wash.* 272:1083–1090.

Padilla, D. K. 1985. Structural resistance of algae to herbivores—A biomechanical approach. *Mar. Biol.* 90:103–109.

Paul, V. 1992. Seaweed chemical defenses on coral reefs. In: *Ecological Roles of Marine Natural Products*, V. Paul (Ed.), pp. 24–50. Ithaca, NY: Comstock Publishing Associates.

Paul, V. J., E. Cruz-Rivera, and R. W. Thacker. 2001. Chemical mediation of seaweed–herbivore interactions: Ecological and evolutionary perspectives. In: *Marine Chemical Ecology*, J. B. McClintock and B. Baker (Eds.), Boca Raton, FL: CRC Press.

Paul, V. J., E. Cruz-Rivera, and R. W. Thacker. 2001. Chemical mediation of macroalgal-herbivore interactions: Ecological and evolutionary perspectives. In: *Marine Chemical Ecology*, J. B. McClintock, and B. J. Baker (Eds.), pp. 227–265. Boca Raton, FL: CRC Press.

Pavia, H., and G. Toth. 2008. Macroalgal models in testing and extending defense theories. In: *Algal Chemical Ecology*, C. D. Amsler (Ed.), pp. 147–172. Berlin, Germany: Springer.

Pavia, H., G. B. Toth, and P. Åberg. 2002. Optimal defense theory: Elasticity analysis as a tool to predict intraplant variation in defenses. *Ecology* 83:891–897.

Pennings, S. C., T. H. Carefoot, M. Zimmer, J. P. Danko, and A. Ziegler. 2000. Feeding preferences of supralittoral isopods and amphipods. *Can. J. Zool.* 78:1918–1929.

Pennings, S. C., M. T. Nadeau, and V. J. Paul. 1993. Selectivity and growth of the generalist herbivore *Dolabella auricularia* feeding upon complementary resources. *Ecology* 74:879–890.

Pennings, S. C., V. J. Paul, D. C. Dunbar, M. T. Hamann, W. A. Lumbang, B. Novack, and R. S. Jacobs. 1999. Unpalatable compounds in the marine gastropod *Dolabella auricularia*: Distribution and effect of diet. *J. Chem. Ecol.* 25:735–755.

Pinn, E. H., A. Rogerson, and R. J. A. Atkinson. 1997. Microbial flora associated with the digestive system of *Upogebia stellata* (Crustacea: Decapoda: Thalassinidea). *J. Mar. Biol. Assoc. UK* 77:1083–1096.

Poore, A. G. B., and N. A. Hill. 2006. Sources of variation in herbivore preference: Among-individual and past diet effects on amphipod host choice. *Mar. Biol.* 149:1403–1410.

Poore, A. G. B., and P. D. Steinberg. 1999. Preference-performance relationships and effects of host plant choice in an herbivorous marine amphipod. *Ecol. Monogr.* 69:443–464.

Poore, A. G. B., A. H. Campbell, R. A. Coleman, J. E. Duffy, G. J. Edgar, V. Jormalainen, P. L. Reynolds et al. 2012. Global patterns in the impact of marine herbivores on benthic primary producers. *Ecol. Lett.* 15:912–922.

Poore, A. G. B., N. A. Hill, and E. E. Sotka. 2008. Phylogenetic and geographic variation in host breadth and composition by herbivorous amphipods in the family Ampithoidae. *Evolution* 62:21–38.

Poore, A. G. B., M. J. Watson, R. de Nys, J. K. Lowry, and P. D. Steinberg. 2000. Patterns of host use among alga- and sponge-associated amphipods. *Mar. Ecol. Prog. Ser.* 208:183–196.

Poore, A. G. B., and P. D. Steinberg. 2001. Host plant adaptation in a herbivorous marine amphipod: Genetic potential not realized in field populations. *Evolution* 55:68–80.

Rasher, D. B., A. S. Hoey, and M. E. Hay. 2013. Consumer diversity interacts with prey defenses to drive ecosystem function. *Ecology* 94:1347–1358.

Raubenheimer, D., W. L. Zemke-White, R. J. Phillips, and K. D. Clements. 2005. Algal macronutrients and food selection by the omnivorous marine fish *Girella tricuspidata*. *Ecology* 86:2601–2610.

Richardson, K. L., G. Gold-Bouchot, and D. Schlenk. 2009. The characterization of cytosolic glutathione transferase from four species of sea turtles: Loggerhead (*Caretta caretta*), green (*Chelonia mydas*), olive ridley (*Lepidochelys olivacea*), and hawksbill (*Eretmochelys imbricata*). *Comp. Biochem. Physiol. C Toxicol. Pharmacol.* 150:279–284.

Rogers, C. N., R. de Nys, and P. D. Steinberg. 2002. Effects of algal diet on the performance and susceptibility to predation of the sea hare *Aplysia parvula*. *Mar. Ecol. Prog. Ser.* 236:241–254.

Rogers, C. N., R. de Nys, T. S. Charlton, and P. D. Steinberg. 2000. Dynamics of algal secondary metabolites in two species of sea hare. *J. Chem. Ecol.* 26:721–744.

Rorandelli, R., M. Gomei, M. Vannini, and S. Cannicci. 2007. Feeding and masking selection in *Inachus phalangium* (Decapoda, Majidae): Dressing up has never been so complicated. *Mar. Ecol. Prog. Ser.* 336:225–233.

Salminen, J. P., and M. Karonen. 2011. Chemical ecology of tannins and other phenolics: We need a change in approach. *Funct. Ecol.* 25:325–338.

Schiel, D. R., and Foster, M. S. 1985. The structure of subtidal algal stands in temperate waters. *Oceanogr. Mar. Biol. Annu. Rev.* 24:265–307.

Schnitzler, I., G. Pohnert, M. Hay, and W. Boland. 2001. Chemical defense of brown algae (*Dictyopteris* spp.) against the herbivorous amphipod *Ampithoe longimana*. *Oecologia* 126(4):515–521.

Siska, E. L., S. C. Pennings, T. L. Buck, and M. D. Hanisak. 2002. Latitudinal variation in palatability of salt-marsh plants: Which traits are responsible? *Ecology* 83(12):3369–3381.

Singer, M. S. 2008. Evolutionary ecology of generalism. In: *Specialization, Speciation, and Radiation: The Evolutionary Biology of Herbivorous Insects*, K. J. Tilmon (Ed.), pp. 29–42. Berkeley, CA: University of California Press.

Sorensen, J. S., and M. D. Dearing. 2006. Efflux transporters as a novel herbivore countermechanism to plant chemical defenses. *J. Chem. Ecol.* 32:1181–1196.

Sotka, E. E. 2003. Genetic control of feeding preference in the herbivorous amphipod *Ampithoe longimana*. *Mar. Ecol. Prog. Ser.* 256:305–310.

Sotka, E. E. 2005. Local adaptation in host use among marine invertebrates. *Ecol. Lett.* 8:448–459.

Sotka, E. E. 2007. Restricted host use by the herbivorous amphipod *Peramphithoe tea* is motivated by food quality and abiotic refuge. *Mar. Biol.* 151:1831–1838.

Sotka, E. E., and J. Gantz. 2013. Preliminary evidence that the feeding rates of generalist marine herbivores are limited by detoxification rates. *Chemoecology* 23:233–240.

Sotka, E. E., J. Forbey, M. Horn, A. G. B. Poore, D. Raubenheimer, and K. E. Whalen. 2009. The emerging role of pharmacology in understanding consumer-prey interactions in marine and freshwater systems. *Integr. Comp. Biol.* 49:291–313.

Sotka, E. E., and M. E. Hay. 2002. Geographic variation among herbivore populations in tolerance for a chemically-rich seaweed. *Ecology* 83:2721–2735.

Sotka, E. E., M. E. Hay, and J. D. Thomas. 1999. Host-plant specialization by a non-herbivorous amphipod: Advantages for the amphipod and costs for the seaweed. *Oecologia* 118:471–482.

Sotka, E. E., and P. L. Reynolds. 2011. Rapid experimental shift in host use traits of a polyphagous marine herbivore reveals fitness costs on alternative hosts. *Evol. Ecol.* 25:1335–1355.

Sotka, E. E., J. P. Wares, and M. E. Hay. 2003. Geographic and genetic variation in feeding preference for chemically defended seaweeds. *Evolution* 57:2262–2276.

Sotka, E. E., and K. E. Whalen. 2008. Herbivore offense in the sea: The detoxification and transport of secondary metabolites. In: *Algal Chemical Ecology*, C. D. Amsler (Ed.), pp. 203–228. Berlin, Germany: Springer.

Stachowicz, J. J., and M. E. Hay. 1999. Reducing predation through chemically mediated camouflage: Indirect effects of plant defenses on herbivores. *Ecology* 80:495–509.

Stachowicz, J. J., and M. E. Hay. 2000. Geographic variation in camouflage specialization by a decorator crab. *Am. Nat.* 156:59–71.

Stachowicz, J. J., J. F. Bruno, and J. E. Duffy. 2007. Understanding the effects of marine biodiversity on communities and ecosystems. *Annu. Rev. Ecol. Evol. Syst.* 38:739–766.

Stegeman, J. J., B. R. Woodin, H. Singh, M. F. Oleksiak, and M. Celander. 1997. Cytochromes p450 (CYP) in tropical fishes: Catalytic activities, expression of multiple CYP proteins and high levels of microsomal p450 in liver of fishes from Bermuda. *Comp. Biochem. Physiol. C Pharmacol. Toxicol. Endocrinol.* 116:61–75.

Steinberg, P. D. 1984. Algal chemical defense against herbivores: Allocation of phenolic compounds in the kelp *Alaria marginata*. *Science* 223:405–406.

Steinberg, P. D. 1986. Chemical defenses and the susceptibility of tropical brown algae to herbivores. *Oecologia* 69:628–630.

Steinberg, P., K. Edyvane, R. de Nys, R. Birdsey, and I. Van Altena. 1991. Lack of avoidance of phenolic-rich brown algae by tropical herbivorous fishes. *Mar. Biol.* 109:335–343.

Steinberg, P. D., and I. van Altena. 1992. Tolerance of marine invertebrate herbivores to brown algal phlorotannins in temperate Australasia. *Ecol. Monograph.* 62:189–222.

Steinberg, P. D., J. A. Estes, and F. C. Winter. 1995. Evolutionary consequences of food-chain length in kelp forest communities. *Proc. Nat. Acad. Sci. USA* 92:8145–8148.

Steneck, R. S., and Watling, L. 1982. Feeding capabilities and limitation of herbivorous mollusks: A functional-group approach. *Mar. Biol.* 68:299–319.

Steneck, R. S., M. H. Graham, B. J. Bourque, D. Corbett, J. M. Erlandson, J. A. Estes, and M. J. Tegner. 2002. Kelp forest ecosystems: Biodiversity, stability, resilience and future. *Environ. Conserv.* 29:436–459.

Stern, J. L., A. E. Hagerman, P. D. Steinberg, and P. K. Mason. 1996. Phlorotannin-protein interactions. *J. Chem. Ecol.* 22:1877–1899.

Strauss, S. Y., H. Sahli, and J. K. Conner. 2005. Toward a more trait-centered approach to diffuse (co)evolution. *New Phytol.* 165:81–89.

Strong, D. R. Jr, J. H. Lawton, and T. R. E. Southwood. 1984. *Insects on Plants: Community Patterns and Mechanisms*. Oxford, UK: Blackwell Science, 313 pp.

Sword, G. A., A. Joern, and L. B. Senior. 2005. Host plant-associated genetic differentiation in the snakeweed grasshopper, *Hesperotettix viridis* (Orthoptera: Acrididae). *Mol. Ecol.* 14:2197–2205.

Tang, K. W., V. Turk, and H. P. Grossart. 2010. Linkage between crustacean zooplankton and aquatic bacteria. *Aquat. Microb. Ecol.* 61:261–277.

Targett, N., and T. Arnold. 2001. Effects of secondary metabolites on digestion in marine herbivores. In: *Marine Chemical Ecology*, B. Baker and K. McClintock (Eds.), pp. 391–411. Boca Raton, FL: CRC Press.

Targett, N. M., A. A. Boettcher, T. E. Targett, and N. H. Vrolijk. 1995. Tropical marine herbivore assimilation of phenolic-rich plants. *Oecologia* 103:170–179.

Targett, N. M., L. D. Coen, A. A. Boettcher, and C. E. Tanner. 1992. Biogeographic comparisons of marine algal polyphenolics: Evidence against a latitudinal trend. *Oecologia* 89:464–470.

Taylor, R. B., and P. J. Brown. 2006. Herbivory in the gammarid amphipod *Aora typica*: Relationships between consumption rates, performance and abundance across ten seaweed species. *Mar. Biol.* 149:455–463.

Taylor, R. B., N. Lindquist, J. Kubanek, and M. E. Hay. 2003. Intraspecific variation in palatability and defensive chemistry of brown seaweeds: Effects on herbivore fitness. *Oecologia* 136:412–423.

Taylor, R. B., E. Sotka, and M. E. Hay. 2002. Tissue-specific induction of herbivore resistance: Seaweed response to amphipod grazing. *Oecologia* 132:68–76.

Taylor, R., and P. Steinberg. 2005. Host use by Australasian seaweed mesograzers in relation to feeding preferences of larger grazers. *Ecology* 86:2955–2967.

Tegner, M. J., and P. K. Dayton. 1991. Sea urchins, El Ninos, and the long-term stability of southern California kelp forest communities. *Mar. Ecol. Prog. Ser.* 77:49–63.

Thompson, J. N. 2005. *The Geographic Mosaic of Coevolution*. Chicago, IL: University of Chicago Press.

Thornber, C., E. Jones, and J. Stachowicz. 2008. Differences in herbivore feeding preferences across a vertical rocky intertidal gradient. *Mar. Ecol. Prog. Ser.* 363:51–62.

Tomas, F., J. M. Abbott, C. Steinberg, M. Balk, S. L. Williams, and J. J. Stachowicz. 2011. Plant genotype and nitrogen loading influence seagrass productivity, biochemistry, and plant–herbivore interactions. *Ecology* 92:1807–1817.

Toth, G. B., and H. Pavia. 2007. Induced herbivore resistance in seaweeds: A meta-analysis. *J. Ecol.* 95:425–434.

Toth, G. B., O. Langhamer, and H. Pavia. 2005. Inducible and constitutive defenses of valuable seaweed tissues: Consequences for herbivore fitness. *Ecology* 86:612–618.

Trowbridge, C. D., and C. D. Todd. 2001. Host-plant change in marine specialist herbivores: Ascoglossan sea slugs on introduced macroalgae. *Ecol. Monograph.* 71:219–243.

Trowbridge, C. D. 1991. Diet specialization limits herbivorous sea slug's capacity to switch among food species. *Ecology* 72:1880–1888.

Tugwell, S., and G. M. Branch. 1992. Effects of herbivore gut surfactants on kelp polyphenol defenses. *Ecology* 73:205–215.

Van Alstyne, K. L., and V. Paul. 1990. The biogeography of polyphenolic compounds in marine macroalgae: Temperate brown algal defenses deter feeding by tropical herbivorous fishes. *Oecologia* 84:158–163.

Van Alstyne, K. L., D. O. Duggins, and M. N. Dethier. 2001a. Spatial patterns in macroalgal chemical defenses. In: *Marine Chemical Ecology*, J. B. McClintock, and B. J. Baker (Eds.), pp. 301–224. Boca Raton, FL: CRC Press.

Van Alstyne, K. L., A. V. Nelson, J. R. Vyvyan, and D. A. Cancilla. 2006. Dopamine functions as an antiherbivore defense in the temperate green alga *Ulvaria obscura*. *Oecologia* 148:304–311.

Van Alstyne, K. L. 2008. Ecological and physiological roles of dimethylsulfoniopropionate and its products in marine macroalgae. In: *Algal Chemical Ecology*, C. D. Amsler (Ed.), pp. 173–194. Berlin, Germany: Springer-Verlag.

Vasconcelos, M. A., T. C. Mendes, W. L. S. Fortes, and R. C. Pereira. 2009. Feeding and decoration preferences of the Epialtidae crab *Acanthonyx scutiformis*. *Braz. J. Oceanogr.* 57:137–143.

Vergés, A., P. D. Steinberg, M. E. Hay, A. G. B. Poore, A. H. Campbell, E. Ballesteros, K. L. Heck et al. 2014. The tropicalization of temperate marine ecosystems: Climate-mediated changes in herbivory and community phase shifts. *Proc. Biol. Sci.* 281:20140846.

Vesakoski, O., C. Boström, T. Honkanen, and V. Jormalainen. 2008. Sexual and local divergence in host exploitation the marine herbivore *Idotea baltica* (Isopoda). *J. Exp. Mar. Biol. Ecol.* 367:118–126.

Vesakoski, O., J. Rautanen, V. Jormalainen, and T. Ramsay. 2009. Divergence in host use ability of a marine herbivore from two habitat types. *J. Evol. Biol.* 22:1545–1555.

Via, S. 1984. The quantitative genetics of polyphagy in an insect herbivore. II. Genetic correlations in larval performance within and among host plants. *Evolution* 38:896–905.

Wägele, H., M. Ballesteros, and C. Avila. 2006. Defensive glandular structures in opisthobranch molluscs—From histology to ecology. *Oceanogr. Mar. Biol. Annu. Rev.* 44:197–276.

Whalen, K. E., E. E. Sotka, J. V. Goldstone, and M. E. Hahn. 2010. The role of multixenobiotic transporters in predatory marine molluscs as counter-defense mechanisms against dietary allelochemicals. *Comp. Biochem. Physiol. C Toxicol. Pharmacol.* 152:288–300.

Wikström, S. A., M. B. Steinarsdóttir, L. Kautsky, and H. Pavia. 2006. Increased chemical resistance explains low herbivore colonization of introduced seaweed. *Oecologia* 148:593–601.

Williams, S. L., and J. E. Smith. 2007. A global review of the distribution, taxonomy, and impacts of introduced seaweeds. *Annu. Rev. Ecol. Evol. Syst.* 38:327–359.

Williamson, J. E., and P. D. Steinberg. 2012. Fitness benefits of size-dependent diet switching in a marine herbivore. *Mar. Biol.* 159:1001–1010.

Winkler, I., and C. Mitter. 2008. The phylogenetic dimension of insect/plant interactions: A summary of recent evidence. In: *Specialization, Speciation, and Radiation: The Evolutionary Biology of Herbivorous Insects*, K. Tillmon (Ed.), pp. 240–263. Berkeley, CA: University of California Press.

Wright, J. T., R. de Nys, A. G. B. Poore, and P. D. Steinberg. 2004. Chemical defense in a marine alga: Heritability and the potential for selection by herbivores. *Ecology* 85:2946–2959.

Zimmer, M., and S. Bartholme. 2003. Bacterial endosymbionts in *Asellus aquaticus* (Isopoda) and *Gammarus pulex* (Amphipoda) and their contribution to digestion. *Limnol. Oceanogr.* 48:2208–2213.

6 The Role of Natural Products in Structuring Microbial Communities of Marine Algae

Jennifer M. Sneed and Melany P. Puglisi

CONTENTS

6.1 INTRODUCTION

Most organisms have a resident microbial community (microbiome) associated with them. The importance of these communities to the health and overall functioning of their hosts is becoming increasingly recognized. This is exemplified by the recent proliferation of studies demonstrating the role of the human microbiome in everything from digestive to mental health (Cho and Blaser 2012, Foster and McVey Neufeld 2013). Similarly, microbiomes residing within marine organisms serve a variety of functions from defense to nutrient cycling (McFall-Ngai et al. 2013, Apprill 2017). Disruption of these microbiomes is often harmful to the host and it has been postulated that many marine diseases are caused by destabilization of the host microbiome (see reviews by Krediet et al. 2013, Egan and Gardiner 2016).

Until recently, our knowledge of the ecological roles of microbes was limited to those organisms that could be cultured under laboratory conditions. It is estimated that only a small fraction of marine microbes are currently culturable and that we have yet to discover the conditions necessary to culture the majority (Amann et al. 1995). However, new advances in sequencing technology have made it possible to examine marine microbial diversity more fully. Bacteria and many archaea can be identified using the 16S rRNA marker gene (Olsen et al. 1986). Targeted sequencing of the 16S rRNA gene using next-generation sequencing techniques can generate millions of sequences from a single sample. This allows for a thorough sampling of the microbial diversity found within a sample and has led to the discovery of microbial taxa that were previously unknown.

Because bacteria play such important roles in the ecology of marine organisms and because their abundance is so high in seawater, it is likely that marine organisms have evolved mechanisms to regulate the bacterial community surrounding them (Singh and Reddy 2014). There is growing evidence that many marine organisms host species-specific bacterial communities on their surfaces and/or within their tissues/bodies. This has been most clearly demonstrated in studies of marine macroalgae (as reviewed in Singh and Reddy 2014). The composition of bacterial communities on the surfaces of macroalgae is often different from that of non-living substrates and in the surrounding seawater (Dobretsov et al. 2006, Sneed and Pohnert 2011, Sneed et al. 2015). Additionally, some macroalgal species maintain a unique bacterial community across varying geographical locations. Lachnit et al. (2009) found that the surface bacterial communities of six species of macroalgae (three brown, two red, and one green) differed significantly from each other and that the composition of the bacterial community was affected more by the host species than by the location from which the algae were collected (North Sea vs. Baltic Sea). Nylund et al. (2010) likewise found that algae collected from two locations within the North Sea maintained a specific surface bacterial community regardless of geography. Similar patterns of bacterial-host specificity have also been described for other sessile marine organisms, including sponges and corals (Rohwer et al. 2002, Taylor et al. 2004, La Rivière et al. 2015). Although these examples indicate that many benthic marine organisms harbor unique bacterial communities, the mechanisms governing this type of regulation are still mostly unknown. Sessile marine organisms produce a wide variety of secondary metabolites and it is likely that these small molecules are involved in such regulation (reviewed in Engel et al. 2002). Here we will review what is known about the role of secondary metabolites in the regulation of macroalgal associated microbial communities within benthic marine environments.

6.2 MICROBIAL CHEMICAL DEFENSES

Mounting evidence suggests that increases in ocean temperatures may result in major disease outbreaks in marine organisms caused by opportunistic or emerging pathogens (Harvell et al. 2002, 2004, 2007, Lafferty et al. 2004, Sokolow 2009, Burge et al. 2013, Egan et al. 2014, Maynard et al. 2015). In addition, changes in oceanic salinity and pH are occurring that may further influence the frequency and severity of disease outbreaks (Hoegh-Guldberg et al. 2007, Sokolow 2009). These factors alter the relationship between the host and pathogen by increasing disease susceptibility in physiologically stressed hosts and causing changes in the growth rate and virulence of pathogens (Harvell et al. 1999, 2002, Hoegh-Guldberg et al. 2007). The loss of biodiversity in marine habitats due to disease outbreaks and other anthropogenic pressures may increase the transmission of infectious diseases, resulting in a negative feedback loop and increasingly common disease in marine environments (Keesing et al. 2010).

Despite the increase in disease in the marine environment, algae appear to be less susceptible to infection compared to marine plants and animals (Engel et al. 2002). This is likely due in part to the effective production of antimicrobial chemical defenses by these organisms. Antimicrobial chemical defenses of macroalgae have been previously reviewed by Engel et al. (2002), Paul and Puglisi (2004), Lane and Kubanek (2008), and Egan et al. (2014). In this section, the prevalence of antimicrobial defenses in marine macroalgae and natural products with antimicrobial activities against environmental organisms will be discussed.

6.2.1 ANTIMICROBIAL EXTRACTS

Most studies on the effects of macroalgal natural products on the growth of individual bacterial species have focused on the search for highly potent antibacterial compounds, in most cases for pharmaceutical use or as antifouling agents (Reichelt and Borowitzka 1984, Freile-Pelegrin and Morales 2004, Salvador et al. 2007, Shanmughapriya et al. 2008). However, the examination of the activity of algal metabolites against ecologically relevant bacterial species is increasing (see review by Goecke et al. 2010). Several broad surveys have shown that crude extracts from marine algae exhibit antimicrobial activities against marine bacteria and fungi (Engel et al. 2006, Puglisi et al. 2007, Lane et al. 2010, Goeke et al. 2012).

Broad studies of common algal species from the tropical Atlantic and Indo-Pacific examined the effects of lipophilic and hydrophilic algal extracts on the growth of a diverse panel of marine microorganisms including *Lindra thalassiae* (pathogenic fungus), *Dendryphiella salina* (saprophytic fungus), *Halophytophthora spinosa* and *Schizochytrium aggregatum* (saprophytic stramenopiles), and *Pseudoaltermonas bacteriolytica* (pathogenic bacterium) (Engel et al. 2006, Puglisi et al. 2007). These studies demonstrated more than 95% of the extracts exhibited antimicrobial activities against at least one ecologically relevant microorganism in the panel. Extracts from 55 out of 103 algal species tested in their studies inhibited the growth of a known algal pathogen, *P. bacteriolytica*. Broad-spectrum activity against at least three assay microbes was observed in 50% of the extracts. Extracts from the green alga *Bryopsis pennata* and the red alga *Portieria hornemannii* from Guam and green algae *Halimeda copiosa* and *Penicillus capitatus* from the Caribbean inhibited the growth of all assay microorganisms.

A similar study in Fiji evaluated extracts from 43 species of red algae for antimicrobial activity against *L. thalassiae*, *D. salina*, and *P. bacteriolytica* (Lane et al. 2010). Extracts were fractionated by reversed-phase chromatography yielding four fractions for bioassays: 1:1 MeOH/H$_2$O, 4:1 MeOH/H$_2$O, MeOH, and Acetone. At least one fraction from all 43 species (69 specimens in total) in the study reduced the growth of *L. thalassiae* and *P. bacteriolytica*. The polar and nonpolar fractions were equally effective against *L. thalassiae* while *P. bacteriolytica* was more susceptible to the polar fractions. The authors also reported intra-specific variability in microbial defenses in algae collected at different locations.

Another survey of algae from the Western Baltic Sea looked at the antimicrobial activities of 16 algae against ten bacterial strains, including five environmental strains reported to be associated with macroalgae (Goeke et al. 2012). Three of the environmental strains used in the panel were isolated from the surface of *Fucus vesiculosus* and *Delesseria sanguinea*: *Bacillus algicola*, *Paenibacillus lautus*, and *Pseudomonas marincola*. Other strains included the marine pathogens *Agricola bacteriolytica* and *Pseudoalteromonas elyakovii*. Antimicrobial activity was reported for 88% of the algal extracts against bacteria. *Bacillus subtilis* was the most susceptible to the algal extracts. This study differs from those conducted in the Atlantic and Indo-Pacific in that some of the algal extracts also stimulated growth of some surface-associated and pathogenic strains.

Dubber and Harder (2008) investigated the antibacterial effects of lipophilic and hydrophilic extracts of three algae—*Mastocarpus stellatus, Laminaria digitata*, and *Ceramium rubrum*—collected over two seasons (November and March) against a panel of environmental and pathogenic bacteria. The assay panel included twelve strains of environmental marine bacteria isolated from the mudflats of the Wendell Sea and seven fish pathogens: *Aeromonas hydrophila* subsp. *hydrophila*, *A. salmonicida, Vibirio (Listonella) anguillarum, Photobacterium damselae* subsp. *damselae, Pseudomonas anguilliseptica, Vibrio alginolyticus*, and *Yersinia ruckeri*. There were no notable seasonal differences between two collection periods for any of the algal species. The hydrophilic extract of *C. rubrum* and the lipophilic extract of *L. digitata* were the most active extracts, inhibiting at least 78% of the bacterial species in the panel. Of the environmental strains, the Gram-negative marine *Vibrionaceae* were less susceptible to the algal extracts than the Gram-positive marine *Bacillaceae*. Algal extracts were effective in reducing the growth of the fish pathogens *L. anguillarum, P. anguilliseptica*, and *A. salmonicida*, suggesting that these may be useful in the control of pathogens in mariculture.

The surface extract from *Bonnemaisonia asparagoides*, rich in halogenated metabolites, was shown to inhibit the growth of 10 of 12 bacteria isolated from the surface of three red algae—*Ceramium virgatum, Polysiphonia fucoides*, and *Rhodomela confervoides*—from the same habitat (Nylund et al. 2010). Extracts, prepared by dipping *B. asparagoides* in a mixture of 1:1 hexanes and 0.5 M NaCl to obtain the surface-associated metabolites, were evaluated for antimicrobial activity in membrane bioassays that mimic the surface of the alga. It was postulated that these compounds may be stored in specialized gland cells at the surface of the alga preventing settlement of bacterial epiphytes.

A similar hexane dipping method was used by Thabard et al. (2011) to prepare surface extracts from *Sargassum polyceratium* for antimicrobial and antifouling investigation. Surface extracts were more effective inhibitors against four marine fouling and pathogenic bacteria (*Halomonas marina, Pseudoalteromonas elyakovii, Polaribacter irgensii*, and *Vibrio aestuarianus*) compared with five terrestrial bacterial strains isolated from estuaries and coastal environments. They also inhibited embryo development in common fouling invertebrates. Results reported in this study suggest that microbial chemical defenses may evolve to adapt to the potential threats in a specific habitat.

6.2.2 Antimicrobial Natural Products

Chemical investigation of the active extracts from these larger surveys has resulted in the isolation and characterization of novel and known compounds that exhibit antimicrobial activity at or below natural concentrations. Chemical investigation of the brown alga *Lobophora variegata* resulted in the isolation of a structurally unusual antifungal macrolide: lobophorolide (**1**) (Kubanek et al. 2003). This compound is a potent inhibitor of the pathogen *Lindra thallasiae* and the saprophyte *Dendryphiella salina*. It was also shown to be active against the human pathogen *Candida albicans*. Mechanism of action studies demonstrated that two lobophorolide molecules cooperate to form a dimerization interface that binds to the actin filament and prevents microfilament stabilization (Blain et al. 2010).

Bioassay-guided fractionation of active crude extracts from the green alga *Penicillus capitatus* resulted in the isolation of two triterpene sulfates: capisterones A and B (**2, 3**) (Puglisi et al. 2004). These compounds selectively inhibited the growth of the *L. thallasiae*. Subsequent investigation of *P. capitatus* reported in a review by Lane and Kubanek (2008) (Engel and Fenical unpublished work) showed that the capisterones are concentrated in the filamentous cap of the alga and cannot be found in measurable quantities in the calcareous base. These compounds are also found in other species in the genus *Penicillus*.

Chemical investigation of the red alga *Callophycus serratus* yielded a suite of antifungal metabolites including the bromophycolides A–I, debromophycolide A, and callophycoic acids/callophycols (Lane et al. 2009). All of the bromophycolides were active against *L. thallasiae* at or below the natural concentration while only callophycoic acids C and G inhibited growth at or near natural concentration.

Desorption electrospray ionization mass spectrometry (DESI-MS) was used to locate the antifungal compounds at the surface of the alga at active concentrations, demonstrating that antifungal metabolites are positioned within the alga to prevent infection by the aggressive pathogen *L. thallasiae*. DESI-MS indicated the presence of bromophycolides in association with distinct surface patches and within the internal algal tissue. This groundbreaking study is among the first examples employing DESI-MS for natural product imaging on biological surfaces.

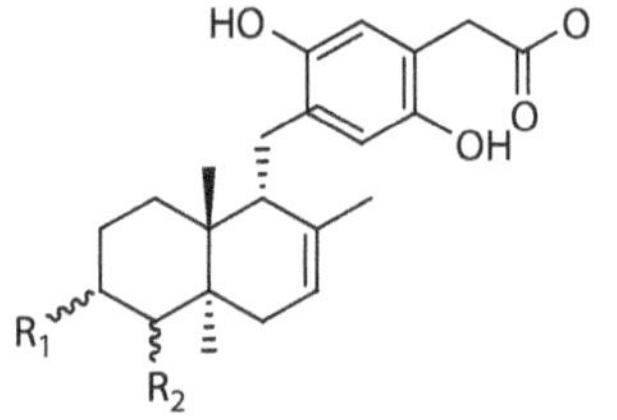

4   **5**

$R_1 = $ —Br $R_1 = $ ····CH$_3$

$R_2 = $ ····CH$_3$ CH$_3$ $R_2 = $ = CH$_2$

Of the 43 species of red algae surveyed in Fiji for antimicrobial activity, the fractions of the crustose red alga *Peyssonnelia* sp. that exhibited potent activity against the bacterium *Pseudoalteromonas bacteriolytica* and the fungus *Lindra thalassiae* were identified for further investigation (Lane et al. 2010). Bioassay-guided fractionation yielded two novel sesquiterpene hydroquinones, peyssonoic acids A–B (**4, 5**). These compounds were reported to be active below natural concentrations.

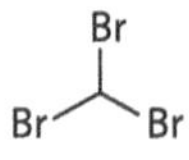

6 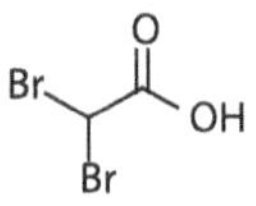   **7**

In another study, the simple halogenated compounds bromoform (**6**) and dibromoacetic acid (**7**) produced by the alga red *Aspragopsis armata* were reported to be antibacterial agents against the marine strains *Vibrio harveyii* and *V. alginolyticus* (Paul et al. 2006). Gas chromatography-mass spectrometry analysis (GC-MS) showed that the compounds were present in both the tetrasporophyte and gametophyte forms of the alga. These brominated compounds are localized in retractable vesicles in *A. armata* that are lost when bromine is eliminated from an artificial culture medium. Subsequent experiments with algae that no longer produced halogenated metabolites showed that epiphytic bacteria will preferentially settle on *A. armata* without brominated compounds. Both compounds inhibited the growth of bacteria isolated from

algae without brominated compounds, indicating that halogenated metabolites of
A. armata may have a role in forming the epiphytic bacterial community.

The carotenoid pigment fucoxanthin (**8**) was identified as a microbial chemical
defense in the brown seaweed *Fucus vesiculosus* (Saha et al. 2011). This metabolite
was isolated from the surface of the alga using the hexanes dipping technique at
concentrations between 0.7 and 9 mg cm^{-2} on the algal surface. Of the 15 bacteria
in the study, *Cytophaga KT0804*, *Bacillus aquimaris*, *Ulvabacter littoralis*,
and Alteromondaceae A1 were susceptible to fucoxanthin at natural surface
concentrations.

Finally, work by Case et al. (2011) demonstrated that furanones produced by
Delisea pulchra inhibit colonization and infection by *Ruegeria* sp. R11, a marine
pathogen causing gall disease in red algae. In controlled experiments with furanone
(−) and furanone (+) *D. pulchra* thalli, *Ruegeria* sp. R11 was able to form biofilms,
invade and bleach the furanone (−) thalli at 24°C but not at temperatures lower
than 19°C temperatures. This result suggests that increasing ocean temperatures can
increase the susceptibility of algae to infection.

6.3   CHEMICALLY MEDIATED BIOFILM FORMATION

Studies like those discussed in Section 6.2 demonstrate that macroalgae are capable
of producing compounds that inhibit the growth of certain microbes, often for the
purpose of defending the alga. The bioassays traditionally used to assess the activity of algal extracts against bacteria were adapted from the pharmaceutical industry
and are biased toward the discovery of inhibitory compounds. The complexity of
microbial interactions within biofilms on algal surfaces makes it imperative to examine the effects of algal secondary metabolites on microbes through a more holistic
lens. Algal biofilm communities are likely structured in part by the production of
compounds that selectively inhibit or promote the growth of specific microbes and/or
impact the ability of microbes to colonize surfaces and maintain their populations
within the biofilm community. In addition to directly affecting the growth of microorganisms, there is also a growing body of evidence that several macroalgae produce
compounds that interfere with the process of communication among bacterial cells
known as quorum sensing (Borchardt et al. 2001, Skindersoe et al. 2008). Quorum
sensing regulates many physiological processes in bacteria including expression of
virulence, biofilm formation, and antibiotic production. Disruption of quorum sensing can therefore greatly impact the formation of biofilm communities and the interactions between host organisms and their associated microbes. Several recent studies
that have evaluated the impacts of algal chemistry on the composition of biofilm
communities are discussed in this section.

9

The red alga *Bonnemaisonia hamifera* has significantly lower abundances of bacteria on its surface compared to a co-occurring species, *Chondrus crispus* (Nylund et al. 2005). Crude extracts of *B. hamifera* demonstrated broad-spectrum activity against a suite of marine and model bacterial strains in disk diffusion assays but had little effect on the attachment behavior of a subset of these bacterial strains (Nylund et al. 2005). The antibacterial activity of *B. hamifera* was attributed to the compound 1,1,3,3-tetrabromo-2-heptanone (**9**) found on the surface of the alga. At ecologically relevant concentrations, this compound selectively inhibited the growth of bacterial strains isolated from red alga that co-occur with *B. hamifera*. Bacteroidetes, firmicutes, and actinobacteria were particularly sensitive to the compound, while proteobacteria were largely resistant (Nylund et al. 2008). In addition to reducing the growth of bacteria in culture, **9** also reduced the number of bacterial cells colonizing artificial surface in the natural environment. The differential activity of this compound against bacterial strains in laboratory assays suggests a role in structuring the composition of the alga's surface microbiome. Molecular analysis terminal restriction fragment length polymorphism [TRFLP] of petri dishes coated with **9** demonstrated significant shifts in the bacterial community composition and increased bacterial evenness compared to controls (Persson et al. 2011). Contrary to laboratory culture assays, molecular evidence showed that differences in communities formed on plates coated with **9** could be attributed mostly to reductions in strains found within the alpha and gamma-proteobacteria.

In an attempt to gain a better understanding of how algal secondary metabolites contribute to the formation of unique bacterial communities on the algal surfaces, Lachnit et al. (2010) developed a delivery system to present marine bacteria with both polar and non-polar compounds within the environment. The system consisted of a hydrogel matrix (Phytagel) with non-polar extract imbedded in it and an apparatus that continuously pumped polar extracts through the hydrogel matrix. Denaturing gradient gel electrophoresis (DGGE) profiles of the bacterial communities on the hydrogels with both non-polar and polar surface extracts of *Fucus vesiculosus* were not significantly different from those found on the alga itself, indicating that the surface compounds play a large role in structuring the bacterial community. However, later tests using 454 pyrosequencing instead of DGGE found different results (Lachnit et al. 2013). They determined that surface extracts from *F. vesiculosus* did not significantly inhibit the growth of 5 marine bacterial strains but did significantly reduce the attachment of *Pseudomonas* sp. isolated from the marine environment. Based on a bioassay-guided fractionation with *Pseudomonas* sp., they isolated two compounds that were responsible for this activity and demonstrated that these compounds also inhibited settlement in the other four marine strains. One of these compounds was identified as fucoxanthin (**8**) and the other was not identified. Fucoxanthin, along with DMSP, and proline isolated from the surface of *F. vesiculosus* had been previously demonstrated to selectively inhibit settlement marine bacterial strains tested at naturally relevant concentrations (Saha et al. 2011, 2012). The two strains that were not affected by fucoxanthin were the only two strains originally isolated from the alga itself. Additionally, these compounds reduced the number of bacterial cells present on an artificial surface placed underwater for 3 days. However, the composition of bacterial communities on surfaces treated with the compounds did not significantly differ from controls and was significantly different from communities residing on the surface of *F. vesiculosus*. The authors point out that this may be a result of a need for a better delivery system to more accurately approximate the surface of the alga and the natural release of compounds from that surface. They also suggest that 3 days may not be long enough to allow for the formation of a stable biofilm community. Other factors such as microbe-microbe interactions and surface structure likely contribute to the biofilm composition. As methods for profiling microbial communities become more sophisticated, we can continue to expect more accurate approximations of the true composition of these communities and gain better understanding of their ecology.

The anti-settlement activity of *F. vesiculosus* surface extracts differentially affected bacterial strains. The effects of surface extracts on those strains that are susceptible varies by location and by season with consistently higher activity in the summer and autumn months compared to the winter and spring months across locations (Saha and Wahl 2013). This corresponds to a higher microfouling pressure caused by an increase in planktonic bacteria during these months. Analysis of the bacterial communities on the surfaces of *F. vesiculosus* treated with varying temperature and light regimes demonstrated that there were correlations between the concentrations of DMSP, proline, and fucoxanthin on the surface of *F. vesiculosis* and the bacterial OTUs present (Saha et al. 2014).

10   **11**

The red alga *Delisea pulchra* is one of the most well-studied examples of the role of secondary metabolites in controlling the composition of an organism's associated microbiome in the marine environment (reviewed in Harder et al. 2012). This alga produces halogenated furanones (**10, 11**) that mimic quorum sensing molecules (AHLs) typically produced by Gram-negative bacteria. The furanones interrupt cell-cell communication between bacteria, which disrupts a variety of bacterial processes including biofilm formation and virulence. Furanones have a wide variety of ecological activities including antiherbivore and antifouling defense. Loss of furanones during periods of increased seawater temperature stress leads to an imbalance in the resident microbiome of *D. pulchra* resulting in a bleaching disease characterized by increased bacterial diversity and a shift in the major components of the associated microbial community (Campbell et al. 2011, Case et al. 2011, Fernandes et al. 2012).

6.4 THE ROLE OF ALGAL PRODUCTS IN STRUCTURING MARINE MICROBIAL COMMUNITIES

Studies discussed so far in this chapter concern the effect of algal extracts and metabolites as chemical defenses and mediators of the microbiome of the host. The final section of this chapter will focus on the growing research area concerning the roles that algae and its chemistry have on the adjacent bacterial communities of hard corals.

Outbreaks of disease on coral reefs have had devastating effects all over the world (Bythell and Shepard 1993, Aronson et al. 1998, Bythell et al. 2000). The earliest experimental evidence of the role of algae in facilitating the spread of disease demonstrated that physical contact with *Halimeda opuntia* can trigger white plague type II in *Montastraea faveolata*. *Aurantimonas coralicida*, a causative agent of the disease, was present on *H. opuntia* populations, suggesting that the alga serves as a reservoir for the pathogen (Nugues et al. 2004). Since then it has been shown that algae appear to be a reservoir for the coral pathogen causing White Syndrome in the Indo-Pacific (Sweet et al. 2013).

Additional studies have utilized transplant experiments to assess the effects of macroalgae on the surface microbiome of adjacent corals. *Dictyota bartayresiana* and *Halimeda opuntia* were shown to increase variance in the coral-associated microbial community of *Porites astreoides* in Curacao (Barrott et al. 2012), while *Galaxaura obtusa, Lobophora variegata, Halimeda tuna,* and *Sargassum polyceratium* changed the coral microbial community of *Porites astreoides* and caused the transfer of microbes from the algae to the coral in the Florida Keys (Thurber et al. 2012). *Halimeda opuntia* and *Dictyota menstrualis* were reported to cause a shift in the coral-associated bacteria of *Montastraea faveolata* and *Porites astreoides* to a profile that closely matched that found on the algae (Morrow et al. 2013).

There is also a growing body of evidence demonstrating that algal chemistry can alter the microbial communities of corals and promote the spread of disease with the potential to cause a phase shift from a coral reef to an algal-dominated habitat. A field study conducted by Morrow et al. (2012) in Belize, the Florida Keys, and St. Thomas, Virgin Islands, reported the polar crude extract embedded in a Phytagel matrix and live algal thalli from *Lobophora variegata* caused shifts to occur in the surface mucus layer bacterial communities of *Montastraea faveolata* and *Porites astreoides*. They also observed variations in the effects of the extracts on the microbial communities depending upon the site. In addition, the crude extract from *Halimeda tuna* induced a shift in the bacterial community of *P. astreoides*. These changes in the bacterial assemblage were determined by DGGE profile analysis. While the authors recognize that there were some limitations to the study not allowing for broad conclusions, the study is one of the first attempts to investigate the effects of algal chemistry on adjacent coral microbiome.

A very recent study by Morrow et al. (2017) examined the effects of the extract from *Lobophora* sp. on coral-associated bacteria and coral larvae. Simple growth assays demonstrated that the organic and aqueous extracts from the Pacific brown macroalga *Lobophora* sp. inhibited the growth of five coral-associated bacterial isolates from three classes including *Bacterioplanes sanyensis* JB47, *Marinobacter* sp. JB49, *Shewenella* p. CO41, *Paracoccus detrificans* JB11, and *Pseudovibrio dentrificans* JB12. To determine the effect of *Lobophora* sp. on the microbial communities of field-collected branches of *Porites cylindrica*, treated coral branches were exposed to aqueous crude extract embedded on Phytagel for 24 hours. The authors reported a shift in the microbial communities associated with coral tissues to *Vibrio* dominance. Further laboratory studies exposing coral larvae from *Acropora millepora* to the extracts demonstrated that the extracts exhibited potent activity inhibiting coral larval settlement and causing mortality well below natural concentrations. These experiments, while conducted in the laboratory, contribute to growing evidence that macroalgae produce chemistry with the potential to alter the surface associated microbial communities of nearby organisms.

6.5   CONCLUSION AND FUTURE DIRECTIONS

Advanced techniques in molecular biology have contributed significantly to recent advances in marine algal-microbial chemical ecology. This area of study is in its infancy compared to the study of herbivore chemical defenses and will continue to expand as researchers develop new field techniques to capture the changes in situ. While there have been several studies identifying specific metabolites that can inhibit the growth of potential algal pathogens, there is a need for field experiments that clearly define the release mechanisms of these compounds when exposed to a pathogen. Further understanding of the role of chemistry in the formation of the algal microbiome, as well as the effects of these metabolites on the microbiome of near neighbors, is essential in the ever-changing marine environment. The understanding gained through investigation of both negative and positive effects of algal metabolites on marine microbial communities will have implications in conservation management of marine ecosystems around the world.

REFERENCES

Amann, R. I., W. Ludwig, and K. H. Schleifer. 1995. Phylogenetic identification and in situ detection of individual microbial cells without cultivation. *Microbiol. Rev.* 59:143–169.

Apprill, A. 2017. Marine animal microbiomes: Toward understanding host–microbiome interactions in a changing ocean. *Front. Mar. Sci.* 4:222.

Aronson, R. B., W. F. Precht, and I. G. Macintyre. 1998. Extrinsic control of species replacement on a Holocene reef in Belize: The role of coral disease. *Coral Reefs* 17:223–230.

Barott, K. L., B. Rodriguez-Brito, J. Janouškovec, K. L. Marhaver, J. E. Smith, P. Keeling, and F. L. Rohwer. 2011. Microbial diversity associated with four functional groups of benthic reef algae and the reef-building coral *Montastraea annularis*. *Environ. Microbiol.* 13:1192–1204.

Barott, K. L., B. Rodriguez-Mueller, M. Youle, K. L. Marhaver, M. J. A. Vermeij, J. E. Smith, F. L. Rohwer. 2012. Microbial to reef scale interactions between the reef-building coral *Montastraea annularis* and benthic algae. *Proc. R. Soc. B* 279:655–1664.

Blain, J. C., Y.-F. Mok, J. Kubanek, and J. S. Allingham. 2010. Two molecules of lobophorolide cooperate to stabilize an actin dimer using both their "ring" and "tail" region." *Chem. Biol.* 17:802–807.

Borchardt, S. A., E. J. Allain, J. J. Michels, G. W. Stearns, R. F. Kelly, and W. F. McCoy. 2001. Reaction of acylated homoserine lactone bacterial signaling molecules with oxidized halogen antimicrobials. *Appl. Environ. Microbiol.* 67:3174–3179.

Burge, C. A., C. J. Kim, J. M. Lyles, and C. D. Harvell. 2013. Special issue oceans and humans health: The ecology of marine opportunists. *Microb. Ecol.* 65:869–879.

Bythell, J., and C. Sheppard. 1993. Mass mortality of Caribbean shallow corals. *Mar. Poll. Bull.* 26:296–297.

Bythell, J. C., Z. M. Hillis-Starr, and C. S. Rogers. 2000. Local variability but landscape stability in coral reef communities following repeated hurricane impacts. *Mar. Ecol. Prog. Ser.* 204:93–100.

Campbell, A. H., T. Harder, S. Nielsen, S. Kjelleberg, and P. D. Steinberg. 2011. Climate change and disease: Bleaching of a chemically defended seaweed. *Global Change Biol.* 17:2958–2970.

Case, R. J., S. R. Longford, A. H. Campbell, A. Low, N. Tujula, P. D. Steinberg, and S. Kjelleberg. 2011. Temperature induced bacterial virulence and bleaching disease in a chemically defended marine macroalga. *Environ. Microbiol.* 13:529–537.

Cho, I., and M. J. Blaser. 2012. The human microbiome: At the interface of health and disease. *Nat. Rev. Genet.* 13:260–270.

Dobretsov, S., H. U. Dahms, T. Harder, and P. Y. Qian. 2006. Allelochemical defense against epibiosis in the macroalga *Caulerpa racemosa* var. turbinata. *Mar. Ecol. Prog. Ser.* 318:165–175.

Dubber, D., and T. Harder. 2008. Extracts of *Ceramium rubrum, Mastocarpus stellatus* and *Laminaria digitata* inhibit growth of marine and fish pathogenic bacteria at ecologically realistic concentrations. *Aquaculture* 274:196–200.

Egan, S., N. D. Fernandes, V. Kumar, M. Gardiner, and T. Thomas. 2014. Bacterial pathogens, virulence and host defence in marine macroalgae. *Environ. Microbiol.* 16:915–938.

Egan, S., and M. Gardiner. 2016. Microbial dysbiosis: Rethinking disease in marine ecosystems. *Front. Microbiol.* 7:991.

Engel, S., P. R. Jensen, and W. Fenical. 2002. Chemical ecology of marine microbial defense. *J. Chem. Ecol.* 28:1971–1985.

Engel, S., M. P. Puglisi, P. R. Jensen, and W. Fenical. 2006. Antimicrobial activities of extracts from tropical Atlantic marine plants against marine pathogens and saprophytes. *Mar. Biol.* 149:991–1002.

Fernandes, N., P. Steinberg, D. Rusch, S. Kjelleberg, and T. Thomas. 2012. Community structure and functional gene profile of bacteria on healthy and diseased thalli of the red seaweed *Delisea pulchra. PLoS One* 7:e50854.

Foster, J. A., and K.-A. McVey Neufeld. 2013. Gut–brain axis: How the microbiome influences anxiety and depression. *Trends Neurosci.* 36:305–312.

Freile-Pelegrin, Y., and J. L. Morales. 2004. Antibacterial activity in marine algae from the coast of Yucatan, Mexico. *Botanica Marina* 47:140–146.

Goecke, F., A. Labes, J. Wiese, and J. F. Imhoff. 2010. Chemical interactions between marine macroalgae and bacteria. *Mar. Ecol. Prog. Ser.* 409:267–299.

Goecke, F., A. Labes, J. Wiese, and J. F. Imhoff. 2012. Dual effect of macroalgal extracts on growth of bacteria in Western Baltic Sea. *Rev. Biol. Mar. Oceanogr.* 1:75–86.

Harder, T., A. H. Campbell, S. Egan, and P. D. Steinberg. 2012. Chemical mediation of ternary interactions between marine holobionts and their environment as exemplified by the red alga *Delisea pulchra. J. Chem. Ecol.* 38:442–450.

Harvell, C. D., K. Kim, J. M. Burkholder, R. R. Colwell, P. R. Epstein, D. J. Grimes, E. E. Hofmann et al. 1999. Emerging marine diseases—Climate links and anthropogenic factors. *Science* 285:1505–1510.

Harvell, C. D., C. E. Mitchell, J. R. Ward, S. Altizer, A. P. Dobson, R. S. Ostfeld et al. 2002. Climate warming and disease risks for terrestrial and marine biota. *Science* 296:2158–2162.

Harvell, C. D., R. Aronson, N. Baron, J. Connell, A. Dobson, S. Ellner, L. Gerber et al. 2004. The rising tide of ocean diseases: Unsolved problems and research priorities. *Front. Ecol. Environ.* 2:375–382.

Harvell, C. D., E. Jordan-Dahlgren, S. Merkel, E. Rosenberg, L. Raymundo, G. Smith, E. Weil, and B. Willis. 2007. Coral disease, environmental drivers, and the balance between coral microbial associates. *Oceanography* 20:172–195.

Hoegh-Guldberg, O., P. J. Mumby, A. J. Hooten, R. S. Steneck, P. Greenfield, E. Gomez, C. D. Harvell et al. 2007. Coral reefs under rapid climate change and ocean acidification. *Science* 318:1737–1742.

Keesing, F., L. K. Belden, P. Daszak, A. Dobson, C. D. Harvell, R. D. Holt, P. Hudson et al. 2010. Impacts of biodiversity on the emergence and transmission of infectious diseases. *Nature* 7324:647.

Krediet, C. J., K. B. Ritchie, V. J. Paul, and M. Teplitski. 2013. Coral-associated microorganisms and their roles in promoting coral health and thwarting diseases. *Proc. R. Soc. Lond. B Biol. Sci.* 280:20122328.

Kubanek, J., P. R. Jensen, P. A. Keifer, M. C. Sullards, D. O. Collins, and W. Fenical. 2003. Seaweed resistance to microbial attack: A targeted chemical defense against marine fungi. *Proc. Natl. Acad. Sci.* 100:6916–6921.

La Rivière, M., J. Garrabou, and M. Bally. 2015. Evidence for host specificity among dominant bacterial symbionts in temperate gorgonian corals. *Coral Reefs* 34:1087–1098.

Lachnit, T., M. Blumel, J. F. Imhoff, and M. Wahl. 2009. Specific epibacterial communities on macroalgae: Phylogeny matters more than habitat. *Aqua. Biol.* 5:181–186.

Lachnit, T., M. Fischer, S. Künzel, J. F. Baines, and T. Harder. 2013. Compounds associated with algal surfaces mediate epiphytic colonization of the marine macroalga *Fucus vesiculosus*. *FEMS Microbiol. Ecol.* 84:411–420.

Lachnit, T., M. Wahl, and T. Harder. 2010. Isolated thallus-associated compounds from the macroalga *Fucus vesiculosus* mediate bacterial surface colonization in the field similar to that on the natural alga. *Biofouling* 26:247–255.

Lafferty, K. D., J. W. Porter, and S.E. Ford, 2004. Are diseases increasing in the ocean? *Annu. Rev. Ecol. Evol. Syst.* 35:31–54.

Lane, A. L., and J. Kubanek. 2008. Secondary metabolite defenses against pathogens and biofoulers. In *Algal Chemical Ecology*, pp. 229–243. Berlin, Germany: Springer.

Lane, A. L., L. Mular, E. J. Drenkard, T. L. Shearer, S. Engel, S. Fredericq, C. R. Fairchild et al. 2010. Ecological leads for natural product discovery: Novel sesquiterpene hydroquinones from the red macroalga *Peyssonnelia* sp. *Tetrahedron* 66:455–461.

Lane, A. L., L. Nyladong, A. S. Galhena, T. L. Shearer, E. P. Stout, R. M. Perry, M. Kwasnik et al. 2009. Desorption electrospray ionization mass spectrometry reveals surface-mediated antifungal chemical defense of a tropical seaweed. *Proc. Natl. Acad. Sci.* 106:7314–7319.

Maynard, J., R. van Hooidonk, C. M. Eakin, M. Poutinen, M. Garren, G. Williams, S. F. Heron et al. 2015. Projections of climate conditions that increase susceptibility and pathogen abundance and virulence. *Nat. Clim. Change* 5(7):688.

McFall-Ngai, M., M. G. Hadfield, T. C. G. Bosch, H. V. Carey, T. Domazet-Lošo, A. E. Douglas, N. Dubilier et al. 2013. Animals in a bacterial world, a new imperative for the life sciences. *Proc. Natl. Acad. Sci.* 110:3229–3236.

Morrow, K. M., R. Ritson-Williams, C. Ross, M. R. Liles, and V. J. Paul. 2012. Macroalgal extracts induce bacterial assemblage shifts and sublethal tissue stress in Caribbean corals. *PLoS One* 7:e44859.

Morrow, K. M., M. R. Liles, V. J. Paul, A. G. Moss, and N. E. Chadwick. 2013. Bacterial shifts associated with coral–macroalgal competition in the Caribbean Sea. *Mar. Ecol. Prog. Ser.* 488:103–117.

Morrow, K. M., K. Bromhall, C. A. Motti, C. B. Munn, and D. G. Bourne. 2017. Allelochemicals produced by brown macroalgae of the *Lobophora* genus are active against coral larvae and associated bacteria, supporting pathogenic shifts to *Vibrio* dominance. *Appl. Environ. Microbiol.* 83:e02391–16.

Nugues, M. M., W. G. Smith, R. J. van Hooidonk, M. I. Seabra, and R. P. P. Bak. 2004. Algal contact as a trigger for coral disease. *Ecol. Lett.* 7:919–923.

Nylund, G. M., G. Cervin, M. Hermansson, and H. Pavia. 2005. Chemical inhibition of bacterial colonization by the red alga *Bonnemaisonia hamifera*. *Mar. Ecol. Prog. Ser.* 302:27–36.

Nylund, G. M., G. Cervin, F. Persson, M. Hermansson, P. D. Steinberg, and H. Pavia. 2008. Seaweed defence against bacteria: A poly-brominated 2-heptanone from the red alga *Bonnemaisonia hamifera* inhibits bacterial colonisation. *Mar. Ecol. Prog. Ser.* 369:39–50.

Nylund, G. M., F. Persson, M. Lindegarth, G. Cervin, M. Hermansson, and H. Pavia. 2010. The red alga *Bonnemaisonia asparagoides* regulates epiphytic bacterial abundance and community composition by chemical defence. *FEMS Microbiol. Ecol.* 71:84–93.

Olsen, G. J., D. J. Lane, S. J. Giovannoni, N. R. Pace, and D. A. Stahl. 1986. Microbial ecology and evolution: A ribosomal RNA approach. *Annu. Rev. Microbiol.* 40:337–365.

Paul, V. J., and M. P. Puglisi. 2004. Chemical mediation of interactions among marine organisms. *Nat. Prod. Rep.* 21:189–209.

Paul, N. A., R. de Nys, and P. D. Steinberg. 2006. Chemical defence against bacteria in the red alga *Asparagopsis armata*: Linking structure with function. *Mar. Ecol. Prog. Ser.* 306:87–101.

Persson, F., R. Svensson, G. M. Nylund, N. J. Fredriksson, H. Pavia, and M. Hermansson. 2011. Ecological role of a seaweed secondary metabolite for a colonizing bacterial community. *Biofouling* 27:579–588.

Puglisi, M. P., L. T. Tan, P. R. Jensen, and W. Fenical. 2004. Capisterones A and B from the tropical green alga *Penicillus capitatus*: Unexpected anti-fungal defenses targeting the marine pathogen *Lindra thallasiae*. *Tetrahedron* 60:7035–7039.

Puglisi, M. P., S. Engel, P. R. Jensen, and W. Fenical. 2007. Antimicrobial activities of extracts from Indo-Pacific marine plants against marine pathogens and saprophytes. *Mar. Biol.* 150:531–540.

Reichelt, J. L., and M. A. Borowitzka. 1984. Antimicrobial activity from marine algae: Results of a large-scale screening program. *Hydrobiologia* 116:158–168.

Rohwer, F., V. Seguritan, F. Azam, and N. Knowlton. 2002. Diversity and distribution of coral-associated bacteria. *Mar. Ecol. Prog. Ser.* 243:1–10.

Saha, M., M. Rempt, B. Gebser, J. Grueneberg, G. Pohnert, and F. Weinberger. 2012. Dimethylsulphopropionate (DMSP) and proline from the surface of the brown alga *Fucus vesiculosus* inhibit bacterial attachment. *Biofouling* 28:593–604.

Saha, M., M. Rempt, K. Grosser, G. Pohnert, and F. Weinberger. 2011. Surface-associated fuco-xanthin mediates settlement of bacterial epiphytes on the rockweed *Fucus vesiculosus*. *Biofouling* 27:423–433.

Saha, M., M. Rempt, S. B. Stratil, M. Wahl, G. Pohnert, and F. Weinberger. 2014. Defence chemistry modulation by light and temperature shifts and the resulting effects on asso-ciated epibacteria of *Fucus vesiculosus*. *PLoS One* 9:e105333.

Saha, M., and M. Wahl. 2013. Seasonal variation in the antifouling defence of the temperate brown alga *Fucus vesiculosus*. *Biofouling* 29:661–668.

Salvador, N., A. G. Garreta, L. Lavelli, and M. A. Ribera. 2007. Antimicrobial activity of Iberian macroalgae. *Scientia Marina* 71:101–113.

Shanmughapriya, S., A. Manilal, S. Sujith, J. Selvin, G. S. Kiran, and K. Natarajaseenivasan. 2008. Antimicrobial activity of seaweeds extracts against multiresistant pathogens. *Ann. Microbiol.* 58:535–541.

Singh, R. P., and C. R. K. Reddy. 2014. Seaweed-microbial interactions: Key functions of seaweed-associated bacteria. *FEMS Microbiol. Ecol.* 88:213–230.

Skindersoe, M. E., P. Ettinger-Epstein, T. B. Rasmussen, T. Bjarnsholt, R. de Nys, and M. Givskov. 2008. Quorum sensing antagonism from marine organisms. *Mar. Biotechnol.* 10:56–63.

Sneed, J. M., and G. Pohnert. 2011. The green alga *Dicytosphaeria ocellata* and its organic extracts alter natural bacterial biofilm communities. *Biofouling* 27:347–356.

Sneed, J. M., R. Ritson-Williams, and V. J. Paul. 2015. Crustose coralline algal species host distinct bacterial assemblages on their surfaces. *ISME J.* 9:2527–2536.

Sokolow, S. 2009. Effects of a changing climate on the dynamics of coral infectious disease: A review of the evidence. *Dis Aquat Organ* 87(1–2):5–18.

Sweet, M. J., J. C. Bythell, and M. M. Nugues. 2013. Algae as reservoirs for coral pathogens. *PLoS One* 8:e69717.

Taylor, M. W., P. J. Schupp, I. Dahllof, S. Kjelleberg, and P. D. Steinberg. 2004. Host specificity in marine sponge-associated bacteria, and potential implications for marine microbial diversity. *Environ. Microbiol.* 6:121–130.

Thabard, M., O. Gross, C. Hellio, and J. Marechal. 2011. *Sargassum polyceratium* (Phaeophyceae, Fucaceae) surface olecule activity towards fouling organisms and embryonic development of benthic species. *Botanica Marina* 54:147–157.

Thurber, R. V., D. E. Burkepile, A. M. S. Correa, A. R. Thurber, A. A. Shantz, R. Welsh, C. Pritchard, and S. Rosales. 2012. Macroalgae decrease growth and alter microbial community structure of the reef-building coral, *Porites astreoides. PLoS One* 7:e44246.

7 New Insights into the Chemical Ecology of *Vibrio cholerae*

Jason R. Graff and David C. Rowley

CONTENTS

7.1 INTRODUCTION

Marine microbes directly impact human health in coastal communities. Pathogenic bacteria and toxic phytoplankton in coastal waters can lead to outbreaks of communicable diseases and acute poisonings through exposure to contaminated water and ingestion of shellfish and other fish stocks. Economic costs can reach well into the hundreds of millions of dollars (US$) for a single outbreak (e.g., cholera) and place human lives and livelihoods at risk. Such extensive economic and human costs emphasize the need for continued research into the ecology of marine pathogens.

A better understanding of their behaviors and interactions with other organisms could lead to more effective means for limiting human contact with these pathogens and for minimizing their detrimental effects on people living in coastal communities.

In a typical 1 mm^3 of surface seawater there are more than 1000 prokaryotes and eukaryotes, resulting in a proximity of less than 100 μm between these organisms (Azam and Malfatti 2007). Chemical interactions between marine microbial groups in close proximity undoubtedly play a role in the success of pathogenic and non-pathogenic species alike. Chemical cues structure many communities (Hay 2009; Weitz 2013) yet little is known regarding the influence of specific secondary metabolites on the behaviors and life histories of marine pathogens and how these interactions might alter the impact these organisms have on human populations. For bacteria-bacteria relationships, intraspecific interactions may include quorum-sensing activities (Gram et al. 2002) while interspecific competition is evident in the high incidence of antagonistic interactions between marine bacteria found on particles (Long and Azam 2001; Grossart et al. 2004). Chemically mediated interactions (antagonistic or protagonistic) likely affect the microscale distributions of bacteria species on particles (DeLong et al. 1993; Long et al. 2003). This is important because the bacteria present within aggregates have implications for human health; particularly for *Vibrio cholerae*, which is often vectored into human food sources and drinking water via particles (Colwell 1996).

In this chapter, we explore the influence of secondary metabolites on the behavior of *V. cholerae* by reviewing previously published work as well as presenting new results. The metabolites considered here include (1) an antibiotic produced by a competing particle-colonizing bacteria and (2) the natural suite of excreted compounds from marine phytoplankton whose blooms have been linked to cholera epidemics (Tamplin et al. 1990; Alfansane et al. 2002). We discuss what the behavioral responses of *V. cholerae* to these metabolites imply for particle colonization by this bacterium and how these results could be further explored and applied in future efforts aimed at mitigating the influence of this pathogen on its human hosts.

7.2 MARINE PARTICLES: A HOTBED OF GROWTH AND INTERACTION

The ocean is a particle-rich environment. Living particles such as phytoplankton and zooplankton and abiotic particles composed of colloids, submicrometer particles, and transparent exopolymers (TEP) (Wells and Goldberg 1991; Azam and Long 2001; Passow 2002; Verdugo et al. 2004) create a heterogeneous environment in which motile microorganisms must navigate. For particle-attaching bacteria, these particles represent potentially hospitable conditions suitable for colonization and growth. High rates of enzyme activities, incorporation of dissolved organic matter, and growth in these organic-rich microenvironments are hypothesized to account for a large portion of the microbial production in the ocean (Fuhrman and Azam 1982; Azam and Long 2001; Kiørboe and Jackson 2001; Hunt et al. 2010). Some marine microbes maximize their utilization of patchy resources (e.g., phytoplankton cells, colloidal particles, and plumes of organic substrates) by employing chemotaxis to aggregate around these microzones (Mitchell et al. 1985; Blackburn et al. 1998; Azam and Malfatti 2007; Stocker et al. 2008; Seymour et al. 2009).

Intraspecific interactions for marine bacteria may include quorum-sensing activities (Gram et al. 2002) while interspecific competition is evident in the high incidence of antagonistic interactions, particularly those attached to particles (Long and Azam 2001; Grossart et al. 2004). These chemically mediated interactions likely affect the microscale distributions of bacteria species (DeLong et al. 1993; Long et al. 2003). This is important because the bacteria present on particles differentially influence element cycling in the ocean due to variations in enzymatic activities (Smith et al. 1992; Martinez et al. 1996; Zaccone et al. 2002) and have implications for human health as these microzones can host pathogenic species (Colwell 1996). Deciphering the mechanisms that drive bacterial diversity and abundance is key to mitigating the potentially harmful effects of pathogens such as *V. cholerae*.

In addition to the dynamics between particle-colonizing bacteria, cell–cell relationships between bacteria and other organisms such as phytoplankton and zooplankton may also be important in the success of either group. This direct association may be species-specific (Schafer et al. 2002; Pinhassi et al. 2004; Jasti et al. 2005; Rooney-Varga et al. 2005). While these relationships are not fully understood, they are thought to have implications for phytoplankton growth (Mayali and Azam 2004; Croft et al. 2005), toxin production by harmful algae (Alavi et al. 2001; Alverca et al. 2002; Bates et al. 2004; Kaczmarska et al. 2005), probiotic associations (Cole et al. 1982; Kodama et al. 2006), and microbial community succession (Rooney-Varga et al. 2005) and may lead to the proliferation, transport, and vectoring of pathogens such as *Vibrio cholerae* (Tamplin et al. 1990).

Despite the potential significance of the interactions between phytoplankton and colonizing bacteria, the mechanisms that regulate the process of colonization of phytoplankton by bacteria have not been thoroughly explored. Similar to the chemically mediated interactions between marine bacteria, phytoplankton exudate chemistry may control bacterial attachment to algal cells. Microalgae excrete chemicals into the surrounding water, the amount and composition of which change over time in relation to the growth stage and health of the phytoplankton (Fogg et al. 1965; Hellebust 1965; Sharp 1977). Phytoplankton exudates can increase the growth of some heterotrophic bacteria while depressing the growth of others (Bell et al. 1974; Naviner et al. 1999) and some phytoplankton may produce metabolites with antibiotic properties (Sieburth 1959). A study in Monterey Bay indicated that diatoms and dinoflagellates that formed dense and often nearly monospecific thin layers and surface slicks had reduced numbers of attached bacteria (Graff et al. 2011). Such dense aggregations may produce high concentrations of inhibitory secondary metabolites, similar to toxin production during harmful algal blooms, which could reduce bacterial attachment to the microalgae. Secondary metabolites excreted by the phytoplankton may serve as cues to bacteria to find or avoid specific phytoplankton. It has been previously observed that bacteria are more commonly found attached to algal cells that are in later stages of growth or at the end of a bloom (Droop and Elson 1966; Vaqué et al. 1989; Smith et al. 1995; Kaczmarska et al. 2005; Grossart et al. 2006) and cells in these stages of growth likely produce and excrete different compounds than during bloom initiation.

Figure 7.1 illustrates some of the interactions between marine bacteria and phytoplankton that will be discussed in this chapter as they have been observed by us and in previous research. These include free-associations between bacteria (e.g., *V. cholerae*) and phytoplankton and the colonization of microalgae, TEPs, and larger

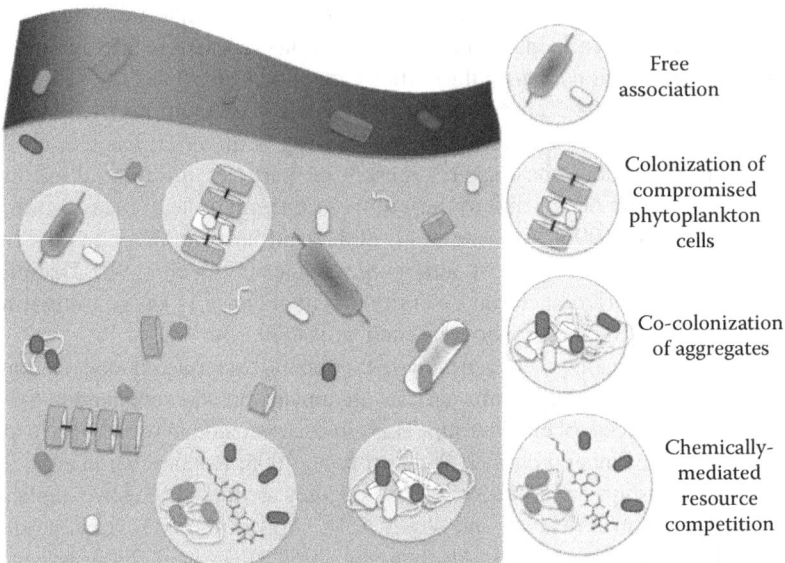

Free association

Colonization of compromised phytoplankton cells

Co-colonization of aggregates

Chemically-mediated resource competition

FIGURE 7.1 Examples of phytoplankton–bacteria and bacteria–bacteria interactions in the pelagic marine environment that we explore in this chapter, with specific emphasis on *Vibrio cholerae*.

aggregates. It is likely that some or all of these associations are chemically mediated relationships that influence colonization dynamics, resource competition, and the growth and proliferation of nearby community members, and can ultimately impact human health.

7.3 PARTICLE ECOLOGY OF *VIBRIO CHOLERAE*

7.3.1 INTERSPECIFIC INTERACTIONS

Vibrio cholerae colonizes marine particles including phytoplankton (Tamplin et al. 1990; Alfansane et al. 2002) and zooplankton (Tamplin et al. 1990; Huq et al. 1993). Interactions with other bacteria or the colonized plankton undoubtedly influence the success of colonization and proliferation of *V. cholerae* on these surfaces. Antagonism between bacteria is most often experimentally demonstrated in the laboratory through antibiotic activity. An example of this with respect to the human pathogen *V. cholerae* is its relationship with the competing Vibrionales SWAT3-wild-type (SWAT3-*wt*) (Long et al. 2005). *Vibrio cholerae* particle colonization is inhibited in the presence of SWAT3-*wt*, a competitor and a producer of the antibiotic andrimid (Long et al. 2005). Andrimid has been isolated from both cultures of marine (Needham et al. 1994; Long et al. 2005) and terrestrial Gammaproteobacteria (Fredenhagen et al. 1987) and has been shown to block the carboxyl transfer reaction of acetyl-CoA carboxylase (Freiberg et al. 2004). *Vibrio cholerae* did not colonize model agar-based particles that were previously colonized by the wild-type SWAT3 but readily colonized particles containing a SWAT3 mutant incapable of producing andrimid. While andrimid potently inhibited the growth of *V. cholerae* in the Long et al. (2005) study, it was unclear how sub-lethal concentrations could act to mediate this competitive interaction. Sub-lethal interactions are likely to be found in naturally occurring environments where it may be difficult to maintain growth-inhibiting concentrations of an antibiotic by competing species due to the effects of rapid diffusion. A growing body of research supports an ideological framework in which antibiotics are thought to be signaling molecules and serve roles other than antibiosis in the environment (Davies et al. 2006; Linares et al. 2006; Mlot 2009). Exploring interactions at more realistic, sub-lethal concentrations is important for understanding the regulation of bacterial diversity, but also because particle attachment and subsequent growth is the major mechanism for the transmission of an infective dose of *V. cholerae* to a human host (Colwell 1996).

7.3.2 Colonization of Phytoplankton

Vibrio cholerae has been found attached to phytoplankton (Tamplin et al. 1990; Alfansane et al. 2002) and cholera outbreaks often coincide with phytoplankton blooms (Epstein 1993; Colwell 1996). This association may be due to the production of chitin by some phytoplankton (Durkin et al. 2009) and the colonization and utilization of chitin by *V. cholerae* (Nalin et al. 1979; Pruzzo et al. 2008). Given the observation that bacteria are more likely to attach to microalgae at the end of a bloom, there may be changes in excreted compounds controlling the timing of colonization by *V. cholerae* as well. If there is indeed a relationship in the timing of the bloom and colonization of post-bloom phytoplankton, this information could be used to minimize human contact with particles hosting the pathogen. The pathogens could enter human food sources as phytoplankton and other particles to which they are attached are consumed by shellfish (Potasman et al. 2002) or when these particles are inadvertently included in water used for drinking other daily activities (Colwell 1996).

7.4 BEHAVIORAL RESPONSE OF *VIBRIO CHOLERAE* TO SECONDARY METABOLITES

In the coming sections, we explore how secondary metabolites from a competing bacterium or from phytoplankton at different stages of growth influence the behavior of *V. cholerae*. Transmitted light, epifluorescence, and video microscopy techniques in conjunction with fluorescent and non-fluorescent staining of samples were used to visualize (1) the bacteria and their behaviors relative to a gradient of a competitor derived metabolites, (2) their interactions with phytoplankton cells, and (3) their response to the excreted products of the microalgae. For behavioral assays we developed a novel microchannel chemotaxis assay (Graff et al. 2013) and microtiter well motility assays to explore the behavioral response of *V. cholerae* to the secondary metabolites of interest.

7.5 ANDRIMID: A COMPETITOR-PRODUCED ANTIBIOTIC

We have recently shown that one species of particle-attaching marine bacteria produces an antibiotic that causes *V. cholerae* to alter its swimming behaviors and in turn is

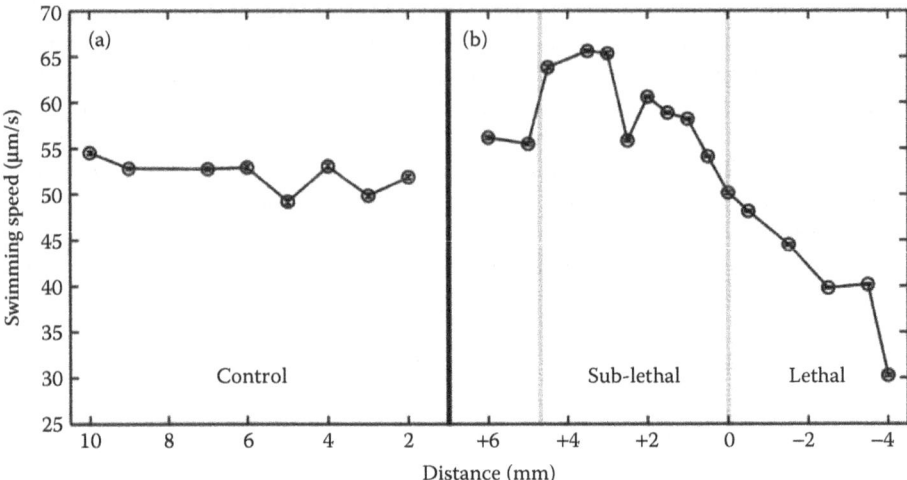

FIGURE 7.2 Mean swimming speeds of *V. cholerae* at distances (a) from a colony of SWAT3-111 (non-andrimid producing mutant) and (b) inside the sub-lethal (+) and lethal (−) zones where 0 mm indicates the delineation of the zone of growth inhibition created by an andrimid-producing colony of wild-type SWAT3. Swimming speeds increase by >10% in the sub-lethal zone and decrease upon exposure to lethal concentrations of andrimid. Plotted error bars are contained within data symbols and represent 1 standard error of the mean. (Reproduced from Graff, J.R. et al., *Front. Microbiol.*, 4, 8, 2013. With permission.)

hypothesized to reduced particle colonization by *V. cholerae* (Graff et al. 2013). For the previously discussed antagonistic relationship between *V. cholerae* and the andrimid producing *Vibrionales* SWAT3 wild-type (SWAT3-*wt*), we hypothesized that andrimid's role in the interaction was to function as a chemical signal that would deter *V. cholerae* from colonizing particles (Figure 7.2). To test this hypothesis, we developed a chemotaxis assay (Graff et al. 2013) that incorporated a polydimethylsiloxane (PDMS) microchannel with a disk or colony diffusion assay (Bauer et al. 1966). By placing a three-sided microchannel on top of an agar surface we were able to replicate the gradient of compound(s) emanating radially from the disk or colony within the channel. We then quantified the behaviors of cells within the channel, including swimming speeds and turning rates, of individual motile bacteria using microvideography and cell tracking methods (Graff et al. 2013). The movements of individual bacteria were tracked as they swam within a gradient of pure andrimid or the suite of metabolites produced by a growing colony of SWAT3-*wt* or the non-andrimid producing mutant SWAT3-111. Growth assays run in parallel using purified andrimid or SWAT3-*wt* provided a direct measurement for the extent of the lethal zone.

We determined that a gradient of diffusible metabolites, including andrimid, from the SWAT3-*wt* induced concentration specific changes in the swimming behavior of *V. cholerae*. The pathogen exhibited increased swimming speeds, turning rates, and run lengths and directed its movements away from the metabolite source (Graff et al. 2013). Purified andrimid elicited a similar response in all behaviors, suggesting it was the principle metabolite affecting the motility behavior of *V. cholerae*. These changes effectively, although not completely, would ultimately result in avoidance of the toxic agent and a reduction in the colonization of surfaces where the competing SWAT3-*wt* was already residing. We expected the swimming patterns of *V. cholerae* to be altered when exposed to lethal concentrations of andrimid (i.e., cells stop swimming as they succumb to the compound and the possible movement of cells away from toxic concentrations). Surprisingly, *V. cholerae* showed alterations in behavior to sub-lethal concentrations of andrimid, either the pure compound or when produced by a colony of SWAT3-*wt*. This suggests that competitive interactions may not require toxic concentrations of compounds to be effective in deterring would-be rivals. More so, such high concentrations would be difficult and costly to maintain in the aquatic environment.

Bacterial motility is frequently modeled as a random walk so that population distribution and changes therein can be approximated similar to the diffusion of molecules (Berg 1993). Measured swimming speeds and run times of tracked bacteria were used to calculate the treatment-specific bacterial diffusion coefficient (D), which is calculated as $D = (v^2\tau)/3$, where D is the diffusivity ($\mu m^2\ s^{-1}$), v is the speed ($\mu m\ s^{-1}$), and τ is the run duration (s) (Berg 1993) (Table 7.1). A high diffusivity implies a high rate of population dispersal, caused by bacteria swimming with high speeds and low turning rates for example, to disperse from a point in space. A behavioral shift toward lower diffusivity, effected through lower swimming speeds or higher turning rates, or both, implies that the population will have a lower dispersal rate and may aggregate at a source of a chemoattractant (Berg 1993; Packer and Armitage 1994). We calculated the diffusivity for *V. cholerae* in our experiments using the mean swimming speeds and run durations from each treatment and zone of interest. The diffusion coefficient for the media and SWAT3-111 controls were 320 ($\pm$ 2.6 Standard Error of the Mean, SE) and 324 ($\pm$ 3.7 SE) $\mu m^2 s^{-1}$, respectively. In the sub-lethal zone, the diffusion coefficient of *V. cholerae* was 25% higher, increasing to 417 ($\pm$ 7.5 SE) and 426 ($\pm$ 2.5 SE) $\mu m^2 s^{-1}$. In the lethal zone, this decreased to 224 ($\pm$ 2.5 SE) and 206 ($\pm$ 1.8 SE) $\mu m^2 s^{-1}$, 31 and 39% lower, respectively.

These results provide a mechanistic explanation of the observations that *V. cholerae* did not colonize particles previously colonized by SWAT3-*wt* but readily attached

TABLE 7.1

Calculated Diffusion Coefficients ($\pm$1 Compounded Standard Error of the Mean, SE) for *V. cholerae*

	Treatment	Diffusion Coefficient (μms^{-1}; $\pm$SE)
Andrimid	Media control	320 (2.6)
	Sub-lethal zone	417 (7.5)
	Lethal zone	224 (2.5)
SWAT3-*wt*	SWAT3-111 control	324 (3.7)
	Sub-lethal zone	426 (2.5)
	Lethal zone	206 (1.8)

Source: Graff, J.R. et al., *Front. Microbiol.*, 4, 8, 2013.

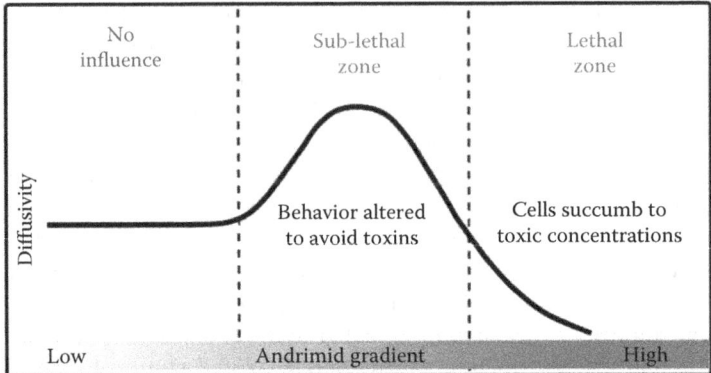

FIGURE 7.3   Conceptual model of *V. cholerae's* swimming behaviors resulting in avoidance of inhospitable living conditions. (Reproduced from Graff, J.R. et al., *Front. Microbiol.*, 4, 8, 2013. With permission.)

to particles colonized by SWAT3-111 (Long et al. 2005). This competitor-induced modulation of movements effectively is negative chemotaxis and results in decreased exposure of *V. cholerae* to higher concentrations of the competitor-derived effector. These results suggest a mechanism by which *V. cholerae* exploits competitor-derived chemical information to limit exposure to the lethal antibiotic (Figure 7.3) by altering its swimming behavior to avoid contact with potentially lethal concentrations of an antibiotic. It also suggests that sub-lethal concentrations of andrimid may serve as interspecific signaling molecules in the marine environment to deter particle colonization and ultimately competition for resources. Ultimately, these types of avoidance behaviors will alter the rates of particle attachment by the pathogen and modify its trajectory through the food web and thus the extent of human exposure.

7.6   PHYTOPLANKTON: HARBINGERS OF CHOLERA?

Phytoplankton potentially play a significant role in cholera outbreaks, harboring the pathogen at a particular stage of a bloom. We investigated the temporal and spatial relationship between diatoms, a type of phytoplankter, as well as the abiotic particulate and dissolved exudates with (1) the bacteria community that was associated with the microalgae when they were isolated from seawater and (2) the particle colonizing human pathogen *V. cholerae*. The particulate fraction of the exudates included the polysaccharides that aggregate to form TEP. The dissolved components included the secondary metabolites present in the cell-free supernatants and dimethylsulfoniopropionate (DMSP), a metabolite produced by many phytoplankton (Keller et al. 1989). We used light, epifluorescence, and video microscopy techniques in conjunction with fluorescent and non-fluorescent staining of samples to visualize the bacteria and their associations and interactions with the phytoplankton cells and with TEP in the cultures. Because bacterial chemotaxis is a mechanism for motile marine bacteria to locate and exploit patchy food resources, such as phytoplankton and other organic particles (Bell and Mitchell 1972; Fenchel 2002;

Stocker et al. 2008; Seymour et al. 2009), we also used microchannel chemotaxis (Graff et al. 2013) and microtiter well motility assays to explore the influence of the cell-free supernatants and DMSP on the swimming behaviors of *V. cholerae*.

7.6.1 Patterns of Bacterial Colonization of Diatoms and Particles in Culture

The natural bacteria community in the diatom cultures rarely colonized phytoplankton cells or cell-associated TEP when the phytoplankton were in the exponential phase of growth (Figure 7.4a–d). The diatoms *Thalassiosira nordenskioeldii* and *Chaetoceros diadema* were more heavily colonized when the cultures became senescent (Figure 7.4e–h). Many bacteria attached to TEP within the senescent cultures (Figure 7.4i and j) and appeared to preferentially colonize TEP, even when the polymers were entangled with the diatom chains (Figure 7.4e and f). Bacteria selectively attached to dead or otherwise compromised cells within the diatom chains, while avoiding healthy autotrophic cells (Figure 7.4g and h).

DAPI labeled *V. cholerae* physically came into contact with and colonized dead or compromised diatom cells and particles from senescent cultures. We often saw the stained *V. cholerae* swimming around and associating with amorphous particles and malformed diatom chains that were morphologically irregular (Figure 7.5a) or diatoms that were physically disrupted (Figure 7.5c). In these cases, multiple bacteria associated with particle surfaces, occasionally remaining on the surfaces for up to 15 minutes before swimming away. *Vibrio cholerae* rarely attached to cells in exponential growth and were most often observed swimming past chains of healthy autotrophic cells with little or no direct interaction with the diatoms (Figure 7.5b). The use of the DAPI on live cultures of *V. cholerae* decreased the leucine incorporation rate of the cells by approximately 20%.

7.6.2 *Vibrio cholerae* Modifies Its Behavior in the Presence of Diatom Supernatants

In the microchannel chemotaxis assays, *V. cholerae* modified its swimming behavior when exposed to diatom culture supernatants collected from exponentially growing diatoms but to a lesser extent when exposed to supernatants from senescent phytoplankton cultures (Table 7.2). There was, however, no pattern suggesting a reaction to a gradient of compounds within the channel. Thus, we combined the data from all locations and replicates within the microchannel for a specific assay. The average swimming speed of *V. cholerae* in the F/2 Media control was 53.6 ($\pm$ 0.09 SE) μm s^{-1}. In contrast, bacteria exposed to supernatants from exponentially growing *T. nordenskioeldii* or *Ch. diadema* increased their average swimming speed by 14% to 61.1 ($\pm$ 0.1 SE) and 63.2 ($\pm$ 0.09 SE) μm s^{-1}, respectively. In chemotaxis assays using senescent phytoplankton supernatants, average swimming speeds were closer to the swimming speeds of cells in the F/2 Media control; 55.5 ($\pm$ 0.08 SE) μm s^{-1} for *T. nordenskioeldii* supernatants and 55.4 ($\pm$ 0.09 SE) μm s^{-1} for *Ch. diadema* supernatants. A comparison of the distribution of swimming speeds from the exponential and senescent supernatant chemotaxis assays show a shift of >10% of observations toward higher swimming speeds in the presence of exponentially growing diatom supernatants (Figure 7.6).

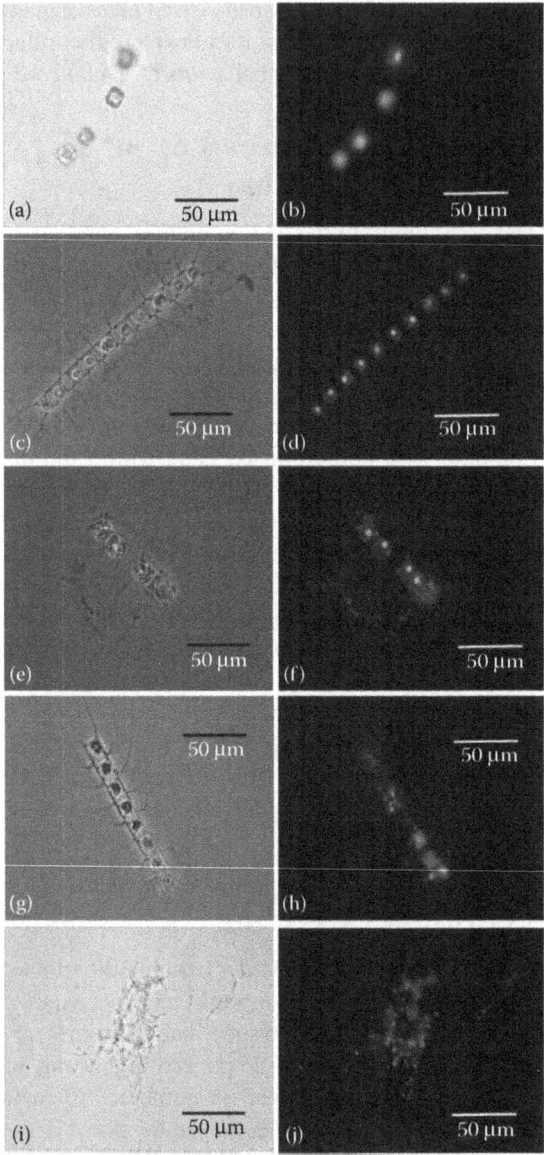

FIGURE 7.4 Brightfield or phase contrast (left) and epifluorescent images (right) of Alcian Blue and SYBR Green stained samples from exponential (a–d) and senescent cultures (e–h) of the marine diatoms *T. nordenskioeldii* (a–b, e–f) and *Ch. diadema* (c–d, g–h). A highly colonized TEP particle from a senescent culture of *T. nordenskioeldii* (i–j). In the left panels, the blue stain indicates the presence of polymers labeled with Alcian Blue. In the right panels, red is chlorophyll autofluorescence and green is SYBR Green staining of bacteria and phytoplankton nuclei (highlighted region in the center of diatom cells). SYBR Green staining of bacteria is prominent on compromised phytoplankton (cells not containing bright autofluorescence from chlorophyll) and on the TEP in the cultures. Even "healthy" cells in the senescent cultures (red autofluorescent cells in E and H) remain mostly free of colonizing bacteria relative to compromised diatom cells.

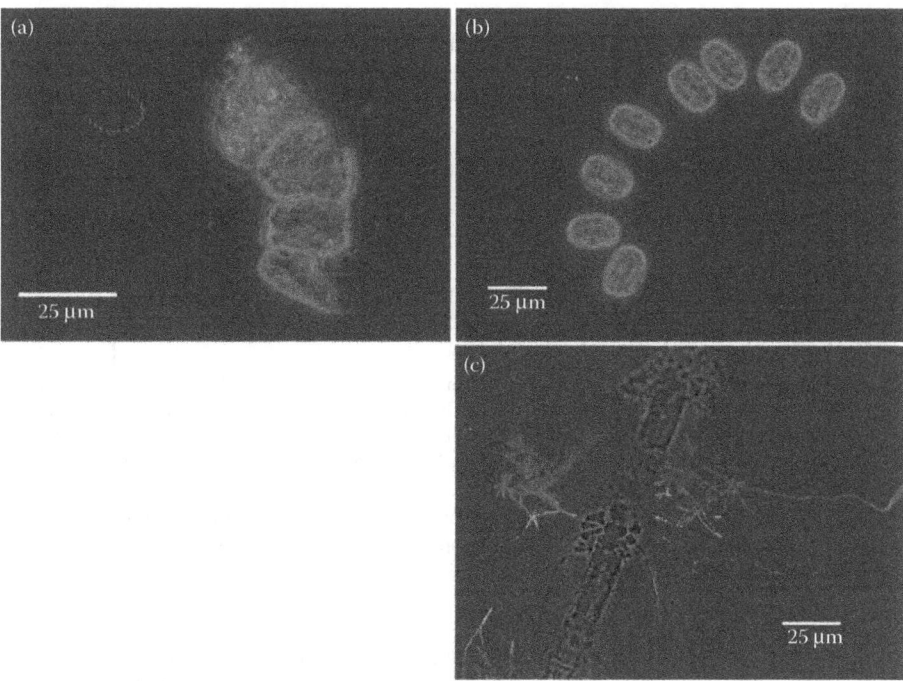

FIGURE 7.5 A projection of 30 seconds of video of DAPI stained *V. cholerae* cells (fluorescent blue tracks) as they swim around and/or interact with (a) a particle of malformed *T. nordenskioeldii* from a senescent culture, (b) an intact, healthy chain of *T. nordenskioeldii*, and (c) a broken, leaky chain of *Ch. diadema*. More interactions occurred around and on particles composed of physically compromised or senescent cells.

TABLE 7.2

Mean Swimming Speed, Turning Rate, Run Length, and Calculated Diffusivity of *V. cholerae* When Exposed to the F/2 Control, and Exponentially Growing and Senescent Diatom Supernatants

Treatment		Speed (μm s^{-1})	Turn Rate (deg s^{-1})	Run Length (μm)	Diffusivity (μm^2 s^{-1})
Control	F/2	53.6 ± 0.09	116.7 ± 0.7	19.3 ± 0.2	344.8
Exponential	*Ch. diadema*	63.2 ± 0.09	109.5 ± 0.6	24.5 ± 0.2	516.1
	T. nordenskioeldii	61.1 ± 0.10	99.4 ± 0.7	25.1 ± 0.3	511.2
Senescent	*Ch. diadema*	55.5 ± 0.08	120.9 ± 0.7	19.8 ± 0.2	366.3
	T. nordenskioeldii	55.4 ± 0.09	101.8 ± 0.6	22.7 ± 0.3	419.2

Note: Values $\pm$ standard error of the mean are reported here.

Background concentrations of
phytoplankton and bacteria are
generally low. Healthy phytoplankton
dominate and colonization is rare.

Compromised or unhealthy phytoplankton are more
common, TEP production increases, and larger
aggregates begin to form. Bacterial colonization of
cells and particles becomes a more common occurrence.

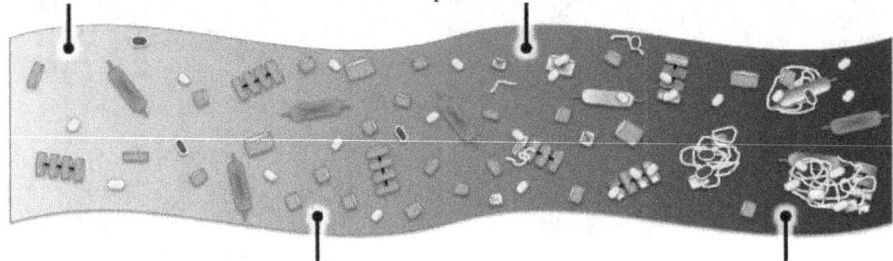

Phytoplankton experience favorable
conditions to "bloom." Bacteria benefit
from excreted or released products from
phytoplankton.

Colonization of aggregates and sinking particles is high.
This likely represents the time with the greatest
probability for colonizing, pathogenic bacteria (e.g.,
V. cholerae) to be vectored into food and water sources.

FIGURE 7.6 Conceptual diagram of the temporal and spatial relationships between *V. cholerae* and other colonizing bacteria with phytoplankton and larger aggregates moving from pre-bloom (far left) to post-bloom (far right) of the microalgal community.

The patterns for other *V. cholerae* behaviors were similar for each diatom species supernatants. Turning rates of *V. cholerae* were lower relative to the F/2 Media control (116.7 [$\pm$ 0.7 SE] deg s^{-1}) in the presence of *T. nordenskioeldii's* and *Ch. diadema's* exponential supernatant, 99.4 ($\pm$ 0.7 SE) deg s^{-1} and 109.5 ($\pm$ 0.6 SE) deg s^{-1}, respectively (Table 7.2). When *V. cholerae* was exposed to the supernatants from senescent cultures their average turning rates increased in comparison to assays using exponential growth supernatants; increasing to 101.8 ($\pm$ 0.6 SE) and 120.9 ($\pm$ 0.7 SE) deg s^{-1} for *T. nordenskioeldii* and *Ch. diadema*, respectively (Table 7.2). Similarly, *V. cholerae's* run lengths were higher in exponential treatments, 25.1 ($\pm$ 0.3 SE) and 24.5 ($\pm$ 0.2 SE) μm, and lower in assays using senescent supernatants, 22.7 ($\pm$ 0.3 SE) and 19.8 ($\pm$ 0.2 SE) μm (Table 7.2).

The direction of swimming and magnitude of bacterial diffusivity indicates that *V. cholerae* was attracted to phytoplankton supernatants. The highest rates of diffusivity were observed in exponential supernatant chemotaxis assays and lower, but still higher than the control, for senescent supernatant assays (Table 7.2). The net direction of the cells down the channel and toward the disc loaded with supernatants is enhanced in supernatant assays in comparison to the F/2 control for both diatom species and both growth phases (data not shown). Thus, increased rates of diffusivity in the direction of the supernatant source indicates an attraction of the *V. cholerae* to the bulk compounds excreted by phytoplankton.

LC-MS results confirmed the presence of DMSP in the exponentially growing *Ch. diadema* culture. In the microtiter well DMSP experiments, the swimming speeds of *V. cholerae* increased relative to the FASW control (54.1 μm s^{-1}) to between 58 and 80 μm s^{-1} (Table 7.3) with the exception of the lowest DMSP concentration tested (10 μM DMSP) where cells swam at average speeds of 54.0 μm s^{-1} (Table 7.3). This change in speed at the higher DMSP concentrations is similar to that seen in response to the phytoplankton exudates. Turning rates, while decreasing in response to all of

TABLE 7.3
Swimming Speeds (μm s⁻¹) of *V. cholerae* in Microtiter Wells

Control/Treatment	Swimming Speed
F/2	51.5
FASW	54.1
Culture Supernatant	62.3
DMSP (μM)	—
10	54.0
50	80.4
125	61.4
250	58.0

Note: Speeds increased in response to whole cell-free supernatants from a culture of *Ch. diadema* and to higher concentrations of DMSP (>10 μM). Standard error of the mean < 0.1 for all treatments.

the microalgal exudates, increased in the wells containing 10 and 250 μM DMSP, 148.6 and 149.9 deg s⁻¹ respectively, but did decrease in the 50 and 125 μM wells to 111.5, 129.4 in comparison to the FASW control (133.1 deg s⁻¹). Run lengths increased in all wells to between 18 and 33 μm in comparison to the FASW control value of 16.8 μm, also similar to the response of *V. cholerae* to phytoplankton exudates.

7.6.3 DISCUSSION

Prior and current observations of the behavior of *V. cholerae* indicate a strong dependence on chemical information for structuring the movements of this human pathogen, ultimately impacting the potential for the cells to colonize particles in the marine environment. Previous work exposed the role of sub-lethal concentrations of a competitor-produced antibiotic for deterring particle colonization by this bacterium (Long et al. 2005; Graff et al. 2013). We now show how the excreted metabolites from phytoplankton can alter the behavior of *V. cholerae* and potentially impact their colonization of microalgae.

The natural community of bacteria preferentially colonized TEP particles with the most extensive colonization of microalgae occurring in senescent cultures. We commonly observed bacteria attached to dead or non-autotrophic phytoplankton cells, with no or few bacteria attached to intact and visually healthy cells within the same diatom chain. *Vibrio cholerae* interacted with and attached to senescent cells or particles within the cultures but not to exponentially growing and healthy, intact diatoms. *Vibrio cholerae* altered its swimming patterns in response to dissolved, cell-free supernatants with similar behavioral changes when exposed to DMSP. Our analysis indicates that the pathogen is attracted to diatom-excreted metabolites. This attraction apparently does not result in colonization of healthy diatoms in exponential growth, suggesting that interactions that control bacterial colonization of phytoplankton occur in the phycosphere or at the phytoplankton cell surface.

The pattern of diatom colonization by the natural community of bacteria in our cultures agrees with previous observations (Droop and Elson 1966; Vaqué et al. 1989; Kaczmarska et al. 2005; Grossart et al. 2006; Graff et al. 2011; Znachor et al. 2012) that healthy and/or exponentially growing phytoplankton are rarely colonized by marine bacteria but become more colonized when cells are senescent or unhealthy. Our observations that TEP is heavily colonized by bacteria also agrees with previous studies that demonstrate high rates of attachment to these particles composed of phytoplankton produced exopolymers (Alldredge et al. 1986; Passow 2002). Vaqué et al. (1989) found the greatest extent of attachment in a microcosm study when phytoplankton and bacteria were at their highest concentrations. They proposed that as the number of organisms in each group increases, the encounter rate of the two groups, and thus colonization rates, increase (Vaqué et al. 1989). However, even in our senescent diatom cultures, the bacteria preferentially colonized TEP particles, the TEP associated with diatom chains, and the bacteria selectively colonized the dead or non-autotrophic cells within diatom chains. While the abundance of the two groups may play a role in this process, the patterns of attachment we observed are consistent with the notion that the physiology or health of a diatom cell is important for regulating bacterial colonization of marine phytoplankton.

The fluorescently labeled *V. cholerae* cells showed similar patterns of interactions with the diatoms. *Vibrio cholerae* did not attach to exponentially growing diatoms and swimming cells of *V. cholerae* rarely came into contact with visually healthy diatom cells or chains. They did, however, colonize and come into contact with broken diatom cells and aggregates within senescent cultures. The DNA/RNA binding DAPI stain could have influenced the bacteria's ability to sense or attach to phytoplankton, and did lower their ability to incorporate leucine, but did not appear to influence their behavior. The bacteria, however, were viable for >5 days, were motile, and were found attached to amorphous particles or diatoms that were physically compromised. These behaviors were not observed when they were introduced into cultures of exponentially growing diatoms or with visibly healthy and intact diatoms present in older cultures.

Because *V. cholerae* is known to colonize and metabolize chitin (Nalin et al. 1979; Pruzzo et al. 2008), we expected to see the bacteria interacting with or colonizing the chitin threads produced by *T. nordenskioeldii*. However, we did not observe this particular pattern of colonization or interaction. The increased activity of chitobiase was found to be negatively correlated with particle aggregation in mesocosm experiments (Smith et al. 1995). It is also possible that other organic substrates that are preferred by *V. cholerae* were present in the cultures and limited their interactions with or attraction to the chitin threads. The interactions between the swimming cells of *V. cholerae* and the senescent diatoms and particles suggest that the transmission of this pathogen is not likely to occur at the onset or peak of a phytoplankton bloom. Rather, it is the senescent sinking cells and particles produced during the termination of a bloom that are likely to be colonized by *V. cholerae*. These particles and aggregates could be vectored through shellfish and other filter feeding organisms that are harvested for human consumption (Potasman et al. 2002).

Vibrio cholerae altered its swimming behavior in the presence of phytoplankton supernatants. Their behavior was most influenced by the supernatants from exponentially growing diatom cultures. The increase in diffusivity and changes in the direction of the bacteria suggest that the bacteria were attracted to the supernatants from both exponentially growing diatom species. Smaller changes in behavior in bacteria exposed to supernatants from senescent cultures also suggest an attraction, but to a lesser degree. The attraction of marine bacteria to phytoplankton super-natants, which can include amino acids, sugars, and polysaccharides (Hellebust 1965; Myklestad 1995, 2000), were observed in other chemotaxis assays (Bell and Mitchell 1972; Stocker et al. 2008; Seymour et al. 2009). Stocker et al. (2008) found similar results using a microfluidics assay designed to mimic an organic plume produced by a phytoplankton colony or particle. Their research demonstrated that *Pseudoalteromonas haloplanktis*, a motile marine bacterium, can quickly sense, locate, and alter their behavior to maximize their exposure to the plume (Stocker et al. 2008). Bell et al. (1974), using capillary chemotaxis assays, also found that motile marine bacterium, specifically pseudomonad HNY and spirillum 7697, were attracted to filtrates from phytoplankton cultures. These results, however, do not explain the lack of attached bacteria to exponentially growing diatoms. An attrac-tion to the source of the exudates should lead to a concentration of motile bacteria, including particle-attaching bacteria, around a phytoplankton cell that should lead to a higher occurrence of attachment.

The response of *V. cholerae* to DMSP, which in many ways was similar to its response to whole phytoplankton exudates, confirmed our speculation that chemosen-sory reactions may be responsible, at least in part, for the behavioral modifications we observed in this bacterium. As DMSP was the only pure molecule/metabolite specifi-cally tested, we cannot rule out the possibility of multiple metabolite interactions with the bacteria that led to our observations in the culture supernatant assays. Seymour et al. (2010) also observed behavioral modification in response to DMSP, indicating it was a chemoattractant for bacteria. The results in this study indicate that DMSP is a chemical signal between marine phytoplankton and bacteria, likely as a chemoat-tractant, although further investigations would need to be carried out to confirm this.

The lack of attached bacteria in exponential cultures, despite an attraction to phy-toplankton exudates, may be due to diatom cell surface chemistry or the chemical conditions near the microalgal cell. Azam and Smith (1991) hypothesized that phy-toplankton would benefit by excreting organic matter to prevent or reduce bacterial attack or to attract bacteria that would remineralize nutrients from the nearby particu-late matter. Their hypothesized model of interaction extends to flagellates that would consume the bacteria and also regenerate nutrients for the phytoplankton (Azam and Smith 1991). The results of this study support such a model. However, the surface of the phytoplankton would provide immediate access to the source of the excreted metabolites and carbon sources, yet particle-attaching bacteria avoid colonizing the microalgae until the cells become unhealthy. The phycosphere is postulated to be a region immediately surrounding a phytoplankton cell that would increase bacterial growth due to the release or leaking of photoassimilated carbon (Bell and Mitchell 1972). The conditions of the phycosphere are likely to be very different for healthy and intact autotrophic phytoplankton compared to senescent or dying cells. The amount and composition of exudates that are released or leaked into the surrounding water are dependent on the phytoplankton species, growth phase, and growth condi-tions (Fogg et al. 1965; Hellebust 1965; Myklestad 2000). Extracellular superoxide production by phytoplankton (Kustka et al. 2005) would likely impact the response of bacteria to the phycosphere, making it an inhospitable surface for colonization. Oxidative bursts in response to pathogens are known in higher plants (Lamb and Dixon 1997). Unfortunately, many of the details of the phycosphere remain elusive due to the physical constraints for directly examining its chemical nature.

7.6.4 CONCLUSION

Investigations into the patterns and specificity of direct associations between phytoplankton and bacteria have increased in recent years. These studies have identified the potential ecological consequences of these associations that range from phytoplankton life cycle and bloom dynamics (Kim et al. 1998; Mayali and Azam 2004; Croft et al. 2005; Mayali et al. 2007) to influences on human health via harmful algal bloom toxin production (Smith et al. 2002; Uribe and Espejo 2003; Bates et al. 2004) and aquatic pathogens (Nalin et al. 1979; Tamplin et al. 1990). We wanted to understand the mechanisms underlying the observed patterns of attachment documented here and elsewhere.

The absence of bacteria attached to autotrophic, intact marine phytoplankton, and more colonization of older, senescent cells indicates that health and/or growth stage is an important variable in this process. In these experiments, bacteria preferentially colonized and interacted with TEP and diatom cells that appeared physically or physiologically compromised. Chemically mediated interactions do influence bacterial colonization of marine particles (Long et al. 2005; Graff et al. 2013) and may play a role in the colonization of phytoplankton. While we were able to demonstrate an influence of secondary metabolites from phytoplankton cultures on the chemotactic behaviors of *V. cholerae*, these changes cannot explain the lack of colonization of exponentially growing diatoms. The interactions regulating attachment may take place within the phycosphere near the phytoplankton cell surface as evidenced by the swimming behavior of *V. cholerae* around diatoms of varying degrees of health. Future studies of this process should focus on the chemical properties of the phycosphere and the cell surface of phytoplankton and would benefit from the ability to quantify the physiological state of individual microalgae under the microscope.

Applying ours and others' observations to the dynamics of phytoplankton blooms in the field, we are able to construct a simple temporal model of bacterial colonization of phytoplankton and larger aggregates (Figure 7.6). This conceptual model depicts the seasonal cycle of a phytoplankton bloom. Pre-bloom conditions are depicted on the left-hand side of the model when phytoplankton are at background concentrations. The diagram continues through the seasonal peak in phytoplankton biomass and ends post-bloom when growth conditions have become unfavorable, TEP production and flocculation is high, and larger aggregates are forming. The data presented in this chapter and elsewhere would suggest that colonization of phytoplankton and aggregates by bacteria, including human pathogens such as *V. cholerae* barring an extremely high incidence of competition (Long et al. 2005; Graff et al. 2013), would be highest at this time. This would also be the time when the potential for transmission of pathogens to humans via particles (Colwell 1996) would be greatest.

7.6.5 Methods for Exploring Bacteria–Phytoplankton Interactions

7.6.5.1 Diatoms and *V. cholerae* Cultures

Chaetoceros diadema (Ehrenberg) Gran and *Thalassiosira nordenskioeldii* Cleve were isolated from Narragansett Bay on January 19, 2010. Phytoplankton species were identified by J. Graff, who has many years of experience identifying the plankton of Narragansett Bay, with the aid of the book *Identifying Marine Phytoplankton* (Tomas 1997). *Chaetoceros diadema* and *Th. nordenskioeldii* are chain-forming diatom species; that is, they form colonies with multiple cells linked together with external structures. For each species, a single colony was picked from the seawater sample with a sterile pipette and washed five times in sterile filtered F/2 medium. The diatoms were cultured under non-axenic conditions in glass culture tubes containing 30 mL of F/2 medium at 12°C–14°C on a 12:12 hour light:dark cycle at 150 µE m^{-2} s^{-1}. Growth of the phytoplankton cultures was measured by tracking *in vivo* fluorescence with a Turner-10 AU fluorometer. All samples used for microscopy and microchannel chemotaxis assays were obtained within five months of the original isolation of the diatoms. For the bacteria, we inoculated 10 mL of ZoBell (ZB) 2216 media with *Vibrio cholerae* (isolate Vc N16961 obtained from Richard Long [University of South Carolina] and initially gifted to R. Long from John Mekalanos [Harvard Medical School]) 16–20 hours prior to use in experiments. The bacteria were cultured at 24°C and shaken at 175 rpm.

7.6.5.2 Bacterial Interaction with and Colonization of Diatoms and Particles

Samples for microscopic investigations of the attachment of the natural bacteria community to diatoms and other particles were collected when phytoplankton cultures were in exponential growth and a minimum of two days after they reached senescence. We added 75 µL of a working solution of Alcian Blue (Passow and Alldredge 1994) to stain TEP particles and 75 µL of 1X working solution of SYBR green to stain bacteria and phytoplankton to a 20 mL aliquot of the phytoplankton cultures. Samples were stained for a minimum of 30 minutes before observations were made. To image the phytoplankton, bacterial cells, and particles, one to two drops of the stained sample were spotted onto a glass microscope slide and covered with a glass coverslip. We viewed the samples using a Nikon E800 compound microscope using bright field, phase, and epifluorescence techniques and imaged cells at 40X with a SPOT Insight CCD camera.

To assess the direct interaction of *V. cholerae* with phytoplankton and particles, we introduced *V. cholerae* into the diatom cultures. *Vibrio cholerae* were pre-labeled with DAPI in order to distinguish them from other bacteria present in the diatom cultures. To stain the *V. cholerae* cells, 1 mL of the bacteria culture was centrifuged for 3 minutes at 7000 rpm and re-suspended in sterile filtered artificial seawater media (FASW). A 100 µL aliquot of re-suspended cells was added to 900 µL of FASW to create a 1:10 dilution. Then, we added 25 µL of a DAPI stock solution (250 µg mL^{-1}) to the diluted and re-suspended cells. Bacteria were checked for motility after 4 hours of staining by placing a drop of the stained culture under a cover slip on a clean glass microscope slide and examining with epifluorescence microscopy. Cells were re-centrifuged, the supernatant with DAPI removed, and the cells re-suspended in 1 mL of FASW. Approximately 100 µL of the stained cells were added to 10 mL aliquots of diatom cultures. We documented the interactions of DAPI labeled *V. cholerae* with phytoplankton and particles in cultures using a Nikon Eclipse 80*i* with epifluorescence and a Retiga 2000R digital camera. Thirty-second videos were captured at >18 fps. The viability of the stained *V. cholerae* was also checked by testing the ability of the cells to incorporate ^{3}H-labeled leucine following the methods in Smith and Azam (1992).

7.6.5.3 Microchannel Chemotaxis Assays Using Diatom Culture Supernatants and *V. cholerae*

To assess the response of *V. cholerae* to phytoplankton exudates, we tracked the movements of *V. cholerae* cells exposed to a concentration gradient of phytoplankton cell-free supernatants. To do this, we utilized a newly developed experimental assay designed to assess the influence of a gradient of secondary metabolites on bacterial motility (Graff et al. 2013). This assay combines poly(dimethylsiloxane) (PDMS) microchannels created using soft-lithography techniques (Whitesides et al. 2001; Sia and Whitesides 2003) with a disc diffusion assay (Burkholder et al. 1966). In these assays, a PDMS microchannel is placed onto an agar surface and solution from the agar fills the microchannel, creating an aquatic microenvironment in which motile planktonic microorganisms can swim (DiLuzio et al. 2005) while being exposed to the gradient of metabolites that have diffused from the paper disk. A more detailed description of this assay—including microchannel fabrication, image acquisition, and data analysis—can be found in Graff et al. (2013).

To obtain the phytoplankton cell-free supernatants, 200 mL of each diatom culture in exponential growth or senescence were filtered through a 0.2 µm pore size disposable filter system with cellulose nitrate membrane (Corning). Cell-free supernatants were immediately frozen at −80°C until just prior to use in each chemotaxis assay. For this series of experiments, 80 µL of culture supernatant was loaded onto a sterile 6 mm paper disc in 20 µL increments, allowing the disc to dry before adding the next 20 µL aliquot. The disc was then placed in the center of a low-nutrient (ZB/8) agar plate and the metabolites were allowed to diffuse for 16–20 hours and then the PDMS microchannel was arranged radially outward from the disc so that the chemical gradient in the agar was aligned with the microchannel. For the control, the same procedure was carried out with F/2 seawater medium.

To prepare *V. cholerae* to be introduced into the microchannel, a 1 mL sub-sample of culture was centrifuged at 7000 rpm for 3 minutes, the supernatant removed, and the cells re-suspended in sterile filtered seawater media. A 1 µL aliquot of the re-suspended cells was spotted near the microchannel opening and the 1 µL of FASW was absorbed into the agar leaving a micro-colony of cells near the channel opening. Just prior to the experiment, the microchannel was gently maneuvered toward the spot of *V. cholerae* until motile cells were visible at the channel entrance. Videos of movements of the cells were collected at multiple distances from the loaded disc at 20X magnification for 60 seconds at a minimum of 15 fps using a Photometrics CoolSNAP HQ camera mounted on a Nikon Eclipse TE2000-E inverted microscope with an automated stage that was used locate specific distances from the paper disc within the channel. Videos were processed in ImageJ (NIH) to remove background noise and a manually identified threshold was applied to identify bacterial cells and determine the Cartesian coordinates of each bacterium in each frame. Swimming tracks of the bacteria were created in a version of Tracker 3D (Menden-Deuer and Grünbaum 2006) that was modified for 2D analysis and optimized to track the movements of motile bacteria. We used a cubic smoothing spline to remove high frequency noise from paths and then sampled swimming statistics from these paths at 0.05 seconds intervals. The methods for organism tracking closely follow previously established methods for tracking marine protists (Menden-Deuer and Grünbaum 2006). Only data for cells tracked longer than one second were included in this analysis. This resulted in hundreds of swimming tracks per video and thousands to tens of thousands of observations of bacterial behaviors. We performed multiple replicates of each experiment and data from identical distances in relation to the chemical source were pooled to increase the number of observations at each of these locations.

7.6.5.4 DMSP Microtiter Well Assays

We investigated the response of *V. cholerae* to DMSP, a molecule produced by many phytoplankton (Keller et al. 1989) and a known chemoattractant for marine bacteria (Seymour et al. 2010). These tests were carried out in microtiter wells with 10, 50, 125, and 250 µM pure DMSP dissolved in FASW as well as a FASW control. Swimming behaviors and analysis were carried out as described previously for the microchannel chemotaxis assays. LC-MS was performed on cell-free supernatants from the culture of *Ch. diadema* and a culture of *Isochrysis galbana*, a known producer of DMSP (Keller et al. 1989). Pure DMSP dissolved in FASW was used as a control to identify the presence of this molecule in phytoplankton cultures.

REFERENCES

Alavi, M., T. Miller, K. Erlandson, R. Schneider, and R. Belas. 2001. Bacterial community associated with *Pfiesteria*-like dinoflagellate cultures. *Environ. Microbiol.* 3:380–396.

Alfansane, A., M. S. Islam, and T. Begum. 2002. Association of *Vibrio cholerae* 01 and 0139 with phytoplankton in a pond of Bangladesh. *Bangladesh J. Bot.* 31:19–22.

Alldredge, A. L., J. J. Cole, and D. A. Caron. 1986. Production of heterotrophic bacteria inhabiting macroscopic organic aggregates (marine snow) from surface waters. *Limnol. Oceanogr.* 31:68–78.

Alverca, E., I. Biegala, G. Kennaway, J. Lewis, and S. Franca. 2002. In situ identification and localization of bacteria associated with *Gyrodinium instriatum* (Gymnodiniales, Dinophyceae) by electron and confocal microscopy. *Eur. J. Phycol.* 37:523–530.

Azam, F., and R. Long. 2001. Oceanography: Sea snow microcosms. *Nature* 414:495.

Azam, F., and D. C. Smith. 1991. Bacterial influence on the variability in the ocean's biogeochemical state: A mechanistic view. In: *Particle Analysis in Oceanography*, S. Demers (Ed.), pp. 213–236. Berlin, Germany: Springer-Verlag.

Azam, F. and F. Malfatti. 2007. Microbial structuring of marine ecosystems. *Nat. Rev. Microbiol.* 5(10):782.

Bates, S. S., J. Gaudet, I. Kaczmarska, and J. M. Ehrman. 2004. Interaction between bacteria and the domoic-acid-producing diatom *Pseudo-nitzschia multiseries* (Hasle) Hasle; can bacteria produce domoic acid autonomously? *Harmful Algae* 3:11–20.

Bauer, A., W. Kirby, J. Sherris, and M. Turck. 1966. Antibiotic susceptibility testing by a standardized single disk method. *Am. J. Clin. Pathol.* 45:493–496.

Bell, W., and R. Mitchell. 1972. Chemotactic and growth responses of marine bacteria to algal extracellular products. *Biol. Bull.* 143:265–277.

Bell, W. H., J. M. Lang, and R. Mitchell. 1974. Selective stimulation of marine bacteria by algal extracellular products. *Limnol. Oceanogr.* 19:833–839.

Blackburn, N., T. Fenchel, and J. Mitchell. 1998. Microscale nutrient patches in planktonic habitats shown by chemotactic bacteria. *Science* 282:2254–2256.

Burkholder, P. R., R. M. Pfister, and F. H. Leitz. 1966. Production of a pyrrole antibiotic by a marine bacterium. *Appl. Microbiol.* 14:649–653.

Cole, J. J., G. E. Likens, and D. L. Strayer. 1982. Photosynthetically produced dissolved organic carbon: An important carbon source for planktonic bacteria. *Limnol. Oceanogr.* 27:1080–1090.

Colwell, R. R. 1996. Global climate and infectious disease: The cholera paradigm. *Science* 274:2025–2031.

Croft, M. T., A. D. Lawrence, E. Raux-Deery, M. J. Warren, and A. G. Smith. 2005. Algae acquire vitamin B12 through a symbiotic relationship with bacteria. *Nature* 438:90–93.

Davies, J., G. Spiegelman, and G. Yim. 2006. The world of sub-inhibitory antibiotic concentrations. *Curr. Opin. Microbiol.* 9:445–453.

DeLong, E. F., D. G. Franks, and A. L. Alldredge. 1993. Phylogenetic diversity of aggregate-attached vs. free-living marine bacterial assemblages. *Limnol. Oceanogr.* 38:924–934.

DiLuzio, W. R., L. Turner, M. Mayer, P. Garstecki, D. B. Weibel, H. C. Berg, and G. M. Whitesides. 2005. *Escherichia coli* swim on the right-hand side. *Nature* 435:1271–1274.

Droop, M. R., and K. G. R. Elson. 1966. Are pelagic diatoms free from bacteria? *Nature* 211:1096–1097.

Durkin, C. A., T. Mock, and E. V. Armbrust. 2009. Chitin in diatoms and its association with the cell wall. *Eukaryot. Cell* 8:1038–1050.

Epstein, P. R. 1993. Algal blooms in the spread and persistence of cholera. *Bio Systems* 31:209–221.

Fenchel, T. 2002. Microbial behavior in a heterogeneous world. *Science* 296:1068–1071.

Fogg, G. E., C. Nalewajko, and W. D. Watt. 1965. Extracellular products of phytoplankton photosynthesis. *Proc. R. Soc. Lond. [Biol.]* 162:517–534.

Fredenhagen, A., S. Y. Tamura, P. T. Kenny, H. Komura, Y. Naya, K. Nakanishi, K. Nishiyama, M. Sugiura, and H. Kita. 1987. Andrimid, a new peptide antibiotic produced by an intracellular bacterial symbiont isolated from a brown planthopper. *J. Am. Chem. Soc.* 109:4409–4411.

Freiberg, C., N. A. Brunner, G. Schiffer, T. Lampe, J. Pohlmann, M. Brands, M. Raabe, D. Häbich, and K. Ziegelbauer. 2004. Identification and characterization of the first class of potent bacterial acetyl-CoA carboxylase inhibitors with antibacterial activity. *J. Biol. Chem.* 279:26066–26073.

Fuhrman, J. A., and F. Azam. 1982. Thymidine incorporation as a measure of heterotrophic bacterioplankton production in marine surface waters: Evaluation and field results. *Mar. Biol.* 66:109–120.

Graff, J. R., J. E. B. Rines, and P. Donaghay. 2011. Enumeration of marine bacteria attached to phytoplankton and factors regulating attachment. *Mar. Ecol. Prog. Ser.* 441:15–24.

Graff, J. R., S. R. Forschner-Dancause, S. Menden-Deuer, R. A. Long, and D. C. Rowley. 2013. *Vibrio cholera* exploits sub-lethal concentrations of a competitor-produced antibiotic to avoid toxic interactions. *Front. Microb.* 4:8.

Gram, L., H. Grossart, A. Schlingloff, and T. Kiørboe. 2002. Possible quorum sensing in marine snow bacteria: Production of acylated homoserine lactones by *Roseobacter* strains isolated from marine snow. *Appl. Environ. Microbiol.* 68:4111–4116.

Grossart, H.-P., G. Czub, and M. Simon. 2006. Algae-bacteria interactions and their effects on aggregation and organic matter flux in the sea. *Environ. Microbiol.* 8:1074–1084.

Grossart, H. P., A. Schlingloff, M. Bernhard, M. Simon, and T. Brinkhoff. 2004. Antagonistic activity of bacteria isolated from organic aggregates of the German Wadden Sea. *FEMS Microbiol. Ecol.* 47:387–396.

Hay, M. 2009. Marine chemical ecology: Chemical signals and cues structure marine populations, communities, and ecosystems. *Ann. Rev. Mar. Sci.* 1:193–212.

Hellebust, J. A. 1965. Excretion of some organic compounds by marine phytoplankton. *Limnol. Oceanogr.* 10:192–206.

Hunt, D., E. Ortega-Retuerta, and C. Nelson. 2010. Connections between bacteria and organic matter in aquatic ecosystems: Linking microscale ecology to global carbon cycling. In: *Eco-DASVIII Symposium Proceedings*, P. F. Kemp (Ed.), pp. 110–128. Waco, TX: ASLO.

Huq, A., E. B. Small, P. A. West, R. Rahman, and R. R. Colwell. 1993. Ecological relationships between *Vibrio cholerae* and planktonic crustacean copepods. *Appl. Environ. Microbiol.* 45: 275–283.

Jasti, S., M. E. Sieracki, N. J. Poulton, M. W. Giewat, and J. N. Rooney-Varga. 2005. Genetic diversity and specificity of bacteria closely associated with *Alexandrium* spp. and other phytoplankton. *Appl. Environ. Microbiol.* 71:3483–3494.

Kaczmarska, I., J. M. Ehrman, S. S. Bates, D. H. Green, C. Leger, and J. Harris. 2005. Diversity and distribution of epibiotic bacteria on *Pseudo-nitzschia multiseries* (Bacillariophyceae) in culture, and comparison with those on diatoms in native seawater. *Harmful Algae* 4:725–741.

Keller, M. D., W. K. Bellows, and R. R. L. Guillard. 1989. Dimethyl sulfide production in marine phytoplankton. (In) Biogenic Sulfur in the Environment. *ACS Symp. Ser.* 393:167–182.

Kim, M., I. Yoshinaga, I. Imai, K. Nagasaki, S. Itakura, and Y. Ishida. 1998. A close relationship between algicidal bacteria and termination of *Heterosigma akashiwo* (Raphidophyceae) blooms in Hiroshima Bay, Japan. *Mar. Ecol. Prog. Ser.* 170:25–32.

Kiørboe, T., and G. Jackson. 2001. Marine snow, organic solute plumes, and optimal chemosensory behavior of bacteria. *Limnol. Oceanogr.* 46:1309–1318.

Kodama, M., G. J. Doucette, and D. H. Green. 2006. Relationships between bacteria and harmful algae. In: *Ecology of Harmful Algae*, Berlin, Germany: Springer.

Kustka, A.B., Y. Shaked, A. J. Milligan, D. W. King, and F. M. Morel. 2005. Extracellular production of superoxide by marine diatoms: Contrasting effects on iron redox chemistry and bioavailability. *Limnol. Oceanogr.* 50(4):1172–1180.

Lamb, C., and R. A. Dixon. 1997. The oxidative burst in plant disease resistance. *Annu. Rev. Plant Physiol. Plant Mol. Biol.* 48:251–275.

Linares, J., I. Gustafsson, F. Baquero, and J. Martinez. 2006. Antibiotics as intermicrobial signaling agents instead of weapons. *Proc. Natl. Acad. Sci. U.S.A.* 103:19484–19489.

Long, R. A., and F. Azam. 2001. Antagonistic interactions among marine pelagic bacteria. *Appl. Environ. Microbiol.* 67:4975–4983.

Long, R. A., A. Qureshi, D. Faulkner, and F. Azam. 2003. 2-n-pentyl-4- quinolinol produced by a marine *Alteromonas* sp. and its potential ecological and biogeochemical roles. *Appl. Environ. Microbiol.* 69:568–576.

Long, R. A., D. C. Rowley, E. Zamora, J. Liu, D. H. Bartlett, and F. Azam. 2005. Antagonistic interactions among marine bacteria impede the proliferation of *Vibrio cholerae*. *Appl. Environ. Microbiol.* 71:8531–8536.

Martinez, J. L., D. C. Smith, G. F. Steward, and F. Azam. 1996. Variability in ectohydrolytic enzyme activities of pelagic marine bacteria and its significance for substrate processing in the sea. *Aquat. Microb. Ecol.* 10:223–230.

Mayali, X., and F. Azam. 2004. Algicidal bacteria in the sea and their impact on algal blooms. *J. Euk. Microbiol.* 51:139–144.

Mayali, X., P. J. S. Franks, and F. Azam. 2007. Bacterial induction of temporary cyst formation by the dinoflagellate *Lingulodinium polyedrum*. *Aquat. Microb. Ecol.* 50:51–62.

Menden-Deuer, S., and D. Grünbaum. 2006. Individual foraging behaviour and population distributions of a planktonic predator agregating to phytoplankton thin layers. *Limnol. Oceanogr.* 51:109–116.

Mitchell, J., A. Okubo, and J. Fuhrman. 1985. Microzones surrounding phytoplankton form the basis for a stratified marine microbial ecosystem. *Nature* 316:58–59.

Mlot, C. 2009. Antibiotics in nature: Beyond biological warfare. *Science* 324:1637–1639.

Myklestad, S. M. 1995. Release of extracellular products by phytoplankton with special emphasis on polysaccharides. *Sci. Total Environ.* 165:155–164.

Myklestad, S. M. 2000. Dissolved organic carbon from phytoplankton. In: *The Handbook of Environmental Chemistry*, Vol. 5, P. Wangersky (Ed.), pp. 111–148. Berlin, Germany: Springer-Verlag.

Nalin, D. R., V. Daya, A. Reid, M. M. Levine, and L. Cisneros. 1979. Adsorption and growth of *Vibrio cholerae* on chitin. *Infect. Immun.* 25:768–770.

Naviner, M., J. P. Berge, P. Durand, and H. Le Bris. 1999. Antibacterial activity of the marine diatom *Skeletonema costatum* against aquacultural pathogens. *Aquaculture* 174:15–24.

Needham, J., M. Kelly, M. Ishige, and R. Andersen. 1994. Andrimid and moiramides AC, metabolites produced in culture by a marine isolate of the bacterium *Pseudomonas fluorescens*: Structure elucidation and biosynthesis. *J. Org. Chem.* 59:2058–2063.

Packer, H., and J. Armitage. 1994. The chemokinetic and chemotactic behavior of *Rhodobacter sphaeroides*: Two independent responses. *J. Bacteriol.* 176:206–212.

Passow, U., and A. L. Alldredge. 1994. Distribution, size and bacterial colonization of transparent exopolymer particles (TEP) in the ocean. *Mar. Ecol. Prog. Ser.* 113:185–198.

Passow, U. 2002. Transparent exopolymer particles (TEP) in aquatic environments. *Prog. Oceanogr.* 55:287–333.

Pinhassi, J., M. M. Sala, H. Havskum, F. Peters, O. Guadayol, A. Malits, and C. Marrase. 2004. Changes in bacterioplankton composition under different phytoplankton regimens. *Appl. Environ. Microb.* 70:6753–6766.

Potasman, I., A. Paz, and M. Odeh. 2002. Infectious outbreaks associated with bivalve shellfish consumption: A worldwide perspective. *Clin. Infect. Dis.* 35:921–928.

Pruzzo, C., L. Vezzulli, and R. R. Colwell. 2008. Global impact of *Vibrio cholerae* interactions with chitin. *Environ. Microbiol.* 10:1400.

Rooney-Varga, J. N., M. W. Giewat, M. C. Savin, S. Sood, M. LeGresley, and J. L. Martin. 2005. Links between phytoplankton and bacterial community dynamics in a coastal marine environment. *Microb. Ecol.* 49:163–175.

Schafer, H., B. Abbas, H. Witte, and G. Muyzer. 2002. Genetic diversity of "satellite" bacteria present in cultures of marine diatoms. *FEMS Microb. Ecol.* 42:25–35.

Seymour, J. R., Marcos, and R. Stocker. 2009. Resource patch formation and exploitation throughout the marine microbial food web. *Am. Nat.* 173:E15–E29.

Seymour, J. R., R. Simó, T. Ahmed, and R. Stocker. 2010. Chemoattraction to dimethylsulfoniopropionate throughout the marine microbial food web. *Science* 329:342–345.

Sharp, J. H. 1977. Excretion of organic matter by marine phytoplankton: Do healthy cells do it? *Limnol. Oceanogr.* 22:381–399.

Sia, S. K., and G. M. Whitesides. 2003. Microfluidic devices fabricated in poly (dimethylsiloxane) for biological studies. *Electrophoresis* 24:3563–3576.

Sieburth, J. M. 1959. Antibacterial activity of antarctic marine phytoplankton. *Limnol. Oceanogr.* 4:419–424.

Smith, D. C., and F. Azam. 1992. A simple, economical method for measuring bacteria protein synthesis rates in seawater using ^{3}H-leucine. *Mar. Microb. Food Webs* 6:107–114.

Smith, D. C., M. Simon, A. L. Alldredge, and F. Azam. 1992. Intense hydrolytic enzyme activity on marine aggregates and implications for rapid particle dissolution. *Nature* 359:139–142.

Smith, D. C., G. F. Steward, R. A. Long, and F. Azam. 1995. Bacterial mediation of carbon fluxes during a diatom bloom in a mesocosm. *Deep-Sea Res II* 42:75.

Smith, E. A., F. H. Mackintosh, F. Grant, and S. Gallacher. 2002. Sodium channel blocking (SCB) activity and transformation of paralytic shellfish toxins (PST) by dinoflagellate-associated bacteria. *Aquat. Microb. Ecol.* 29:1–9.

Stocker, R., J. R. Seymour, A. Samadani, D. E. Hunt, and M. F. Polz. 2008. Rapid chemotactic response enables marine bacteria to exploit ephemeral microscale nutrient patches. *Proc. Natl. Acad. Sci. U.S.A.* 105:4209–4213.

Tamplin, N. L., A. L. Gauzens, A. Huq, D. A. Sack, and R. R. Colwell. 1990. Attachment of *Vibrio cholerae* serogroup O1 to zooplankton and phytoplankton of Bangladesh waters. *Appl. Environ. Microbiol.* 56:1977–1980.

Tomas, C. R. (Ed.). 1997. *Identifying Marine Phytoplankton.* San Diego, CA: Academic press.

Uribe, P., and R. T. Espejo. 2003. Effect of associated bacteria on the growth and toxicity of *Alexandrium catenella. Appl. Environ. Microbiol.* 69:659–662.

Vaqué, D., C. M. Duarte, and C. Marrasé. 1989. Phytoplankton colonization by bacteria: Encounter probability as a limiting factor. *Mar. Ecol. Prog. Ser.* 54:137–140.

Verdugo, P., A. Alldredge, F. Azam, D. Kirchman, U. Passow, and P. Santschi. 2004. The oceanic gel phase: A bridge in the DOM-POM continuum. *Mar. Chem.* 92:67–85.

Weitz, M., K. Duncan, N. V. Patin, and J. R. Jensen. 2013. Antagonistic interactions mediated by marine bacteria: The role of small molecules. *J. Chem. Ecol.* 39:879–891.

Wells, M., and E. Goldberg. 1991. Occurrence of small colloids in sea water. *Nature* 353:342–344.

Whitesides, G. M., E. Ostuni, S. Takayama, X. Jiang, and D. E. Ingber. 2001. Soft lithography in biology and biochemistry. *Annu. Rev. Biomed. Eng.* 3:335–373.

Zaccone, R., G. Caruso, and C. Calě. 2002. Heterotrophic bacteria in the northern Adriatic Sea: Seasonal changes and ectoenzyme profile. *Mar. Environ. Res.* 54:1–19.

Znachor, P., K. Šimek, and J. Nedoma. 2012. Bacterial colonization of the freshwater planktonic diatom *Fragilaria crotonensis. Aquat. Microb. Ecol.* 66: 87–94.

8 Interactions Between Microorganisms as Modulators of Natural Product Biosynthesis

Amy L. Lane and Taylor A. Lundy

CONTENTS

8.1 INTRODUCTION

Microorganisms are remarkable chemists. They utilize enzymes as synthetic tools to yield not only primary metabolites essential for life, but also secondary metabolites with a wide range of biological functions and diverse chemical structures (Bugni and Ireland 2004; Gunatilaka 2006; Hughes and Fenical 2010; Rateb and Ebel 2011). These natural products have broad-reaching applications, including leads for pharmaceutical development (Clardy et al. 2006; Fischbach and Walsh 2009), probes to elucidate molecular mechanisms of biological processes (Carlson 2010), and inspirations for organic syntheses (Nicolaou 2005). Over the past several years, the discovery of novel metabolites from both marine and terrestrial microorganisms has become increasingly challenging due to high rates of re-isolation of known compounds (Gaudencio and Pereira 2015). Further, low production yields—sometimes less than a milligram from over one hundred liters of fermentation (Gomes et al. 2010)—commonly make metabolite identification arduous and limit practical applications. Innovative strategies for metabolite discovery and titer enhancement are needed in order for society to fully reap the rewards of Nature's chemical diversity.

8.1.1 BIOSYNTHETIC PATHWAYS GREATLY OUTNUMBER
              OBSERVED NATURAL PRODUCTS

Genome sequencing has become faster and more affordable during the 20 years since completion of the first bacterial genome in Fleischmann et al. (1995). As a result, genome sequencing has emerged as a prominent resource for evaluating the biosynthetic capabilities of microorganisms. Bioinformatics analyses of these genetic data have revealed that phylogenetically diverse microorganisms devote surprisingly large portions of their genetic material to secondary metabolite biosynthesis. Bioinformatics analyses revealed that obligate marine actinomycete bacteria of the genus *Salinispora* commonly commit greater than 10% of their genetic material to natural product pathways (Udwary et al. 2007; Ziemert et al. 2014).

Analogous analyses of the *Aspergillus nidulans* genome also revealed impressive biosynthetic capacity; this terrestrial filamentous fungus is predicted to produce over fifty different polyketides and nonribosomal peptides (von Doehren 2009). The abundance of genomic data appearing in publicly available databases over the past decade supports that these examples are not unique isolated cases; the capacity of microorganisms to yield unique natural products appears far greater than recognized in the pre-genomics era (Cimermancic et al. 2014).

Intriguingly, only a small fraction of natural products predicted by bioinformatics analyses are typically observed for microorganisms grown in routine laboratory cultures of single species (Corre and Challis 2009; Craney et al. 2013). Predicted biosynthetic pathways for which the corresponding natural product(s) are unknown are referred to as orphan pathways (Gross et al. 2007). Two possible explanations for the abundance of orphan pathways are that (1) most biosynthetic genes are poorly expressed under laboratory conditions, or that (2) biosynthetic enzymes from orphan pathways are incapable of catalysis (Corre and Challis 2009). The heterologous expression of orphan biosynthetic pathways and the engineering of pathway regulatory genes have yielded structurally unprecedented, bioactive natural products (Van Lanen and Shen 2006; Corre and Challis 2009; Gross 2009; Zotchev et al. 2012; Yaegashi et al. 2014). This suggests that many orphan pathway enzymes are catalytically competent, yet inadequately expressed under typical laboratory conditions. As a consequence, a tremendous wealth of Nature's chemical diversity remains silently locked within microbial genomes. A natural product renaissance may be realized through the activation of these untapped biosynthetic capabilities. However, the activation of pathways to reveal corresponding chemical structures poses significant challenges. These challenges are being tackled by complementary genetic manipulation, media manipulation, and biotic interaction approaches that offer promise for unlocking Nature's immense chemistry toolbox (see Section 8.1.2 below).

8.1.2 STRATEGIES FOR BIOSYNTHETIC PATHWAY UPREGULATION

8.1.2.1 Genetic Manipulation

Both targeted and untargeted genetic manipulation approaches have been applied to activate or upregulate biosynthetic pathways. Targeted approaches are generally directed at genetic manipulation of a specific pathway that was predicted through bioinformatics analyses of genome sequence data. Targeted manipulations include the heterologous expression of biosynthetic genes in hosts optimized for natural product production (Yamanaka et al. 2014), the introduction of activating regulatory genes (Karray et al. 2010; Martin and Liras 2010), and/or the elimination of suspected repressor genes of pathway transcription (Chen et al. 2011). For example, heterologous expression and regulatory gene manipulation of a silent gene cluster from the marine actinomycete *Saccharomonospora* sp. CNQ-490 revealed taromycin A (**1**) (Figure 8.1), a novel antibiotic (Yamanaka et al. 2014). A drawback of targeted approaches is that several aspects of biosynthetic pathway regulation remain poorly understood, and hence some pathways remain poorly expressed after extensive manipulation.

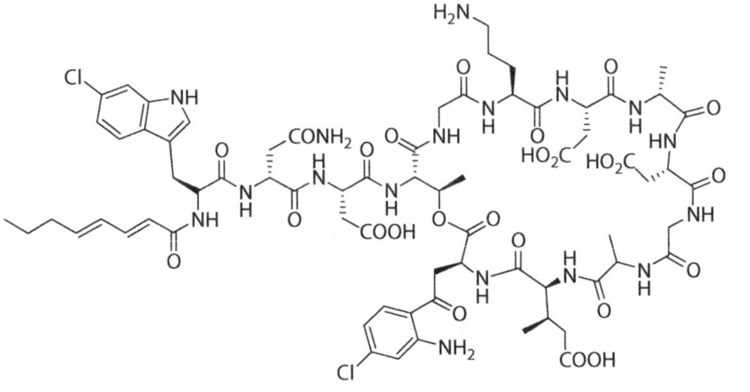

Taromycin A (**1**)

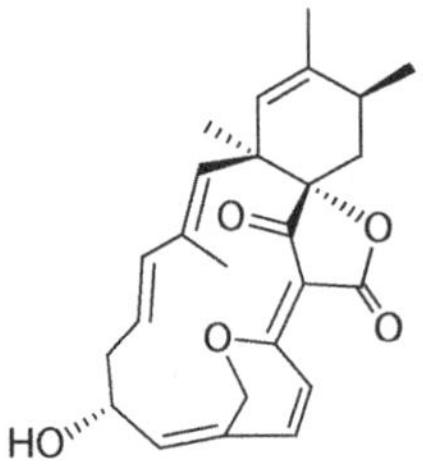

Spirohexenolide A (**2**)

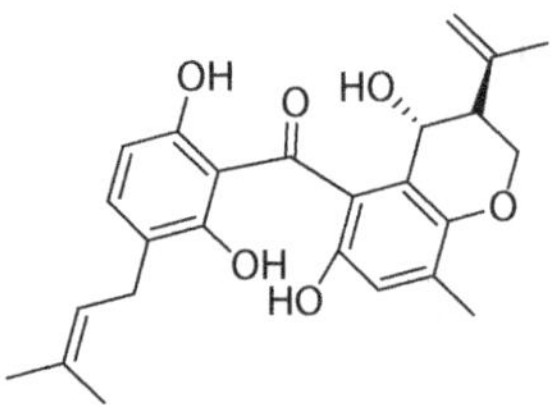

Pre-shamixanthone (**3**)

FIGURE 8.1 Structures of novel natural products discovered through genetic manipulation (**1–2**) or through media manipulation (**3**).

Untargeted approaches for genetic manipulation entail randomly mutating microorganisms to elicit metabolite production. For example, structure elucidation of the spirohexenolides (e.g., **2**, Figure 8.1) from *Streptomyces platensis* was enabled largely by dramatic yield improvements resulting from both chemical- and UV-based mutagenesis of this actinomycete (Kang et al. 2009). A drawback of untargeted approaches is that they typically require screening of relatively large numbers of mutant microorganisms to find one offering the desired chemical profile.

8.1.2.2 Media and Growth Condition Manipulation

A variety of media and growth condition manipulations have also been used to alter the expression of biosynthetic pathways. These modifications include changes to nutrient availability (Bode et al. 2002; Rigali et al. 2008; Sarkar et al. 2012), cultivation temperature or aeration changes (Sidebottom et al. 2013), the addition of small organic molecules (Pettit 2010; Yoon and Nodwell 2014), and the addition of inorganic substances such as rare earth elements (Tanaka et al. 2010).

As one example, the production of pre-shamixanthone (**3**, Figure 8.1), a novel polyketide from *Aspergillus nidulans*, was triggered by nitrogen limitation (Sarkar et al. 2012). The effects of media or growth condition changes on natural product production must generally be determined empirically, which necessitates the implementation of a large number of experiments with relatively low individual success rates.

8.1.2.3 Biotic Interactions

Natural product biosynthetic pathways presumably evolved through natural selection (Williams et al. 1989; Stone and Williams 1992). Pathway expression likely confers a competitive advantage to the producing organism in the context of the complex biotic and abiotic milieu in which it lives (Williams et al. 1989). It is plausible that microorganisms reduce the energetic cost of biosynthetic pathways by activating them only when metabolite production offers a fitness advantage. For example, some microbial natural products act as chemical cues to trigger biofilm formation and other cooperative processes between microorganisms (Linares et al. 2006; O'Brien and Wright 2011). The production of such positive chemical cues only when proximal to targeted microorganisms could offer the producing organism significant benefits over constitutive biosynthesis. Benefits of induced metabolites might include reduced production costs as well as diminished opportunities for organisms from untargeted genera to recognize and respond to cues that are present only transiently (Agrawal and Karban 1999).

In addition to the potential advantages of dynamic biosynthesis of cues favoring cooperation, the induced biosynthesis of deterrent cues may also offer significant advantages. For example, some biosynthetic pathways may have evolved as responses to microbial competition (Wietz et al. 2013). In such cases, natural products could potentially act as on-demand weapons to overcome competing organisms and thus reduce microbial competition for limited resources. By initiating biosynthesis of these chemical defenses only as needed, organisms may reap the advantages mentioned previously for positive cues as well as reduced potential for autotoxicity (Agrawal and Karban 1999).

In the remainder of this chapter, we aim to highlight progress during the past decade (2005–2015) in revealing interspecies interactions as determinants of natural product production, as well as what is known and unknown about the ecological functions of these induced metabolites. We survey reports of metabolites whose production dramatically increases in response to interspecies interactions (Section 8.2), highlight experimental approaches and technical innovations that facilitated these discoveries (Section 8.3), and evaluate current understanding of physiochemical factors that trigger biotic interaction-mediated biosynthesis (Section 8.4). Rather than comprehensively review these topics, we discuss selected seminal examples in detail to illustrate these concepts. Whenever possible, examples are drawn from marine-derived microorganisms. However, in many cases, critical work toward understanding biotic interaction-driven regulation of natural product biosynthesis was conducted in terrestrial systems. Thus, we also present terrestrial examples with the goal of suggesting likely parallels between terrestrial and marine microbial chemical ecology and laying the framework for analogous marine explorations. We conclude by highlighting future opportunities (Section 8.5). Future studies may better connect current realization of biotic interaction as a driver of natural product biosynthesis with the ecological roles of these dynamic secondary metabolites. Ultimately, understanding the ecological functions of these metabolites may logically guide efforts toward activating the full potential of Nature's genetically encoded chemical arsenal for application in myriad scientific endeavors.

8.2 MICROBIAL METABOLITES WHOSE PRODUCTION IS STIMULATED IN RESPONSE TO INTERSPECIES INTERACTIONS

Review of literature from the past decade (January 2005–February 2015) revealed at least 55 different natural products whose production is either initiated or dramatically increased in response to the co-cultivation of microorganisms (Table 8.1). A total of 22 of these metabolites were identified from marine-derived microorganisms, while 33 were derived from terrestrial organisms. It appears that induced metabolites have been discovered at increasing rates during most of the past decade (Table 8.1). This trend likely reflects both the growing scientific focus on biotic interaction as a strategy for enhancing metabolic diversity as well as improved analytical techniques used for the detection of these metabolites (Section 8.3) (Figure 8.2).

Induced metabolites presented in Table 8.1 (Oh et al. 2005; Zhu and Lin 2006; Park et al. 2009, 2011; Li et al. 2011; Zhu et al. 2011, 2014; Zuck et al. 2011; Cho and Kim 2012; Chagas et al. 2013; Konig et al. 2013; Ola et al. 2013; Rateb et al. 2013; Wang et al. 2013; Huang et al. 2014; Carlson et al. 2015) span a variety of biosynthetic classes including polyketides, nonribosomal peptides, and terpenoids. Commonly observed metabolite classes may reflect the types of biosynthetic pathways most frequently activated in response to biotic interaction. Alternatively, it is possible that commonly recognized classes of induced metabolites reflect sampling biases introduced by analytical or bioassay methods employed in the detection of these compounds. For example, liquid chromatography-mass spectrometry (LC/MS) with positive mode electrospray ionization (ESI) is a common method for the detection of induced metabolites (Section 8.3.1.1). This approach biases for the detection of molecules featuring readily ionizable groups, such as cyclic peptides (e.g., **11**; **17–18**; **26–27**) (Guo et al. 2009), while potentially overlooking metabolites that fail to ionize well in positive mode (e.g., carboxylic acids).

TABLE 8.1

Natural Products Reported during the Past Decade (2005–Early 2015) Whose Production Dramatically Increased in Response to Biotic Interaction

Co-cultured Species	Natural Habitat	Induced Metabolite(s)	Producer of Induced Natural Products	References
Libertella sp. CNL-523 and unidentified bacterium CNJ-328	Marine	libertellenones A-D (**4–7**)	*Libertella* sp. CNL-523	Oh et al. (2005)
unidentified endophytic fungus Nos-1924 and unidentified endophytic fungus Nos-3893	Marine	marinamides A-B (**8–9**); 6-methylsalicylic acid (**10**); cyclo(Phe-Phe) (**11**)	unknown	Zhu and Lin (2006) and Zhu et al. (2013)
Emericella sp. CNL-878 and *Salinispora arenicola* CNH-665	Marine	emericellamides A and B (**12–13**)	*Emericella* sp. CNL-878	Oh et al. (2007)
Phomopsis sp. K38 and *Alternaria* sp. E33	Marine	(−)-byssochlamic acid bisdiimide (**14**); 8-hydroxy-3-methyl-9-oxo-9H-xanthene-1-carboxylic acid methyl ether (**15**), ethyl 5-ethoxy-2-formyl-3-hydroxy-4-methylbenzoate (**16**) cyclo (D-Pro-L-Tyr-L-Pro-L-Tyr) (**17**); cyclo (Gly-L-Phe-L-Pro-L-Tyr) (**18**)	*Phomopsis* sp. K38 and *Alternaria* sp. E33	Li et al. (2010), Li et al. (2011), Wang et al. (2013), Huang et al. (2014)
Aspergillus sp. FSY-01 and *Aspergillus* sp. FSY-02	Marine	aspergicin (**19**); neoaspergillic acid (**20**); ergosterol (**21**)	unknown	Zhu et al. (2011)
Streptomyces cinnabarinus PK209 and *Alteromonas* sp. KNS-16	Marine	lobocompactol (**22**)	*Streptomyces cinnabarinus* PK209	Cho and Kim (2012)
Actinokineospora sp. EG49 and *Nocardiopsis* sp. RV163	Marine	N-(2-hydroxyphenyl)-acetamide (**23**); 1,6-dihydroxyphenazine (**24**); 5a,6,11a,12-tetrahydro-5a,11a-dimethyl[1,4]benzoxazino[3,2-b][1,4]benzoxazine (**25**)	*Nocardiopsis* sp. RV163[a]	Dashti et al. (2014)
Aspergillus fumigatus KMC-901 and *Sphingomonas* sp. KMK-001	Terrestrial	glionitrin A-B (**26–27**)	*Aspergillus fumigatus* KMC-901[a]	Park et al. (2009, 2011)

(Continued)

TABLE 8.1 (Continued)
Natural Products Reported during the Past Decade (2005–Early 2015) Whose Production Dramatically Increased in Response to Biotic Interaction

Co-cultured Species	Natural Habitat	Induced Metabolite(s)	Producer of Induced Natural Products	References
Streptomyces coelicolor and Bacillus subtilis	Terrestrial	streptorubin (**28**); undecylprodiginine (**29**)	Streptomyces coelicolor	Yang et al. (2009)
Ovadendron sulphureoochraceum, Ascochyta pisi, Emericellopsis minima, Cylindrocarpon destructans, and Fusarium oxysporum	Terrestrial	lateritin (**30**)	Fusarium oxysporum[a]	Pettit et al. (2010)
Aspergillus nidulans and Streptomyces rapamycinicus	Terrestrial	orsellinic acid (**31**); lecanoric acid (**32**); F-9775A (**33**); F-9775B (**34**)	Aspergillus nidulans	Nützmann et al. (2011)
Streptomyces endus S-522 and Tsukamurella pulmonis TP-B0596	Terrestrial	alchivemycin A (**35**)	Streptomyces endus S-522	Onaka et al. (2011)
Aspergillus fumigatus and Streptomyces peucetius	Terrestrial	fumiformamide (**36**); N,N′-((1Z,3Z)-1,4-bis(4-methoxyphenyl)buta-1,3-diene-2,3-diyl)diformamide (**37**)	Aspergillus fumigatus	Zuck et al. (2011)
Alternaria tenuissima and Nigrospora sphaerica	Terrestrial	stemphyperylenol (**38**) and other polyketides	Alternaria tenuissima	Chagas et al. (2013)
Streptomyces coelicolor and Amycolatopsis sp. AA4	Terrestrial	acyl-desferrioxamines (**39**)	Streptomyces coelicolor	Traxler et al. (2013)
Aspergillus fumigatus and Streptomyces rapamycinicus	Terrestrial	fumicyclines A and B (**40–41**)	Aspergillus fumigatus	Konig et al. (2013)
Alternaria tenuissima and Nigrospora sphaerica	Terrestrial	stemphyperylenol (**38**) and other polyketides	Alternaria tenuissima	Chagas et al. (2013)

(Continued)

TABLE 8.1 (*Continued*)

Natural Products Reported during the Past Decade (2005–Early 2015) Whose Production Dramatically Increased in Response to Biotic Interaction

Co-cultured Species	Natural Habitat	Induced Metabolite(s)	Producer of Induced Natural Products	References
Aspergillus fumigatus MBC-F1-10 and *Streptomyces bullii*	Terrestrial	ergosterol (**18**); brevianamide F (**42**); spirotryprostatin (**43**); 6-methoxyspirotryprostatin B (**44**); fumitremorgin B (**45**); fumitremorgin C and dihydroxy derivatives (**46–47**); verruculogen (**48**); 11-*O*-methylpseurotin A-A2 (**49–50**)	*Aspergillus fumigatus* MBC-F1-10	Rateb et al. (2013)
Streptomyces sp. CJ-5 and *Tsukamurella pulmonis* TP-B0596	Terrestrial	chojalactones A–C (**51–53**)	*Streptomyces* sp. CJ-5	Hoshino et al. (2015)
Streptomyces sp. B033 and strains of the phylum *Proteobacteria*	Terrestrial	resistomycin (**54**)	*Streptomyces* sp. B033	Carlson et al. (2015)
Fusarium tricinctum and *Bacillus subtilis*	Terrestrial	macrocarpon C (**55**), (−)-citreoisocoumarin (**56**), 2-(carboxymethylamino)benzoic acid (**57**), (−)-citreoisocoumarinol (**58**)	*Fusarium tricinctum*[a]	Ola et al. (2013)

[a] Indicates most likely producer of natural product, but that the true producer has not been experimentally confirmed.

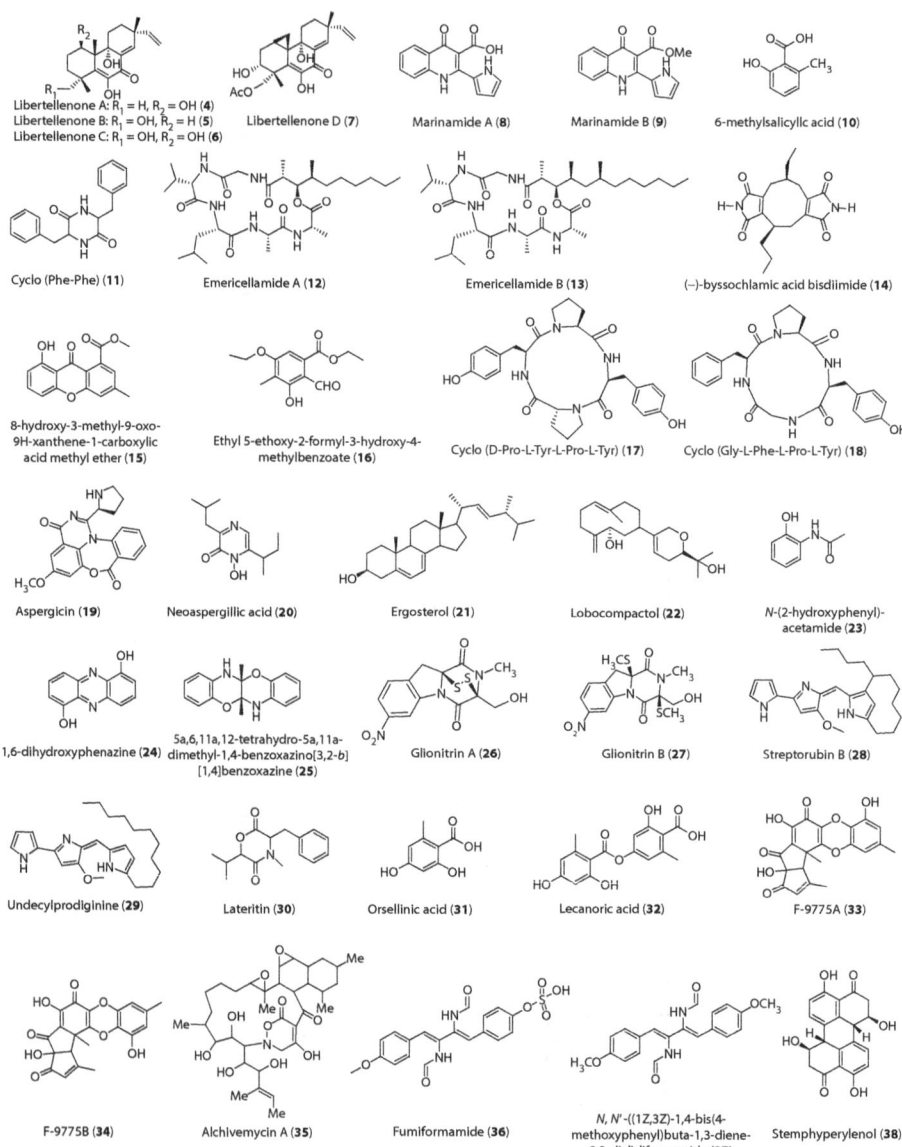

FIGURE 8.2 Structures of metabolites whose production was reported to dramatically increase in response to biotic interaction. Table 8.1 provides additional details about these metabolites. *(Continued)*

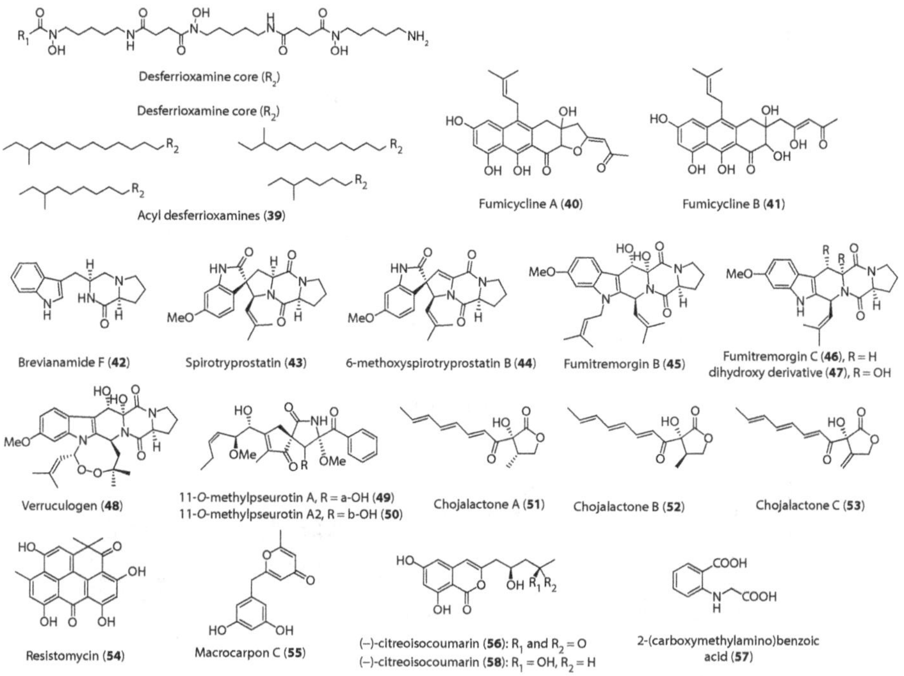

FIGURE 8.2 Structures of metabolites whose production was reported to dramatically increase in response to biotic interaction. Table 8.1 provides additional details about these metabolites.

8.3 DETECTION OF INDUCED NATURAL PRODUCTS

Natural product identification from a single microbial species is analogous to seeking a needle in a haystack because trace metabolites of interest are commonly found in complex mixtures with more abundant metabolites and media components. In the case of microbial co-cultures, this challenge is amplified and comparable to seeking a single needle that differentiates multiple haystacks. To tackle this challenge, researchers during the past decade have applied several innovative experimental approaches to probe metabolic differences between isolated microorganisms and those engaged in biotic interactions. These approaches include the comparison of metabolite profiles by analytical techniques (Section 8.3.1), the comparison of biosynthetic gene expression by molecular biology techniques (Section 8.3.2), and the comparison of biological activities by either ecology- or medically-motivated assays (Section 8.3.3; Figure 8.3). Application of these strategies has enhanced recognition of induced metabolites and, in some cases, enabled the conclusive determination of which co-cultivation partner was responsible for the biosynthesis of induced metabolites.

8.3.1 COMPARISON OF METABOLITE PROFILES BY ANALYTICAL CHEMISTRY APPROACHES

During the past decade, several analytical methods have been employed to detect changes in metabolite profiles in response to biotic interaction. These methods include

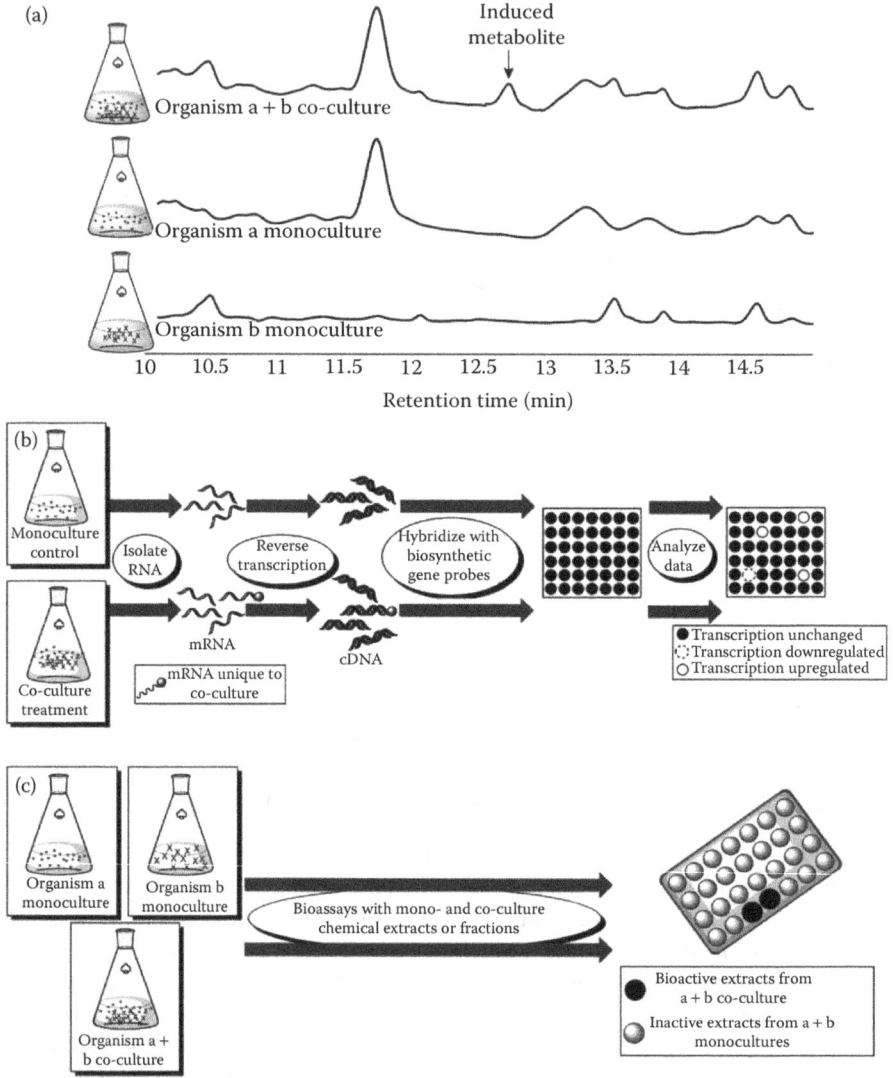

FIGURE 8.3 Overview of selected experimental approaches for evaluating biosynthetic pathway responses to interactions between microorganisms. (a) Analytical chemistry approaches include the comparison of metabolite profiles between mono- and co-cultures by LC/MS or LC/UV; (b) Molecular biology approaches include the use of microarrays to evaluate changes in biosynthetic gene transcription in response to biotic interactions; (c) Biological assays of chemical extracts from mono- and co-cultures facilitate the discovery of metabolites whose production changes in response to biotic interactions.

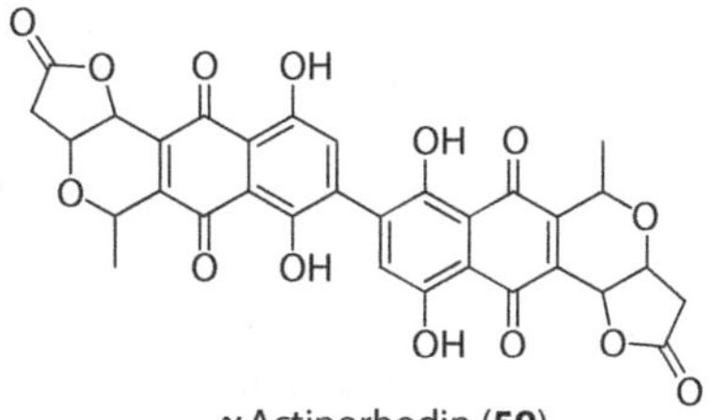

γ Actinorhodin (**59**)

FIGURE 8.4 γ Actinorhodin, a pigment produced by some *Streptomyces* spp. Changes in production of this metabolite are readily assessed colorimetrically. (From Traxler, M.F., et al., *mBio*, 4, e00459–13, 2013; Onaka, H. et al., *Appl. Environ. Microbiol.*, 77, 400–406, 2011.)

LC/MS and LC/UV (Sections 8.3.1.1; Figure 8.3a), MS imaging (Section 8.3.1.2; Figure 8.5), and NMR spectroscopy (Section 8.3.1.3). Such analytical techniques can be daunting for screening copious co-cultures; therefore, researchers have also evaluated phenotypic changes between mono- and co-cultures as an initial rapid indicator of potential metabolic changes. These phenotypic changes commonly include alterations to colony morphology or color. Colorimetric approaches have most often been applied to rapidly screen for the modulation of known pathways, such as those encoding the production of γ Actinorhodin, (**59**, Figure 8.4) and undecylprodiginine (**29**) pigments by some *Streptomyces* bacteria (Onaka et al. 2011; Traxler et al. 2013).

8.3.1.1 Comparative Metabolite Profiling by Liquid Chromatography–Mass Spectrometry

Oh et al. (2007) provide one example of the utility of LC/MS in the evaluation of biotic interaction-induced natural product biosynthesis among marine microorganisms. Metabolite profiles were compared by LC/MS between monocultures and co-cultures of more than 50 combinations of marine sediment-derived bacteria and fungi. These analyses revealed that the chemical profiles of individual monocultures of the obligate marine actinomycete bacterium *Salinispora arenicola* CNH-665 and the fungus *Emericella* sp. CNL-878 differed markedly from the chemical profile of the co-culture. Two upregulated metabolites were isolated and identified as the novel emericellamides A–B (**12–13**), molecules of likely hybrid polyketide-nonribosomal peptide biosynthetic origin. To establish which co-cultivation partner was responsible for emericellamide biosynthesis, high sensitivity LC/MS was applied to reveal trace quantities of these metabolites from chemical extracts of pure *Emericella* sp. but absent from pure *S. arenicola*. Quantification of **12–13** demonstrated that the production of each of these metabolites by *Emericella* sp. increased by approximately 100-fold in response to biotic interaction with *S. arenicola*.

Both emericellamides were moderately cytotoxic (IC_{50} ~ 20–40 μM) toward a human colon cancer cell line and inhibited the growth of methicillin-resistant *Staphylococcus aureus* (MIC ~5 μM). Based upon this antibacterial activity, it may be speculated that *Emericella* sp. deploys emericellamides as chemical defenses to thwart *S. arenicola*. However, without evaluation of the effects of emericellamides at realistic natural concentrations on *S. arenicola*, the ecological implications of induced emericellamide biogenesis remain unknown.

8.3.1.2 Spatially Resolved Mass Spectrometry for Metabolite Profiling

MS innovations, including desorption electrospray ionization (DESI) MS (Cooks et al. 2006) and matrix-assisted laser desorption ionization (MALDI) MS imaging (Figure 8.5) (Cornett et al. 2007), offer advantages over conventional MS for the study of interspecies interactions. These advantages include the detection of metabolites directly from living cells with minimal sample preparation and the pinpointing of metabolite locations on agar plates at micrometer scales.

While the capabilities of MS imaging hold immense appeal in chemical ecology, challenges currently limit their widespread use. Namely, the generation of MS data

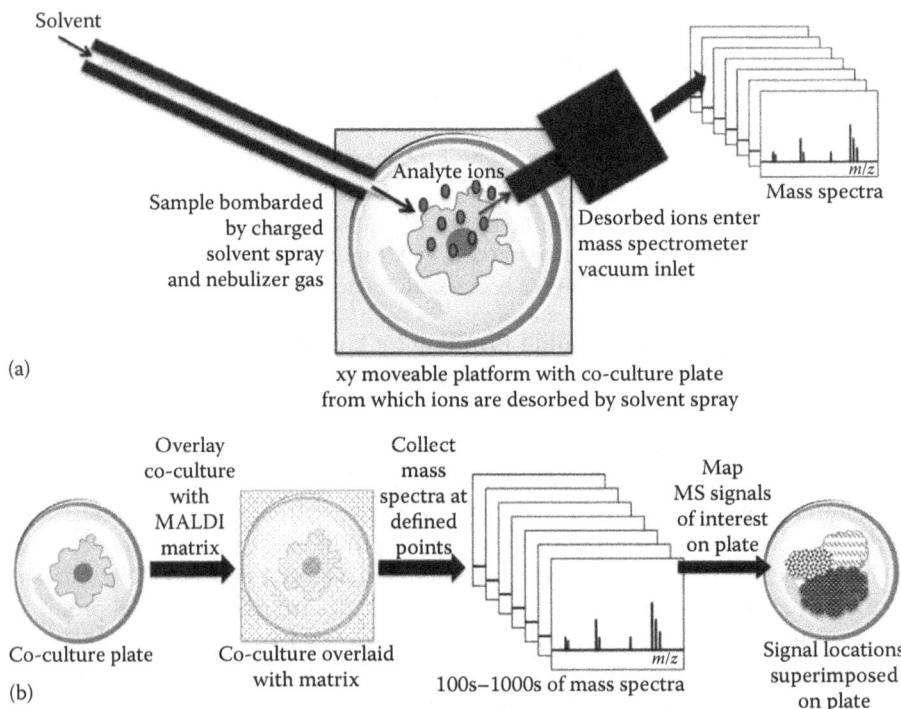

FIGURE 8.5 Overview of spatially resolved mass spectrometry (MS) methods for tracking metabolic responses of microorganisms to biotic interaction. (a) In desorption electrospray ionization (DESI) MS, samples are mounted at atmospheric pressure on an xy-moveable stage and subjected to a charged solvent spray, with desorbed ions then entering the mass spectrometer via a vacuum inlet. Movement of the stage yields spectra at targeted locations across the sample. (b) In matrix-assisted laser desorption ionization (MALDI) MS imaging, samples are coated with a MALDI matrix and then bombarded by the MALDI laser at targeted raster points to desorb analyte ions from the samples. Mass spectra generated at these points are compiled to map locations of analytes on the original sample. (From [a] Cooks, R.G. et al., *Science*, 311, 1566–1570, 2006; [b] Cornett, D.S. et al., *Nat. Methods*, 4, 828–833, 2007.)

across surfaces may yield 100s–1000s of spectra, the interpretation of which is non-trivial (Angel and Caprioli 2013). This challenge is magnified by the inherent complexity of untargeted mass spectra for mixtures such as those derived from cells. As a consequence of this complexity, signals for novel secondary metabolites may be overlooked in a barrage of more prominent signals. Hence, spatially resolved MS is most readily applicable for the targeted detection of specific metabolites of interest, although recent seminal studies have successfully tracked untargeted metabolites as well. Work by Yang et al. (2009) and Traxler et al. (2013) highlights the emerging potential of MS imaging to probe the effects of biotic interaction on terrestrial microbial metabolism in situ (Sections 9.3.2.1 and 9.3.2.2).

To our knowledge, no studies have applied spatially resolved MS to evaluate the responses of marine microorganisms to biotic interaction. However, DESI MS was previously utilized to determine the locations and relative concentrations of chemical defense metabolites on macroalgal surfaces (Lane et al. 2009; Andras et al. 2012). These marine macroalgal studies, together with the terrestrial microorganism studies highlighted in the following sections, suggest that MS imaging should be suitable for tracking marine microorganism responses to biotic interactions.

8.3.1.2.1   Matrix-Assisted Laser Desorption Ionization Mass Spectrometry

Yang et al. (2009) applied MALDI time of flight (TOF) imaging MS to reveal metabolic changes resulting from interaction between the model bacteria *Bacillus subtilis* and *Streptomyces coelicolor*. Focusing both on known metabolites and selected unidentified metabolites, molecular ions were compared temporally and spatially between monocultures and co-cultures of bacteria grown in close proximity on agar plates. These experiments revealed that co-cultivation resulted in striking decreases in titers of some *S. coelicolor* metabolites, including the common *Streptomyces* pigment prodiginine. Hypothesizing that the excreted *B. subtilis* metabolite bacillaene (**60**, Figure 8.6) was responsible for silencing these *S. coelicolor* biosynthetic pathways, a *B. subtilis* mutant lacking an essential bacillaene biosynthetic gene was co-plated with *S. coelicolor*. Imaging MS revealed that the bacillaene-deficient *B. subtilis* mutant was incapable of attenuating *S. coelicolor* biosynthesis, and implicated bacillaene as an effector of *S. coelicolor* biosynthesis. These results highlighted the synergistic potential of MALDI imaging MS and genetic manipulation for illuminating chemical communication between microorganisms and suggested the importance of chemical crosstalk as a driver of biosynthetic pathway activity.

8.3.1.2.2   Nanospray Desorption Electrospray Ionization Mass Spectrometry

Efforts by Traxler et al. (2013) demonstrated that nanospray DESI (nanoDESI) MS in conjunction with MALDI MS imaging is a powerful combination for the evaluation of biosynthetic pathway induction. Plating of *S. coelicolor* in close proximity to ~20 other actinomycete strains revealed five strains that dramatically altered the gross morphology or pigmentation of *S. coelicolor*. Due to these drastic phenotypic changes, these five strains were selected as likely candidates for inducing metabolic changes of *S. coelicolor*. Hence, these strains were individually co-plated with *S. coelicolor*, and these co-cultures were evaluated by nanoDESI at multiple time points to yield tandem MS data. These molecular fragmentation MS data were

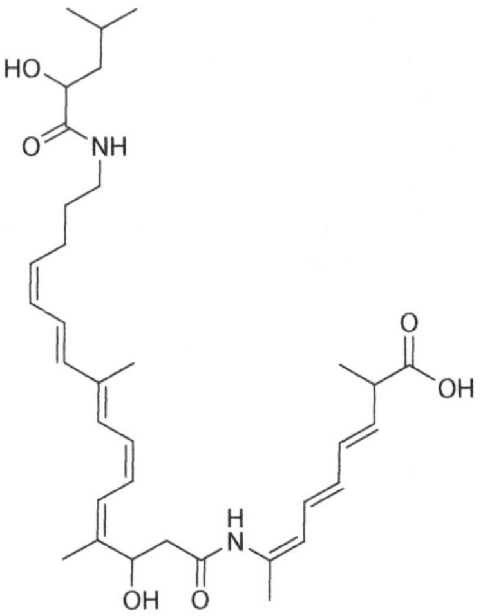

Bacillaene (**60**)

FIGURE 8.6 Bacillaene, a metabolite produced by *Bacillus subtilis*. This metabolite causes dramatic shifts in *S. coelicolor* secondary metabolism as revealed by imaging MALDI MS. (From Yang, Y.L. et al., *Nat. Chem. Biol.*, 1–3, 2009.)

computationally mined to locate fragmentation patterns observed across one or more co-culture treatments but absent from monoculture controls.

Data for structurally related metabolites, based on similar fragmentation patterns, were further evaluated. These evaluations aimed to determine whether induced metabolites were unique to the interaction of *S. coelicolor* with one other strain or were induced by the interaction of *S. coelicolor* with multiple strains. This metabolite networking approach revealed dramatic increases of multiple acyl-desferrioxamines (e.g., **39**) during interactions between *S. coelicolor* and several other actinomycetes. Intriguingly, nanoDESI revealed that both the time course of acyl-desferrioxamine production and the exact chemical structures of these molecules varied with the co-cultivation partner.

Two pieces of information strongly implicated *S. coelicolor* as the producer of the induced acyl-desferrioxamines. First, MALDI imaging MS revealed a concentration gradient of these compounds centered around *S. coelicolor* (Traxler et al. 2013). Second, bioinformatics analyses of the genome sequences of organisms co-cultured with *S. coelicolor* revealed that some co-cultivation partners lacked genes necessary for acyl-desferrioxamine biosynthesis.

Siderophores, such as desferrioaximines, sequester environmental iron that is limited in many environments (Hider and Kong 2010). Hence, Traxler et al. (2013) hypothesized that desferrioaximine biosynthesis may be induced in response to competition for iron. Supporting this hypothesis, MALDI imaging MS revealed that *S. coelicolor* produced dramatically less acyl-desferrioxamines when co-cultured with mutant actinomycetes deficient in siderophore pathways. This suggested that *S. coelicolor* applies acyl-desferrioxamines as a dynamic chemical tool to improve its fitness in competition for iron.

8.3.1.3 Comparative Metabolite Profiling by NMR Spectroscopy

Combining LC-based approaches with ^{1}H NMR metabolite fingerprinting offers benefits in the detection and identification of induced metabolites over LC/MS alone. Namely, ^{1}H NMR universally detects all sufficiently abundant metabolites that contain hydrogen atoms, while LC/MS and LC/UV overlook metabolites that ionize poorly or that do not absorb UV radiation, respectively (Crews et al. 1998). Further, ^{1}H NMR generally provides more structural details than LC/MS or LC/UV. Despite these advantages of NMR, it remains less routinely applied in metabolite profiling than LC/MS and LC/UV, likely because of its decreased sensitivity relative to UV or MS detection. However, recent advancements in high field NMR spectroscopy have made significant strides toward improving the limit of detection of NMR and hence enhanced its capabilities for probing microbial metabolism (Lane et al. 2015).

Dashti et al. (2014) illustrated the utility of merging ^{1}H NMR metabolite fingerprinting with LC/UV to detect metabolic changes. They identified metabolites whose production was initiated by the biotic interaction of two sponge-derived actinomycetes, *Actinokineospora* sp. EG49 and *Nocardiopsis* sp. RV163. LC/UV analyses revealed that co-cultivation of these two actinomycetes resulted in dramatic metabolic profile changes, including both the up- and down-regulation of several natural products. Following partial chromatographic purification of metabolites from large-scale cultures, ^{1}H NMR metabolite fingerprints were compared between mono- and co-cultures. This revealed the novel natural product 5a,6,11a,12-tetrahydro-5a,11a-dimethyl[1,4]benzoxazino[3,2-*b*][1,4]benzoxazine **(25)**, as well as previously reported natural products *N*-(2-hydroxyphenyl)-acetamide **(23)** and 1,6-dihydroxyphenazine **(24)**.

These induced metabolites were evaluated for inhibition of selected protozoans, bacteria, and fungi (Dashti et al. 2014). Most relevant to ascribing an ecological function to these metabolites, **24** significantly inhibited the growth of *Actinokineospora* sp. in disk diffusion assays but did not inhibit *Nocardiopsis* sp. Thus, it is plausible that *Nocardiopsis* sp. employs **24** as a chemical defense against other sponge-associated actinomycetes including *Actinokineospora* sp. However, without conclusive evidence revealing which organism produces this compound and its natural abundance, the ecological significance of this finding remains unclear.

8.3.2 Evaluation of Microbial Metabolic Changes by Molecular Biology Approaches

The number of complete prokaryote genomes available in in the National Center for Biotechnology Information (NCBI) database has grown to nearly 4,000 during the twenty years since release of the first bacterial genome (Fleischmann et al. 1995). This DNA sequence information affords unprecedented opportunities to evaluate responses of microorganisms to biotic interaction and other stimuli by using a wealth of molecular biology tools. Studies such as those by Schroeckh et al. (2009) (Section 8.3.2.1) and Seyedsayamdost (2014) (Section 8.3.2.2) highlight selected opportunities enabled by genome sequencing and molecular biology.

8.3.2.1 Biosynthetic Gene Transcription Arrays

Bioinformatics analyses of the *Aspergillus nidulans* genome suggested that this terrestrial filamentous fungus harbored ~50 biosynthetic pathways, the majority of which were apparently silent under typical fermentation conditions (Galagan et al. 2005; von Doehren 2009). Schroeckh et al. (2009) applied this information to develop a molecular biology-based approach to explore the hypothesis that interactions between *A. nidulans* and actinomycetes activate these silent biosynthetic pathways. To evaluate biosynthetic gene transcription changes in response to biotic interaction, a "secondary metabolism array" including probes for genes from each predicted *A. nidulans* biosynthetic pathway was developed. RNA from co-cultivation treatments and monoculture controls was subjected to reverse transcription to yield cDNA corresponding to transcriptionally active genes. This cDNA was detected using the secondary metabolism array (Figure 8.3b), which revealed *A. nidulans* pathways whose transcription was initiated or ceased in response to biotic interaction with over 50 different actinomycetes.

The array revealed that interaction between *A. nidulans* and one actinomycete, *Streptomyces rapamycinicus*, activated an orphan polyketide-encoding gene cluster from *A. nidulans* (Schroeckh et al. 2009). To help connect this activated pathway to its corresponding metabolites, *A. nidulans* was genetically engineered to delete essential genes from this orphan polyketide pathway. Both this mutant and wild type *A. nidulans* were grown in co-culture with *S. rapamycinicus*. Comparative metabolite profiling of these two cultures linked the orphan polyketide pathway to biosynthesis of orsellinic acid (**31**) and related metabolites (**32–34**). Intriguingly, biosynthesis of these polyketides by *A. nidulans* was triggered only through intimate contact with *S. rapamycinicus*, rather than through a diffusible chemical signal from the bacterium. Subsequent studies revealed that the *A. nidulans* pathway encoding **31–34** is activated by fungal histone acetylation (Nützmann et al. 2011); the specific physiochemical feature of *S. rapamycinicus* that triggers histone acetylation and concomitant *A. nidulans* biosynthetic activation remains cryptic.

8.3.2.2 Reporter-Based Evaluation of Biosynthesis

Seyedsayamdost (2014) developed an innovative reporter-based method to evaluate the activation of poorly expressed polyketide and nonribosomal peptide biosynthetic pathways of the bacterium *Burkholderia thailandensis* E264 in response to external stimuli. *B. thailandensis* mutants were created, in which targeted biosynthetic genes were linked to *lacZ* reporters. This afforded *B. thailandensis* fluorescence only upon expression of the biosynthetic gene-*lacZ* fusion. Screening in a 384-well plate format, a library of >500 small molecules were evaluated for elicitation of each pathway, as evidenced by LacZ fluorescence output. This revealed that some antibiotics at sub-inhibitory concentrations modulated *B. thailandensis* biosynthetic pathway expression, a result confirmed by subsequent isolation of the induced natural products.

Relative to comparative metabolite profiling (Section 8.3.1), reporter assays offer potential for expedited evaluation of responses of biosynthetic pathways to external stimuli. This makes it particularly attractive as a tool for the high throughput screening of biosynthetic pathway activation. To our knowledge, fluorescent reporter-based strategies have not yet been applied for the evaluation of biosynthetic responses to direct interspecies interactions, although they should be directly applicable in such studies. Importantly, reporter strategies are limited to microorganisms amenable to genetic manipulation. While genetic manipulation is facile for some well-characterized and biosynthetically rich microorganisms (Gust et al. 2003), others are not amenable to such modifications (Alqahtani et al. 2015).

8.3.3   BIOACTIVITY-BASED EVALUATION OF METABOLIC RESPONSES TO BIOTIC INTERACTION

In addition to analytical chemistry and molecular biology approaches, bioactivity-centric strategies have also proven useful for tracking biosynthetic responses to microbial interactions. The Clardy and Kolter groups (2012) provided an excellent overview of methods for several bioactivity-guided approaches. Such approaches are particularly suitable for studies aimed at the identification of induced metabolites with specific biological activities of either medical or ecological interest. Bioactivity-guided approaches often require less costly experimental resources than the analytical chemistry-centric approaches and may offer higher throughputs due to their suitability for microtiter plate formats with quickly evaluated data outputs. The combination of bioactivity-guided approaches and complementary analytical chemistry strategies offers much opportunity for expediting the detection and identification of dynamic bioactive metabolites.

8.3.3.1   Induced Production of Bioactive Metabolites from Complex Ensembles of Fungi

A study by Pettit et al. (2010) is unique amongst bioactivity-guided approaches in that the co-cultivation of more than two microorganisms triggered biosynthetic activation. They investigated five filamentous fungi that individually produced no metabolites inhibitory toward cancer cell lines. However, the co-cultivation of these five fungi resulted in production of an anticancer metabolite. Through bioassay-guided fractionation, the induced anticancer natural product was identified as the previously reported metabolite lateritin (**30**). *Fusarium oxysporum* was reported as the most plausible producer of lateritin since this was the sole detectable fungus at the end of fermentation. Based on these data, it is tempting to speculate that **30** or other unidentified metabolites act as chemical defenses of *F. oxysporum*. However, **30** was not evaluated for effects on co-cultured fungi, so its ecological function remains cryptic.

8.4   SIGNALS TRIGGERING BIOSYNTHETIC PATHWAY ACTIVATION DURING BIOTIC INTERACTION

For most reports of induced natural product biosynthesis (Table 8.1), the physical and/or chemical signal(s) from one organism that trigger biosynthetic pathway activation of another remain unknown. Most studies of biotic interaction-triggered biosynthesis suggest that these metabolic responses are not broad stress responses to all interactions, but instead responses to interaction with specific microorganisms or small groups of related microorganisms (e.g., Sections 8.3.1.1, 8.3.1.2.2, 8.3.2.1). This suggests that the physical and/or chemical signal(s) that trigger biosynthetic induction are likewise unique to small subsets of microorganisms.

8.4.1 SMALL MOLECULES AS MEDIATORS OF BIOSYNTHETIC PATHWAY ACTIVITY

The addition of some natural products, such as quorum sensing molecules and antibiotics, to microbial growth media has been shown to influence natural product biosynthesis (Igarashi et al. 2001; Takano 2006; Hsiao et al. 2009; Kalivoda et al. 2010). Likewise, recent work revealed that *B. subtilis* utilized bacillaene to manipulate the metabolome of *S. coelicolor* (Yang et al. 2009) (Section 8.3.1.2.1), and that gluconic acid from *Pseudomonas* sp. thwarted γ actinorhodin biosynthesis by *S. coelicolor* (Galet et al. 2014). These observations suggest small molecules as important drivers of biosynthetic pathway activity during microbial interactions. To better understand the ecological relevance of these observations, experiments in contexts more closely representative of interspecies interactions in nature are needed (Bernier and Surette 2013).

While the previous examples highlight the effects of diffusible small molecule chemical signals on natural product biosynthesis, in many cases the physical and/or chemical factors that trigger biosynthetic pathway activation or repression appear more complex. For example, while interaction between *S. rapamycinicus* and *A. nidulans* has been demonstrated to induce polyketide production by *A. nidulans*, this biosynthetic activation appears not to be triggered by a diffusible chemical signal but instead by a cryptic feature afforded exclusively by close contact between the two microorganisms (Schroeckh et al. 2009; Nützmann et al. 2011) (Section 8.3.2.1). Recent studies also suggest that complex physiochemical signals may trigger natural product biosynthesis, as highlighted in recent work by Onaka et al. (2011) (Section 8.4.2).

8.4.2 MYCOLIC ACID-PRODUCING BACTERIA INDUCE *STREPTOMYCES* BIOSYNTHESIS

Co-cultivation of 400 different bacterial strains with *S. lividans* revealed that only co-cultivation with *Tsukamurella pulmonis* TP-B0596 yielded visually discernable upregulation of γ actinorhodin (**59**) and undecylprodiginine (**29**) pigments by *S. lividans* (Onaka et al. 2011). *T. pulmonis* TP-B0596, a member of the *Corynebacteriaceae* family, also modulated biosynthetic pathways among other representative *Streptomyces* species. Likewise, closely related *Corynebacteriaceae* representatives also stimulated *Streptomyces* biosynthesis, while more distant relatives failed to elicit biosynthesis.

A hallmark feature of *Corynebacteriaceae* bacteria is the production of cell membrane mycolic acids, prompting the hypothesis that mycolic acids induced *Streptomyces* biosynthesis (Onaka et al. 2011). Supporting this hypothesis, mutants lacking only an essential mycolic acid biosynthetic gene failed to activate *Streptomyces* biosynthesis. However, the addition of either pure mycolic acids, chemical extracts, or filter-sterilized exudates from wild type *Corynebacteriaceae* family bacteria failed to trigger *Streptomyces* biosynthesis. This suggested that the context in which mycolic acids were presented to *Streptomyces* was critical in dictating biosynthetic activation. It is also feasible that mycolic acids acted in concert with labile elicitor metabolites unsuitable for extraction, or that both mycolic acids and non-chemical signals were required for biosynthetic induction.

Interspecies interactions between *Corynebacteriaceae* family bacteria and *Streptomyces* species have now revealed multiple novel metabolites whose biosynthesis is triggered or dramatically upregulated in response to interaction. Alchivemycin A (**35**) is a novel antibiotic whose production by *S. endus* S-522 was induced by biotic interaction with mycolic acid-producing bacteria (Onaka et al. 2011). Analogously, chojalactones A–C (**51–53**) were also recently isolated from the soil-derived *Streptomyces* sp. CJ-5 through this experimental approach (Hoshino et al. 2015).

Together, these studies suggest *Corynebacteriaceae–Streptomyces* associations as promising model systems for future investigations of the ecological implications of corynebacteria-induced biosynthesis. Thus far, *Corynebacteriaceae–Streptomyces* interactions have been exclusively evaluated among terrestrial microorganisms. Members of both groups are also recognized as co-occurring members of marine microbial communities (Gontang et al. 2010), suggesting that analogous biotic interactions may also be relevant in marine environments.

8.5 OPPORTUNITIES AND OUTLOOK

Genome sequencing has shown that a tremendous amount of Nature's genetically encoded natural product potential remains untapped. The lack of expression of many biosynthetic genes in laboratory cultures is one cause of the discrepancy between genetic potential and observed chemistry. A growing body of evidence supports that interspecies interactions dictate the expression of biosynthetic pathways (Sections 8.1 through 8.4). These findings suggest induced natural products as potential mediators of interspecies interactions. However, studies of these metabolites as determinants of competitive fitness remain remarkably scarce.

Study of the ecological functions of dynamic metabolites is anticipated to help address fundamental questions in microbial ecology, including how chemical signals structure microbial communities such as those associated with marine sediments and sponges. Knowledge of the ecological roles of induced metabolites may also be used to develop rational strategies for activating silent biosynthetic pathways, and ultimately holds potential to facilitate discovery of the next generation of natural products. Realization of these goals necessitates collaborative efforts spanning from ecology to microbiology, chemistry, and molecular biology. Three current interdisciplinary opportunities for enhancing understanding of the ecological functions of induced natural products are: (1) implementation of technologies that emulate natural microbial habitats; (2) utilization of cultivation-independent approaches for evaluating biosynthetic dynamics in nature; and (3) genetic manipulation of microbial secondary metabolic pathways (Sections 8.5.1 through 8.5.3).

8.5.1 Emulating Natural Habitats to Reveal Ecological Functions of Induced Natural Products

It is estimated that less than 1% of bacteria can be cultured via commonly laboratory techniques (McCraig et al. 2001). In an effort to facilitate the cultivation of marine microorganisms, researchers have focused on the development of laboratory culture conditions that more closely emulate nature. For example, the iChip microfluidic device was recently developed to promote the cultivation of previously "uncultur-able" microorganisms (Kaeberlein et al. 2002; Nichols et al. 2010). Using the iChip, individual microbial cells from environmental samples may be introduced into sepa-rate chambers that permit chemical exchange with the external environment. By incubating the iChip in nature or immersed by components from native environ-ments (e.g., seawater), microbial propagation is promoted by exposure to conditions that emulate those faced in nature. This technology was recently applied to initiate a laboratory culture of the soil dwelling β-proteobacterium *Eleftheria terrae*, which yielded the novel antibiotic teixobactin (**61**, Figure 8.7) (Ling et al. 2015).

Although nature-emulating culture conditions such as those provided by the iChip have not yet seen widespread application for probing interspecies interac-tions, their success in the cultivation of new microorganisms suggests their poten-tial in chemical ecology. It is envisioned that the execution of ecological assays using the iChip and related innovations would allow the functions of natural prod-ucts to be probed in contexts more environmentally realistic than provided by today's common agar plate and liquid culture-based methods. Further, by enhanc-ing cultivable microbial diversity, such microfluidic devices provide opportunities for exploring chemically mediated interactions between previously inaccessible

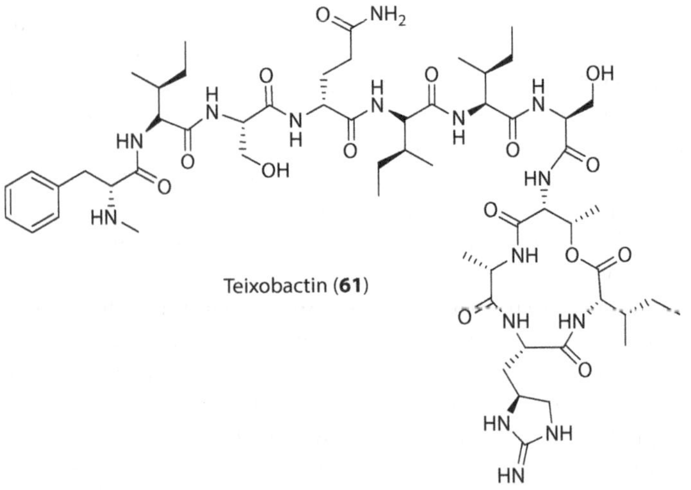

Teixobactin (**61**)

FIGURE 8.7 Structure of teixobactin, a natural product produced by the β-proteobacterium *E. terrae*. Laboratory cultivation of this bacterium was facilitated by utilization of the iChip microfluidic device, which exposes microbial cells to chemicals from their native habitats.

representatives of microbial communities. This should afford better understanding of how induced natural products mediate associations among microorganisms more reflective of nature than previously possible.

8.5.2 CULTIVATION-INDEPENDENT MICROBIAL CHEMICAL ECOLOGY

Despite recent cultivation innovations (Section 8.5.1), it remains impossible to fully emulate Nature in the laboratory. The relative abundances of diverse microbial genera, specific flow conditions, and other environmental variables remain impossible to replicate in the laboratory. This fact poses inherent limitations on understanding of the ecological functions of induced secondary metabolites. Cultivation-independent field-based approaches offer emerging opportunities to overcome these challenges and enrich understanding of chemically mediated interactions between microorganisms.

Metagenomics and metatranscriptomics are the analyses of mixtures of genomic DNA and corresponding messenger RNA sequences, respectively, from ensembles of organisms (Moran 2009; Su et al. 2012). Hence, these approaches allow both the assessment of all genes within a community as well as determination of which genes are transcriptionally active. These meta-analyses are representative of cultivation-independent techniques that offer promise for addressing fundamental chemical ecology questions. For example, by conducting metagenome and metatranscriptome analyses in parallel, genetically encoded biosynthetic potential may be compared with corresponding gene transcription for ensembles of microorganisms within natural environments such as marine sediments. Such studies may help to address the question of whether biosynthetic genes expressed at low levels in the laboratory are more active in Nature and offer clues about the evolutionary impetus for maintenance of these pathways.

Cultivation-independent meta-analyses also offer opportunities to probe the induction of natural product biosynthesis. For example, such approaches should offer opportunities to track responses of biosynthetic genes to environmental manipulation, such as changes in microbial titers or chemical signals in situ. Through such experiments, the triggers of biosynthesis in nature could be explored with fewer inherent biases than plausible through cultivation-dependent laboratory approaches. Suggesting the promise of cultivation-independent field approaches in chemical ecology, Penn et al. (2014) applied metatranscriptomics to probe diurnal gene expression dynamics within the prokaryotic community associated with a freshwater harmful microalgal bloom. This study revealed that the most highly expressed polyketide synthase-nonribosomal peptide synthetase (PKS/NRPS) hybrid biosynthetic gene cluster within the community corresponded to a cryptic natural product.

Bioinformatics analyses of genes and transcripts can also often predict structural features of corresponding nonribosomal peptide, polyketide, and other microbial natural products (Medema et al. 2011). Hence, these data could ultimately be applied to guide the implementation of high sensitivity analytical techniques to detect and monitor corresponding metabolites directly from environmental samples. These advancements would certainly usher a wealth of opportunity for chemical ecology.

8.5.3 PROBING THE ECOLOGICAL FUNCTIONS OF SECONDARY METABOLITES THROUGH GENETIC MANIPULATION

Investigations of natural products as mediators of interspecies interactions often employ assays in which chemical extracts or pure natural products are added to microbial cultures and resulting changes in microbial growth/survival or behavior (e.g., biofilm formation) are measured (Lane et al. 2010). These chemical addition-based approaches cannot fully represent the temporal and spatial distribution of metabolites deployed as chemical communication signals in situ. Hence, experimental approaches that more realistically allow the tracking of actual chemical communication between microorganisms are needed to probe the roles of secondary metabolites as determinants of the outcomes of interactions between microorganisms.

Modern molecular biology offers promising avenues for improving capabilities to probe natural products as mediators of ecological interactions. Methods are now available for targeted genetic manipulation of diverse marine microbial genera (Sanchez et al. 2012; Kieler et al. 2013; Jensen et al. 2015; Zhang et al. 2015). These genetic manipulation approaches permit the targeted elimination of selected biosynthetic genes. Such gene elimination mutations are analogous to resetting Nature's evolutionary clock back to a time prior to an organism's acquisition of targeted biosynthetic genes. Hence, the biotic interactions between wild type and biosynthetic gene elimination mutant microorganisms with targeted species can be compared, thus allowing in situ assessment of how the presence of a targeted biosynthetic pathway dictates the outcome of an interspecies interaction. Such strategies have thus far seen limited application in understanding roles of chemical crosstalk between organisms as governors of ecological interactions (Yang et al. 2009). They offer considerable future promise, particularly in conjunction with experimental approaches that emulate nature in laboratory settings (Section 8.5.1).

8.6 CONCLUSION

The studies presented in Sections 8.1 through 8.5 highlight interspecies interactions as mediators of natural product biosynthesis. These biotic interactions hold substantial promise to overcome practical natural product discovery challenges associated with the enormous prevalence of transcriptionally silent biosynthetic genes in single species laboratory cultures. To fully realize this promise, the natural functions of metabolites whose biosynthesis is triggered by biotic interaction must be better understood. By revealing the ecological functions of such metabolites, fundamental insights into roles of chemistry as governors of microbial communities may be achieved. In addition to bolstering the field of chemical ecology, such knowledge is also anticipated to offer significant benefits to drug discovery, chemical synthesis, and the immense variety of other scientific endeavors that draw inspiration from Nature's chemical wealth.

REFERENCES

Agrawal, A. A., and R. Karban. 1999. Why induced defenses may be favored over constitutive strategies in plants. In *The Ecology and Evolution of Inducible Defenses*, R. Tollrian and C. D. Harvell (Eds.). Princeton, NJ: Princeton University Press.

Alqahtani, N., S. Porwal, E. D. James, J. A. Karty, A. L. Lane, and R. Viswanathan. 2015. Synergism between genome sequencing, tandem mass spectrometry and bio-inspired synthesis reveals insights into nocardioazine B biogenesis. *Org. Biomol. Chem.* 13(26):7177–7192.

Andras, T. D., T. S. Alexander, A. Gahlena, R. M. Parry, F. M. Fernandez, J. Kubanek, M. D. Wang, and M. E. Hay. 2012. Seaweed allelopathy against coral: Surface distribution of a seaweed secondary metabolite by imaging mass spectrometry. *J. Chem. Ecol.* 38(10):1203–1214.

Angel, P. M., and R. M. Caprioli. 2013. Matrix-assisted laser desorption ionization imaging mass spectrometry: In situ molecular mapping. *Biochemistry* 52(22):3818–3828.

Bernier, S. P., and M. G. Surette. 2013. Concentration-dependent activity in natural environments. *Front. Microbiol.* 4:1–14.

Bode, H. B., B. Bethe, R. Hofs, and A. Zeeck. 2002. Big effects from small changes: Possible ways to explore nature's chemical diversity. *ChemBioChem* 3(7):619–627.

Bugni, T. S., and C. M. Ireland. 2004. Marine-derived fungi: A chemically and biologically diverse group of microorganisms. *Nat. Prod. Rep.* 21(1):143.

Carlson, E. E. 2010. Natural products as chemical probes. *ACS Chem. Biol.* 5(7):639–653.

Carlson, S., U. Tanouye, S. Omarsdottir, and B. T. Murphy. 2015. Phylum-specific regulation of resistomycin production in a *Streptomyces* sp. via microbial coculture. *J. Nat. Prod.* 78(3):381–387.

Chagas, F. O., L. G. Dias, and M. T. Pupo. 2013. A mixed culture of endophytic fungi increases production of antifungal polyketides. *J. Chem. Ecol.* 39(10):1335–1342.

Chen, Y., M. Yin, G. P. Horsman, and B. Shen. 2011. Improvement of the enediyne antitumor antibiotic C-1027 production by manipulating its biosynthetic pathway regulation in *Streptomyces globisporus*. *J. Nat. Prod.* 74(3):420–424.

Cho, J. Y., and M. S. Kim. 2012. Induction of antifouling diterpene production by *Streptomyces cinnabarinus* PK209 in co-culture with marine-derived *Alteromonas* sp. KNS-16. *Biosci. Biotechnol. Biochem.* 76(10):1849–1854.

Cimermancic, P., M. H. Medema, J. Claesen, K. Kurita, L. C. Wieland Brown, K. Mavrommatis, A. Pati et al. 2014. Insights into secondary metabolism from a global analysis of prokaryotic biosynthetic gene clusters. *Cell* 158:412–421.

Clardy, J., M. A. Fischbach, and C. T. Walsh. 2006. New antibiotics from bacterial natural products. *Nat. Biotechnol.* 24:1541–1550.

Cooks, R. G., Z. Ouyang, Z. Takats, and J. M. Wiseman. 2006. Ambient mass spectrometry. *Science* 311(5767):1566–1570.

Cornett, D. S., M. L. Reyzer, P. Chaurand, and R. M. Caprioli. 2007. MALDI imaging mass spectrometry: Molecular snapshots of biochemical systems. *Nat. Methods* 4:828–833.

Corre, C., and G. L. Challis. 2009. New natural product biosynthetic chemistry discovered by genome mining. *Nat. Prod. Rep.* 26(8):977–986.

Craney, A., S. Ahmed, and J. Nodwell. 2013. Towards a new science of secondary metabolism. *J. Antibiot.* 66(7):387–400.

Crews, P., J. Rodriguez, and M. Jaspars. 1998. *Organic Structure Analysis*. Oxford University Press.

Dashti, Y., T. Grkovic, U. Abdelmohsen, U. Hentschel, and R. Quinn. 2014. Production of induced secondary metabolites by a co-culture of sponge-associated actinomycetes, *Actinokineospora* sp. EG49 and *Nocardiopsis* sp. RV163. *Mar. Drugs* 12(5):3046–3059.

Fischbach, M. A., and C. T. Walsh. 2009. Antibiotics for emerging pathogens. *Science* 325(5944):1089–1093.

Fleischmann, R. D., M. D. Adams, O. White, R. A. Clayton, E. F. Kirkness, A. R. Kerlavage, C. J. Bult et al. 1995. Whole-genome random sequencing and assembly of *Haemophilus-influenzae* Rd. *Science* 269(5223):496–512.

Galagan, J. E., S. E. Calvo, C. Cuomo, L. J. Ma, J. R. Wortman, S. Batzoglou, S. I. Lee et al. 2005. Sequencing of *Aspergillus nidulans* and comparative analysis with *A. fumigatus* and *A. oryzae*. *Nature* 438(7071):1105–1115.

Galet, J., A. Deveau, L. Hôtel, P. Leblond, P. Frey-Klett, and B. Aigle. 2014. Gluconic acid-producing *Pseudomonas* sp. prevent γ-actinorhodin biosynthesis by *Streptomyces coelicolor* A3(2). *Arch. Microbiol.* 196(9):619–627.

Gaudencio, S. P., and F. Pereira. 2015. Dereplication: Racing to speed up the natural products discovery process. *Nat. Prod. Rep.* 32(6):779–810.

Gomes, P. B., M. Nett, H. M. Dahse, I. Sattler, K. Martin, and C. Hertweck. 2010. Bezerramycins A–C, antiproliferative phenoxazinones from *Streptomyces griseus* featuring carboxy, carboxamide or nitrile substituents. *Eur. J. Org. Chem.* 2010(2):231–235.

Gontang, E. A., S. P. Gaudencio, W. Fenical, and P. R. Jensen. 2010. Sequence-based analysis of secondary metabolite biosynthesis in marine actinobacteria. *Appl. Environ. Microbiol.* 76(8):2487–2499.

Gross, H. 2009. Genomic mining—A concept for the discovery of new bioactive natural products. *Curr. Opin. Drug Discov. Devel.* 12(2):207–219.

Gross, H., V. O. Stockwell, M. D. Henkels, B. Nowak-Thompson, J. E. Loper, and W. H. Gerwick. 2007. The genomisotopic approach: A systematic method to isolate products of orphan biosynthetic gene clusters. *Chem. Biol.* 14(1):53–63.

Gunatilaka, A. A. L. 2006. Natural products from plant-associated microorganisms: Distribution, structural diversity, bioactivity, and implications of their occurrence. *J. Nat. Prod.* 69(3):509–526.

Guo, Y. C., S. X. Cao, X. K. Zong, X. C. Liao, and Y. F. Zhao. 2009. ESI-MSn study on the fragmentation of protonated cyclic-dipeptides. *Spectrosc. Int. J.* 23(3–4):131–139.

Gust, B., G. L. Challis, K. Fowler, T. Kieser, and K. F. Chater. 2003. PCR-targeted *Streptomyces* gene replacement identifies a protein domain needed for biosynthesis of the sesquiterpene soil odor geosmin. *Proc. Natl. Acad. Sci. U. S. A.* 100(4):1541–1546.

Hider, R. C., and X. Kong. 2010. Chemistry and biology of siderophores. *Nat. Prod. Rep.* 27:637–657.

Hoshino, S., T. Wakimoto, H. Onaka, and I. Abe. 2015. Chojalactones A–C, cytotoxic butanolides isolated from *Streptomyces* sp. cultivated with mycolic acid containing bacterium. *Org. Lett.* 17(6):1501–1504.

Hsiao, N. H., S. Nakayama, M. E. Merlo, M. de Vries, R. Bunet, S. Kitani, T. Nihira, and E. Takano. 2009. Analysis of two additional signaling molecules in *Streptomyces coelicolor* and the development of a butyrolactone-specific reporter system. *Chem. Biol.* 16(9):951–960.

Huang, S., W. Ding, C. Li, and D. G. Cox. 2014. Two new cyclopeptides from the co-culture broth of two marine mangrove fungi and their antifungal activity. *Pharmacogn. Mag.* 10(40):410–414.

Hughes, C. C., and W. Fenical. 2010. Antibacterials from the Sea. *Chem. Eur. J.* 16(42):12512–12525.

Igarashi, Y., Y. Kan, K. Fujii, T. Fujita, K. I. Harada, H. Naoki, H. Tabata, H. Onaka, and T. Furumai. 2001. Goadsporin, a chemical substance which promotes secondary metabolism and morphogenesis in streptomycetes. II. Structure determination. *J. Antibiot.* 54(12):1045–1053.

Jensen, P. R., B. S. Moore, and W. Fenical. 2015. The marine actinomycete genus *Salinispora*: A model organism for secondary metabolite discovery. *Nat. Prod. Rep.* 32(5):738–751.

Kaeberlein, T., K. Lewis, and S. S. Epstein. 2002. Isolating "uncultivable" microorganisms in pure culture in a simulated natural environment. *Science* 296:1127–1129.

Kalivoda, E. J., N. A. Stella, M. A. Aston, J. E. Fender, P. P. Thompson, R. P. Kowalski, and R. M. Q. Shanks. 2010. Cyclic AMP negatively regulates prodigiosin production by *Serratia marcescens*. *Res. Microbiol.* 161(2):158–167.

Kang, M. J., B. D. Jones, A. L. Mandel, J. C. Hammons, A. G. DiPasquale, A. L. Rheingold, J. J. La Clair, and M. D. Burkart. 2009. Isolation, structure elucidation, and antitumor activity of spirohexenolides A and B. *J. Org. Chem.* 74(23):9054–9061.

Karray, F., E. Darbon, H. C. Nguyen, J. Gagnat, and J. L. Pernodet. 2010. Regulation of the biosynthesis of the macrolide antibiotics spiramycin in *Streptomyces ambofaciens*. *J. Bacteriol.* 192(21):5813–5821.

Kieler, J. B., K. L. Duong, W. S. Moye-Rowley, and J. S. Klutts. 2013. Targeted gene deletion in *Aspergillus fumigatus* using microbial machinery and a recyclable marker. *J. Microbiol. Methods* 95(3):373–378.

Konig, C. C., K. Scherlach, V. Schroeckh, F. Horn, S. Nietzsche, A. A. Brakhage, and C. Hertweck. 2013. Bacterium induces cryptic meroterpenoid pathway in the pathogenic fungus *Aspergillus fumigatus*. *ChemBioChem* 14(8):938–942.

Lane, A. L., P. E. Mandelare, and Y. H. Ban. 2015. New developments in NMR methodologies with special roles in drug discovery. In *Applications of NMR Spectroscopy*, A. U. Rahman and I. Choudhary (Eds.). Bentham Science.

Lane, A. L., L. Mular, E. J. Drenkard, T. L. Shearer, S. Engel, S. Fredericq, C. R. Fairchild et al. 2010. Ecological leads for natural product discovery: Novel sesquiterpene hydroquinones from the red macroalga *Peyssonnelia* sp. *Tetrahedron* 66:455–461.

Lane, A. L., L. Nyadong, A. S. Galhena, T. L. Shearer, E. P. Stout, R. M. Parry, M. Kwasnik et al. 2009. Desorption electrospray ionization mass spectrometry reveals surface-mediated antifungal chemical defense of a tropical seaweed. *Proc. Natl. Acad. Sci.* 106(18):7314–7319.

Li, C. Y., W. J. Ding, C. L. Shao, Z. G. She, Y. C. Lin. 2010. A new diimide derivative from the co-culture broth of two mangrove fungi (strain no. E33 and K38). *J. Asian Nat. Prod. Res.* 12:809–813.

Li, C., J. Zhang, C. Shao, W. Ding, Z. She, and Y. Lin. 2011. A new xanthone derivative from the co-culture broth of two marine fungi (strain No. E33 and K38). *Chem. Nat. Compd.* 47(3):383–384.

Linares, J. F., I. Gustafsson, F. Baquero, and J. L. Martinez. 2006. Antibiotics as intermicrobial signaling agents instead of weapons. *Proc. Natl. Acad. Sci. U. S. A.* 103(51):19484–19489.

Ling, L. L., T. Schneider, A. J. Peoples, A. L. Spoering, I. Engels, B. P. Conlon, A. Mueller et al. 2015. A new antibiotic kills pathogens without detectable resistance. *Nature* 1–19.

Martin, J. F., and P. Liras. 2010. Engineering of regulatory cascades and networks controlling antibiotic biosynthesis in *Streptomyces*. *Curr. Opin. Microbiol.* 13(3):263–273.

McCraig, A. E., S. J. Grayston, J. I. Prosser, and L. A. Glover. 2001. Impact of cultivation on characterization of species composition of soil bacterial communities. *FEMS Microbiol Ecol.* 35:37–48.

Medema, M. H., K. Blin, P. Cimermancic, V. de Jager, P. Zakrzewski, M. A. Fischbach, T. Weber, E. Takano, and R. Breitling. 2011. antiSMASH: Rapid identification, annotation and analysis of secondary metabolite biosynthesis gene clusters in bacterial and fungal genome sequences. *Nucleic Acids Res.* 39(Web Server):W339–W346.

Moran, M. A. 2009. Metatranscriptomics: Eavesdropping on complex microbial communities. *Microbe* 4:329–335.

Nichols, D., N. Cahoon, E. M. Trakhtenberg, L. Pham, A. Mehta, A. Belanger, T. Kanigan, K. Lewis, and S. S. Epstein. 2010. Use of ichip for high-throughput in situ cultivation of "uncultivable" microbial species. *Appl. Environ. Microbiol.* 76(8):2445–2450.

Nicolaou, K. C. 2005. Joys of molecules. 1. Campaigns in total synthesis. *J. Org. Chem.* 70:7007–7027.

Nützmann, H. W., Y. Reyes-Dominguez, K. Scherlach, V. Schroeckh, F. Horn, A. Gacek, J. Schümann, C. Hertweck, J. Strauss, and A. A. Brakhage. 2011. Bacteria-induced natural product formation in the fungus *Aspergillus nidulans* requires Saga/Ada-mediated histone acetylation. *Proc. Natl. Acad. Sci.* 108(34):14282–14287.

O'Brien, J., and G. D. Wright. 2011. An ecological perspective of microbial secondary metabolism. *Curr. Opin. Biotechnol.* 22(4):552–558.

Oh, D. C., P. R. Jensen, C. A. Kauffman, and W. Fenical. 2005. Libertellenones A–D: Induction of cytotoxic diterpenoid biosynthesis by marine microbial competition. *Bioorganic Med. Chem.* 13(17):5267–5273.

Oh, D. C., C. A. Kauffman, P. R. Jensen, and W. Fenical. 2007. Induced production of emericellamides A and B from the marine-derived fungus *Emericella* sp. in competing coculture. *J. Nat. Prod.* 70(4):515–520.

Ola, A. R., D. Thomy, D. Lai, H. Brotz-Oesterhelt, and P. Proksch. 2013. Inducing secondary metabolite production by the endophytic fungus *Fusarium tricinctum* through coculture with *Bacillus subtilis*. *J. Nat. Prod.* 76(11):2094–2099.

Onaka, H., Y. Mori, Y. Igarashi, and T. Furumai. 2011. Mycolic acid-containing bacteria induce natural-product biosynthesis in *Streptomyces* species. *Appl. Environ. Microbiol.* 77(2):400–406.

Park, H. B., Y. J. Kim, J. S. Park, H. O. Yang, K. R. Lee, and H. C. Kwon. 2011. Glionitrin B, a cancer invasion inhibitory diketopiperazine produced by microbial coculture. *J. Nat. Prod.* 74(10):2309–2312.

Park, H. B., H. C. Kwon, C. H. Lee, and H. O. Yang. 2009. Glionitrin A, an antibiotic-antitumor metabolite derived from competitive interaction between abandoned mine microbes. *J. Nat. Prod.* 72:248–252.

Penn, K., J. Wang, S. C. Fernando, and J. R. Thompson. 2014. Secondary metabolite gene expression and interplay of bacterial functions in a tropical freshwater cyanobacterial bloom. *ISME J.* 8(9):1866–1878.

Pettit, R. K. 2010. Small-molecule elicitation of microbial secondary metabolites. *Microbial Biotechnol.* 4(4):471–478.

Pettit, R. K., G. R. Pettit, J. P. Xu, C. A. Weber, and L. A. Richert. 2010. Isolation of human cancer cell growth inhibitory, antimicrobial lateritin from a mixed fungal culture. *Planta Medica* 76(5):500–501.

Rateb, M. E., I. Hallyburton, W. E. Houssen, A. T. Bull, M. Goodfellow, R. Santhanam, M. Jaspars, and R. Ebel. 2013. Induction of diverse secondary metabolites in *Aspergillus fumigatus* by microbial co-culture. *RSC Adv.* 3:14444.

Rateb, M. E., and R. Ebel. 2011. Secondary metabolites of fungi from marine habitats. *Nat. Prod. Rep.* 28(2):290–344.

Rigali, S., F. Titgemeyer, S. Barends, S. Mulder, A. W. Thomae, D. A. Hopwood, and G. P. van Wezel. 2008. Feast or famine: The global regulator DasR links nutrient stress to antibiotic production by *Streptomyces*. *EMBO Rep.* 9(7):670–675.

Sanchez, J. F., A. D. Somoza, N. P. Keller, and C. C. C. Wang. 2012. Advances in *Aspergillus* secondary metabolite research in the post-genomic era. *Nat. Prod. Rep.* 29:351–571.

Sarkar, A., A. N. Funk, K. Scherlach, F. Horn, V. Schroeckn, P. Chankhamjon, M. Westermann et al. 2012. Differential expression of silent polyketide biosynthesis gene clusters in chemostat cutlures of *Aspergillus nidulans*. *J. Biotechnol.* 160:64–71.

Schroeckh, V., K. Scherlach, H. W. Nutzmann, E. Shelest, W. Schmidt-Heck, J. Schuemann, K. Martin, C. Hertweck, and A. A. Brakhage. 2009. Intimate bacterial-fungal interaction triggers biosynthesis of archetypal polyketides in *Aspergillus nidulans*. *Proc. Natl. Acad. Sci. U. S. A.* 106(34):14558–14563.

Seyedsayamdost, M. R. 2014. High-throughput platform for the discovery of elicitors of silent bacterial gene clusters. *Proc. Natl. Acad. Sci.* 111(20):7266–7271.

Sidebottom, A. M., A. R. Johnson, J. A. Karty, D. J. Trader, and E. E. Carlson. 2013. Integrated metabolomics approach facilitates discovery of an unpredicted natural product suite from *Streptomyces coelicolor* M145. *ACS Chem. Biol.* 8(9):2009–2016.

Stone, M. J., and D. H. Williams. 1992. On the evolution of functional secondary metabolites. *Mol. Microbiol.* 6(1):29–34.

Su, C., L. Lei, Y. Duan, K. Q. Zhang, and J. Yang. 2012. Culture-independent methods for studying environmental microorganisms: Methods, application, and perspective. *Appl. Microbiol. Biotechnol.* 93(3):993–1003.

Takano, E. 2006. γ-Butyrolactones: *Streptomyces* signalling molecules regulating antibiotic production and differentiation. *Curr. Opin. Microbiol.* 9(3):287–294.

Tanaka, Y., T. Hosaka, and K. Ochi. 2010. Rare earth elements activate the secondary metabolic biosynthetic gene clusters in *Streptomyces coelicolor* A3(2). *J. Antibiot.* 63(8):477–481.

Traxler, M. F., J. D. Watrous, T. Alexandrov, P. C. Dorrestein, and R. Kolter. 2013. Interspecies interactions stimulate diversification of the *Streptomyces coelicolor* secreted metabolome. *mBio* 4 (4):e00459-13-e00459-13.

Udwary, D. W., L. Zeigler, R. N. Asolkar, V. Singan, A. Lapidus, W. Fenical, P. R. Jensen, and B. S. Moore. 2007. Genome sequencing reveals complex secondary metabolome in the marine actinomycete *Salinispora tropica*. *Proc. Natl. Acad. Sci.* 104(25):10376–10381.

Van Lanen, S. G., and B. Shen. 2006. Microbial genomics for the improvement of natural product discovery. *Curr. Opin. Microbiol.* 9(3):252–260.

von Doehren, H. 2009. A survey of nonribosomal peptide synthetase (NRPS) genes in *Aspergillus nidulans*. *Fungal Genet. Biol.* 46:S45–S52.

Wang, J., W. Ding, C. Li, S. Huang, Z. She, and Y. Lin. 2013. A new polysubstituted benzaldehyde from the co-culture broth of two marine fungi (Strains Nos. E33 and K38). *Chem. Nat. Compd.* 49(5):799–802.

Wietz, M., K. Duncan, N. V. Patin, and P. R. Jensen. 2013. Antagonistic interactions mediated by marine bacteria: The role of small molecules. *J. Chem. Ecol.* 39:879–891.

Williams, D. H., M. J. Stone, P. R. Hauck, and S. K. Rahman. 1989. Why are secondary metabolites (natural products) biosynthesized? *J. Nat. Prod.* 52 (6):1189–1208.

Yaegashi, J., B. R. Oakley, and C. C. C. Wang. 2014. Recent advances in genome mining of secondary metabolite biosynthetic gene cluster and the development of heterologous expression systems in *Aspergillus nidulans*. *J. Ind. Microbiol. Biotechnol.* 41:433–442.

Yamanaka, K., K. A. Reynolds, R. D. Kersten, K. S. Ryan, D. J. Gonzalez, V. Nizet, P. C. Dorrestein, and B. S. Moore. 2014. Direct cloning and refactoring of a silent lipopeptide biosynthetic gene cluster yields the antibiotic taromycin A. *Proc. Natl. Acad. Sci.* 111(5):1957–1962.

Yang, Y. L., Y. Xu, P. Straight, and P. C. Dorrestein. 2009. Translating metabolic exchange with imaging mass spectrometry. *Nat. Chem. Biol.* 1–3.

Yoon, V., and J. R. Nodwell. 2014. Activating secondary metabolism with stress and chemicals. *J. Ind. Microbiol. Biotechnol.* 41(2):415–424.

Zhang, Y., L. Xuegong, D. H. Bartlett, and X. Xiao. 2015. Current developments in marine microbiology: High-pressure biotechnology and the genetic engineering of piezophiles. *Curr. Opin. Biotechnol.* 33:157–164.

Zhu, F., G. Chen, X. Chen, M. Huang, and X. Wan. 2011. Aspergicin, a new antibacterial alkaloid produced by mixed fermentation of two marine-derived mangrove epiphytic fungi. *Chem. Nat. Compd.* 47(5):767–769.

Zhu, F., G. Y. Chen, J. S. Wu, and J. H. Pan. 2014. Structure revision and cytotoxic activity of marinamide and its methyl ester, novel alkaloids produced by co-cultures of two marine-derived mangrove endophytic fungi. *Nat. Prod. Res.* 27(21):1960–1964.

Zhu, F., and Y. Lin. 2006. Marinamide, a novel alkaloid and its methyl ester produced by the application of mixed fermentation technique to two mangrove endophytic fungi from the South China Sea. *Chinese Sci. Bull.* 51(12):1426–1430.

Ziemert, N., A. Lechner, M. Wietz, N. Millan-Aguinaga, K. L. Chavarria, and P. R. Jensen. 2014. Diversity and evolution of secondary metabolism in the marine actinomycete genus *Salinispora. Proc. Natl. Acad. Sci.* 111(12):E1130–E1139.

Zotchev, S. B., O. N. Sekurova, and L. Katz. 2012. Genome-based bioprospecting of microbes for new therapeutics. *Curr. Opin. Biotechnol.* 23(6):941–947.

Zuck, K. M., S. Shipley, and D. J. Newman. 2011. Isolation of N-formyl alkaloids from *Aspergillus fumigatus* by co-culture with *Streptomyces peucetius. J. Nat. Prod.* 74:1653–1657.

9 Predator–Prey Interactions in the Marine Plankton
The Role of Signals, Cues, and Defensive Compounds

Emily K. Prince and Kelsey L. Poulson-Ellestad

CONTENTS

9.1   INTRODUCTION

Many of earth's biogeochemical processes are driven by phytoplankton: humble, single-celled microbes with a crucial role in nutrient cycling (Falkowski 2012; Falkowski et al. 2008). However, understanding the large-scale processes driven by phytoplankton involves understanding the ecological interactions that influence phytoplankton abundance and community composition (Strom 2008). Chemical signals and cues are the primary mechanism for communication in marine planktonic systems. Indeed, phytoplankton use chemical compounds to detect predators (Selander et al. 2015), defend against predation (Selander et al. 2006), find mates (Gillard et al. 2013), and compete for resources (Legrand et al. 2003). Zooplankton, consumers of phytoplankton, also rely on chemical cues as mechanisms for detecting and recognizing prey (Roberts et al. 2011), immobilizing prey (Sheng et al. 2010), detecting hosts (Mordue et al. 2009), finding mates (Heuschele and Kiorboe 2012), and avoiding predators (Cohen and Forward 2005). Because microalgae are responsible for approximately 50% of global carbon fixation (Field et al. 1998) and zooplankton are the primary link between these primary producers and higher trophic levels (Calbet and Landry 2004; Turner 2004), understanding how compounds mediate planktonic predator–prey interactions is crucial.

Recent advances in chemical ecology have greatly enhanced our understanding of chemical communication in the sea. Improved chemical techniques have allowed for more efficient identification of compounds, propelling the field from a "black box" approach, where ecological effects were assessed by crude extracts and exudates, to a more mechanistic approach, where the roles of specific compounds and the factors that influence their production can now be assessed (e.g., Van Wagoner et al. 2008; Pohnert, 2012; Gillard et al. 2013). In addition, molecular tools have allowed chemical ecologists to determine the molecular underpinnings of chemical communication in order to understand how these compounds affect cellular (e.g., Lauritano et al. 2011; Yang et al. 2011; Poulson-Ellestad et al. 2014); community (e.g., Bricelj et al. 2005; Vardi et al. 2008); and, ultimately, ecosystem processes (e.g., Falkowski 2012).

Here, we consider the chemical ecology of plankton predator–prey interactions in marine systems, focusing on phytoplankton and their micro- and mesozooplankton predators. We investigate the production of compounds that are beneficial to the producer (e.g., defenses) as well as compounds that benefit the organism detecting the molecule rather than the producer (e.g., cues). Our review considers interactions both from the perspective of the prey species and the zooplankton predator species. However, in planktonic systems, lines between competition and predation can be blurred, and mixotrophic algae can be predators, prey, or competitors to other protists depending on the circumstances (Flynn et al. 2012), so we place interactions in a community and ecosystem context. Phytoplankton prey use cues to perceive predation risk (e.g., Selander et al. 2015) and can respond with chemical, morphological, or behavioral defenses (reviewed in Van Donk et al. 2011). Predators can also use cues to detect prey (e.g., Breckels et al. 2011), and, in some cases, these predators produce their own toxins to immobilize prey species (e.g., Sheng et al. 2010). We consider the implications of these interactions in structuring communities and ecosystems as well as driving evolution. For instance, phytoplankton and zooplankton face selective pressures to produce fewer cues indicating their presence, to evolve more efficient defenses or weapons, and to evolve resistance to defenses or predatory toxins. Therefore, chemical signaling between predators and prey not only have implications for individual organisms but also can change the structure of communities and ecosystems.

Our goal in this chapter is not to provide a comprehensive review of the current state of the field. Instead, we focus on several important systems and the ecologically relevant compounds that shape them. We have chosen a series of case studies of chemically mediated predator–prey interactions that illustrate the ecosystem-wide and evolutionary implications of chemical ecology in the plankton. We focus on harmful algal bloom (HAB)-forming algae, including *Karenia brevis*, *Karlodinium* spp., and *Alexandrium* spp., as these dinoflagellates produce compounds with ecosystem-wide effects. We delve more deeply into the chemical ecology of *Alexandrium* spp. because it has been a model organism for understanding the evolutionary pressures associated with secondary metabolite production in the plankton. We also hope to shed light on some of the more contradictory and controversial findings in planktonic predator–prey interactions. For example, we explore the importance of diatom-produced polyunsaturated aldehydes (PUAs), both for the producing organisms and because of their importance in global nutrient cycling. We also consider how dimethysulfionypropionate (DMSP) and related compounds are used in planktonic systems, addressing how these compounds influence higher trophic levels and even global climate. Finally, in this chapter we hope to draw attention to fruitful areas of future research, emphasizing the questions and approaches that will advance the field.

9.2 THE DEFENSIVE AND OFFENSIVE CHEMICAL STRATEGIES OF HARMFUL ALGAL BLOOMS

Between 60 and 80 species of phytoplankton produce HABs; among these bloom-forming species, flagellates (mostly dinoflagellates) make up the vast majority (Smayda 1997). HAB species produce a number of toxins with diverse structures, including polyketides, alkaloids, non-ribosomal peptides, and isoprenoids (Figure 9.1) (Hay and Kubanek 2002). These primarily neurotoxic compounds can have dramatic effects on marine ecosystems, causing adverse effects on other phytoplankton (Legrand et al. 2003), zooplankton (Turner 2014), marine invertebrates, fish, marine mammals, seabirds (Landsberg 2002), and even human health (Van Dolah 2000). Paradoxically, although a number of HAB toxins have been well characterized, their physiological and ecological roles continue to be debated. HAB species, including *Karenia brevis* (e.g., Waggett et al. 2012), *Prymnesium parvum* (e.g., Sopanen et al. 2008),

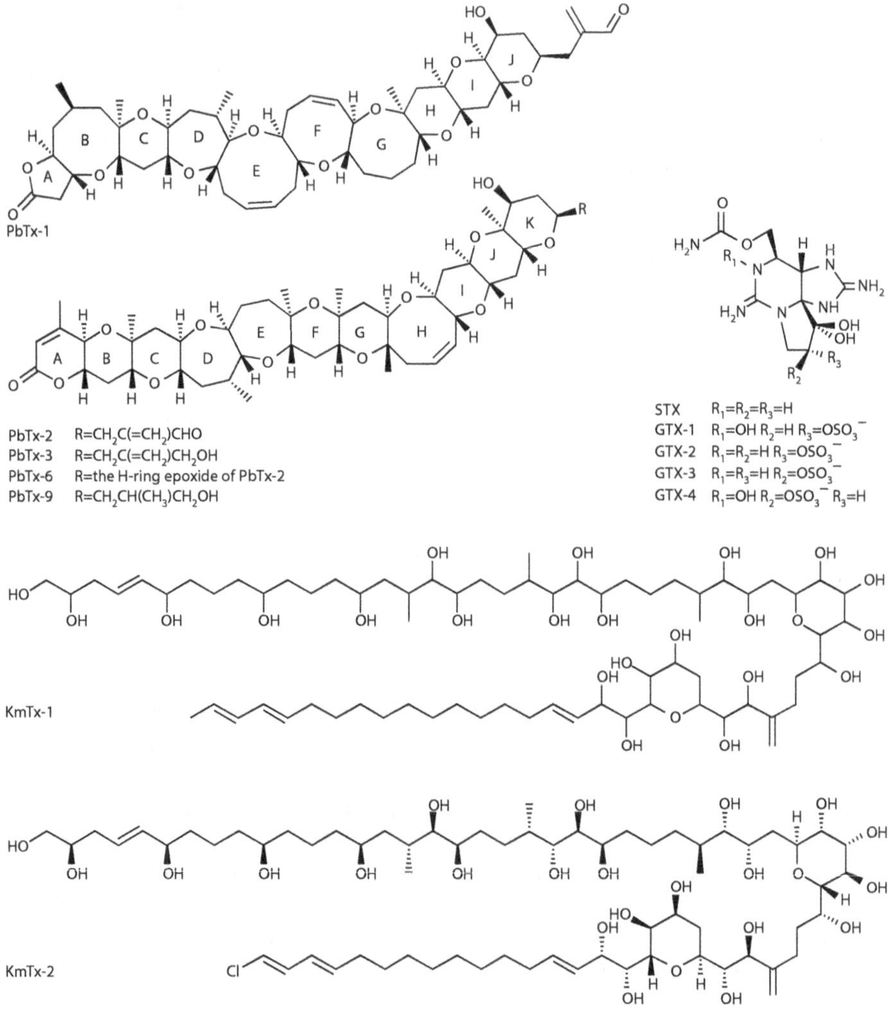

FIGURE 9.1 Structures of toxins produced by harmful algal bloom-forming dinoflagellates, including brevetoxins (PbTx), which are produced by *Karenia brevis*; saxitoxin (STX) and the related gonyautoxins (GTX), which are produced by members of the genus *Alexandrium*; and karlotoxins (KmTx), produced by *Karlodinium* spp.

Karlodinium venificum (e.g., Adolf et al. 2007), *Heterosigma akashiwo* (e.g., Clough and Strom 2005), *Pseudo-nitzschia* spp. (e.g., Bargu et al. 2006), *Dinophysis* spp. (e.g., Kozlowsky-Suzuki et al. 2006), and others appear defended from copepod and protist grazing; however, the particular identities of the defensive compounds are often unknown. In many cases, the presence of chemical defenses are deduced by observing reduced grazing rates or harmful effects on grazers, including reduced fecundity (Barreiro et al. 2006; Turner et al. 2012), cell lysis (Tillmann and John 2002), and death (Zou et al. 2010), but these defenses may or may not be the same as the infamous toxins isolated from many HAB species. In some studies, the production of known neurotoxins has been correlated with reduced grazing, suggesting that an ecological role for these toxins includes defense from predation (e.g., Waggett et al. 2012). In several cases, however, the relationship is not as clear. Studies may conflate the effects of nutritional inadequacy with toxicity and the impacts of noxious (i.e., non-toxic) defenses with toxins. Nonetheless, particular HAB compounds that have been implicated as grazer defenses include brevetoxins (Waggett et al. 2012), okadaic acid (Kozlowsky-Suzuki et al. 2006), domoic acid (Bargu et al. 2006), saxitoxins (STX) (Selander et al. 2006), and karlotoxins (Adolf et al. 2007), as well as several unidentified compounds (e.g., Lewitus et al. 2006; Zou et al. 2010; Tang and Gobler 2012; John et al. 2015). In the next section, we discuss the roles of algal natural products in the bloom ecology of three HAB-forming dinoflagellates: *Karenia brevis*, *Karlodinium* spp., and *Alexandrium* spp.

9.2.1 KARENIA BREVIS

Efforts to understand the ecology of polyketide neurotoxins, brevetoxins (Figure 9.1) produced by the red tide dinoflagellate, *Karenia brevis*, illustrate the difficulty in determining the ecological roles of HAB toxins. Although brevetoxins are potent agonists of vertebrate sodium channels, these compounds may not have detrimental effects to invertebrate grazers. For instance, several studies (e.g., Prince et al. 2006; Breier and Buskey 2007) have suggested that *K. brevis* is nutritionally inadequate but not toxic to copepods, while other studies (e.g., Kubanek et al. 2007; Turner et al. 2012) have indicated that *K. brevis* is chemically defended from zooplankton, but compounds other than brevetoxins are responsible. Still other studies report that brevetoxin-producing strains are more toxic to grazers than strains that do not produce brevetoxins (Hong et al. 2012; Waggett et al. 2012), which suggests that in some cases brevetoxins themselves can adversely affect zooplankton. However, of the studies cited here, only Kubanek et al. (2007) tested purified brevetoxins directly, and these authors observed no effect of brevetoxins on rotifer grazing. However, purified brevetoxins may cause sublethal behavioral changes in copepods, with some copepod species affected more than others (Cohen et al. 2007). *Karenia brevis* is known to produce other, partially-characterized bioactive compounds, which inhibit the growth of competing phytoplankton (Prince et al. 2010), raising the possibility that the production of brevetoxins may correlate with the production of other bioactive molecules, or brevetoxins may act additively or synergistically with these molecules to impact grazers. Bioassays, which can directly test the effect of ingested brevetoxins, or brevetoxin mixtures, on copepods, must be developed if these questions are ever to be answered definitively.

Karenia brevis blooms occur seasonally in the Gulf of Mexico, impacting the entire coastal community. For example, *K. brevis* blooms alter the composition of the phytoplankton community (West et al. 1996), although this may be through allelopathic compounds other than brevetoxins (Prince et al. 2010; Poulson-Ellestad et al. 2014). The effects of *K. brevis* on higher trophic levels, however, are primarily due to brevetoxins (Landsberg 2002). *K. brevis* blooms often result in massive fish kills, both because copepods act as vectors of the toxins (Tester et al. 2000) and because fish may absorb waterborne brevetoxins directly across the gills (Landsberg 2002). Marine mammals, such as manatees and dolphins, have also died from brevetoxin poisoning (Flewelling et al. 2005). Breathing aerosolized brevetoxins results in respiratory distress in humans (Hoagland et al. 2014). Additionally, people consuming bivalves that have accumulated brevetoxins can suffer from neurotoxic shellfish poisoning (NSP), characterized by gastrointestinal and neurological symptoms, which often require hospitalization (Watkins et al. 2008).

9.2.2 *KARLODINIUM* SPP.

The HAB forming mixotrophic dinoflagellates in the genus *Karlodinium* produce several polyketide toxins called karlotoxins. These toxins are reportedly used as a defense against predation (Adolf et al. 2007) and as an allelopathic weapon against competitors (Adolf et al. 2006). However, the primary role of karlotoxins may be in prey capture. Sheng et al. (2010) showed that karlotoxins, as well as karlotoxin-producing *Karlodinium veneficum*, slowed down or immobilized the algal prey species *Storeatula major.* No effects on *S. major* were observed for a non-toxic (and non-predatory) strain of *K. veneficum.* Once *S. major* cells were immobilized, *K. veneficum* captured and consumed the *S. major* cells (Sheng et al. 2010). *K. veneficum* exemplifies the blurry line between predator–prey interactions and competitive interactions in planktonic systems, as this dinoflagellate deploys karlotoxins as weapons in multiple interactions.

Chemically mediated predation may be a widespread strategy among *Karlodinium* spp. In fact, these dinoflagellates are not limited to consuming unicellular algae. *Karlodinium armiger* does not produce karlotoxins but has been reported to produce unidentified compounds with neurotoxic activity (Garcés et al. 2006). At low cell concentrations (below 1000 cells/mL), *K. armiger* is ingested by the copepod *Acartia tonsa* (Berge et al. 2012). However, at higher concentrations, *K. armiger* cells immobilize the copepod and ingest it. Similar effects were detected when *K. armiger* cells were fed other metazoans, including nematodes and polychaete larvae (Berge et al. 2012). Interestingly, strains of *K. veneficum* reported to produce karlotoxins did not prey on copepods, supporting the hypothesis that metazoan predation is not mediated by karlotoxins (Berge et al. 2012). Additionally, *K. armiger* does not exude molecules to immobilize prey into the surrounding media, as filtrates of *K. armiger* have no effect on copepods (Berge et al. 2012). Berge et al. (2012) hypothesized that because *K. armiger* causes erratic movements and the inability to swim in *A. tonsa,* a fast-acting neurotoxin, which could be transferred into the copepod nervous system through the dinoflagellate feeding tube or extrusosomes, may be involved. As of yet no bioactive molecule has been isolated or identified.

In contrast, the structures of karlotoxins have been elucidated (Figure 9.1) (Van Wagoner et al. 2008; Peng et al. 2010), and the mechanism by which they act has been characterized (Deeds et al. 2015). Karlotoxins generate pores in cell membranes containing sterols, although they do not bind to all sterols with equal affinity (Deeds and Place 2006). Thus, karlotoxins have the potential to alter community composition during *K. veneficum* blooms. Karlotoxins do not interact strongly with gymnodinosterol, found in *Karlodinium* spp., or with amphisterol, found in

Alexandrium spp., so it is perhaps unsurprising that karlotoxins did not impact the growth rate of *Karlodinium venificum* or *Alexandrium carterae* (Adolf et al. 2006). However, karlotoxin exposure did result in cell lysis of *Oxyrrhis marina*, a heterotrophic dinoflagellate with membranes dominated by brassicasterol and cholesterol (Deeds and Place 2006). Brassicasterol also dominates the membrane of *K. veneficum's* prey, *S. major* (Adolf et al. 2006). These species-specific effects of karlotoxins impact potential grazers, potential prey, and potential competitors, which could alter the composition of the plankton community during *K. veneficum* blooms. Blooms of *Karlodinium* spp. also have detrimental effects on higher trophic levels and are associated with fish kills (Place et al. 2012; Lim et al. 2014). This is likely because karlotoxins lead to non-selective permeabilization of fish cell membranes, resulting in cell lysis (Deeds et al. 2015). Because not all species of fish are equally affected by *Karlodinium* spp., blooms of this organism may also impact the relative abundance of different fish species (Hallett 2012).

9.2.3 *ALEXANDRIUM* SPP.

One of the best-studied, chemically defended HABs are dinoflagellates in the genus *Alexandrium*. Many *Alexandrium* species produce potent alkaloid neurotoxins, STX, and the related gonyautoxins (GTX), which are collectively known as paralytic shellfish toxins (PSTs) (Figure 9.1). The PSTs produced by *Alexandrium* species have effects on the plankton community as well as higher trophic levels. They also adversely impact human health and economies. PSTs primarily act as a defense against copepod grazing (Selander et al. 2006), and the evolutionary implications of this interaction are explored in Section 9.3. However, smaller zooplankton, including rotifers (Wang et al. 2005) and ciliates (Kamiyama et al. 2005) seem to be unaffected by PSTs, suggesting that blooms of *Alexandrium* spp. may shift size distribution of zooplankton to smaller organisms. This is especially relevant, since grazing by microzooplankton, not copepods, has the potential to control *Alexandrium minutum* blooms (Calbet et al. 2003). Therefore, it would also benefit *Alexandrium* spp. to produce defensive compounds (other than PSTs) that can act against smaller grazers. In fact, numerous studies have demonstrated that *Alexandrium* spp. produce a variety of defensive compounds. For example, strains of *Alexandrium tamarense* are reported to produce reactive oxygen species that cause significant mortality in ciliates and heterotrophic dinoflagellates (Flores et al. 2012). Additionally, strains of *Alexandrium fundyense* that produce an uncharacterized defensive compound were not grazed by a heterotrophic dinoflagellate, whereas strains of *A. fundyense* that do not produce the defensive compound were grazed (John et al. 2015). However, when the strains were cultured together, neither were grazed, indicating that facilitation among the strains is possible, and these defensive molecules may help maintain phenotypic and genotypic diversity in the population (John et al. 2015).

The PSTs produced by *Alexandrium* spp. accumulate in higher trophic levels. Many studies have shown that copepods can accumulate PSTs (reviewed in Turner 2014), but all size classes of the zooplankton community contain PSTs, suggesting that there are multiple vectors by which PSTs reach higher trophic levels (Petitpas et al. 2014). PSTs can have devastating effects in secondary and tertiary consumers. They have been detected in endangered Harbor seals (Jensen et al. 2015) and seabirds (Shearn-Bochsler et al. 2014). They are also found in North Atlantic right whales, a species believed to consist of only 400 individuals (Doucette et al. 2012) and have been responsible for humpback whale fatalities (Geraci et al. 1989). Humans are also affected by PSTs, usually by eating shellfish that have accumulated the toxins. Overall, shellfish poisoning accounts for around 1% of foodborne illness in the United States, and around 6% of people with paralytic shellfish poisoning die (Isbister and Kiernan 2005).

9.3 INDUCIBLE DEFENSES OF *ALEXANDRIUM*: A MODEL SYSTEM FOR CHEMICAL CO-EVOLUTION

In contrast to many HAB species, where chemical defenses are either produced constitutively or the pattern of defense production is uncharacterized, the chemical defenses of *Alexandrium* spp. are the best-studied example of inducible defenses in the marine phytoplankton (illustrated in Figure 9.2). When grazing pressure is variable, it may benefit an organism to produce a defense only when the threat of predation is present (Rhoades 1979). Because inducible chemical, morphological, and behavioral defenses in planktonic systems have been reviewed elsewhere (Van Donk et al. 2011), we focus on how the genus *Alexandrium* has been and should continue to be a model system to understand the ecological and evolutionary pressures influencing the induction of planktonic chemical defenses.

9.3.1 INDUCED DEFENSES IN *ALEXANDRIUM*

As described previously, *Alexandrium* spp. produce a suite of neurotoxins, PSTs, some of which act to defend cells against copepod grazers. Additionally, the production of these compounds is enhanced by the presence of a predator. For instance, exposing *Alexandrium minutum* cells to the copepod *Acartia tonsa*, increased the production of GTXs 5-fold in *A. minutum* cells (Selander et al. 2006, 2015). The degree of induction is specific to the cues produced by particular copepod species.

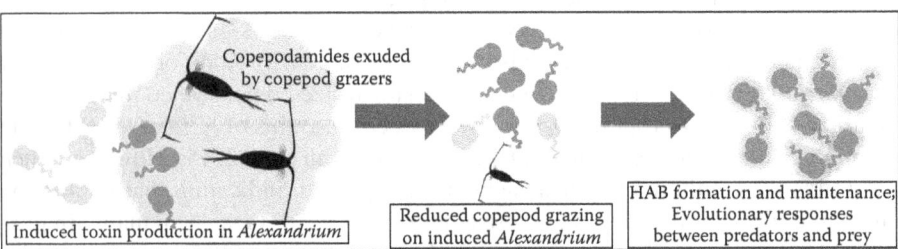

Copepodamides exuded by copepod grazers

Induced toxin production in *Alexandrium*

Reduced copepod grazing on induced *Alexandrium*

HAB formation and maintenance; Evolutionary responses between predators and prey

FIGURE 9.2 Induction of chemical defenses in *Alexandrium* spp. Copepodamides are released from grazing copepods and induce the production of PSTs in non-grazed *Alexandrium* cells. Cells that have enhanced production of PSTs are no longer a preferred food source for copepod grazers, thereby reducing grazing pressure on *Alexandrium* populations and promoting bloom formation and/or maintenance.

As an example, when exposed to chemical cues from grazing *Centropages typicus* copepods, *A. minutum* cells increased GTX production by up to 20 times, which, in turn, reduced grazing on *A. minutum* cells (Figure 9.2) (Bergkvist et al. 2008). Another copepod species, *Pseudocalanus* sp., did not induce the production of GTXs (Bergkvist et al. 2008), indicating species specificity of these predator–prey interactions.

Although almost all studies of induced defenses in *Alexandrium* spp. focus on copepod grazers, the response of *A. fundyense* to grazers from multiple phyla was investigated in a recent study (Senft-Batoh et al. 2015a). The authors found that copepods, ascidians, and mollusks all induce PST production in *A. fundyense*, either through compounds released by the grazer or feeding-related cues, such as damaged *A. fundyense* cells. However, neither of the two protist grazers, a heterotrophic dinoflagellate and a ciliate, caused an increase in PST production in *A. fundyense* (Senft-Batoh 2015a). Not all grazers are adversely affected by PSTs, so the differential induction of PSTs by different phyla may be explained by their evolutionary history with *Alexandrium* spp. (explored in Section 9.3.2.2).

To explore how *Alexandrium* spp. respond to cues from grazers, transcriptional responses of *A. minutum* to *C. typicus* have also been measured, demonstrating that *Alexandrium* cells sense and respond to surrounding predators by increasing toxin production per cell and upregulating the expression of genes involved in protein folding and modification (Yang et al. 2011). Differential expressions of several serine/threonine kinases have also been observed in *A. tamarense* in response to the presence of grazing copepods (Wohlrab et al. 2010), but whether these genes are involved directly in toxin production is unknown. Overall, studies of this system that directly tie changes in gene and protein expression to physiological or ecological roles are still lacking.

9.3.2 EVOLUTIONARY IMPLICATIONS

Predator–prey interactions are likely to exert strong selective pressure on both phytoplankton and zooplankton (Smetacek 2001; Dam 2013). A recent report of convergent evolution in the chemical signals between terrestrial and marine systems suggests that lessons learned in terrestrial systems may be able to be applied in the marine environment (Rasher et al. 2015). In terrestrial systems, the co-evolution of plant–herbivore interactions is theorized to be an important driver of chemical diversity in plants (Ehrlich and Raven 1964; Vermeij 1994). Further, an increase in the number and diversity of plant secondary metabolites in response to herbivory has been demonstrated in some systems (Becerra et al. 2009). Additionally, the use of genetically modified crops producing insecticidal proteins from *Bacillus thuringiensis* has enabled researchers to see that herbivores develop resistance to new plant defenses within a few years of their introduction (Tabashnik et al. 2013). In terrestrial systems, the genetic underpinnings of predator resistance to chemically defended prey have been identified in some cases (Geffeney et al. 2005), but, in the planktonic environment, co-evolution of predators and prey has just begun to be investigated (Bricelj et al. 2005; Dam 2013). Chemical signals are likely to be essential in this process because they are ubiquitous in the planktonic environments, serving as a way to detect both predators and prey, as a defense against predation, and as a tool to capture and ingest prey. Thus, one would expect strong selection for the production of toxins and defensive metabolites and for resistance to those same compounds. Selection should work against the release of compounds that allow prey to be detected by predators and vice versa. Additionally, toxins and defensive metabolites should drive resistance in predators or prey. The importance of evolutionary relationships is clear—placing planktonic predator–prey interactions in an evolutionary context may be key to understanding complex and contradictory reports in these systems.

9.3.2.1 Predator-Evolved Resistance to *Alexandrium* Chemical Defenses

Blooms of *Alexandrium* are globally distributed (Anderson et al. 2012) but have historically been common at northern latitudes on the East Coast of North America, including New Brunswick and Maine. Recently, the range of these dinoflagellates has expanded, with blooms reported as far south as Connecticut and New Jersey (Anderson 1997). Several researchers have taken advantage of the opportunity provided by this expanding range to study the impacts of exposure history on predator–prey interactions, as selection pressure for resistance to PSTs should be highest at northern latitudes and decrease in more southern populations.

The only example of evolved resistance to algal chemical defenses where the mechanism of resistance is known occurs in the clam *Mya arenaria*. *M. arenaria* can be found along the east coast of North America, both in regions where blooms of PST producing dinoflagellates are common and in areas with no records of PST producing dinoflagellates. When fed diets of PST-producing *A. tamarense,* clams from areas facing frequent blooms experienced almost no mortality, whereas naïve clams experienced up to 40% mortality (Bricelj et al. 2005). Because PSTs act by blocking voltage-gated sodium channels (Cestèle and Catterall 2000), Bricelj et al. (2005) investigated how the amino acid sequence in the sodium channels in clams resistant to STX differed from those susceptible to STX. They found that the mutation of a single amino acid from glutamic acid to aspartic acid between the two clam populations provided resistant clams with a sodium channel that was 1,000 times less sensitive to STX than in the susceptible population. Subsequent studies have found that clams with the susceptible genotype were eliminated from the population at higher rates than resistant clams, indicating strong selection for resistant clams in the presence of PST-producing dinoflagellates (Bricelj et al. 2010). Because resistant clams accumulated approximately 10 times as much STX as susceptible clams (Bricelj et al. 2005), the evolution of resistance in clams may lead to increased intoxication and death in the animals that prey on these clams.

Copepods also adapt to toxic *Alexandrium* spp. For example, when populations of the copepod *Acartia hudsonica* from Maine (co-occurring with *Alexandrium*) and from New Jersey (naïve populations) were fed diets including the toxic *Alexandrium fundyense*, both survivorship and egg production of the population from New Jersey declined dramatically, but survivorship and egg production of the population from Maine was unchanged (Colin and Dam 2004). However, resistance to diets containing *Alexandrium* spp. does appear to have a cost. Avery and Dam (2007) identified five distinct *A. hudsonica* phenotypes, finding that the phenotypes producing the highest number of eggs on a standard diet experienced a dramatic reduction in fitness when fed a toxic diet. In contrast, copepods with the highest fitness when fed a toxic diet were slightly less fit (relative to the other phenotypes) when fed the standard diet. In copepods, evolution of resistance to PSP toxins may occur quickly, with naïve populations performing significantly better on toxic diets after only three generations (Colin and Dam 2004). Conversely, resistance to toxic diets may also be lost within a few generations (Avery and Dam 2007). Similar results have been observed for populations of the copepod *Acartia tonsa* grazing on the harmful dinoflagellate *Cochlodinium polykrikoides* (Jiang et al. 2011). The rapid loss of resistance to harmful algae by *Acartia* spp. also suggests that resistance to these compounds is costly.

In contrast to clams, the mechanism by which copepods evolve resistance to PSTs is unknown. The presence of discrete copepod phenotypes suggests the action of a relatively simple genetic mechanism (Avery and Dam 2007). While mutant isoforms of sodium channels have been detected in *A. hudsonica*, the mutations are not in the sodium binding domain of the channel, they do not appear to confer resistance to PSTs, and no geographic differences in sodium channel expression could be detected in *A. hudsonica* (Finiguerra et al. 2014b). Additionally, all *A. hudsonica* individuals that were tested expressed both the wild-type and the mutant sodium channel isoform in varying proportions, but expression of the mutant isoform was not induced by exposure to toxic *A. fundyense* (Finiguerra et al. 2014a). The proportion of mutant isoforms in the population of *A. hudsonica* did not increase, even after exposure to toxic food for four generations. Interestingly, fitness of *A. hudsonica* did increase over four generations with *A. fundyense* in the diet, emphasizing that adaptation occurred, but the mutant sodium channel was not responsible (Finiguerra et al. 2014a).

The evolutionary response of copepods to toxic *Alexandrium* spp. could also have ecosystem-wide consequences, but in quite a different way than the evolutionary responses of clams. Unlike clams, resistant *A. hudsonica* exposed to *Alexandrium* do not accumulate more toxins than susceptible *A. hudsonica*, even though resistant populations ingest significantly more of the toxic dinoflagellate. Although the populations have similar depuration rates for PSTs, toxin retention efficiency was significantly higher for susceptible *A. hudsonica* populations than for resistant populations (Dam and Haley 2011). Importantly, the evolution of resistance in copepods may mitigate the effects of harmful *Alexandrium* blooms, as grazing by resistant copepods is more likely to control blooms than grazing by susceptible copepods (Figure 9.2) (Colin and Dam 2007).

Only a few studies investigate whether zooplankton have evolved resistance to dinoflagellates other than *Alexandrium* spp., including *Karenia brevis*, common in the Gulf of Mexico. Observed patterns of resistance suggest zooplankton populations regularly exposed to *K. brevis* are more resistant to this dinoflagellate. For example, Walsh and O'Neil (2014) found that Gulf of Mexico copepod populations ingested *K. brevis* at higher rates than copepod populations from the Chesapeake Bay, where *K. brevis* blooms do not occur. Similar results have been observed for rotifer grazers. Kubanek et al. (2007) reported that a sympatric rotifer—*Brachionus ibericus*—ingested *K. brevis* as well as *K. brevis* extracts coated on palatable food, but an allopatric rotifer—*Brachionus plicatilis*—did not ingest it until the rotifers were acclimated to their new diets. In contrast, Turner et al. (2012) found that the allopatric copepod *Centropages helgolandicus* was actually less sensitive to diets of *K. brevis* than a co-occurring species (*Temora stylifera*), highlighting that more work is required to understand the influence of exposure history on grazing interactions in the plankton.

Overall, studies suggest that mesozooplankton can evolve resistance to defensive compounds produced by phytoplankton. However, these studies have focused on a very few phytoplankton species and have relied almost exclusively on copepods as model grazers (but see Senft-Batoh et al. 2015a). Microzooplankton grazing is responsible for the majority of phytoplankton mortality in the oceans (Calbet and Landry 2004), so conclusions drawn about evolution from reports focusing solely on mesozooplankton will necessarily be limited. Future studies should consider whether microzooplankton, such as the model heterotrophic dinoflagellate *Oxyrrhis marina*, also evolve resistance to phytoplankton-defensive compounds. Additionally, studies are limited regarding the evolution of or resistance to chemical defenses and do not consider the myriad of other chemically-mediated interactions between zooplankton and phytoplankton. For example, have zooplankton evolved to detect cues from phytoplankton prey? Do toxins, used to subdue prey, such as karlotoxins, vary with different prey species or habitat, and can differences be attributed to evolutionary history? Future studies should attempt to decipher how evolution shapes all aspects of chemically mediated predator–prey interactions.

9.3.2.2 *Alexandrium* Detection of Predator Cues

The evolution of marine phytoplankton chemical defenses in response to zooplankton has, to our knowledge, never been addressed directly. However, a few studies considering the dinoflagellate *Alexandrium* spp. and its copepod predators have found that *Alexandrium* spp. may respond differently to copepods with which it has co-evolved than with copepods with which the dinoflagellate does not share evolutionary history. In the presence of chemical cues from the copepod *Acartia tonsa*, *Alexandrium minutum* more than doubled its concentration of PSTs, which reduced *A. tonsa* grazing on induced *A. minutum* cells relative to the control *A. minutum* (Selander et al. 2006). Bergkvist et al. (2008) found that cues from *Acartia clausi* and *Centropages typicus* caused an increase in PST content in *A. minutum*, while cues from *Pseudocalanus* sp. did not. Additionally, Wohlrab et al. (2010) also found that only certain species of copepods induce the toxin production in *Alexandrium tamarense*. In both studies, the co-evolutionary history of the copepod grazers and the *Alexandrium* spp. may explain the differential induction of PSTs. While *A. clausi* and *C. typicus* co-occur with *A. minutum*, *Pseudocalanus* sp. is a cold-water species and is unlikely to have historically co-occurred with *A. minutum* (Bergkvist et al. 2008). All of the copepods used by Wohlrab et al. (2010) co-occurred with *A. tamarense*. However, in this case PST induction was highest for the copepod most tolerant to the toxins, *Calanus helgolandicus*. No PST induction was observed for the copepod least able to tolerate the toxins, *Oithona similis*, suggesting that *A. tamarense* may have evolved to upregulate PST production only when it is needed.

Experiments thus far have focused on pairwise-interactions between *Alexandrium* spp. and a single grazer. In naturally occurring plankton assemblages,

Alexandrium spp. are likely to be exposed to a mixture of mesozooplankton and microzooplankton. Investigating the potential trade-offs that *Alexandrium* spp. must make in natural systems is likely to shed light on the selective pressure of different grazers in planktonic communities. For instance, *A. fundyense* does not increase toxin production in response to protist grazers (Senft-Batoh et al. 2015a), perhaps because protists are adversely affected by reactive oxygen species produced by *Alexandrium* spp. rather than PSTs (Flores et al. 2012; Senft-Batoh et al. 2015a), meaning that inducing PST production is an unnecessary defense against protist grazing.

The mechanism by which *Alexandrium* spp. perceive different predators is unknown; however, Wohlrab et al. (2010) speculated that *Alexandrium* cells must have cell-surface receptors that bind to specific cues released by copepods, resulting in a signaling cascade that changes *Alexandrium* spp. gene expression. The evolution of these receptors may account for the ability of *Alexandrium* spp. to respond differently to different species of grazers. A recent study has shown that even different populations of the same species of copepods can induce different responses in *Alexandrium fundyense*, perhaps because of evolutionary history (Senft-Batoh et al. 2015b). *Acartia hudsonica* from Maine have historically been exposed to *A. fundyense* blooms, while *A. hudsonica* from New Jersey have not. Interestingly, cues from Maine populations of *A. hudsonica* induce PST production in *A. fundyense*, while cues from New Jersey populations of *A. hudonica* do not (Senft-Batoh et al. 2015). The mechanism by which *A. fundyense* discriminates between these two populations is unknown but may be related to different cues released by different populations of *A. hudsonica*.

Recently, the cues released by copepods that cause induction of PSTs in *Alexandrium minutum* have been identified (Figure 9.3), which will allow many questions about the co-evolutionary relationship between predators and prey to be directly addressed in this system. A suite of eight taurolipids, known as copepodamides, isolated from the copepods *Centropages typicius* and *Calanus finmarchicus*, induce the production of toxins by the dinoflagellate *Alexandrium minutum* (Selander et al. 2015). Moreover, natural populations of copepods released copepodamides,

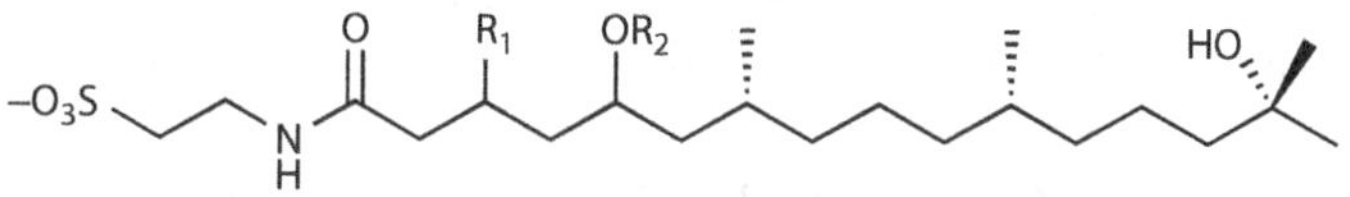

A $R_1 = -CH_3$ $R_2 =$ Docosahexaenoic acid (DHA)
B $R_1 = -CH_3$ $R_2 =$ Eicosapentaenoic acid (EPA)
C $R_1 = -CH_3$ $R_2 =$ Stearidonic acid (SDA)
D $R_1 = =CH_2$ $R_2 =$ Docosahexaenoic acid (DHA)
E $R_1 = =CH_2$ $R_2 =$ Eicosapentaenoic acid (EPA)
F $R_1 = =CH_2$ $R_2 =$ Stearidonic acid (SDA)
G $R_1 = -CH_3$ $R_2 = -H$
H $R_1 = =CH_2$ $R_2 = -H$

FIGURE 9.3 Structure of the recently discovered copapodamides, produced by calanoid copepods. The letters A–H indicate the specific copepodamide, with the corresponding R-groups indicated.

and copepodamides have been quantified in field samples (Selander et al. 2015). The potencies of these molecules are dependent on the presence of a methyl group at the RI position, with copepodamides A–C an order of magnitude more potent in inducing toxin production than copepodamides with a methylene group in the same position (Figure 9.3) (Selander et al. 2015). Researchers found that several copepod species, including *Calanus* sp., *C. typicus*, and *Pseudocalanus* sp. produce copepodamides; however, the make-up of the copepodamide suite varied according to copepods species. Copepod species that elicit the strongest response in *A. minutum* produced greater concentrations of copepodamides A–C (Selander et al. 2015). Thus, the relative concentrations of particular copepodamides in this mixture modulate their potency, possibly allowing algal cells to discriminate between copepod species and detect environmental patches containing copepod grazers.

The magnitude of the induction effect is also dependent upon whether copepods are actively feeding. Exposure to chemical cues from actively feeding copepods results in an even greater increase in toxicity of *A. minutum* (Selander et al. 2006; Bergkvist et al. 2008), *A. fundyense* (Senft-Batoh 2015a, 2015b), and *A. tamarense* cells (Wohlrab et al. 2010) than exposure to cues from starved copepods, suggesting that *Alexandrium* cells are also sensitive to alarm cues emanating from damaged conspecifics. However, as it is thought that copepodamides are involved in digestion and their release rapidly decreases when copepods are starved (Selander et al. 2015), it may be that alarm cues or copepod metabolic transformation products work in tandem with copepodamides, or that these kairomones are only produced in sufficient quantities by grazing copepods.

Clearly, the specificity and context dependency of the responses of *Alexandrium* spp. to various copepod chemical cues suggests that these cuing systems are sophisticated and information-rich languages, demonstrating that chemical variation in suites of released cues allows for the evolution of specific responses in receiving organisms (Meinwald 2010). The receptors that bind to the copepodamides may also be able to be identified, and experiments could be designed to test whether expression of these receptors has evolved in response to copepodamides from different copepod species. Experiments may also be designed to detect whether copepods exposed to PST producing *Alexandrium* spp. are selected to release lower concentrations of copapodamides or copapodamides that are less potent inducers of PSTs.

9.4 POLYUNSATURATED ALDEHYDES: ACTIVATED DEFENSES AND NUTRIENT CYCLING

Activated defenses are those compounds that are produced or modified upon wounding, usually as a result of enzymatic conversion of a precursor metabolite into a noxious or toxic compound. For example, terrestrial plants of the order Capparales enzymatically degrade glucosinolates upon herbivore attack, producing toxic nitriles, isothiocyanoates, thiocyanates, and other noxious compounds in a defense known as the "mustard-oil bomb" (Rask et al. 2000; Wittstock et al. 2004). A potential ecological and physiological advantage to employing activated defenses includes minimizing self-toxicity (Pohnert 2000); however, the pressures that would select for a single-celled organism to produce compounds that impact grazers only after cell destruction are unclear (Thornton 2002; Flynn and Irigoien 2009). Regardless, there are several examples of toxic and teratogenic compounds that are produced by phytoplankton only after cell lysis. Additionally, there are several well-studied examples of activated chemical defenses of marine phytoplankton. Among the best studied are the PUAs.

9.4.1 Polyunsaturated Aldehydes as Activated Defenses

Oxylipins are a group of lipid compounds, including PUAs, produced as a result of enzymatic oxidative cascades of unsaturated fatty acids. Oxylipins and related compounds are often cited as examples of diatom-activated defenses, as studies have shown that copepods reared on some diatoms experienced low fecundity (Ban et al. 1997) and had offspring with high rates of birth defects (Poulet et al. 1994; Ianora et al. 2004). While many organisms, including terrestrial plants, animals, and fungi, produce oxylipins, diatom oxylipins appear to be produced from C20 and C16 membrane lipids, while other organisms produce these compounds through a slightly different pathway (Pohnert 2002; d'Ippolito et al. 2009; Ianora and Miralto 2010; Cutignano et al. 2011). The enzymatically-driven cascade that produces oxylipins is rapidly initiated by cell lysis and membrane disruption (Pohnert 2000), and, although the particular biosynthetic pathways vary, many lineages of diatoms produce oxylipins (Cutignano et al. 2011).

One of the best-known classes of diatom oxylipins is the PUAs, which have been implicated in the detrimental effects of diatom diets on copepod grazers. Examples of PUAs include the compounds decadienal, decatrienal (Miralto et al. 1999), octadienal, and heptadienal among others (Figure 9.4) (Pohnert 2000). Once ingested by copepods, diatom PUAs have been shown to decrease the fitness of copepods by causing reduced fecundity, egg viability, and larval birth defects, while not necessarily deterring ingestion (Figure 9.5) (Miralto et al. 1999; Pohnert 2002; Leflaive and Ten-Hage 2009). Many laboratory-based studies have demonstrated that PUAs cause apoptosis in the cells of copepod gonads and nauplii (Poulet et al. 1994; Romano et al. 2003; Buttino et al. 2004). This pattern of reduced fecundity also has been demonstrated in field populations of copepods feeding on a natural *Skeletonema* bloom (Miralto et al. 1999). Thus, this pattern has been dubbed the "insidious effects" of diatom diet on copepod populations since these compounds impact future populations of copepods (Miralto et al. 1999), potentially benefiting diatom populations by reducing future grazing pressures (Figure 9.5). Thus far, investigations of PUAs as defenses against grazers have focused exclusively on copepods, so their effects on microzooplankton are entirely unknown. Future work should shed light on the potential defensive roles of these molecules against microzooplankton.

Although the impacts of PUAs on copepod physiology have been demonstrated, several researchers have cautioned that the utility of a defense that acts on future generations of grazers is of an unknown benefit to current generations of prey species. Other studies have questioned the selective pressures that would result in insidious defenses of diatoms. For example, Flynn and Irigoien (2009) argued there is no benefit for the diatom individuals that are currently being grazed. Moreover, several field-based studies have found little influence of diatom toxicity on copepod reproduction

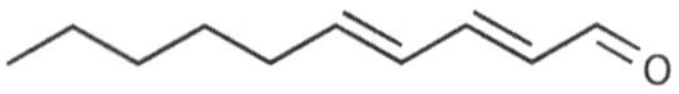

(2E,4E)-Decadienal

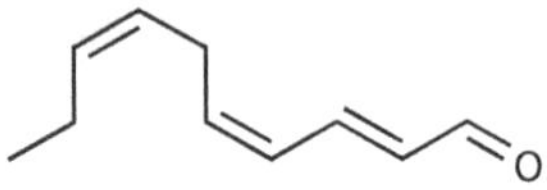

(2E,4Z,7Z)-Decatrienal

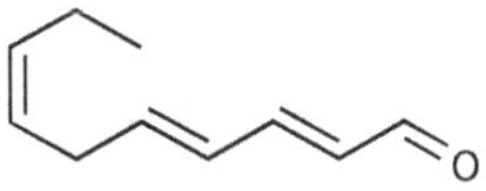

(2E,4E,7Z)-Decatrienal

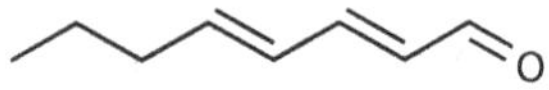

(2E,4E)-Octadienal

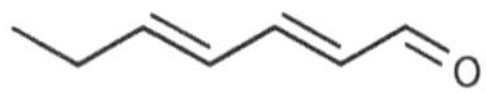

(2E,4E)-Heptadienal

FIGURE 9.4   Structure of common diatom-produced PUAs.

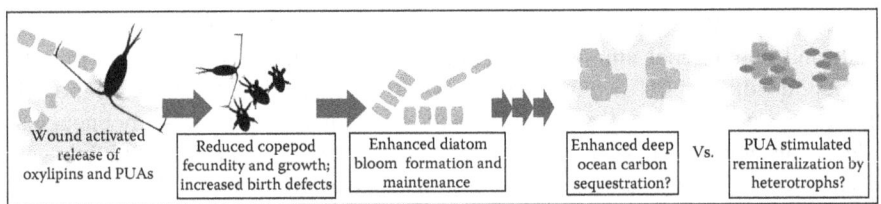

FIGURE 9.5   Influence of diatom-produced PUAs and oxylipins on copepod populations. While feeding, copepods break open diatom cells, promoting the production of PUAs and oxylipins. Exposure to these natural products causes birth defects in subsequent generations of copepods and release future diatom populations from grazing pressure and may ultimately stimulate the sequestration of diatom-fixed carbon. Alternatively, the production of PUAs and oxylipins by degrading diatoms may stimulate bacterial metabolism, enhancing carbon remineralization in the upper water column.

(e.g., Irigoien et al. 2002), or have found contrasting results on the effects of PUAs on copepod grazers. In some cases studies found little to no correlation between PUA concentration and copepod fitness (Poulet et al. 2006; Wichard et al. 2008; Jónasdóttir et al. 2009; Koski et al. 2012), suggesting that other compounds, perhaps other oxylipins, contribute to the diatom-copepod paradox (Pohnert et al. 2002; Dutz et al. 2008; Ianora et al. 2008; Amin et al. 2011), although this has not been demonstrated. These contrasting results have stimulated a large body of work in which researchers have struggled to explain the variable effects of PUAs on copepod physiology.

Extreme variability in PUA production by different diatoms and in response to environmental factors makes it difficult to characterize the role of PUAs and other oxylipins as defenses against grazers. Several studies have described the variability in PUA and oxylipin production across diatom species (e.g., Wichard et al. 2005; Lamari et al. 2013) and strains (e.g., Gerecht et al. 2011). For example, Lamari et al. (2013) found that different species of *Pseudo-nitzschia* were distinguishable based upon oxylipin profiles and suggested the possibility for these pathways to rapidly evolve in response to interactions between copepods and diatoms. Nutrient status is also important, as researchers have found that the production of the PUAs heptadienal, octadienal, and octatrienal by *Skeletonema marinoi* varied with nutrient supply and increased under silicate and phosphorus limitation, while PUA production decreased under nitrogen limitation (Ribalet et al. 2009). Growth stage also influenced PUA production, as cultures in stationary phase produced more PUAs than those in earlier growth stages (Vidoudez and Pohnert 2008; Ribalet et al. 2007b). Gerecht et al. (2011) found that oxylipin profiles of *S. marinoi* varied among strains. These differences were based upon variable enzyme activities, as opposed to levels of precursor lipid molecules (Gerecht et al. 2011). If researchers are to fully understand the role of PUAs as activated defenses, the environmental and even evolutionary context of the molecules must be considered.

Differences in evolutionary history between grazers and diatoms may help explain the contradictory results in studies attempting to assess the effects of PUA-producing diatoms on copepod reproduction, as there appear to be large levels of variation in the response of copepods to these oxylipins. For instance, at least one species of benthic harpacticoid copepod, *Tisbe holothuriae*, was reported to be resistant to PUAs in a recent study (Taylor et al. 2012). Another recent study showed that the expression of detoxification and stress response genes is reduced in *Calanus helgolandicus* copepods when they are fed a diet of oxylipin producing *S. marinoi* (Lauritano et al. 2011). However, it appears that history of exposure is an additional modulating factor in the relationship between diatom PUAs and copepod grazers. Copepods with a history of exposure to diatoms that produce PUAs upregulate the expression of genes likely to be involved in detoxifying these compounds when fed PUA-producing diatoms. In contrast, copepods without a history of exposure to these diatoms show decreased expression of genes coding for detoxification and stress-related enzymes (Lauritano et al. 2011, 2012). These more recent studies show that in order to understand the ecological role(s) of PUAs, the specific context of diatom-copepod interactions must be considered.

9.4.2 Community and Ecosystem Roles of Polyunsaturated Aldehydes

PUAs have been reported to impact the community and ecosystem in ways that go beyond defending against predation (Ianora et al. 2011). They have been reported to affect phytoplankton cells, including the very cells that produce them. The diatom *Thalassiosira weissflogii* undergoes an apoptosis-like process when exposed to PUAs (Casotti et al. 2005). Casotti et al. (2005) suggested that this may be an intraspecific signal regulating bloom dynamics, but it is unclear how such a response would be beneficial to individual cells (Nedelcu et al. 2011 for a detailed exploration of programmed cell death in unicellular organisms). Other studies have reported that diatoms exposed to sublethal PUA concentrations upregulated enzymes involved in nitric oxide signaling and programed cell death (PCD) (Vardi et al. 2008), and exposure to low concentrations of PUAs protect diatoms from subsequent PUA exposure, preventing diatoms from undergoing PCD (Vardi et al. 2006). PUAs have also been reported to inhibit the growth of a wide range of phytoplankton cells, with some species, such as the prymnesiophyte *Isochrysis galbana*, more affected than other species, like the PUA-producing diatom *Skeletonema marinoi* (Ribalet et al. 2007a). The different susceptibilities may be related to avoidance of auto-toxicity, but other factors, such as cell size and cell wall composition may also be important (Ribalet et al. 2008). For both of these studies, the concentrations of PUAs used were tens to hundreds of times higher than the average dissolved concentration found in seawater (0.005–2.5 nM) (Vidoudez et al. 2011a; Ribalet et al. 2014), but there are likely small patches where PUAs can accumulate to high concentrations, such as in the immediate vicinity of diatom aggregates (Ribalet et al. 2007a; Edwards et al. 2015). PUA production by diatoms may also play a role in seasonal plankton blooms. In the North Atlantic, spring diatom blooms are followed by an increase in zooplankton, primarily copepods (Valiela 1995). Although PUAs have not been implicated in the process, copepods' selective avoidance of diatoms in favor of motile ciliates and dinoflagellates may allow these large diatom blooms to occur, and for ungrazed biomass to sink (Ianora et al. 2004), potentially sequestering carbon in the ocean—a process with major implications for nutrient cycling and climate change. However, to this point, PUA production has not been directly tied to the formation or magnitude of spring diatom blooms.

A number of studies have addressed the indirect effects of PUAs on the bacterial community. For example, some isolates of marine bacteria are inhibited by PUAs, but others are not (Ribalet et al. 2008). In natural assemblages of bacteria, Balestra et al. (2011) reported that while PUAs slightly altered the relative abundance and metabolic activity of some bacterial clades, they did not change the overall abundance of total bacteria in samples from the NW Mediterranean Sea. When PUAs or PUA-producing diatoms were added to the more realistic mesocosm environment, neither had any effect on bacterial community structure, but metabolic response of bacteria was not assessed (Paul et al. 2012). A more recent study has indicated that while the average concentrations of PUAs in seawater used by Paul et al. (2012) do not affect the bacterial community, higher concentrations, such as those found in particulate organic matter, do (Edwards et al. 2015). These authors report that PUAs stimulate bacterial respiration and population growth through an unknown mechanism. Importantly, the increased respiration also increases the rate of remineralization of organic matter, which may actually decrease the amount of particulate organic matter that is sequestered in the deep sea below the mixing layer (Edwards et al. 2015), which is in contrast to previous hypotheses (Ianora et al. 2004). Untangling the effects of PUAs on the build-up of diatom biomass, carbon remineralization, and oceanic carbon sequestration will be important if the effects of PUAs on climate processes are to be understood (Figure 9.5).

9.5 SENSING PREY: DIMETHYL SULFIDE/DIMETHYL SULFONIOPROPIONATE AND TRITROPHIC INTERACTIONS

The zwitterionic metabolite dimethyl sulfoniopropionate (DMSP) and its degradation products, dimethyl sulfide (DMS) and acrylic acid (Figure 9.6), have been studied for their potential roles as grazer deterrents, as cues used by zooplankton to detect prey, and as important components in the global sulfur cycle. Many algal taxa, notably prymnesiophytes and dinoflagellates, produce DMSP and the intracellular enzyme DMSP lyase, which cleaves this molecule into DMS and acrylic acid in response to handling or cell damage by grazers (Wolfe and Steinke 1996; Wolfe et al. 2000; Alcolombri et al. 2015). The ecological and physiological functions of DMSP, DMS, and acrylic acid are still being debated, but recent work has begun to characterize the intraspecific variability in expression of the DMSP lyase genes, (members of the Alma gene family), as well as to determine the presence of this gene family across plankton lineages (Alcolombri et al. 2015). Upon release in the atmosphere, DMS can alter oceanic cloud cover with implications for climate change (Charlson et al. 1987; Watson and Liss 1998). Understanding the molecular and ecological forces driving the production and release of these compounds will be essential to obtain a more complete picture of plankton community ecology and to determine the biogeochemical cycles of marine systems (Strom 2008).

9.5.1 DIMETHYL SULFONIOPROPIONATE AS PART OF AN ACTIVATED DEFENSE SYSTEM

The role of DMSP in activated defenses against grazing has been most studied in the coccolithopohore *Emiliania huxleyi*. Researchers have hypothesized that acrylic acid may be toxic to protist grazers, such that when protists consume *E. huxleyi*, acrylic acid may accumulate to toxic levels in food vacuoles (Wolfe and Steinke 1996; Wolfe et al. 1997). Further research showed that protist grazers consumed fewer *E. huxleyi* strains with high DMSP lyase activity than *E. huxleyi* strains with low DMSP lyase activity (Wolfe and Steinke 1996; Strom et al. 2003a), providing indirect evidence that DMS or acrylic acid could act as defensive molecules. Surprisingly, the additions of exogenous DMS and acrylate (the conjugate base of acrylic acid) did not reduce

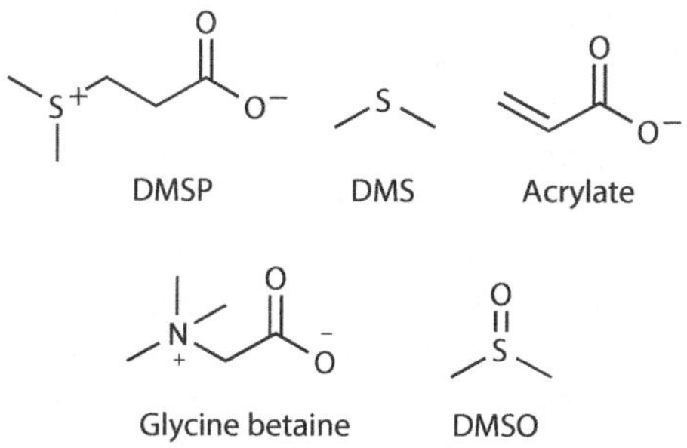

FIGURE 9.6 Structure of dimethylsulfoniopropionate (DMSP) and the structurally related glycine betaine. The products of DMSP cleavage, dimethyl sulfide (DMS) and acrylate, are also shown. DMS is oxidized to dimethylsulfoxide (DMSO) in the atmosphere.

grazing of heterotrophic protists on *E. huxleyi*. However, exogenous DMSP and the structurally related glycine-betaine did reduce grazing at concentrations of 20 μM (Strom et al. 2003b). This concentration could be released from *E. huxleyi* since it falls between the nanomolar concentrations found in seawater and the approximately 100 μM found inside *E. huxleyi* cells (Seymour et al. 2010).

There is little evidence that DMSP itself acts as a toxin against protist grazers, but rather as a cue that indicates unpalatable or toxic prey (Strom et al. 2003b). Indeed, grazers that were most sensitive to the presence of DMSP also tended to reject *E. huxleyi* strains with high DMSP lyase activities (Strom et al. 2003a, 2003b). However, DMSP lyase activity does not necessarily correlate with cellular DMSP concentrations (Wolfe and Steinke 1996; Strom et al. 2003b), which indicates that DMSP production alone is not an accurate indicator of prey palatability or quality. Cumulatively, these studies suggest that multiple signals, including prey nutritional quality, morphology, and cell surface chemistry, may play a role in influencing protist grazing on *E. huxleyi*. Additionally, virtually no studies have investigated the impact of very high concentrations of acrylic acid on protist physiology so the role of acrylic acid as a defensive compound is still unestablished (Wolfe et al. 1997; Breckels et al. 2011). Although concentrations of acrylic acid in protist food vacuoles are expected to be near 70 mM for each ingested *E. huxelyi* cell, previous studies have tested the impacts of only 1 mM acrylate on grazing (e.g., Strom et al. 2003b). Finally, it has also been hypothesized that DMSP (and related glycine betaine; Figure 9.6) may also act as a satiation cue to grazers (Strom et al. 2003b) instead of indicating unpalatable prey. Thus, the defensive roles of DMSP and its degradation products are still unknown.

9.5.2 DIMETHYL SULFONIOPROPIONATE AS A CUE FOR GRAZERS

Both heterotrophic protists and copepods exhibit positive chemotaxis to a number of pure compounds as well as metabolite mixtures, including algal exudates and primary metabolites (reviewed in Breckels et al. 2011; Heuschele and Selander 2014). In addition, both *O. marina* and the heterotrophic nanoflagellate *Neobodo designis* exhibit positive chemotaxis to DMSP, with *O. marina* responding strongly and rapidly to all concentrations of the cue tested, from 2 to 500 μM. *O. marina* also swam towards all concentrations of DMS tested (Seymour et al. 2010). In a similar test of *O. marina* chemotaxis, Breckels et al. (2011) also reported positive chemotaxis towards DMSP at concentrations ranging from 10 nM to 100 μM. However, no significant response to DMS was detected. The presence of acrylate at concentrations ranging from 1 μM to 100 μM resulted in negative chemotaxis at of *O. marina* (Breckels et al. 2011). Investigations of copepod responses to DMS have produced mixed results. When tethered *Temora longicornis* copepods were exposed to a plume of DMS, search behavior increased (Steinke et al. 2006). However, a subsequent study failed to find any response of the copepod *Calanus helgolandicus* (Breckels et al. 2013), suggesting that the response to DMS may be taxon specific.

The observations that zooplankton respond positively to either DMS or DMSP are surprising, since both compounds have been implicated in zooplankton feeding deterrence (see previous; Wolfe et al. 1997; Strom et al. 2003b). In bulk seawater, DMSP concentrations are in the nanomolar range, while concentrations inside cells can be as much as 100 μM, resulting in a patchy distribution of DMSP on spatial scales relevant to microbes (Seymour et al. 2010). Seymour et al. (2010) speculate that the bulk additions of DMSP in early studies created a uniform distribution of DMSP, so positive chemotaxis toward higher concentrations like those exuded from cells was not possible. However, it is less clear why grazers (including ciliates and heterotrophic dinoflagellates) would have higher feeding rates on *Emiliana huxleyi* with low DMSP-lyase activity (Strom et al. 2003a), as these strains should release lower concentrations of DMS and acrylate upon grazing. Overall, the role of DMSP, DMS, and acrylate for phytoplankton is unclear. The primary role of one or all of these compounds may be to serve as a defense against predation. Alternatively, these compounds may have evolved to perform different physiological roles but have been co-opted by zooplankton and used as a cue to indicate prey. Others have suggested that microbes including *O. marina* could directly use DMSP itself as a sulfur source (Breckels et al. 2011). The conflicting effects of DMSP and related compounds may indicate that these compounds actually have multiple roles, and their function and production varies with species and strain of both grazer and prey (Figure 9.7) (Strom et al. 2003a, 2013b; Strom 2008; Seymour et al. 2010).

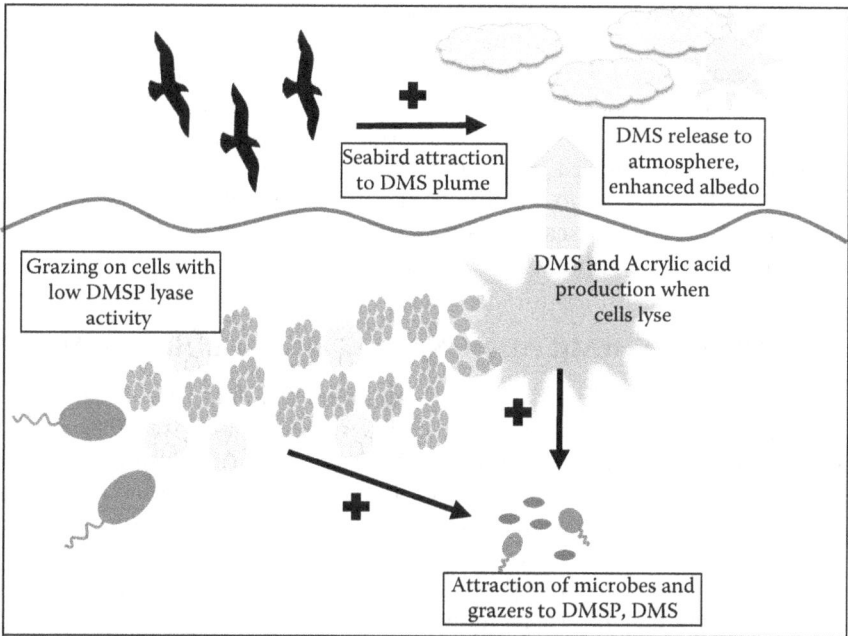

FIGURE 9.7 Multiple impacts of dimethylsulfoniopropionate (DMSP) and related compounds dimethylsulfoxide (DMS) and acrylic acid on the behavior of marine organisms. As cells of the coccolithophore *Emiliania huxleyi* lyse, DMSP is cleaved by DMSP-lyase to DMS and acrylic acid. Microzooplankton preferentially feed on *Emiliania huxleyi* cells with low DMSP-lyase activities, suggesting a defensive role for these compounds. As cells lyse, DMSP and DMS also attract heterotrophic bacteria and other microbes. DMS escapes from surface waters and is used as a cue by procellariiform seabirds to locate potential patches of prey items. DMS is also oxidized in the atmosphere where it stimulates cloud formation.

9.5.3 Dimethyl Sulfide as a Cry for Help

Regardless of the role of DMSP, DMS, and acrylate in the plankton, their production has implications for higher trophic levels. Some procellariiform seabirds have been shown to detect atmospheric DMS, even at picomolar concentrations, and were attracted to DMS-scented slicks on the ocean surface even from thousands of kilometers away (Figure 9.6) (Nevitt et al. 1995; Nevitt 2000, 2011). Because DMS is produced by phytoplankton when they are grazed by crustaceans, it is not surprising that the procellariiform seabirds that detected this compound had diets that consisted primarily of crustaceans (Savoca and Nevitt 2014). The procellariiform seabirds that do not detect DMS eat a wider selection of food, including cephalopods and fish. Savoca and Nevitt (2014) proposed that the relationship between phytoplankton and seabirds is a tritrophic mutualism, and by producing DMS, plankton benefit from seabirds. The benefit comes not only from the removal of grazing pressure, but also because seabird feces are rich in iron and phytoplankton are iron limited in the Southern Ocean, where these studies are focused (Savoca and Nevitt 2014).

DMS and related compounds may act as foraging cues for other organisms as well. For example, the release of DMSP attracted several species of reef fish (DeBose et al. 2008). These fish usually feed on planktivorous prey, so it is logical that DMSP might serve as a reliable cue leading to productive feeding areas. Loggerhead sea turtles also seem to detect DMS; juveniles exposed to DMS-scented air spent more time with their noses out of the water than those exposed to control scents (Endres and Lohmann 2012). Further tests of sea turtle foraging under natural conditions could clarify exactly how sea turtles use DMS. A number of other animals have been shown to be attracted to DMS or DMSP, including basking sharks, African penguins, and Northern fur seals; however, these studies should also be conducted under natural foraging conditions (Nevitt 2011).

9.6 PLANKTON CHEMICAL SIGNALING ON A GLOBAL SCALE

Planktonic chemical signaling has global implications, in large part because it has the potential to impact the earth's climate. The reverse is also true. The decreasing ocean pH and warming temperatures that will accompany rising pCO_2 levels may alter or disrupt chemical signaling in the plankton. Research into the effects of climate change and ocean acidification on chemical signaling in general is in the early stages. However, marine phytoplankton are crucial for large-scale ecosystem processes, so understanding how processes in the plankton are impacted by changing climate patterns may be crucial to the health of marine ecosystems.

9.6.1 Dimethyl Sulfide and Climate Change

The release of DMS from phytoplankton into the atmosphere has global implications for climate change. Once in the atmosphere, DMS is oxidized into a variety of molecules, including dimethylsulfoxide (DMSO) (Figure 9.6). These molecules serve as nuclei for cloud condensation, resulting in increased cloud cover over the ocean (Figure 9.7) (Charlson et al. 1987). DMS concentration is positively correlated with chlorophyll concentration, and satellite observations have shown that chloro-phyll concentrations and cloud cover are also positively correlated, providing some direct evidence for the link between phytoplankton-DMS release and cloud cover (Vogt and Liss 2009). Because the clouds reflect sunlight, increased cloud cover has a cooling effect on the earth's climate (Charlson et al. 1987). In fact, Watson and Liss (1998) calculate that, overall, marine biota cool the earth by 6°C. Only one third of this effect is caused by the uptake of CO_2 by marine phytoplankton; the other 4°C is caused by the release of DMS into the atmosphere.

Because DMS plays a major role in global climate regulation, feedback mecha-nisms that might alter its production should be carefully researched. Nearly half of procellariiform seabirds are listed as vulnerable or endangered, and their decline is likely to reduce iron fertilization in the open ocean, which may result in lower phytoplankton biomass and decreased DMS production (Savoca and Nevitt 2014). Field experiments of iron fertilization in the Southern Ocean have provided mixed results, with some reporting massive increases of DMS and others reporting no change or a pulse of DMS (reviewed in Vogt and Liss 2009). However, these large-scale fertilization events are likely to differ from the lower-concentration continu-ous iron input provided by seabirds. The effects of iron fertilization by seabird feces on DMS production has not been assessed, so it is unclear how seabird population declines might impact DMS concentrations and ocean cloud cover. The effects of ocean acidification, a result of increased atmospheric CO_2, are also likely to have an effect on oceanic DMS production. Decreasing ocean pH is predicted to adversely affect organisms with calcium carbonate shells, including the coccolithophorid *Emiliana huxleyi* (Doney et al. 2009). Because *E. huxleyi* is a major producer of DMS, increasing CO_2 in the atmosphere may decrease DMS production, leading to a positive feedback loop that results in increased global warming (Vogt and Liss 2009).

9.6.2 OCEAN ACIDIFICATION AND CHEMICAL SIGNALING

Increased global temperature and lowered ocean pH are likely to alter chemical signaling in planktonic systems. In fact, ocean acidification has been shown to dis-rupt the perception of chemical signals by marine organisms. For example, larval clownfish were less able to recognize the chemical cues that indicate suitable settle-ment sites and could no longer distinguish between the smell of parent and non-parent conspecifics when reared in a pH of 7.8 (Munday et al. 2009). Later work showed that low pH interfered with fish neurotransmitter function, which altered the perception of chemical signals (Nilsson et al. 2012). Although the effects of ocean acidification on plankton chemical perception are largely uncharacterized, one recent study investigated how a lower pH changes the responses of the copepod *Centropages typicus* to algal cues. Generally, *C. typicus* was less attracted to cues from the dinoflagellate *Prorocentrum minimum* and the diatom *Pseudonitzschia delicatissima* at low pH, although the effects varied widely with algal species and concentration of algal cells (Maibam et al. 2015). The mechanism for and implica-tions of the change in copepod perception of algal chemical cues have not been investigated.

9.6.3   CLIMATE CHANGE AND HARMFUL ALGAL BLOOMS

Ocean acidification is likely to change the production, as well as the perception, of chemical compounds. For example, the dinoflagellate *Karlodinium veneficum* produces increased amounts of karlotoxin-1 and decreased amounts of karlotoxin-2 in high pCO_2 conditions (Fu et al. 2010). Karlotoxin-1 has ten times the hemolytic potency of karlotoxin-2 (Bachvaroff et al. 2009) and is more effective at immobilizing prey (Sheng et al. 2010). Several studies have investigated the effect of decreasing pH on growth rate and toxin production of *Alexandrium* spp. High pCO_2 promoted an increased growth rate in both *Alexandrium catenella* (Tatters et al. 2013) and *Alexandrium fundyense* (Hattenrath-Lehmann et al. 2015) but had no effect on *Alexandrium tamarense* (Van de Waal et al. 2014). The effects of CO_2 levels on toxin production are less clear. High pCO_2 increases toxin production in *A. catenella* (Tatters et al. 2013) but decreases the toxicity of *A. tamarense* (Van de Waal et al. 2014). High pCO_2 increases toxicity of *A. fundyense* collected from Northport Bay, New York, but not in the same species collected from the Bay of Fundy, Canada (Hattenrath-Lehmann et al. 2015). The change in toxicity is often explained by the shift to more (Hattenrath-Lehmann et al. 2015) or less (Van de Waal et al. 2014) toxic analogues in the PST suite of compounds. Overall, these results indicate that the effect of CO_2 on toxin production varies not only by species of *Alexandrium*, but also by strain. However, studies generally suggest that increasing pCO_2 will increase the growth rate of *Alexandrium* spp. and may result in an overall increase of phytoplankton less efficient at concentrating CO_2 (i.e., dinoflagellates) relative to phytoplankton with efficient carbon concentrating mechanisms (i.e., diatoms) (Hattenrath-Lehmann et al. 2015). Because many toxin-producing HAB species are dinoflagellates (Smayda 1997), increasing ocean pCO_2 concentrations may promote HAB formation.

In addition to ocean acidification, rising temperatures have implications for toxin-producing HAB species. Changes in the phytoplankton community will likely vary by region; in tropical and mid-latitude seas limited by nutrients, increased stratification resulting from decreased mixing may result in decreased nutrient flux and an overall decrease in phytoplankton. In contrast, increased stratification in high-latitude light-limited areas should result in phytoplankton remaining in high light surface waters leading to an overall increase in phytoplankton (Hallegraeff 2010). Rising temperatures may result in HAB expansion by a variety of mechanisms. For example, warming temperatures may allow range expansion of warm-water toxic species, such as the ciguatera producing dinoflagellate *Gambierdiscus toxicus* (Moore et al. 2008). Other species may expand ranges northward, as has been seen for several HAB species that have become increasingly common in the North Sea and along the coast of Norway (Hallegraeff 2010). Not only will the shifting ranges of HAB species cause economic disruptions and human health problems, but also may alter marine community composition with potentially dramatic effects for organisms that do not share an evolutionary history with these HAB species. For example, both copepods and clams naïve to blooms of *Alexandrium* spp. experience much lower fitness during blooms than conspecific populations with a shared history (Bricelj et al. 2005; Avery and Dam 2007; Colin and Dam 2007; Bricelj et al. 2010). It is also likely that longer periods of warm weather will result in extended blooms for some species. For example, *Alexandrium* spp. in the Puget Sound bloom when temperatures exceed 13°C. Historically, this has resulted in a 68-day window during which blooms can occur. However, if average temperatures increase by 2°C, *Alexandrium* spp. blooms could last twice as long; if temperatures increase by 6°C, bloom length is expected to triple (Moore et al. 2008). Increasing HABs are likely to result in increasing adverse health outcomes for humans; however, a paucity of long-term historical data sets makes this change difficult to assess (Moore et al. 2008).

9.7 CONCLUSIONS AND FUTURE DIRECTIONS

The systems explored in this chapter illustrate the importance of chemically mediated predator–prey interactions in the plankton. Both predators and prey use chemical compounds to sense the presence of each other. Prey can respond by upregulating or activating defensive compounds, while predators may use toxic compounds to immobilize prey, making them easier to consume. The critical role of plankton in carbon fixation and biogeochemical cycling is recognized (Field et al. 1998), but the fact that these processes are dependent on ecological interactions within the phytoplankton community is easily overlooked (Strom 2008). Understanding the mechanisms that control the abundance and distribution of organisms within these communities may help us understand how planktonic communities respond to perturbations, such as ocean acidification and climate change, with major implications for marine communities, ecosystems, and even global climate.

9.7.1 EVOLUTIONARY CONSIDERATIONS

In this chapter, we highlight the complexity of chemically-mediated interactions in the plankton. Single molecules often have multiple functions, and, as is the case for DMSP and related compounds, the relative importance of each role (i.e., as a defense or as a cue for predators) may be difficult to determine (Strom 2008). Similarly, the adverse effects of PUAs on copepods may be the primary role of these compounds, or PUAs may serve some other function while their impact on grazers is simply a "happy coincidence" from the perspective of the diatom. In order to determine the selective pressure leading to the production of these multi-functional compounds, it is critical to place interactions in an evolutionary context. Where do selective pressures lie? In which situations do organisms benefit from the production of these compounds? Answers to these questions will provide insights into how the function of these molecules is likely to shift in a changing ecosystem, as well as explain why the patterns of production and response to signaling molecules are variable across taxa. Researchers considering the evolutionary context of signaling molecules have already made strides in explaining, for example, why the effects of PSTs (Avery and Dam 2007) or PUAs (Lauritano et al. 2012) on grazers varies dramatically.

9.7.2 IDENTIFYING PLANKTON METABOLITES: CHALLENGES AND SOLUTIONS

Chemical signals are often a blend of many different molecules in a particular ratio, instead of a single compound. In some cases, the relevant molecules in the mixture have been characterized (e.g., Selander et al. 2015), but more often crude extracts or exudates are used to elicit responses while the ecologically relevant molecules remain unknown. Experiments with crude extracts and exudates are useful for shedding light on ecological relationships but isolating and elucidating the structure of signaling molecules can illuminate a relationship in greater detail. Without the ability to test the effects of each compound individually at ecologically appropriate concentrations, knowledge of how multiple compounds act additively or synergistically on the physiology of a predator or prey species remains unknown. Determining the chemical structure of defenses will also allow researchers to determine their cellular targets. Structure elucidation of bioactive compounds will aid in determining the biosynthetic pathways involved in their production and could suggest the metabolic costs required to make them. Placing signaling molecules in an evolutionary context by determining the selective pressures that contribute to the production and maintenance of these compounds requires the identification of the bioactive molecules. Thus, structure elucidation is necessary if the factors, both biotic and abiotic, that influence compound production are to be determined.

Identifying the compounds mediating predator–prey interactions can be challenging, especially when molecules are produced at low concentrations, are unstable, act additively or synergistically, or include the involvement of a complex blend of metabolites (Kuhlisch and Pohnert 2015). It is likely that new techniques will need to be developed to expand our ability to detect molecules with high sensitivity in situ. However, advances in system approaches to biology, most notably metabolomics, have the potential to help scientists quickly home in on relevant compounds while minimizing the fractionation steps required to isolate them, and to detect multiple molecules that may act additively or synergistically (Kuhlisch and Pohnert 2015). Metabolomic approaches, which attempt to identify and quantify the metabolites produced by an organism as completely as possible, have been used successfully to answer other questions in planktonic systems. For example, a diatom sex pheromone was identified using this approach (Gillard et al. 2013), and the effect of *Karenia brevis* allelopathy on the metabolism of competing species was characterized (Poulson-Ellestad et al. 2014). In our minds, several areas mentioned in this chapter would benefit from the application of metabolomic techniques. While the defenses described are mostly made up of secondary metabolites, other cues are likely complex mixtures of primary metabolites (Hay 2009), which would be amenable to metabolomic characterization (e.g., Poulson-Ellestad et al. 2016). Additionally, many defensive compounds produced by phytoplankton remain unknown (e.g., Kubanek et al. 2007; Turner et al. 2012), and, if bioactive compounds can be identified, perhaps patterns of expression could be determined. Identifying novel compounds could also help determine whether currently unidentified compounds interact with known toxins (e.g., brevetoxins), possibly explaining some of the contradictory results explored in Section 9.2.1. The danger of metabolomic approaches for chemical ecologists remains that metabolites will be characterized but will not be placed in an ecological context (Prince and Pohnert 2010). Researchers must confirm the roles of differentially expressed metabolites, (e.g., physiological responses to stress, defenses, cues to locate prey) through further manipulative experiments. Without an attempt to understand the ecological or physiological role of variably expressed compounds, studies employing metabolomic approaches run the risk of simply generating a list of compounds that correlate with a biological metric, an approach as limited as testing for the ecological roles of crude extracts.

Molecular tools are increasingly common in investigations of chemical ecology because they allow researchers to manipulate the production of natural products. For bacteria, genes encoding the enzymes necessary for secondary metabolites can be eliminated, and the changes in ecological interactions can be observed (Jensen et al. 2015; Lane and Taylor, this book). However, this technique is limited to known metabolites from organisms with well-characterized genomes, which are often lacking in planktonic systems. However, rapidly advancing techniques such as meta-genomic, -transcriptomic, and -proteomic strategies will enable hypotheses about the ecological roles of natural products to be developed and tested in situ. These meta "omic" techniques have the potential to allow investigation of mixed community assemblages with incompletely sequenced genomes. If well-annotated libraries of planktonic communities can be developed, metagenomics techniques will allow scientists to determine the genetic and, therefore, metabolic potential of a community (Kujawinski 2011). For example, questions such as whether the genes coding for enzymes that responsible for PUA production or detoxification are present in a natural community could be answered. Other techniques, such as transcriptomics and metabolomics, will allow scientists to determine which genes are expressed and metabolites produced under which conditions (Kujawinski 2011). For example, these techniques could help to address what conditions lead to PUA production and when the enzymes that lead to their detoxification are expressed. Some techniques are already being applied to answer ecological questions in planktonic systems, including measures of planktonic diversity (de Vargas et al. 2015) and how biotic and abiotic factors predict plankton community structure (Lima-Mendez et al. 2015). These same techniques can be used to detect novel secondary metabolites and to determine how their patterns of expression relate to community interactions.

9.7.3   Roles of Metabolites in Natural Communities

Future experiments designed to decipher chemically-mediated predator prey interactions must determine the ecological and evolutionary meaning of those interactions. To accomplish this, researchers must go beyond pairwise interactions in a laboratory and attempt to clarify the roles of signaling molecules on a community and in the field. Laboratory experiments involving only two species can certainly be useful, and results are relatively easy to interpret. However, these experiments are inherently limited, especially in planktonic systems where the lines between primary producer and primary consumer are often blurred, making it difficult to distinguish between predator and competitor (Flynn et al. 2012). Compounds with allelopathic roles against competitors may also be used for predation, as is the case for karlotoxins (Adolf et al. 2006; Sheng et al. 2010). Compounds identified because of their roles as defensive metabolites, such as PUAs, have also been found to impact competitors and change the microbial community (Ianora and Miralto 2010). Bioassays designed only to detect one interaction will often miss conditions where molecules have more than one role, or change roles depending on the environment. Incorporating community-wide interactions into ecological experiments is not without challenges. Researchers must think carefully about what data to collect, and indirect interactions may cause unexpected or confusing results that are difficult to interpret. Regardless, these types of studies provide valuable data on the cascading effects of chemical cues throughout communities.

A number of approaches may be taken to enable a community-level understanding of chemically mediated interactions. One of the simplest approaches is to include a wider variety of prey and predators, both at the species and population levels. For example, investigations of zooplankton have been dominated by calanoid copepods, despite the fact that microzooplankton are responsible for the majority of consumption in planktonic marine ecosystems (Calbet and Landry 2004). However, apart from experiments with the model heterotrophic dinoflagellate *Oxhrrhis marina* (reviewed in Breckels et al. 2011; Roberts et al. 2011), the chemical ecology of heterotrophic and mixotrophic dinoflagellates has been woefully understudied. A more difficult, but worthwhile, approach is to use experiments that consider many or all of the members of a community when studying predator–prey interactions. Many of the experiments investigating the roles of PUAs have been conducted in mesocosms (e.g., Vidoudez et al. 2011b; Jónasdóttir et al. 2011; Paul et al. 2012). Mesocosm studies have the benefit of allowing investigators to not only consider a whole community, but also implement controls and increase replication. Other studies investigating PUAs have measured predation and grazer success in the field (e.g., Irigoien et al. 2002), providing data that is unlikely to have been altered by artificial conditions, but without the benefit of controls. Similar approaches could be used to measure other compounds in natural populations. The use of large-scale manipulations of signaling molecules in natural conditions, similar to the iron-fertilization experiments, has also been proposed (Pohnert 2010). The metabolomic approaches described previously can also be used for field studies, and, as the number of annotated genomes expands, metagenomic analysis of planktonic ecosystems will be valuable.

9.7.4 CONCLUSION

Phytoplankton play pivotal roles in ecosystems by fixing carbon and driving biogeochemical nutrient cycles (Falkowski 2012). Zooplankton are the critical link between phytoplankton primary production and higher trophic levels (Calbet and Landry 2004; Turner 2004). The compounds produced by both phytoplankton and zooplankton often determine the predator–prey interactions in the marine plankton, and these interactions shape the plankton community. Changes in plankton community structure can have drastic consequences across the ecosystem, as in the formation of HABs (Landsberg 2002), as well as across the planet, as in the promotion of cloud formation by DMS (Charlson et al. 1987). Global climate change is likely to alter chemical communication in the plankton, although how and to what extent remains unclear. Given the importance of these processes to the health of marine ecosystems, it is crucial to understand both the chemical and ecological aspects of chemically-mediated communication in the marine plankton.

REFERENCES

Adolf, J. E., T. R. Bachvaroff, D. N. Krupatkina, H. Nonogaki, P. J. P. Brown, A. J. Lewitus, H. R. Harvey, and A. R. Place. 2006. Species specificity and potential roles of *Karlodinium micrum* toxin. *Afr. J. Mar. Sci.* 28(2):415–419.

Adolf, J. E., Danara Krupatkina, Tsvetan Bachvaroff, and Allen R. Place. 2007. Karlotoxin mediates grazing by *Oxyrrhis marina* on Strains of *Karlodinium veneficum*. *Harmful Algae* 6(3):400–412.

Alcolombri, U., S. Ben-Dor, E. Feldmesser, Y. Levin, D. S. Tawfik, and A. Vardi. 2015. Identification of the algal dimethyl sulfide–releasing enzyme: A missing link in the marine sulfur cycle. *Science* 348(6242):1466–1469.

Amin, R. M., M. Koski, U. Bamstedt, and C. Vidoudez. 2011. Strain-related physiological and behavioral effects of *Skeletonema marinoi* on three common planktonic copepods. *Mar. Biol.* 158(9):1965–1980.

Anderson, D. M. 1997. Bloom dynamics of toxic *Alexandrium* species in the Northeastern US. *Limnol. Oceanogr.* 42(5):1009–1022.

Anderson, D. M., T. J. Alpermann, A. D. Cembella, Y. Collos, E. Masseret, and M. Montresor. 2012. The globally distributed genus *Alexandrium*: Multifaceted roles in marine ecosystems and impacts on human health. *Harmful Algae* 14:10–35. doi:10.1016/j.hal.2011.10.012.

Avery, D. E., and H. G. Dam. 2007. Newly discovered reproductive phenotypes of a marine copepod reveal the costs and advantages of resistance to a toxic dinoflagellate. *Limnol. Oceanogr.* 52(5):2099–2108.

Bachvaroff, T. R., J. E. Adolf, A. R. Place, and others. 2009. Strain variation in *Karlodinium veneficum* (Dinophyceae): Toxin profiles, pigments, and growth characteristics. *J. Phycol.* 45(1):137–153.

Balestra, C., L. Alonso-Sáez, J. M. Gasol, R. Casotti. 2011. Group-specific effects on coastal bacterioplankton of polyunsaturated aldehydes produced by diatoms. *Aquat. Microb. Ecol.* 63(2):123–131.

Ban, S., C. Burns, and J. Castel. 1997. The paradox of diatom-copepod interactions. *Oceanograph. Lit. Rev.* 3(45):500.

Bargu, S., K. Lefebvre, and M. W. Silver. 2006. Effect of dissolved domoic acid on the grazing rate of krill *Euphausia pacifica*. *Mar. Ecol. Prog. Ser.* 312:169–175.

Barreiro, A., C. Guisande, M. Frangopulos, A. Gonzalez-Fernandez, S. Munoz, D. Perez, S. Magadan, I. Maneiro, I. Riveiro, and P. Iglesias. 2006. Feeding strategies of the copepod *Acartia clausi* on single and mixed diets of toxic and non-toxic strains of the dinoflagellate *Alexandrium minutum*. *Mar. Ecol. Prog. Ser.* 316:115–125.

Becerra, J. X., K. Noge, and D. Lawrence Venable. 2009. Macroevolutionary chemical escalation in an ancient plant–herbivore arms race. *Proc. Natl. Acad. Sci.* 106(43):18062–18066.

Berge, T., L. K. Poulsen, M. Moldrup, N. Daugbjerg, and P. J. Hansen. 2012. Marine microalgae attack and feed on metazoans. *ISME J.* 6(10):1926–1936.

Bergkvist, J., E. Selander, and H. Pavia. 2008. Induction of toxin production in dinoflagellates: The grazer makes a difference. *Oecologia* 156(1):147–154.

Breckels, M. N., N. W. F. Bode, E. A. Codling, and M. Steinke. 2013. Effect of grazing-mediated dimethyl sulfide (DMS) production on the swimming behavior of the copepod *Calanus helgolandicus*. *Mar. Drugs* 11(7):2486–2500.

Breckels, M. N., E. C. Roberts, S. D. Archer, G. Malin, and M. Steinke. 2011. The role of dissolved infochemicals in mediating predator-prey interactions in the heterotrophic dinoflagellate *Oxyrrhis marina*. *J. Plankton Res.* 33(4):629–639.

Breier, C. F., and E. J. Buskey. 2007. Effects of the red tide dinoflagellate, *Karenia brevis*, on grazing and fecundity in the copepod *Acartia tonsa*. *J. Plankton Res.* 29(2):115–126.

Bricelj, V. M., S. P. MacQuarrie, J. A. E. Doane, and L. B. Connell. 2010. Evidence of selection for resistance to paralytic shellfish toxins during the early life history of soft-shell clam, *Mya arenaria*, populations. *Limnol. Oceanogr.* 55(6):2463–2475. doi:10.4319/lo.2010.55.6.2463.

Bricelj, V. M., L. Connell, K. Konoki, S. P. MacQuarrie, T. Scheuer, W. A. Catterall, and V. L. Trainer. 2005. Sodium channel mutation leading to saxitoxin resistance in clams increases risk of PSP. *Nature* 434(7034):763–767.

Buttino, I., M. do Espirito Santo, A. Ianora, and A. Miralto. 2004. Rapid assessment of copepod (*Calanus helgolandicus*) embryo viability using fluorescent probes. *Mar. Biol.* 145(2):393–399.

Calbet, A., and M. R. Landry. 2004. Phytoplankton growth, microzooplankton grazing, and carbon cycling in marine systems. *Limnol. Oceanogr.* 49(1):51–57.

Calbet, A., D. Vaqué, J. Felipe, M. Vila, M. Montserrat Sala, M. Alcaraz, and M. Estrada. 2003. Relative grazing impact of microzooplankton and mesozooplankton on a bloom of the toxic dinoflagellate *Alexandrium minutum*. *Mar. Ecol. Prog. Ser.* 259:303–309.

Casotti, R., S. Mazza, C. Brunet, V. Vantrepotte, A. Ianora, and A. Miralto. 2005. Growth inhibition and toxicity of the diatom aldehyde 2-trans, 4-trans-decadienal on *Thalassiosira weissflogii* (Bacillariophyceae). *J. Phycol.* 41:7–20.

Cestèle, S., and W. A. Catterall. 2000. Molecular mechanisms of neurotoxin action on voltage-gated sodium channels. *Biochimie* 82(9):883–892.

Charlson, R. J., J. E. Lovelock, M. O. Andreae, and S. G. Warren. 1987. Oceanic phytoplankton, atmospheric sulphur, cloud albedo and climate. *Nature* 326(6114):655–661.

Clough, J., and S. Strom. 2005. Effects of *Heterosigma akashiwo* (Raphidophyceae) on protist grazers: Laboratory experiments with ciliates and heterotrophic dinoflagellates. *Aquat. Microb. Ecol.* 39(2):121–134.

Cohen, J. H., and R. B. Forward Jr. 2005. Photobehavior as an inducible defense in the marine copepod *Calanopia americana*. *Limnol. Oceanogr.* 50(4):1269–1277.

Cohen, J. H., P. A. Tester, and R. B. Forward. 2007. Sublethal effects of the toxic dinoflagellate *Karenia brevis* on marine copepod behavior. *J. Plankton Res.* 29(3):301–315.

Colin, S. P., and H. G. Dam. 2004. Testing for resistance of pelagic marine copepods to a toxic dinoflagellate. *Evol. Ecol.* 18(4):355–377.

Colin, S. P., and H. G. Dam. 2007. Comparison of the functional and numerical responses of resistant versus non-resistant populations of the copepod *Acartia hudsonica* fed the toxic dinoflagellate *Alexandrium tamarense*. *Harmful Algae* 6(6):875–882.

Cutignano, A., N. Lamari, E. Manzo, G. Cimino, and A. Fontana. 2011. Lipoxygenase products in marine diatoms: A concise analytical method to explore the functional potential of oxylipins. *J. Phycol.* 47(2):233–243.

Dam, H. G. 2013. Evolutionary adaptation of marine zooplankton to global change. *Ann. Rev. Mar. Sci.* 5(1):121211–172229.

Dam, H. G., and S. T. Haley. 2011. Comparative dynamics of paralytic shellfish toxins (PST) in a tolerant and susceptible population of the copepod *Acartia hudsonica*. *Harmful Algae* 10(3):245–253.

DeBose, J. L., S. C. Lema, and G. A. Nevitt. 2008. Dimethylsulfoniopropionate as a foraging cue for reef fishes. *Science* 319(5868):1356–1356.

Deeds, J. R., R. E. Hoesch, A. R. Place, and J. P. Y. Kao. 2015. The cytotoxic mechanism of karlotoxin 2(KmTx 2) from *Karlodinium veneficum* (Dinophyceae). *Aquat. Toxicol.* 159:148–155. doi:10.1016/j.aquatox.2014.11.028.

Deeds, J. R., and A. R. Place. 2006. Sterol-specific membrane interactions with the toxins from *Karlodinium micrum* (Dinophyceae)—A strategy for self-protection? *Afr. J. Mar. Sci.* 28(2):421–425.

de Vargas, C., S. Audic, N. Henry, J. Decelle, F. Mahé, R. Logares, E. Lara et al. 2015. Eukaryotic plankton diversity in the sunlit ocean. *Science* 348(6237):1261605.

d'Ippolito, G., N. Lamari, M. Montresor, G. Romano, A. Cutignano, A. Gerecht, G. Cimino, and A. Fontana. 2009. 15S-Lipoxygenase metabolism in the marine diatom *Pseudonitzschia delicatissima*. *New Phytol.* 183(4):1064–1071.

Doney, S. C., V. J. Fabry, R. A. Feely, and J. A. Kleypas. 2009. Ocean acidification: The other CO_2 problem. *Annu. Rev. Mar. Sci.* 1:169–192.

Doucette, G. J., C. M. Mikulski, K. L. King, P. B. Roth, Z. Wang, L. F. Leandro, S. L. DeGrasse et al. 2012. Endangered North Atlantic right whales (*Eubalaena glacialis*) experience repeated, concurrent exposure to multiple environmental neurotoxins produced by marine algae. *Environ. Res.* 112:67–76.

Dutz, J., and M. Koski. 2008. Copepod reproduction is unaffected by diatom aldehydes or lipid composition. *Limnol. Oceanogr.* 53(1):225–235.

Edwards, B. R., K. D. Bidle, and B. A. Van Mooy. 2015. Dose-dependent regulation of microbial activity on sinking particles by polyunsaturated aldehydes: Implications for the carbon cycle. *Proc. Natl. Acad. Sci.* 112(19):5909–5914.

Ehrlich, P. R., and P. H. Raven. 1964. Butterflies and plants: A study in coevolution. *Evolution* 18(4):586–608.

Endres, C. S., and K. J. Lohmann. 2012. Perception of dimethyl sulfide (DMS) by loggerhead sea turtles: A possible mechanism for locating high-productivity oceanic regions for foraging. *J. Exp. Biol.* 215(20):3535–3538.

Falkowski, P. G. 2012. Ocean science: The power of plankton. *Nature* 483(7387):S17–S20.

Falkowski, P. G., T. Fenchel, and E. F. Delong. 2008. The microbial engines that drive earth's biogeochemical cycles. *Science* 320(5879):1034–1039.

Field, C. B., M. J. Behrenfeld, J. T. Randerson, and P. Falkowski. 1998. Primary production of the biosphere: Integrating terrestrial and oceanic components. *Science* 281(5374):237–240.

Finiguerra, M., D. E. Avery, and H. G. Dam. 2014a. No evidence for induction or selection of mutant sodium channel expression in the copepod *Acartia husdsonica* challenged with the toxic dinoflagellate *Alexandrium fundyense*. *Ecol. Evol.* 4(17):3470–3481.

Finiguerra, M., D. E. Avery, and H. G. Dam. 2014b. Sodium channel expression in the copepod *Acartia hudsonica* as a function of exposure to paralytic shellfish toxin (PST). *Harmful Algae* 39:75–80.

Flewelling, L. J., J. P. Naar, J. P. Abbott, D. G. Baden, N. B. Barros, G. D. Bossart, M. D. Bottein et al. 2005. Brevetoxicosis: Red tides and marine mammal mortalities. *Nature* 435(7043):755–756.

Flores, H. S., G. H. Wikfors, and H. G. Dam. 2012. Reactive oxygen species are linked to the toxicity of the dinoflagellate *Alexandrium* spp. to protists. *Aquat. Microb. Ecol.* 66(2):199–209.

Flynn, K. J., and X. Irigoien. 2009. Aldehyde-induced insidious effects cannot be considered as a diatom defence mechanism against copepods. *Mar. Ecol. Prog. Ser.* 377:79–89.

Flynn, K. J., D. K. Stoecker, A. Mitra, J. A. Raven, P. M. Glibert, P. L. Hansen, E. Granéli, and J. M. Burkholder. 2012. Misuse of the phytoplankton–zooplankton dichotomy: The need to assign organisms as mixotrophs within plankton functional types. *J. Plankton Res.* 35:3–11.

Fu, F. X., A. R. Place, N. S. Garcia, and D. A. Hutchins. 2010. CO_2 and phosphate availability control the toxicity of the harmful bloom dinoflagellate *Karlodinium veneficum*. *Aquat. Microb. Ecol.* 59(1):55.

Garcés, E., M. Fernandez, A. Penna, K. Van Lenning, A. Gutierrez, J. Camp, and M. Zapata. 2006. Characterization of NW mediterranean *Karlodinium* spp. (Dinophyceae) strains using morphological, molecular, chemical, and physiological methodologies. *J. Phycol.* 42(5):1096–1112.

Geffeney, S. L., E. Fujimoto, E. D. Brodie, and P. C. Ruben. 2005. Evolutionary diversification of TTX-resistant sodium channels in a predator–prey interaction. *Nature* 434(7034):759–763.

Geraci, J. R., D. M. Anderson, R. J. Timperi, D. J. St. Aubin, G. A. Early, J. H. Prescott, and C. A. Mayo. 1989. Humpback whales (*Megaptera novaeangliae*) fatally poisoned by dinoflagellate toxin. *Can. J. Fish. Aquat. Sci.* 46(11):1895–1898.

Gerecht, A., G. Romano, A. Ianora, G. d'Ippolito, A. Cutignano, and A. Fontana. 2011. Plasticity of oxylipin metabolism among clones of the marine diatom *Skeletonema marinoi* (bacillariophyceae). *J. Phycol.* 47(5):1050–1056.

Gillard, J., J. Frenkel, V. Devos, K. Sabbe, C. Paul, M. Rempt, D. Inzé, G. Pohnert, M. Vuylsteke, and W. Vyverman. 2013. Metabolomics enables the structure elucidation of a diatom sex pheromone. *Angew. Chem. Int. Ed.* 52(3):854–857.

Hallegraeff, G. M. 2010. Ocean climate change, phytoplankton community responses, and harmful algal blooms: A formidable predictive challenge. *J. Phycol.* 46(2):220–235.

Hallett, C. S. 2012. Brief report on fish community responses to the *Karlodinium veneficum* Bloom of May 2012, Including the Response of the Fish Community Index of Estuarine Condition. http://researchrepository.murdoch.edu.au/13102/.

Hattenrath-Lehmann, T. K., J. L. Smith, R. B. Wallace, L. R. Merlo, F. Koch, H. Mittelsdorf, J. A. Goleski, D. M. Anderson, and C. J. Gobler. 2015. The effects of elevated CO_2 on the growth and toxicity of field populations and cultures of the saxitoxin-producing dinoflagellate, *Alexandrium fundyense. Limnol. Oceanogr.* 60(1):198–214.

Hay, M. E. 2009. Marine chemical ecology: Chemical signals and cues structure marine populations, communities, and ecosystems. *Annu. Rev. Mar. Sci.* 1:193.

Hay, M. E., and J. Kubanek. 2002. Community and ecosystem level consequences of chemical cues in the plankton. *J. Chem. Ecol.* 28(10):2001–2016.

Heuschele, J., and T. Kiorboe. 2012. The smell of virgins: Mating status of females affects male swimming behaviour in *Oithona davisae. J. Plankton Res.* 34(11):929–935.

Heuschele, J., and E. Selander. 2014. The chemical ecology of copepods. *J. Plankton Res.* 36(4):895–913.

Hoagland, P., D. Jin, A. Beet, B. Kirkpatrick, A. Reich, S. Ullmann, L. E. Fleming, and G. Kirkpatrick. 2014. The human health effects of Florida red tide (FRT) blooms: An expanded analysis. *Environ. Int.* 68:144–153.

Hong, J., S. Talapatra, J. Katz, P. A. Tester, R. J. Waggett, and A. R. Place. 2012. Algal toxins alter copepod feeding behavior. *PLoS One* 7(5):e36845.

Ianora, A., M. G. Bentley, G. S. Caldwell, R. Casotti, A. D. Cembella, J. Engström-Öst, C. Halsband et al. 2011. The relevance of marine chemical ecology to plankton and ecosystem function: An emerging field. *Mar. Drugs* 9(9):1625–1648.

Ianora, A., R. Casotti, M. Bastianini, C. Brunet, G. d'Ippolito, F. Acri, A. Fontana, A. Cutignano, J. T. Turner, and A. Miralto. 2008. Low reproductive success for copepods during a bloom of the non-aldehyde-producing diatom *Cerataulina pelagica* in the North adriatic sea. *Mar. Ecol.* 29(3):399–410.

Ianora, A., and A. Miralto. 2010. Toxigenic effects of diatoms on grazers, phytoplankton and other microbes: A review. *Ecotoxicology* 19(3):493–511.

Ianora, A., A. Miralto, S. A. Poulet, Y. Carotenuto, I. Buttino, G. Romano, R. Casotti et al. 2004. Aldehyde suppression of copepod recruitment in blooms of a ubiquitous planktonic diatom. *Nature* 429(6990):403–407.

Irigoien, X., R. P. Harris, H. M. Verheye, P. Joly, J. Runge, M. Starr, D. Pond et al. 2002. Copepod hatching success in marine ecosystems with high diatom concentrations. *Nature* 419(6905):387–389.

Isbister, G. K., and M. C. Kiernan. 2005. Neurotoxic marine poisoning. *Lancet Neurol.* 4(4):219–228.

Jensen, S. K., J. P. Lacaze, G. Hermann, J. Kershaw, A. Brownlow, A. Turner, and A. Hall. 2015a. Detection and effects of harmful algal toxins in Scottish harbour seals and potential links to population decline. *Toxicon* 97:1–14.

Jensen, P. R., B. S. Moore, and W. Fenical. 2015b. The marine actinomycete genus *Salinispora*: A model organism for secondary metabolite discovery. *Nat. Prod. Rep.* 32:738–751.

Jiang, X., D. J. Lonsdale, and C. J. Gobler. 2011. Rapid gain and loss of evolutionary resistance to the harmful dinoflagellate *Cochlodinium polykrikoides* in the copepod *Acartia tonsa. Limnol. Oceanogr.* 56(3):947–954.

John, U., U. Tillmann, J. Hülskötter, T. J. Alpermann, S. Wohlrab, and D. B. Van de Waal. 2015. Intraspecific facilitation by allelochemical mediated grazing protection within a toxigenic dinoflagellate population. *Proc. R. Soc. Lond. B Biol. Sci.* 282(1798):20141268.

Jónasdóttir, S. H., J. Dutz, M. Koski, L. Yebra, H. H. Jakobsen, C. Vidoudez, G. Pohnert, and J. C. Nejstgaard. 2011. Extensive cross-disciplinary analysis of biological and chemical control of *Calanus finmarchicus* reproduction during an aldehyde forming diatom bloom in mesocosms. *Mar. Biol.* 158(9):1943–1963.

Jónasdóttir, S. H., A. W. Visser, and C. Jespersen. 2009. Assessing the role of food quality in the production and hatching of *Temora longicornis* eggs. *Mar. Ecol. Prog. Ser.* 382:139–150.

Kamiyama, T., M. Tsujino, Y. Matsuyama, and T. Uchida. 2005. Growth and grazing rates of the tintinnid ciliate *Favella taraikaensis* on the toxic dinoflagellate *Alexandrium tamarense. Mar. Biol.* 147(4):989–997.

Koski, M., L. Yebra, J. Dutz, S. H. Jónasdóttir, C. Vidoudez, H. H. Jakobsen, G. Pohnert, and J. C. Nejstgaard. 2012. The effect of egg versus seston quality on hatching success, naupliar metabolism and survival of *Calanus finmarchicus* in mesocosms dominated by *Phaeocystis* and diatoms. *Mar. Biol.* 159(3):643–660.

Kozlowsky-Suzuki, B., P. Carlsson, A. Rühl, and E. Granéli. 2006. Food selectivity and grazing impact on toxic *Dinophysis* spp. by copepods feeding on natural plankton assemblages. *Harmful Algae* 5(1):57–68.

Kubanek, J., T. W. Snell, and C. Pirkle. 2007. Chemical defense of the red tide dinoflagellate *Karenia brevis* against rotifer grazing. *Limnol. Oceanogr.* 52(3):1026–1035.

Kuhlisch, C., and G. Pohnert. 2015. Metabolomics in chemical ecology. *Nat. Prod. Rep.* 32(7):937–955.

Kujawinski, E. B. 2011. The impact of microbial metabolism on marine dissolved organic matter. *Annu. Rev. Mar. Sci.* 3:567–599.

Lamari, N., M. V. Ruggiero, G. d'Ippolito, W. H. C. F. Kooistra, A. Fontana, and M. Montresor. 2013. Specificity of lipoxygenase pathways supports species delineation in the marine diatom genus *Pseudo-Nitzschia. PLoS One* 8(8):e73281.

Landsberg, J. H. 2002. The effects of harmful algal blooms on aquatic organisms. *Rev. Fish. Sci.* 10(2):113–390.

Lauritano, C., M. Borra, Y. Carotenuto, E. Biffali, A. Miralto, G. Procaccini, and A. Ianora. 2011. Molecular evidence of the toxic effects of diatom diets on gene expression patterns in copepods. *PLoS One* 6(10):e26850.

Lauritano, C., Y. Carotenuto, A. Miralto, G. Procaccini, and A. Ianora. 2012. Copepod population-specific response to a toxic diatom diet. *PLoS One* 7(10):e47262.

Leflaive, J., and L. Ten-Hage. 2009. Chemical interactions in diatoms: Role of polyunsaturated aldehydes and precursors. *New Phytol.* 184(4):794–805.

Legrand, C., K. Rengefors, G. O. Fistarol, and E. Graneli. 2003. Allelopathy in phytoplankton-biochemical, ecological and evolutionary aspects. *Phycologia* 42(4):406–419.

Lewitus, A. J., M. S. Wetz, B. M. Willis, J. M. Burkholder, M. W. Parrow, and H. B. Glasgow. 2006. Grazing activity of *Pfiesteria piscicida* (Dinophyceae) and susceptibility to ciliate predation vary with toxicity status. *Harmful Algae* 5(4):427–434.

Lima-Mendez, G., K. Faust, N. Henry, J. Decelle, S. Colin, F. Carcillo, S. Chaffron et al. 2015. Determinants of community structure in the global plankton interactome. *Science* 348(6237):1262073.

Lim, H. C., C. P. Leaw, T. H. Tan, N. F. Kon, L. H. Yek, K. S. Hii, S. T. Teng et al. 2014. A bloom of *Karlodinium australe* (Gymnodiniales, Dinophyceae) associated with mass mortality of cage-cultured fishes in West Johor Strait, Malaysia. *Harmful Algae* 40:51–62.

Maibam, C., P. Fink, G. Romano, M. C. Buia, E. Butera, and V. Zupo. 2015. *Centropages typicus* (Crustacea, Copepoda) reacts to volatile compounds produced by planktonic algae. *Mar. Ecol.* 36(3):819–834. doi:10.1111/maec.12254.

Meinwald, J. 2010. Natural products as molecular messengers. *J. Nat. Prod.* 74(3):305–309.

Miralto, A., G. Barone, G. Romano, S. A. Poulet, A. Ianora, G. L. Russo, I. Buttino et al. 1999. The insidious effect of diatoms on copepod reproduction. *Nature* 402(6758):173–176.

Moore, S. K., V. L. Trainer, N. J. Mantua, M. S. Parker, E. A. Laws, L. C. Backer, and L. E. Fleming. 2008. Impacts of climate variability and future climate change on harmful algal blooms and human health. *Environ. Health* 7(2):S4.

Mordue Luntz, A. J., and M. A. Birkett. 2009. A review of host finding behaviour in the parasitic sea louse, *Lepeophtheirus salmonis* (Caligidae: Copepoda). *J. Fish Dis.* 32(1):3–13.

Munday, P. L., D. L. Dixson, J. M. Donelson, G. P. Jones, M. S. Pratchett, G. V. Devitsina, and K. B. Døving. 2009. Ocean acidification impairs olfactory discrimination and homing ability of a marine fish. *Proc. Natl Acad. Sci.* 106(6):1848–1852.

Nedelcu, A. M., W. W. Driscoll, P. M. Durand, M. D. Herron, and A. Rashidi. 2011. On the paradigm of altruistic suicide in the unicellular world. *Evolution* 65(1):3–20.

Nevitt, G. A. 2000. Olfactory foraging by Antarctic procellariiform seabirds: Life at high reynolds numbers. *Biol. Bull.* 198(2):245–253.

Nevitt, G. A. 2011. The Neuroecology of dimethyl sulfide: A global-climate regulator turned marine infochemical. *Integr. Comp. Biol.* 51(5):819–825.

Nevitt, G. A., R. R. Veit, and P. Kareiva. 1995. Dimethyl sulphide as a foraging cue for Antarctic procellariiform seabirds. *Nature* 376(6542):680–682.

Nilsson, G. E., D. L. Dixson, P. Domenici, M. I. McCormick, C. Sørensen, S. A. Watson, and P. L. Munday. 2012. Near-future carbon dioxide levels alter fish behaviour by interfering with neurotransmitter function. *Nat. Clim. Change* 2(3):201–204.

Paul, C., A. Reunamo, E. Lindehoff, J. Bergkvist, M. A. Mausz, H. Larsson, H. Richter et al. 2012. Diatom derived polyunsaturated aldehydes do not structure the planktonic microbial community in a mesocosm study. *Mar. Drugs* 10(4):775–792.

Peng, J., A. R. Place, W. Yoshida, C. Anklin, and M. T. Hamann. 2010. Structure and absolute configuration of karlotoxin-2, an ichthyotoxin from the marine dinoflagellate *Karlodinium veneficum. J. Am. Chem. Soc.* 132(10):3277–3279.

Petitpas, C. M., J. T. Turner, J. R. Deeds, B. A. Keafer, D. J. McGillicuddy, P. J. Milligan, V. Shue, K. D. White, and D. M. Anderson. 2014. PSP toxin levels and plankton community composition and abundance in size-fractionated vertical profiles during spring/summer blooms of the toxic dinoflagellate *Alexandrium fundyense* in the Gulf of Maine and on Georges Bank, 2007, 2008, and 2010: 2. *Deep Sea Res. Part II Top. Stud. Oceanogr.* 103:350–367.

Place, A. R., H. A. Bowers, T. R. Bachvaroff, J. E. Adolf, J. R. Deeds, and J. Sheng. 2012. *Karlodinium veneficum*—The little dinoflagellate with a big bite. *Harmful Algae* 14:179–195.

Pohnert, G. 2000. Wound-activated chemical defense in unicellular planktonic algae. *Angew. Chem. Int. Ed.* 39(23):4352–4354.

Pohnert, G. 2002. Phospholipase A2 activity triggers the wound-activated chemical defense in the diatom *Thalassiosira rotula. Plant Physiol.* 129(1):103–111.

Pohnert, G. 2010. Chemical noise in the silent ocean. *J. Plankton Res.* 32(2):141–144.

Pohnert, G. 2012. How to explore the sometimes unusual chemistry of aquatic defence chemicals. *Chem. Ecol. Aquat. Syst.* 184–195.

Pohnert, G., O. Lumineau, A. Cueff, S. Adolph, C. Cordevant, M. Lange, and S. Poulet. 2002. Are volatile unsaturated aldehydes from diatoms the main line of chemical defence against copepods? *Mar. Ecol. Prog. Ser.* 245(1):33–45.

Poulet, S. A., A. Ianora, A. Miralto, and L. Meijer. 1994. Do diatoms arrest embryonic development in copepods? *Mar. Ecol. Prog. Ser.* 111(1):79–86.

Poulet, S. A., T. Wichard, J. B. Ledoux, B. Lebreton, J. Marchetti, C. Dancie, D. Bonnet, A. Cueff, P. Morin, and G. Pohnert. 2006. Influence of diatoms on copepod reproduction. I. Field and laboratory observations related to *Calanus helgolandicus* egg production. *Mar. Ecol. Prog. Ser.* 308:129–142.

Poulson-Ellestad, K. L., C. M. Jones, J. Roy, M. R. Viant, F. M. Fernández, J. Kubanek, and B. L. Nunn. 2014. Metabolomics and proteomics reveal impacts of chemically mediated competition on marine plankton. *Proc. Natl. Acad. Sci.* 111(24):9009–9014.

Poulson-Ellestad, K. L., E. L. Harvey, M. D. Johnson, and T. J. Minter. 2016. Evidence for strain-specific exometabolomic responses of the coccolithophore *Emiliania huxleyi* to grazing by the dinoflagellate *Oxyrrhis marina*. *Front. Mar. Sci.* 3:1–13. doi:10.3389/fmars.2016.00001.

Prince, E. K., L. Lettieri, K. J. McCurdy, and J. Kubanek. 2006. Fitness consequences for copepods feeding on a red tide dinoflagellate: Deciphering the effects of nutritional value, toxicity, and feeding behavior. *Oecologia* 147(3):479–488. doi:10.1007/s00442-005-0274-2.

Prince, E. K., and G. Pohnert. 2010. Searching for signals in the noise: Metabolomics in chemical ecology. *Anal. Bioanal. Chem.* 396(1):193–197.

Prince, E. K., K. L. Poulson, T. L. Myers, R. D. Sieg, and J. Kubanek. 2010. Characterization of allelopathic compounds from the red tide dinoflagellate *Karenia brevis*. *Harmful Algae* 10(1):39–48.

Rasher, D. B., E. Paige Stout, S. Engel, T. L. Shearer, J. Kubanek, and M. E. Hay. 2015. Marine and terrestrial herbivores display convergent chemical ecology despite 400 million years of independent evolution. *Proc. Natl. Acad. Sci.* 112(39):12110–12115. doi:10.1073/pnas.1508133112.

Rask, L., E. Andréasson, B. Ekbom, S. Eriksson, B. Pontoppidan, and J. Meijer. 2000. Myrosinase: Gene family evolution and herbivore defense in brassicaceae. *Plant Mol. Evol.* 42(1):93–113.

Rhoades, D. F. 1979. Evolution of plant chemical defense against herbivores. *Herbivores: Their Interaction with Secondary Plant Metabolites*, G. A. Rosenthal, and D. H. Janzen (Eds.), pp. 3–54. New York: Academic Press.

Ribalet, F., M. Bastianini, C. Vidoudez, F. Acri, J. Berges, A. Ianora, A. Miralto et al. 2014. Phytoplankton cell lysis associated with polyunsaturated aldehyde release in the Northern Adriatic Sea. *PLoS One* 9(1):e85947.

Ribalet, F., J. A. Berges, A. Ianora, and R. Casotti. 2007a. Growth inhibition of cultured marine phytoplankton by toxic algal-derived polyunsaturated aldehydes. *Aquat. Toxicol.* 85(3):219–227.

Ribalet, F., L. Intertaglia, P. Lebaron, and R. Casotti. 2008. Differential effect of three polyunsaturated aldehydes on marine bacterial isolates. *Aquat. Toxicol.* 86(2):249–255.

Ribalet, F., C. Vidoudez, D. Cassin, G. Pohnert, A. Ianora, A. Miralto, and R. Casotti. 2009. High plasticity in the production of diatom-derived polyunsaturated aldehydes under nutrient limitation: Physiological and ecological implications. *Protist* 160(3):444–451.

Ribalet, F., T. Wichard, G. Pohnert, A. Ianora, A. Miralto, and R. Casotti. 2007b. Age and nutrient limitation enhance polyunsaturated aldehyde production in marine diatoms. *Phytochemistry* 68(15):2059–2067.

Roberts, E. C., C. Legrand, M. Steinke, and E. C. Wootton. 2011. Mechanisms underlying chemical interactions between predatory planktonic protists and their prey. *J. Plankton Res.* 33(6):833–841.

Romano, G., G. L. Russo, I. Buttino, A. Ianora, and A. Miralto. 2003. A marine diatom-derived aldehyde induces apoptosis in copepod and sea urchin embryos. *J. Exp. Biol.* 206(19):3487–3494.

Savoca, M. S., and G. A. Nevitt. 2014. Evidence that dimethyl sulfide facilitates a tritrophic mutualism between marine primary producers and top predators. *Proc. Natl. Acad. Sci.* 111(11):4157–4161.

Selander, E., J. Kubanek, M. Hamberg, M. X. Andersson, G. Cervin, and H. Pavia. 2015. Predator lipids induce paralytic shellfish toxins in bloom-forming algae. *Proc. Natl. Acad. Sci.* 112(20):6395–6400.

Selander, E., P. Thor, G. Toth, and H. Pavia. 2006. Copepods induce paralytic shellfish toxin production in marine dinoflagellates. *Proc. R. Soc. Lond. B Biol. Sci.* 273(1594):1673–1680.

Senft-Batoh, C. D., H. G. Dam, S. E. Shumway, and G. H. Wikfors. 2015a. A multi-phylum study of grazer-induced paralytic shellfish toxin production in the dinoflagellate *Alexandrium fundyense*: A new perspective on control of algal toxicity. *Harmful Algae* 44:20–31.

Senft-Batoh, C. D., H. G. Dam, S. E. Shumway, G. H. Wikfors, and C. D. Schlichting. 2015b. Influence of predator–prey evolutionary history, chemical alarm-cues, and feeding selection on induction of toxin production in a marine dinoflagellate. *Limnol. Oceanogr.* 60(1):318–328.

Seymour, J. R., R. Simó, T. Ahmed, and R. Stocker. 2010. Chemoattraction to dimethylsulfoniopropionate throughout the marine microbial food web. *Science* 329(5989):342–345.

Shearn-Bochsler, V., E. W. Lance, R. Corcoran, J. Piatt, B. Bodenstein, E. Frame, and J. Lawonn. 2014. Fatal paralytic shellfish poisoning in Kittlitz's Murrelet (*Brachyramphus brevirostris*) nestlings, Alaska, USA. *J. Wildl. Dis.* 50(4):933–937.

Sheng, J., E. Malkiel, J. Katz, J. E. Adolf, and A. R. Place. 2010. A dinoflagellate exploits toxins to immobilize prey prior to ingestion. *Proc. Natl. Acad. Sci.* 107(5):2082–2087.

Smayda, T. J. 1997. Harmful algal blooms: Their ecophysiology and general relevance to phytoplankton blooms in the sea. *Limnol. Oceanogr.* 42(5 part 2):1137–1153.

Smetacek, V. 2001. A watery arms race. *Nature* 411(6839):745–745.

Sopanen, S., M. Koski, P. Uronen, P. Kuuppo, S. Lehtinen, C. Legrand, and T. Tamminen. 2008. *Prymnesium parvum* exotoxins affect the grazing and viability of the calanoid copepod *Eurytemora affinis*. *Mar. Ecol. Prog. Ser.* 361:191–202.

Steinke, M., J. Stefels, and E. Stamhuis. 2006. Dimethyl sulfide triggers search behavior in copepods. *Limnol. Oceanogr.* 51(4):1925–1930.

Strom, S. L. 2008. Microbial ecology of ocean biogeochemistry: A community perspective. *Science* 320(5879):1043–1045.

Strom, S. L., G. Wolfe, J. Holmes, H. Stecher, C. Shimeneck, S. Lambert, and E. Moreno. 2003a. Chemical defense in the microplankton I: Feeding and growth rates of heterotrophic protists on the DMS-producing phytoplankter *Emiliania huxleyi*. *Limnol. Oceanogr.* 48(1):217–229.

Strom, S. L., G. Wolfe, A. Slajer, S. Lambert, and J. Clough. 2003b. Chemical defenses in the microplankton II: Inhibition of protist feeding by B-dimethylsulfoniopropionate (DMSP). *Limnol. Oceanogr.* 48(1):230–237.

Tabashnik, B. E., T. Brévault, and Y. Carrière. 2013. Insect resistance to Bt Crops: Lessons from the first billion acres. *Nat. Biotechnol.* 31(6):510–521.

Tang, Y. Z., and C. J. Gobler. 2012. Lethal effects of Northwest Atlantic Ocean isolates of the dinoflagellate, *Scrippsiella trochoidea*, on Eastern Oyster (*Crassostrea virginica*) and Northern Quahog (*Mercenaria mercenaria*) larvae. *Mar. Biol.* 159(1):199–210.

Tatters, A. O., L. J. Flewelling, F. Fu, A. A. Granholm, and D. A. Hutchins. 2013. High CO_2 promotes the production of paralytic shellfish poisoning toxins by *Alexandrium catenella* from Southern California waters. *Harmful Algae* 30:37–43.

Taylor, R. L., G. S. Caldwell, P. J. W. Olive, and M. G. Bentley. 2012. The harpacticoid copepod *Tisbe holothuriae* is resistant to the insidious effects of polyunsaturated aldehyde-producing diatoms. *J. Exp. Mar. Bio. Ecol.* 413:30–37.

Tester, P. A., J. T. Turner, and D. Shea. 2000. Vectorial transport of toxins from the dinoflagellate *Gymnodinium breve* through copepods to fish. *J. Plankton Res.* 22(1):47–62.

Thornton, D. C. O. 2002. Individuals, clones or groups? Phytoplankton behaviour and units of selection. *Ethol. Ecol. Evol.* 14(2):165–173.

Tillmann, U., and U. John. 2002. Toxic effects of *Alexandrium* spp. on heterotrophic dinoflagellates: An allelochemical defence mechanism independent of PSP-toxin content. *Mar. Ecol. Prog. Ser.* 230:47–58.

Turner, J. T. 2004. The importance of small planktonic copepods and their roles in pelagic marine food webs. *Zool. Stud.* 43(2):255–266.

Turner, J. T. 2014. Planktonic marine copepods and harmful algae. *Harmful Algae* 32:81–93.

Turner, J. T., V. Roncalli, P. Ciminiello, C. Dell'Aversano, E. Fattorusso, L. Tartaglione, Y. Carotenuto et al. 2012. Biogeographic effects of the Gulf of Mexico red tide dinoflagellate *Karenia brevis* on mediterranean copepods. *Harmful Algae* 16:63–73.

Valiela, I. 1995. *Marine Ecological Processes*. New York: Springer Science & Business Media.

Van de Waal, D. B., T. Eberlein, U. John, S. Wohlrab, and B. Rost. 2014. Impact of elevated pCO$_2$ on paralytic shellfish poisoning toxin content and composition in *Alexandrium tamarense*. *Toxicon* 78:58–67.

Van Dolah, F. M. 2000. Marine algal toxins: Origins, health effects, and their increased occurrence. *Environ. Health Perspect.* 108(Suppl 1):133.

Van Donk, E., A. Ianora, and M. Vos. 2011. Induced defences in marine and freshwater phytoplankton: A review. *Hydrobiologia* 668(1):3–19.

Van Wagoner, R. M., J. R. Deeds, M. Satake, A. A. Ribeiro, A. R. Place, and J. L. C. Wright. 2008. Isolation and characterization of karlotoxin 1, a new amphipathic toxin from *Karlodinium veneficum*. *Tetrahedron Lett.* 49(45):6457–6461.

Vardi, A., K. D. Bidle, C. Kwityn, D. J. Hirsh, S. M. Thompson, J. A. Callow, P. Falkowski, and C. Bowler. 2008. A diatom gene regulating nitric-oxide signaling and susceptibility to diatom-derived aldehydes. *Curr. Biol.* 18(12):895–899.

Vardi, A., F. Formiggini, R. Casotti, A. De Martino, F. Ribalet, A. Miralto, and C. Bowler. 2006. A stress surveillance system based on calcium and nitric oxide in marine diatoms. *PLoS Biol.* 4(3):411.

Vermeij, G. J. 1994. The evolutionary interaction among species: Selection, escalation, and coevolution. *Annu. Rev. Ecol. Syst.* 25:219–236.

Vidoudez, C., R. Casotti, M. Bastianini, and G. Pohnert. 2011a. Quantification of dissolved and particulate polyunsaturated aldehydes in the adriatic sea. *Mar. Drugs* 9(4):500–513.

Vidoudez, C., J. C. Nejstgaard, H. H. Jakobsen, and G. Pohnert. 2011b. Dynamics of dissolved and particulate polyunsaturated aldehydes in mesocosms inoculated with different densities of the diatom *Skeletonema marinoi*. *Mar. Drugs* 9(3):345–358.

Vidoudez, C., and G. Pohnert. 2008. Growth phase specific release of polyunsaturated aldehydes by the diatom *Skeletonema marinoi*. *J. Plankton Res.* 30(11):1305.

Vogt, M., and P. S. Liss. 2009. Dimethylsulfide and climate. *Surface Ocean—lower Atmosphere Processes*. Geophysical Monograph Series 198:197–232.

Waggett, R. J., D. Hardison, and P. A. Tester. 2012. Toxicity and nutritional inadequacy of *Karenia brevis*: Synergistic mechanisms disrupt top-down grazer control. *Mar. Ecol. Prog. Ser.* 444:15–30.

Walsh, B. M., and J. M. O'Neil. 2014. Zooplankton community composition and copepod grazing on the West Florida shelf in relation to blooms of *Karenia brevis*. *Harmful Algae* 38:63–72.

Wang, L., T. Yan, R. Yu, and M. Zhou. 2005. Experimental study on the impact of dinoflagellate *Alexandrium* species on populations of the rotifer *Brachionus plicatilis*. *Harmful Algae* 4(2):371–382.

Watkins, S. M., A. Reich, L. E. Fleming, and R. Hammond. 2008. Neurotoxic shellfish poisoning. *Mar. Drugs* 6(3):431–455.

Watson, A. J., and P. S. Liss. 1998. Marine biological controls on climate via the carbon and sulphur geochemical cycles. *Phil. Trans. R. Soc. B Biol. Sci.* 353(1365):41–51.

West, T. L., H. G. Marshall, and P. A. Tester. 1996. Natural phytoplankton community responses to a bloom of the toxic dinoflagellate *Gymnodinium breve* Davis off the North Carolina coast. *Castanea* 356–368.

Wichard, T., S. A. Poulet, A. L. Boulesteix, J. B. Ledoux, B. Lebreton, J. Marchetti, and G. Pohnert. 2008. Influence of diatoms on copepod reproduction. II. Uncorrelated effects of diatom-derived α, β, γ, δ-unsaturated aldehydes and polyunsaturated fatty acids on calanus helgolandicus in the field. *Prog. Oceanogr.* 77(1):30–44.

Wichard, T., S. A. Poulet, C. Halsband-Lenk, A. Albaina, R. Harris, D. Liu, and G. Pohnert. 2005. Survey of the chemical defence potential of diatoms: Screening of fifty species for α, β, γ, δ-unsaturated aldehydes. *J. Chem. Ecol.* 31(4):949–958.

Wittstock, U., N. Agerbirk, E. J. Stauber, C. E. Olsen, M. Hippler, T. Mitchell-Olds, J. Gershenzon, and H. Vogel. 2004. Successful herbivore attack due to metabolic diversion of a plant chemical defense. *Proc. Natl. Acad. Sci. U. S. A.* 101(14):4859–4864.

Wohlrab, S., M. H. Iversen, and U. John. 2010. A molecular and co-evolutionary context for grazer induced toxin production in *Alexandrium tamarense. PLoS One* 5(11):e15039.

Wolfe, G. V., M. Levasseur, G. Cantin, and S. Michaud. 2000. DMSP and DMS dynamics and microzooplankton grazing in the labrador sea: Application of the dilution technique. *Deep Sea Res. Part I Oceanogr. Res. Pap.* 47(12):2243–2264.

Wolfe, G. V., and M. Steinke. 1996. Grazing-activated production of dimethyl sulfide (DMS) by two clones of *Emiliania huxleyi. Limnol. Oceanogr.* 41(6):1151–1160.

Wolfe, G. V., M. Steinke, and G. O. Kirst. 1997. Grazing-activated chemical defence in a unicellular marine alga. *Nature* 387(6636):894–897.

Yang, I., E. Selander, H. Pavia, and U. John. 2011. Grazer-induced toxin formation in dino-flagellates: A transcriptomic model study. *Eur. J. Phycol.* 46(1):66–73. doi:10.1080/09670262.2011.552194.

Zou, Y., Y. Yamasaki, Y. Matsuyama, K. Yamaguchi, T. Honjo, and T. Oda. 2010. Possible involvement of hemolytic activity in the contact-dependent lethal effects of the dino-flagellate *Karenia mikimotoi* on the rotifer *Brachionus plicatilis. Harmful Algae* 9(4):367–373.

10 Chemical Ecology of Marine Invertebrate Larval Settlement

Tilmann Harder, Jan Tebben, Mareen Möller, and Peter J. Schupp

CONTENTS

10.1 INTRODUCTION

Most benthic marine invertebrates have dispersive larvae and sessile or comparatively immobile juvenile stages. These two life stages are linked by a settlement event, during which larvae explore marine surfaces, attach onto substrates, and undergo drastic morphological changes into the benthic form. These events, commonly referred to as larval settlement and metamorphosis, are driven by environmental cues that convey information about settlement sites and aid motile larvae to choose spots supporting the benthic mode of life.

The common usage and definitions of larval settlement and larval metamorphosis often vary depending on the animal and habitat under investigation (Pawlik 1992). The term metamorphosis describes the irreversible transition of larvae, involving a conspicuous and abrupt change in the animal's body structure through cell growth and differentiation, resulting in the loss of relevant body structures for planktonic life. The term settlement has been used for larvae exiting the plankton and exploring the benthic substrate but is also used to collectively describe the overall larval transition from planktonic to benthic life stage including morphogenetic events (Crisp and Meadows 1962; Hadfield and Paul 2001; Pawlik 1992).

In this chapter, we consider the latter definition most appropriate. This is because a clear distinction between initial surface exploration and subsequent metamorphosis is rarely possible from literature data. Instead, the efficiency of settlement cues (which are signals that mediate the pelago-benthic transition) is most commonly determined by scoring percentages of larval metamorphosis in bioassays.

Many invertebrate larvae settle in response to cues from conspecifics (adults, juveniles, or larvae) or other organisms (heterospecifics) and are commonly distinguished as gregarious or associative larval settlement. It is understood that this distinction reflects the invertebrate's adaptation to particular life strategies, such as reproduction, mutualism, parasitism, and food preferences. The ecology of marine surface colonization by invertebrate larvae has received continuous attention during the past 40 years in seminal review papers (Crisp and Meadows 1962; Hadfield and Paul 2001; Pawlik 1992). Marine biofilms have long been recognized as a source of larval settlement cues for a broad range of marine invertebrate larvae (Zobell and Allen 1935). For a comprehensive coverage of marine biofilms, and particularly their bacterial component on larval settlement, the reader is referred to (Hadfield 2011).

Despite 40 years of research into the chemical ecology of surface colonization by benthic marine invertebrates the list of chemically identified settlement cues is surprisingly short.

In this chapter, we reviewed the literature for sources and origins of larval settlement cues, their identification to compound classes, and where applicable to structurally elucidated compounds (Table 10.1). The table covers a phylogenetically narrow range of invertebrates, dominated by species of maritime (e.g., fouling barnacles and tube worms) and commercial interest (e.g., mussels and sea urchins), and focuses on cues that presumably exist in situ and render larval surface colonization in ecologically realistic fashion. We therefore neglected artificial inducers of larval settlement and metamorphosis, such as many neurotransmitters, which play economically important roles in commercial aquaculture species.

The most seminal examples of structurally identified settlement cues stem from invertebrates with distinctive gregarious or associative larval settlement behavior. Contrary to our understanding of gregarious or associative settlement cues, comparatively little is known about the settlement signals present in or emanating from marine biofilms. This is likely due to the complex qualitative and quantitative compositional community structure of microorganisms in biofilms and hence a vast possible chemical complexity that marine biofilms offer to surface-exploring larvae.

This chapter provides an updated state of knowledge on benthic marine larval settlement cues and their biological origin (Table 10.1), with focus on the latest research into biofilm-mediated larval settlement. Given the emerging appreciation that any surface in the marine environment, animate or inanimate, is colonized by microorganisms, we place particular emphasis on the difficulties in differentiating between settlement cues associated with animals or plants and those originating from the microbiota associated with macroorganisms. Finally, we highlight the need to integrate ecology, microbiology, chemistry, and physics for a better understanding of the distribution of surface-active molecules in situ and their ecological relevance and plausibility.

TABLE 10.1

Larval Settlement Cues in Marine Invertebrate Phyla

Recipient			Specificity		Origin		Compound or Compound Class	References
Phylum	Class	Genus Species	Intraspecific (Gregarious)	Interspecific (Associative)	Biofilm			
Porifera	Porifera		x	x	x			Khalaman et al. (2014), Maldonado (2006), Whalan et al. (2008)
Cnidaria	Scyphozoa			x	x			Muller and Leitz (2002), Takahashi and Takeda (2015)
		Cassiopea andromeda Cassiopea xanmachana		x	x	Decaying mangrove leaves	Peptides	Fitt (1991), Fleck and Fitt (1999), Hofmann and Brand (1987), Hofmann et al. (1996)
	Anthozoa			x	x	Crustose coralline algae (CCA)		Gleason and Hofmann (2011), Muller and Leitz (2002), Takahashi and Takeda (2015)
		Leptastrea purpurea		x		CCA	Luminaolide	Kitamura et al. (2009)
		Pseudosiderastrea tayamai		x		CCA	11-deoxyfistularin-3	Kitamura et al. (2007)

(Continued)

TABLE 10.1 (Continued)
Larval Settlement Cues in Marine Invertebrate Phyla

| Recipient | | | Specificity | | Origin | | Compound or | |
Phylum	Class	Genus Species	Intraspecific (Gregarious)	Interspecific (Associative)		Biofilm	Compound Class	References
		Acropora millepora, Porites asteroides, Orbicella franksi, Acropora palmata		x	CCA	x	Tetrabromopyrrole	Negri et al. (2001), Sneed et al. (2014), Tebben et al. (2011)
		Agaricia humilis		x	CCA	x	High molecular weight carbohydrates	Morse and Morse (1991)
		Acropora palmata		x	CCA	x	GLW-Amide (Hym-248)	Erwin and Szmant (2010), Iwao et al. (2002)
	Hydrozoa			x		x		Kroiher and Berking (1999), Leitz and Wagner (1993), Muller and Leitz (2002), Takahashi and Takeda (2015)
		Coryne uchidai		x	Sargassum tortile		δ-tocopherol and its epoxide	Kato et al. (1975)
		Hydractinia echinata		x		x	Metamorphosin A	Leitz et al. (1994)

(Continued)

TABLE 10.1 (*Continued*)
Larval Settlement Cues in Marine Invertebrate Phyla

Phylum	Class	Genus Species	Intraspecific (Gregarious)	Interspecific (Associative)	Origin	Biofilm	Compound or Compound Class	References
Annelida	Polychaeta		x	x		x		Qian (1999)
		Hydroides elegans	x	x		x	6,9-heptadecadiene 12-octadecenoic acid	Hadfield (2011), Hung et al. (2009)
		Phragmatopoma californica	x				Free Fatty Acids	Jensen and Morse (1988, 1990), Pawlik (1986, 1990, 1992), Zimmer and Butman (2000)
Mollusca	Bivalvia		x	x		x		Soo and Todd (2014)
		Crassostrea virginica	x				Thyroxine containing protein	Fitt and Coon (1992), Tamburri et al. (1992), Veitch and Hidu (1971), Zimmerfaust and Tamburri (1994)
		Crassostrea gigas	x				Glycoprotein	Vasquez et al. (2013)
		Pecten maximus		x	*Delesseria sanguinea*		Jacarone	Chevolot et al. (1991), Yvin et al. (1985)
		Phestilla sibogae		x	Coral (Porites sp.)			Hadfield and Scheuer (1985), Ritson-Williams et al. (2009)

(Continued)

TABLE 10.1 (Continued)
Larval Settlement Cues in Marine Invertebrate Phyla

Phylum	Recipient Class	Genus Species	Specificity Intraspecific (Gregarious)	Specificity Interspecific (Associative)	Origin	Origin Biofilm	Compound or Compound Class	References
		Adalaria proxima		x	Electra oilosa		Watersoluble small arginine- or lysine-containing peptide	Lambert et al. (1997)
		Crepidula fornicata	x	x	CCA		Dibromomethane	Taris et al. (2010)
		Cymbula nigra					GABA	Lairnek et al. (2008), Rivera-Ingraham et al. (2011)
		Alderia modesta		x	Vaucheria longicaulis		Water-soluble and surface-associated carbohydrates	Krug and Manzi (1999)
	Gastropoda		x	x		x		Courtois de Vicose et al. (2010), Hadfield and Scheuer (1985), Ritson-Williams et al. (2009), Rodriguez et al. (1993)
		Phestilla sibogae		x	Coral (Porites sp.)			Hadfield and Pennington (1990), Hadfield and Scheuer (1985), Ritson-Williams et al. (2009)

(Continued)

TABLE 10.1 (Continued)
Larval Settlement Cues in Marine Invertebrate Phyla

Recipient			Specificity		Origin		Compound or Compound Class	References
Phylum	Class	Genus Species	Intraspecific (Gregarious)	Interspecific (Associative)	Biofilm			
		Haliotis sp.	x	x	x	Lithothamnion, Lithophyllum, Hildenbranchia (CCA)		Miller and Hadfield (1990), Morse et al. (1979, 1980), Shepherd and Turner (1985), Slattery (1992), Young and Braithwaite (1980)
		Adalaria proxima		x		*Electra oilosa*	Watersoluble small arginine- or lysine-containing peptide	Lambert et al. (1997)
		Crepidula fornicata	x	x		CCA	Dibromomethane	Taris et al. (2010)
		Alderia modesta		x		*Vaucheria longicaulis*	Watersoluble and surface-associated carbohydrates	Krug and Manzi (1999)
Echino-dermata	Asteroidea			x	x	CCA		Johnson and Sutton (1994), Metaxas et al. (2008)
		Acanthaster planci		x	x	CCA	Tyrosine (only in combination with cca)	Johnson and Sutton (1994), Johnson and Cartwright (1996)
	Holothuroidea		x	x	x			Mercier et al. (2000), Young and Chia (1982)

(Continued)

TABLE 10.1 (*Continued*)
Larval Settlement Cues in Marine Invertebrate Phyla

| | Recipient | | Specificity | | Origin | | Compound or | References |
Phylum	Class	Genus Species	Intraspecific (Gregarious)	Interspecific (Associative)	Biofilm		Compound Class	
	Echinoidea	*Stichopus japonicus*		x	x			Ito and Kitamura (1997)
		Dendraster excentricus	x	x	x			Antonette Juinio-Menez and Bangi (2010), Dworjanyn and Pirozzi (2008), Highsmith (1982)
		Holopneustes purpurascens	x				Peptide	Burke (1984), Highsmith (1982)
		Holopneustes purpurascens H. inflatus H. erythrogramma Centrostephanus rodgersii		x		*Delisea pulchra*	Floridoside and isethionic acid complex, histamine	Swanson et al. (2006), Swanson et al. (2007), Swanson et al. (2004), Williamson et al. (2000)
		Pseudocentrotus depressus		x		Coralline Algae	Histamine	Swanson et al. (2004, 2012)
				x		Coralline Algae	Free fatty acids	Kitamura et al. (1994), Rahim et al. (2004) *(Continued)*

TABLE 10.1 (Continued)
Larval Settlement Cues in Marine Invertebrate Phyla

| | Recipient | | Specificity | | Origin | | Compound or | |
Phylum	Class	Genus Species	Intraspecific (Gregarious)	Interspecific (Associative)		Biofilm	Compound Class	References
		Anthocidaris crassispina		x	Coralline Algae		Free fatty acids	Kitamura et al. (1993,1994), Rahim et al. (2004)
		Strongylocentrotus intermedius *S. mirabilis* *S. droebachiensis*		x	Algae		(several) glycoglycerolipids	Pearce and Scheibling (1990), Takahashi et al. (2002)
		Leodia sexiesperforata					Tyrosine	Heyland et al. (2004)
Bryozoa			x	x		x		Brancato and Woollacott (1982), Keough (1984, 1986), Mihm et al. (1981)
Brachiopoda Phoronida			x	x	Adult shells, CCA	x		Pennington et al. (1999)
		Phoronis sp.	x	x		x	Cl⁻	Herrmann (1979), Santagata (2004)
Ascidiacea	Ascidiacea		x	x		x		Szewzyk et al. (1991), Wieczorek and Todd (1997)

(Continued)

TABLE 10.1 (Continued)
Larval Settlement Cues in Marine Invertebrate Phyla

Phylum	Recipient Class	Genus Species	Specificity Intraspecific (Gregarious)	Specificity Interspecific (Associative)	Origin Biofilm	Compound or Compound Class	References
		Ciona savignyi	x	x		Urochordamine A	Fusetani (1997), Tsukamoto et al. (1993, 1999)
		Halocynthia roretzi	x			Lumichrome, anthosamines A and B, diketopipemzine 3, urochordamine,	Tsukamoto et al. (1993, 1995, 1999)
Arthropoda	Cirripedia		x	x	x		Puglisi et al. (2014), Thiyagarajan (2010)
		Semibalanus balanoides	x			Arthropodin, series of acidic proteins	Crisp and Meadows (1962), Knightjones (1953), Larman (1984), Larman and Gabbott (1975), Larman et al. (1982), Pawlik (1992), Yule and Walker (1985)
		Amphibalanus amphitrite	x			Water-borne peptides, Protein Complex	Dreanno et al. (2006a, 2006b, 2006c), Kato-Yoshinaga et al. (2000), Tegtmeyer and Rittschof (1988)

10.2   LARVAL SETTLEMENT MEDIATED BY MARINE BIOFILMS

Marine biofilms have long been recognized as a source of both inductive and inhibitive settlement cues for a broad range of marine invertebrate larvae (Zobell and Allen 1935). Yet, exactly how larvae perceive information, which signals transduce information, the source of these signals and their ecological plausibility, are still issues addressed by ongoing research.

Marine biofilms are assemblages of microorganisms and develop on surfaces immersed by seawater. These surfaces include inorganic substratum (rocks and sediment) and the exterior of marine animals, algae, and plants. The surface-associated microbial assemblages are complex and include many different groups of bacteria, diatoms, fungi, protozoa, viruses, and other small organisms, all of which are embedded in extracellular polymeric substances secreted by many of these organisms. The characteristics of biofilms are finely shaped by their qualitative and quantitative composition, which in turn is highly susceptible to and dependent on environmental conditions in the water column (temperature, light, pH, nutrients) and near the substratum (e.g., surface material, surface energy). Given their sensitivity to environmental conditions, marine biofilms literally transcribe local environmental information onto marine surfaces. Presumably, invertebrate larvae have evolved modes to assess this information during their exploration of marine surfaces prior to attachment and metamorphosis. It is beyond the scope of this chapter to cover the influence of environmental effects on biofilm characteristics; instead the reader is referred to detailed specialty reviews on this subject, for example (Kjelleberg and Givskov 2007).

Here, we review the past 20 years of research on biofilm-mediated induction of larval settlement with particular emphasis on characterized biofilm attributes shown to trigger behavioral responses of invertebrate larvae. The experimental strategies to investigate and correlate natural biofilm attributes with larval settlement have included the manipulation of biofilms by (1) elevated temperature, antibiotics, or UV energy (Anderson and Epifanio 2009; Bao et al. 2007; Lau et al. 2005; Li et al. 2014), (2) the addition of biofilm grazers (Dahms and Qian 2005; Shimeta et al. 2012), (3) the development of natural biofilms on different substratum types (Anderson and Epifanio 2009; Chung et al. 2010; Faimali et al. 2004; Huggett et al. 2009), or the harvest of biofilms at different (4) ages (Bao et al. 2007; Campbell et al. 2011; Chung et al. 2010; Faimali et al. 2004; Toupoint et al. 2012; Wang et al. 2012; Wieczorek and Todd 1997; Yang et al. 2014), (5) density (Bao et al. 2007; Campbell et al. 2011), and (6) location or tidal level (Hung et al. 2007; Qian et al. 2003).

Together these studies have unequivocally demonstrated that the effects of natural biofilms on larval settlement are susceptible to these experimental manipulations. This experience formed the basis for the widely acknowledged paradigm that natural biofilms present a characteristic proxy of the local environment and thus signal information to invertebrate larvae when exploring marine surfaces (Lee et al. 2014; Nielsen et al. 2015; Webster et al. 2004; Whalan and Webster 2014).

10.2.1 LARVAL SETTLEMENT CUES FROM BACTERIAL COMMUNITIES VERSUS INDIVIDUAL STRAINS

A key objective has been to decouple the larval response to the complex composition of microbes in natural biofilms from that to individual groups or microbial members of biofilms. In bacterial culture-dependent laboratory experiments larvae settled in response to a variety of monospecies bacterial films (Bao et al. 2007; Dobretsov and Qian 2006; Huggett et al. 2006; Lau et al. 2003; Sebesvari et al. 2013; Sharp et al. 2015; Tran and Hadfield 2011; Unabia and Hadfield 1999; Yang et al. 2013). Similarly, monospecies films of benthic diatoms were shown to trigger larval settlement (Dahms et al. 2004; Harder et al. 2002a; Jouuchi et al. 2007; Leise et al. 2009; Tung and Alfaro 2011). The shared conclusion of these studies was that phylogenetic affiliation of defined bacteria or diatoms was not a reliable proxy of their property to trigger larval settlement (Huang and Hadfield 2003; Lau et al. 2002; Tran and Hadfield 2011).

While marine invertebrate larvae clearly distinguish between bacterial biofilms from different environments (see above), for a long time it remained unclear if they receive an integrated cue from the quantitative and/or qualitative community composition of bacteria in natural biofilms, or if instead they perceive only a few or even just a single inductive or inhibitive bacterial species in natural biofilms. A recent study used a multivariate statistical approach to investigate if larval settlement of sea urchins correlated with the microbial community composition (Nielsen et al. 2015). This study showed a strong correlation of larval settlement with the community composition in natural biofilms as highlighted by canonical analysis of principle components, a constrained ordination technique. Using this technique, the importance of operational taxonomic units (OTUs) within natural bacterial communities relative to larval settlement was investigated. Surprisingly, larval settlement correlated not only with the epiphytic bacterial community composition but also with the relative abundance of a few OTUs within these communities (Nielsen et al. 2015), thus providing first evidence that both qualitative and quantitative features of bacterial biofilm communities provide an integrated settlement cue to sea urchin larvae.

In a series of papers, Hadfield and colleagues proposed that other than affiliation to certain bacterial species, particular bacterial genes were required to stimulate larval settlement in larvae of the polychaete *Hydroides elegans* (Huang et al. 2012). Transposon mutational analysis of the biofilm bacterium *Pseudoalteromonas luteoviolacea* was used to identify and suggest four genes required for metamorphic induction of larvae and to encode functions related to bacterial cell adhesion and cell secretion systems. Their study suggested that other factors than those relating to physical characteristics of biofilms may be critical to the inductive capacity of bacteria to stimulate larval settlement. This concept was confirmed by (Shikuma et al. 2014), who identified arrays of contractile phage tail-like structures that could be responsible for triggering settlement of tubeworm larvae. These secretion systems are bacterial virulence-associated nano-machines composed of proteins that are evolutionarily related to components of bacteriophage tails. Although the mechanisms of delivery of and response to metamorphosis inducing factors from these secretion systems are not yet understood, the idea that bacteria may facilitate the delivery of settlement cues in ways that twist the classical notion of specialized larval receptors for chemical cues deserves attention. The recent availability of the draft genome sequence of *P. luteoviolacea* (Asahina and Hadfield 2015) may be useful in elucidating the mechanisms involved in the settlement and metamorphosis of *H. elegans* and other marine invertebrates by *P. luteoviolacea*.

10.2.2 ECOLOGICAL CONSIDERATIONS TO ISOLATE AND TEST LARVAL SETTLEMENT CUES OF BACTERIAL ORIGIN

In accordance with the effect of natural biofilms on larval settlement, the properties of individual bacteria and diatoms to influence larval settlement strongly depends on their metabolic activity. Several studies have demonstrated that only viable and metabolically active bacteria or diatoms produce larval settlement cues (Harder et al. 2002a; Huggett et al. 2006; Lau et al. 2003; Unabia and Hadfield 1999). Yet, despite various attempts by many investigators over half a century, the bioassay-guided isolation and purification of single-molecule inducers of larval settlement delivered only a hand full of bacteria-derived compounds. This discrepancy suggests that the general concept of species-specific microbial metabolites that cause different benthic invertebrate species to recruit to specific locations either requires better isolation and assay techniques and/or a critical revision of our conceptual thinking about induction of larval settlement. The recent discovery of specific bacterial genes and delivery features, such as secretion systems (see the previous paragraph), points towards other or alternative modes of action how biofilm bacteria transcribe surface information to surface-exploring invertebrate larvae. Future studies will need to test, differentiate, and verify these signaling modes in ecologically plausible manner that is with ecologically relevant bacteria under realistic environmental concentrations.

Surprisingly, the bacterial isolates that have been demonstrated to cause settlement and metamorphosis in vitro have rarely been quantified in situ. One study aimed to ecologically verify the metamorphosis inducing effect of tetrabromopyrrole (TBP) (Negri et al. 2001; Tebben et al. 2011) and concluded that neither the concentration of this compound nor the cell numbers of bacteria producing this compound were sufficient to elicit larval metamorphosis in situ (Tebben et al. 2015).

However, the quantitative analysis of bacterial gene sequences for defined bacterial strains may overlook functionally redundant, but genetically different bacteria capable of producing the same settlement cues. While in the aforementioned study the absence of the causative molecule in situ was obvious, the direct detection and quantification of settlement inducing factors other than small organic molecules (e.g., phage-like structures) is by far more difficult and requires detailed knowledge of the gene clusters responsible for cue production. Future studies will need to quantify relevant concentrations of a cue directly, demonstrate the sufficient abundance of the cue-producing bacteria, or provide evidence for the presence of specific gene cluster encoding the production of the cue in situ.

A general uncertainty of previous studies has been how to test a chemically separated putative settlement cue in ecologically plausible manner. While the ecological rationale behind a settlement cue is to inform surface-exploring larvae about the suitability of a location as settlement site, it still remains speculative if settlement cues are *a priori* surface-bound and thus need to be tested as such, or if larvae principally respond to soluble settlement cues under laboratory assay conditions as long as the cue concentration is equal to its concentration on the surface. Adding to this ambiguity is whether larvae may require other tactile stimuli that convey proximity to a surface.

Several investigators have used the extracts or exudates of monospecies bacterial cultures as well as those of natural biofilms in laboratory settlement assays. While some managed to stimulate larval settlement with organic extracts of natural biofilms (Chung et al. 2010; Hung et al. 2009), others have shown that no cues could be extracted from larval settlement inducing bacterial culture media (Huang and Hadfield 2003; Sharp et al. 2015; Tebben et al. 2011). Only very few studies have addressed these issues in more detail. By videotaping the trajectories of swimming polychaete larvae (*H. elegans*), a recent study convincingly demonstrated that only those larvae that contacted the biofilmed surface slowed down and crawled over the surface to eventually settle (Hadfield et al. 2014), concluding that soluble settlement cues were not produced by biofilms or perceived by larvae under these conditions.

One study argued that distinction of bacterial biofilms into inductive versus non-inductive categories was unreliable and strongly depended on the amount of bacterial EPS (Harder et al. 2002b). When tested either dissolved in water or adsorbed to a substratum, separated bacterial metabolites only triggered larval settlement when immobilized to surfaces. Interestingly, the same metabolites were isolated from bacteria that did not affect larval settlement as single-species films (Harder et al. 2002b). Once adsorbed and concentrated onto artificial surfaces, these metabolites triggered the same magnitude of larval settlement as those isolated from inductive bacterial biofilms. These findings support the aforementioned notion that other bacterial traits than the production of inductive metabolites, such as the quantity and composition of EPS, play a pivotal role in rendering bacteria inductive to larval settlement.

The function of bacterial EPS as a facilitator of larval settlement induction therefore seems to support binding and concentration of cues to sufficient detection threshold concentrations. It is important to recognize how putative settlement cues, which are typically employed in still water laboratory assays at high concentrations, can reach such concentrations under the rapid turbulent flow conditions that occur in virtually all marine environments. The prospect for most benthic larval species to respond to soluble cues for settlement is indeed small, and has yet unequivocally demonstrated in only a few instances, such as sea urchins whose larvae must descend into an algal lawn to find inductive concentrations of histamine (Swanson et al. 2004, 2012), and the coral-eating nudibranch Phestilla (Hadfield and Koehl 2004; Koehl and Hadfield 2004), whose larvae descend into broad stretches of a coral species.

In summary, these results exemplify the need for future work to address under which environmental circumstances soluble settlement cues are ecologically meaningful and which organisms are likely to employ such cues. On the other hand, it clearly emerges that other bacterial traits such as EPS properties and specific phage-like bacterial secretion systems either support or directly trigger larval settlement in modes of action that have not been considered in prior studies and may indeed explain the low number of identified soluble settlement cues of bacterial origin. One consequence of these studies is the strong need to integrate larval ecology, microbiology, chemistry, and the physics of fluids to better assess the ecological relevance and plausibility of surface-active molecules in situ under turbulent flow conditions and the role of non-chemical modes that trigger settlement responses in invertebrate larvae.

10.3 LARVAL SETTLEMENT ON SURFACES OF OTHER ORGANISMS

The affinity of invertebrate larvae to specific sites and surfaces reflects the organism's adaptation to a particular ecological niche. While there is rich support for the role of marine biofilms on inanimate substratum to steer settlement of benthic marine invertebrates (Hadfield 2011), there is abundant evidence that invertebrate larvae also settle in response to other organisms, for example, gregariously on conspecifics or associatively on heterospecifics (Hadfield and Paul 2001).

Recent studies imply that the precise categorization into hetero- and conspecific settlement is less obvious. For example, (Dunn and Young 2014) showed that parasitic larvae of the nemertean barnacle *Carcinonemertes errans* preferentially settle on the Dungeness crab *Metacarcinus magister*. While larvae respond to cues associated with the crab's exoskeleton and migrate under the abdominal flap to prey on the crab's eggs, *C. errans* larvae settle significantly higher on already infested crabs, suggesting the parasite can sense and settle both associatively (on uninfected crabs) as well as gregariously. Consequently, the ability of *C. errans* to respond to both cues types increases its chance to identify and infest prey items. This example is typical, both for the lack of chemically characterized cues as well as for the complex behavioral response to multiple signals.

In a seminal review paper on larval settlement in response to bacterial biofilms, Hadfield described a similar case when referring to "the confusing barnacles" (Hadfield 2011). Larvae of the barnacle *Amphibalanus amphitrite* settle associatively to bacterial biofilms and gregariously to a surface bound settlement-inducing protein complex (SIPC) as well as in response to seawater conditioned with conspecifics (Elbourne and Clare 2010). The SIPC has been found in adult barnacle shells (Matsumura et al. 1998), cuticles (Dreanno et al. 2006b), and the footprints of surface-exploring cyprids (Dreanno et al. 2006a) and was identified as a α2-macroglobulin-like glycoprotein (Pagett et al. 2012). This broad range of settlement responses to different cue sources raised the question if barnacle larvae perceived the same cue from all settlement-inducing substrates or detected different cues through several specific receptors. It was also considered that some invertebrates produce a genetically mixed batch of larvae, some of which respond to different cues than others (Hadfield 2011).

There are no clear answers to these questions and they may well depend on the phylum under investigation. The overarching question derived from these observations is if larval settlement is commonly triggered by multiple signals. Historically, the concept of larval settlement has been concerned with correlating larval behavior with a specific cue, and much effort has been made to identify molecules that explain larval settlement preferences of different invertebrate phyla to the same environmental substrate, for example, that of corals to crustose coralline algae (CCA).

However, much of the literature seems to contradict this concept. For instance, the aforementioned barnacle *A. amphitrite* settles in response to surface-bound SIPC as well as water-soluble conspecific cues and unknown cues associated with natural biofilms. Female larvae of the parasitic rhizocephalan barnacle *Loxothylacus texanus* settle in response to carbohydrate or glycoprotein cues in the epicuticle layer of its host *Callinectes sapidus*, whereas male larvae settle gregariously in response to conspecific females (Boone et al. 2003). The Australian sea urchin *Holopneustes purpuracens* settles in response to histamine produced by the red algae *Delisea pulchra* (Swanson et al. 2004). These urchins also settle at high rates and with great specificity to several coralline algae co-occurring in the same habitat although these algae lack the required threshold concentrations of histamine (Swanson et al. 2004). Coral larvae settle in response to (organically insoluble) polymeric sugars (Morse and Morse 1991) but also settle in response to ethanolic and methanolic extracts and fractions of CCA (Harrington et al. 2004; Heyward and Negri 1999; Siboni et al. 2014; Tebben et al. 2015).

While it remains open if larvae indeed have different receptors for multiple cues or, alternatively, carry receptors specific to multiple agonists belonging to the same chemical compound class, these observations suggest that many invertebrate larvae respond to multiple substrates and cues, respectively. Some invertebrates seem to be capable of settling both associatively and gregariously, which maximizes their chance of locating suitable habitats that specifically support the adult stage's life strategy and adaptation to its respective ecological niche.

10.3.1 CHEMICALLY COMPLEX SOURCES OF LARVAL SETTLEMENT CUES

Generally, the substrates identified as natural sources of larval settlement cues are conglomerates consisting of multiple species of micro- and macro-organisms (e.g., natural biofilms, algae, carapace surfaces). Thus, the identification of natural settlement inducers has not only been challenged by difficulties to decouple complex behavioral larval responses to multiple substrates (see above), but also by the chemical complexity of these conglomerates. Contrary to microbial films and especially their bacterial constituents, natural plant or animal surfaces comprise an even more complex molecular signature originating from the host and its epibiotic biofilm. Analytically, it is hard to decouple putative settlement cues produced by macroscopic hosts from those produced by host-associated microorganisms, because no chemical separation technique can selectively focus on either one of these putative cue sources.

Rarely has the source of a settlement cue been inferred from the cue's chemical identity. Larval settlement cues of mangrove sponges (*Tedania ignis*) were traced to tannins present in mangrove roots, with tannic acid demonstrated to induce larval settlement of *T. ignis* in vitro (Hunting et al. 2010). In other cases, the settlement cue quantity isolated from the natural substratum allowed inferences about its origin. For example, extracts of the red alga *Delisea pulchra* contain more than 100 μg per g algal dry weight of histamine, the settlement cue of the urchin settlement *Holopneustes purpurascens* (Swanson et al. 2006). This concentration is too high to be produced by epiphytic bacteria on the algal surface and thus is of algal origin.

However, in most studies the origin of putative settlement cues has been less obvious. To address this issue many investigators have manipulated the epibiotic microbial part of multispecies substrates chemically (antibiotics, organic solvents) and/or physically (UV energy). While in some studies these treatments removed the settlement-inducing capacity of the substrate (Anderson and Epifanio 2009; Huang and Hadfield 2003; Unabia and Hadfield 1999), others found that larval settlement was not affected by such treatments, suggesting that invertebrate larvae employed in these studies (Heyward and Negri 1999; Tebben et al. 2015) responded entirely to cues associated with the macroscopic host.

10.4 ECOLOGICAL PLAUSIBILITY OF CHARACTERIZED LARVAL SETTLEMENT CUES

The following case study illustrates the pitfalls when assigning a defined ecological function to single bioactive compounds isolated from complex, habitat-specific cue sources for coordinated larval metamorphosis and settlement. The larvae of scleractinian corals have been consistently shown to settle in response to cues associated with CCA and/or their associated bacteria (Morse et al. 1988; Negri et al. 2001; Sneed et al. 2014; Tebben et al. 2015; Tebben et al. 2011; Webster et al. 2004). CCA are key reef-building species that cement loose coral rubble with a hard carbonate skeleton and thus provide a suitable site for attachment of corals. There is ample evidence that CCA-associated bacteria are strong triggers for coral larval settlement (Negri et al. 2001). Recently, TBP was identified in four closely related *Pseudoalteromonas* strains associated with CCA (Sneed et al. 2014; Tebben et al. 2011). The observation that exposure of coral larvae to TBP resulted in metamorphosed but not attached larvae led to the speculation that TBP had a net negative effect on coral recruitment (Tebben et al. 2011, 2015). The gene expression profiles of coral larvae exposed to either CCA extracts or TBP differed significantly (Siboni et al. 2014), suggesting that TBP did not match and mimic the cue signature inherent to CCA. While densities of at least 7,000 TBP-producing bacteria mm^{-2} crustose coralline algal surface were required to elicit statistically significant levels of larval metamorphosis of acroporid hard coral larvae in vitro (Tebben et al. 2011), it was shown that less than 1 cell mm^{-2} of these bacteria actually occurred in situ, thus challenging the notion of these bacteria as ecologically plausible mediators of coral larval settlement in reef ecosystems (Tebben et al. 2015). Instead, glycosaminoglycans (Morse and Morse 1991) and glycoglycerolipids (Tebben et al. 2015) have been identified as crustose coralline algal-derived inducers of coral larval settlement whereas the role and function of bacteria remains questionable. This study emphasizes the need to verify and compare laboratory larval settlement assay-guided studies using environmental bacterial isolates under ecologically plausible or at least comparable field conditions.

10.5 CONCLUSION

Despite considerable efforts in the past 15 years, only very few compounds that elicit settlement and metamorphosis in marine invertebrate larvae have been fully identified. On the contrary, the past 5 years of research into the chemical ecology of surface colonization in marine benthic systems have seen clear advances with regard to the role of complex bacterial communities as settlement cues, the identification of prominent and phylogenetically closely affiliated bacterial groups as producers of settlement cues, and the genetic requirements and morphological features of bacteria to induce larval settlement. However, our understanding of the chemical ecology of marine benthic surface colonization is far less advanced than studies of chemically mediated plant–herbivore or predator–prey interactions, or indeed of deterrents of larval settlement (Steinberg et al. 2001).

As this present chapter highlights, this is largely because few inducers of invertebrate larval settlement have been chemically characterized, quantified in situ, and shown to induce settlement at naturally occurring concentrations. Instead, our current knowledge of putative settlement cues is often restricted to their presence at unknown concentrations in crude extracts of biological tissues. The demonstration that purified cues in these extracts mediate larval settlement in situ at ecologically plausible concentrations has remained largely unexplored, with the notable exception of Raimondi and Morse (2000) and Tebben et al. (2015). This lack of information remains a critical gap in our understanding of chemical mediation of surface colonization in the marine environment. To advance the field of surface colonization in marine benthic systems, especially under the paradigm to demonstrate the ecological plausibility and efficacy of putative settlement cues, the following issues warrant closer attention and consideration in the future:

- To verify if bacteria and their respective inductive metabolites identified to stimulate larval settlement in vitro occur at relevant natural densities and concentrations, respectively, to trigger larval settlement behavior in situ.
- While the classical petri dish assay design has been useful to quickly screen larval behavior in response to putative settlement cues in stagnant water, other standardized test designs that consider the dilution of putative cues via diffusion or water flow must be employed in future studies.
- The bioassay-guided purification of larval settlement cues from crude biological matrices or extracts is prone to identify false positives as cues. This is because any biologically active compound present in settlement inducing crude extracts may be amplified during this isolation strategy until a behavioral larval response is observed. Since many invertebrate larvae show metamorphosis-like stress responses to many bioactive compounds and neurotransmitters (Beiras and Widdows 1995; Coon et al. 1986; Erwin and Szmant 2010), it will be important to distinguish these behavioral patterns from true settlement responses. It should become common practice to verify isolated putative settlement cues in situ at ecologically plausible threshold concentration.
- So far, our understanding of a correlation between microbial community structure in natural biofilms and the capacity of these communities to trigger larval behavioral responses is scarce, with the exception of Nielsen et al. (2015). However, the widespread paradigm in ecology that community structure determines function has recently been challenged by the high complexity of

microbial communities. The link between community structure and function in complex bacterial communities has been successfully decoupled in seminal studies of soil (Purahong et al. 2014) and marine epiphytic communities (Burke et al. 2011) using metagenomic sequence analyses. Common findings of these studies were that functions were distributed widely across a variety of taxa or phylogenetic groups of bacteria. The observation of similarity with respect to functional genes, but not species, suggests that the key level at which to address the role of structure and function of bacterial communities as larval settlement cues may not be at the level of "species" (by means of rRNA taxonomy), but rather at the functional level of genes.

The research of surface colonization in marine benthic systems began as a discipline largely motivated by maritime (antifouling) and blue biotechnological (aquaculture) interests. While arguably neither of these applications still rely on an active research stream into larval surface colonization, other fields have emerged that clearly ask for detailed knowledge and expertise in signal mediated colonization of marine benthic surfaces. Many marine populations are responding to global change with changes in distribution, abundance, physiology, and phenology (Poloczanska et al. 2013). In this context it will be increasingly important to understand if and to what extent the drivers for surface colonization are affected by large-scale climate changes and increased frequency of extreme events.

REFERENCES

Anderson, J. A., and C. E. Epifanio. 2009. Induction of metamorphosis in the Asian shore crab *Hemigrapsus sanguineus*: Characterization of the cue associated with biofilm from adult habitat. *J. Exp. Mar. Biol. Ecol.* 382(1):34–39.

Antonette Juinio-Menez, M., and H. G. P. Bangi. 2010. Extrinsic and intrinsic factors affecting the metamorphic rate of *Tripneustes gratilla* (Echinodermata: Echinoidea). *Mar. Ecol. Prog. Ser.* 402:137–145.

Asahina, A. Y., and M. G. Hadfield. 2015. Draft genome sequence of *Pseudoalteromonas luteoviolacea* HI1, determined using Roche 454 and PacBio single-molecule real-time hybrid sequencing. *Genome Announc.* 3(1):e01590–14.

Bao, W.-Y., J.-L. Yang, C. G. Satuito, and H. Kitamura. 2007. Larval metamorphosis of the mussel *Mytilus galloprovincialis* in response to *Alteromonas* sp. 1: Evidence for two chemical cues? *Mar. Biol.* 152(3):657–666.

Beiras, R., and J. Widdows. 1995. Induction of metamorphosis in larvae of the oyster *Crassostrea gigas* using neuroactive compounds. *Mar. Biol.* 123(2):327–334.

Boone, E. J., A. A. Boettcher, T. D. Sherman, and J. J. O'Brien. 2003. Characterization of settlement cues used by the rhizocephalan barnacle *Loxothylacus texanus*. *Mar. Ecol. Prog. Ser.* 252:187–197.

Brancato, M. S., and R. M. Woollacott. 1982. Effect of microbial films on settlement of bryozoan larvae (*Bugula simplex*, *B. stolonifera*, and *B. turrita*). *Mar. Biol.* 71(1):51–56.

Burke, C., P. Steinberg, D. Rusch, S. Kjelleberg, and T. Thomas. 2011. Bacterial community assembly based on functional genes rather than species. *Proc. Natl. Acad. Sci. USA* 108(34): 14288–14293.

Burke, R. D. 1984. Pheromonal control of metamorphosis in the Pacific sand dollar, *Dendraster excentricus*. *Science* 225(4660):442–443.

Campbell, A. H., D. W. Meritt, R. B. Franklin, E. L. Boone, C. T. Nicely, and B. L. Brown. 2011. Effects of age and composition of field-produced biofilms on oyster larval setting. *Biofouling* 27(3):255–265.

Chevolot, L., J. C. Cochard, and J. C. Yvin. 1991. Chemical induction of larval metamorphosis of *Pecten maximus* with a note on the nature of naturally occurring triggering substances. *Mar. Ecol. Prog. Ser.* 74(1):83–89.

Chung, H. C., O. O. Lee, Y.-L. Huang, S. Y. Mok, R. Kolter, and P.-Y. Qian. 2010. Bacterial community succession and chemical profiles of subtidal biofilms in relation to larval settlement of the polychaete *Hydroides elegans*. *Isme J.* 4(6):817–828.

Coon, S. L., D. B. Bonar, and R. M. Weiner. 1986. Chemical production of clutchless oyster spat using epinephrine and norepinephrine. *Aquaculture* 58(3–4):255–262.

Courtois de Vicose, G., M. Viera, A. Bilbao, and M. Izquierdo. 2010. Larval settlement of *Haliotis tuberculata coccinea* in response to different inductive cues and the effect of larval density on settlement, early growth, and survival. *J. Shellfish Res.* 29(3):587–591.

Crisp, D. J., and P. S. Meadows. 1962. Chemical basis of gregariousness in cirripedes. *Proc. R. Soc. Ser. B Biol. Sci.* 156(965):500-+.

Dahms, H. U., S. Dobretsov, and P. Y. Qian. 2004. The effect of bacterial and diatom biofilms on the settlement of the bryozoan *Bugula neritina*. *J. Exp. Mar. Biol. Ecol.* 313(1):191–209.

Dahms, H. U., and P. Y. Qian. 2005. Exposure of biofilms to meiofaunal copepods affects the larval settlement of *Hydroides elegans* (Polychaeta). *Mar. Ecol. Prog. Ser.* 297:203–214.

Dobretsov, S., and P. Y. Qian. 2006. Facilitation and inhibition of larval attachment of the bryozoan *Bugula neritina* in association with mono-species and multi-species biofilms. *J. Exp. Mar. Biol. Ecol.* 333(2):263–274.

Dreanno, C., R. R. Kirby, and A. S. Clare. 2006a. Smelly feet are not always a bad thing: The relationship between cyprid footprint protein and the barnacle settlement pheromone. *Biol. Lett.* 2(3):423–425.

Dreanno, C., R. R. Kirby, and A. S. Clare. 2006b. Locating the barnacle settlement pheromone: Spatial and ontogenetic expression of the settlement-inducing protein complex of *Balanus amphitrite*. *Proc. R. Soc. B Biol. Sci.* 273(1602):2721–2728.

Dreanno, C., K. Matsumura, N. Dohmae, K. Takio, H. Hirota, R. R. Kirby, and A. S. Clare. 2006c. An alpha(2)-macroglobulin-like protein is the cue to gregarious settlement of the barnacle *Balanus amphitrite*. *Proc. Natl. Acad. Sci. USA* 103(39):14396–14401.

Dunn, P. H., and C. M. Young. 2014. Larval settlement of the nemertean egg predator *Carcinonemertes errans* on the Dungeness crab, *Metacarcinus magister*. *Invertebr. Biol.* 133(3):201–212.

Dworjanyn, S. A., and I. Pirozzi. 2008. Induction of settlement in the sea urchin *Tripneustes gratilla* by macroalgae, biofilms and conspecifics: A role for bacteria? *Aquaculture* 274(2–4):268–274.

Elbourne, P. D., and A. S. Clare. 2010. Ecological relevance of a conspecific, waterborne settlement cue in *Balanus amphitrite* (Cirripedia). *J. Exp. Mar. Biol. Ecol.* 392(1–2):99–106.

Erwin, P. M., and A. M. Szmant. 2010. Settlement induction of *Acropora palmata* planulae by a GLW-amide neuropeptide. *Coral Reefs* 29(4):929–939.

Faimali, M., F. Garaventa, A. Terlizzi, M. Chiantore, and R. Cattaneo-Vietti. 2004. The interplay of substrate nature and biofilm formation in regulating *Balanus amphitrite* Darwin, 1854 larval settlement. *J. Exp. Mar. Biol. Ecol.* 306(1):37–50.

Fitt, W. K. 1991. Natural metamorphic cues of larvae of a tropical jellyfish. *Am. Zool.* 31(5):A106.

Fitt, W. K., and S. L. Coon. 1992. Evidence for ammonia as a natural cue for recruitment of oyster larvae to oyster beds in a Georgia salt marsh. *Biol. Bull.* 182(3):401–408.

Fleck, J., and W. K. Fitt. 1999. Degrading mangrove leaves of *Rhizophora mangle* Linne provide a natural cue for settlement and metamorphosis of the upside down jellyfish *Cassiopea xamachana* Bigelow. *J. Exp. Mar. Biol. Ecol.* 234(1):83–94.

Fusetani, N. 1997. Marine natural products influencing larval settlement and metamorphosi of benthic invertebrates. *Curr. Org. Chem.* 1(2):127–152.

Gleason, D. F., and D. K. Hofmann. 2011. Coral larvae: From gametes to recruits. *J. Exp. Mar. Biol. Ecol.* 408(1–2):42–57.

Hadfield, M. G., and M. A. R. Koehl. 2004. Rapid behavioral responses of an invertebrate larva to dissolved settlement cue. *Biol. Bull.* 207(1):28–43.

Hadfield, M. G., and V. J. Paul. 2001. Natural chemical cues for settlement and metamorphosis of marine invertebrate larvae. In *Marine Chemical Ecology*, J. B. McClintock and B. J. Baker (Eds.). Boca Raton, FL: CRC Press.

Hadfield, M. G., and J. T. Pennington. 1990. Nature of the metamorphic signal and its internal transduction in larvae of the nudibranch *Phestilla sibogae*. *Bull. Mar. Sci.* 46(2):455–464.

Hadfield, M. G., and D. Scheuer. 1985. Evidence for a soluble metamorphic inducer in *Phestilla*: ecological, chemical and biological data. *Bull. Mar. Sci.* 37(2):556–566.

Hadfield, M. G. 2011. Biofilms and marine invertebrate larvae: What bacteria produce that larvae use to choose settlement sites. *Annu. Rev. Mar. Sci.* 3:453–470.

Hadfield, M. G., B. T. Nedved, S. Wilbur, and M. A. R. Koehl. 2014. Biofilm cue for larval settlement in *Hydroides elegans* (Polychaeta): Is contact necessary? *Mar. Biol.* 161(11):2577–2587.

Harder, T., C. Lam, and P. Y. Qian. 2002a. Induction of larval settlement in the polychaete *Hydroides elegans* by marine biofilms: An investigation of monospecific diatom films as settlement cues. *Mar. Ecol. Prog. Ser.* 229:105–112.

Harder, T., S. C. K. Lau, H. U. Dahms, and P. Y. Qian. 2002b. Isolation of bacterial metabolites as natural inducers for larval settlement in the marine polychaete *Hydroides elegans* (Haswell). *J. Chem. Ecol.* 28(10):2029–2043.

Harrington, L., K. Fabricius, G. De'ath, and A. Negri. 2004. Recognition and selection of settlement substrata determine post-settlement survival in corals. *Ecology* 85(12):3428–3437.

Herrmann, K. 1979. Larvalentwicklung und Metamorphose von *Phoronis psammophila* (Phoronida, Tentaculata). *Helgol. wiss. Meeresunters.* 32:550–581.

Heyland, A., A. M. Reitzel, and J. Hodin. 2004. Thyroid hormones determine developmental mode in sand dollars (Echinodermata: Echinoidea). *Evol. Dev.* 6(6):382–392.

Heyward, A. J., and A. P. Negri. 1999. Natural inducers for coral larval metamorphosis. *Coral Reefs* 18(3):273–279.

Highsmith, R. C. 1982. Induced settlement and metamorphosis of sand dollar (*Dendraster ecentricus*) larvae in predator-free sites: Adult sand dollar beds. *Ecology* 63(2):329–337.

Hofmann, D. K., and U. Brand. 1987. Induction of metamorphosis in the symbiotic scyphozoan *Cassiopea andromeda*: Role of marine bacteria and of biochemicals. *Symbiosis* 4(1–3):99–116.

Hofmann, D. K., W. K. Fitt, and J. Fleck. 1996. Checkpoints in the life-cycle of *Cassiopea* spp.: Control of metagenesis and metamorphosis in a tropical jellyfish. *Int. J. Dev. Biol.* 40(1):331–338.

Huang, S. Y., and M. G. Hadfield. 2003. Composition and density of bacterial biofilms determine larval settlement of the polychaete *Hydroides elegans*. *Mar. Ecol. Prog. Ser.* 260:161–172.

Huang, Y., S. Callahan, and M. G. Hadfield. 2012. Recruitment in the sea: Bacterial genes required for inducing larval settlement in a polychaete worm. *Sci. Rep.* 2:228.

Huggett, M. J., B. T. Nedved, and M. G. Hadfield. 2009. Effects of initial surface wettability on biofilm formation and subsequent settlement of *Hydroides elegans*. *Biofouling* 25(5):387–399.

Huggett, M. J., J. E. Williamson, R. de Nys, S. Kjelleberg, and P. D. Steinberg. 2006. Larval settlement of the common Australian sea urchin *Heliocidaris erythrogramma* in response to bacteria from the surface of coralline algae. *Oecologia* 149(4):604–619.

Hung, O. S., O. O. Lee, V. Thiyagarajan, H. P. He, Y. Xu, H. C. Chung, J. W. Qiu, and P. Y. Qian. 2009a. Characterization of cues from natural multi-species biofilms that induce larval attachment of the polychaete *Hydroides elegans*. *Aquat. Biol.* 4(3):253–262.

Hung, O. S., V. Thiyagarajan, R. Zhang, Rudolf S. S. Wu, and P. Y. Qian. 2007. Attachment of *Balanus amphitrite* larvae to biofilms originating from contrasting environments. *Mar. Ecol. Prog. Ser.* 333:229–242.

Hung, O. S., O. O. Lee, V. Thiyagarajan, H.-P. He, Y. Xu, H. C. Chung, J.-W. Qiu, and P.-Y. Qian. 2009b. Characterization of cues from natural multi-species biofilms that induce larval attachment of the polychaete *Hydroides elegans*. *Aquat. Biol.* 4(3):253–262.

Hunting, E. R., H. G. van der Geest, A. J. Krieg, M. B. L. van Mierlo, and R. W. M. van Soest. 2010. Mangrove-sponge associations: A possible role for tannins. *Aquat. Ecol.* 44(4):679–684.

Ito, S., and H. Kitamura. 1997. Induction of larval metamorphosis in the sea cucumber *Stichopus japonicus* by periphitic diatoms. *Hydrobiologia* 358:281–284.

Iwao, K., T. Fujisawa, and M. Hatta. 2002. A cnidarian neuropeptide of the GLWamide family induces metamorphosis of reef-building corals in the genus *Acropora*. *Coral Reefs* 21(2):127–129.

Jensen, R. A., and D. E. Morse. 1988. The bioadhesive of *Phragmatopoma californica* tubes: A silk-like cement containing L-DOPA. *J. Comp. Physiol. B Biochem. Syst. Environ. Physiol.* 158(3):317–324.

Johnson, C. R., and D. C. Sutton. 1994. Bacteria on the surface of crustose coralline algae induce metamorphosis of the crown-of-thorns starfish *Acanthaster planci*. *Mar. Biol.* 120(2):305–310.

Johnson, L. G., and C. M. Cartwright. 1996. Thyroxine-accelerated larval development in the crown-of-thorns starfish, *Acanthaster planci*. *Biol. Bull.* 190(3):299–301.

Jouuchi, T., C. G. Satuito, and H. Kitamura. 2007. Sugar compound products of the periphytic diatom *Navicula ramosissima* induce larval settlement in the barnacle, *Amphibalanus amphitrite*. *Mar. Biol.* 152(5):1065–1076.

Kato-Yoshinaga, Y., M. Nagano, S. Mori, A. S. Clare, N. Fusetani, and K. Matsumura. 2000. Species specificity of barnacle settlement-inducing proteins. *Comp. Biochem. Physiol. A Mol. Integr. Physiol.* 125(4):511–516.

Kato, T., A. S. Kumanireng, I. Ichinose, Y. Kitahara, Y. Kakinuma, M. Nishihira, and M. Kato. 1975. Active components of *Sargassum* tortile effecting the settlement of swimming larvae of *Coryne uchidai*. *Experientia* 31(4):433–434.

Keough, M. J. 1984. Kin-recognition and the spatial-distribution of larvae of the Bryozoan *Bugula neritina* (L). *Evolution* 38(1):142–147.

Khalaman, V. V., N. M. Korchagina, and A. Yu Komendantov. 2014. The impact of water-borne cues from conspecifics and other species on the larvae of *Halichondria panacea* Pallas, 1766 (Porifera: Demospongiae). *Russ. J. Mar. Biol.* 40(1):36–42.

Kitamura, H., S. Kitahara, and H. B. Koh. 1993. The induction of larval settlement and metamorphosis of two sea urchins, *Pseudocentrotus depressus* and *Anthocidaris crassispina*, by free fatty acids extracted from coralline red alga *Corallina pilulifera*. *Mar. Biol.* 115(3):387–392.

Kitamura, M., T. Koyama, Y. Nakano, and D. Uemura. 2007. Characterization of a natural inducer of coral larval metamorphosis. *J. Exp. Mar. Biol. Ecol.* 340(1):96–102.

Kitamura, M., P. J. Schupp, Y. Nakano, and D. Uemura. 2009. Luminaolide, a novel metamorphosis-enhancing macrodiolide for scleractinian coral larvae from crustose coralline algae. *Tetrahedron Lett.* 50(47):6606–6609.

Kjelleberg, S., and M. Givskov. 2007. The biofilm mode of life. In *The Biofilm Mode of Life, Mechanisms and Adaptations*, S. Kjelleberg and M. Givskov (Eds.). Norwich, UK: Horizon Bioscience.

Knightjones, E. W. 1953. Laboratory experiments on gregariousness during setting in *Balanus balanoides* and other barnacles. *J. Exp. Biol.* 30(4):584–598.

Koehl, M. A. R., and M. G. Hadfield. 2004. Soluble settlement cue in slowly moving water within coral reefs induces larval adhesion surfaces. *J. Mar. Syst.* 49(1–4):75–88.

Kroiher, M., and S. Berking. 1999. On natural metamorphosis inducers of the cnidarians *Hydractinia echinata* (Hydrozoa) and *Aurelia aurita* (Scyphozoa). *Helgoland Mar. Res.* 53(2):118–121.

Krug, P. J., and A. E. Manzi. 1999. Waterborne and surface-associated carbohydrates as settlement cues for larvae of the specialist marine herbivore *Alderia modesta*. *Biol. Bull.* 197(1):94–103.

Lairnek, P., S. Clark, M. Stewart, F. Pfeffer, C. Wanichanon, P. Hanna, and P. Sobhon. 2008. The presence of GABA in gastropod mucus and its role in inducing larval settlement. *J. Exp. Mar. Biol. Ecol.* 354(2):182–191.

Lambert, W. J., C. D. Todd, and J. D. Hardege. 1997. Partial characterization and biological activity of a metamorphic inducer of the dorid nudibranch *Adalaria proxima* (Gastropoda: Nudibranchia). *Invertebr. Biol.* 116(2):71–81.

Larman, V. N. 1984. Protein extracts from some marine animals which promote barnacle settlement: Possible relationship between a protein component of arthropod cuticle and actin. *Comp. Biochem. Physiol. B Biochem. Mol. Biol.* 77(1):73–81.

Larman, V. N., and P. A. Gabbott. 1975. Settlement of cyprid larvae of *Balanus balanoides* and *Elminius modestus* induced by extracts of adult barnacles and other marine animals. *J. Mar. Biol. Assoc. U.K.* 55(1):183–190.

Larman, V. N., P. A. Gabbott, and J. East. 1982. Physico-chemical properties of the settlement factor proteins from the barnacle *Balanus balanoides*. *Comp. Biochem. Physiol. B Biochem. Mol. Biol.* 72(3):329–338.

Lau, S. C. K., T. Harder, and P. Y. Qian. 2003. Induction of larval settlement in the serpulid polychaete *Hydroides elegans* (Haswell): Role of bacterial extracellular polymers. *Biofouling* 19(3):197–204.

Lau, S. C. K., K. K. W. Mak, F. Chen, and P. Y. Qian. 2002. Bioactivity of bacterial strains isolated from marine biofilms in Hong Kong waters for the induction of larval settlement in the marine polychaete *Hydroides elegans*. *Mar. Ecol. Prog. Ser.* 226:301–310.

Lau, S. C. K., V. Thiyagarajan, S. C. K. Cheung, and P. Y. Qian. 2005. Roles of bacterial community composition in biofilms as a mediator for larval settlement of three marine invertebrates. *Aquat. Microb. Ecol.* 38(1):41–51.

Lee, O. O., H. C. Chung, J. Yang, Y. Wang, S. Dash, H. Wang, and P.-Y. Qian. 2014. Molecular techniques revealed highly diverse microbial communities in natural marine biofilms on polystyrene dishes for invertebrate larval settlement. *Microb. Ecol.* 68(1):81–93.

Leise, E. M., S. J. Froggett, J. E. Nearhoof, and L. B. Cahoon. 2009. Diatom cultures exhibit differential effects on larval metamorphosis in the marine gastropod *Ilyanassa obsoleta* (Say). *J. Exp. Mar. Biol. Ecol.* 379(1–2):51–59.

Leitz, T., K. Morand, and M. Mann. 1994. Metamorphosin A: A novel peptide controlling development of the lower metazoan *Hydractinia echinata* (Coelenterata, Hydrozoa). *Dev. Biol.* 163(2):440–446.

Leitz, T., and T. Wagner. 1993. The marine bacterium Alteromonas espejiana induces metamorphosis of the hydroid *Hydractinia echinata*. *Mar. Biol.* 115(2):173–178.

Li, Y.-F., X.-P. Guo, J.-L. Yang, X. Liang, W.-Y. Bao, P.-J. Shen, Z.-Y. Shi, and J.-L. Li. 2014. Effects of bacterial biofilms on settlement of plantigrades of the mussel *Mytilus coruscus*. *Aquaculture* 433:434–441.

Maldonado, M. 2006. The ecology of the sponge larva. *Can. J. Zool.* 84(2):175–194.

Matsumura, K., M. Nagano, and N. Fusetani. 1998. Purification of a larval settlement-inducing protein complex (SIPC) of the barnacle, *Balanus amphitrite*. *J. Exp. Zool.* 281(1):12–20.

Mercier, A., S. C. Battaglene, and J. F. Hamel. 2000. Settlement preferences and early migration of the tropical sea cucumber *Holothuria scabra*. *J. Exp. Mar. Biol. Ecol.* 249(1):89–110.

Metaxas, A., R. E. Scheibling, M. C. Robinson, and C. M. Young. 2008. Larval development, settlement, and early post-settlement behavior of the tropical sea star *Oreaster
reticulatus*. *Bull. Mar. Sci.* 83(3):471–480.

Mihm, J. W., W. C. Banta, and G. I. Loeb. 1981. Effects of adsorbed organic and primary
fouling films on bryozoan settlement. *J. Exp. Mar. Biol. Ecol.* 54(2):167–179.

Miller, S. E., and M. G. Hadfield. 1990. Developmental arrest during larval life and life-span
extension in a marine mollusc. *Science* 248(4953):356–358.

Morse, D. E., N. Hooker, A. N. C. Morse, and R. A. Jensen. 1988. Control of larval metamorphosis and recruitment in sympatric agariciid corals. *J. Exp. Mar. Biol. Ecol.* 116(3):193–217.

Morse, D. E., and A. N. C. Morse. 1991. Enzymatic characterization of the morphogen recognized by *Agaricia humilis* (Scleractinian coral) larvae. *Biol. Bull.* 181(1):104–122.

Morse, D. E., M. Tegner, H. Duncan, N. Hooker, G. Trevelyan, and A. Cameron. 1980.
Induction of settling and metamorphosis of planktonic molluscan (*Haliotis*) larvae. III:
Signaling by metabolites of intact algae is dependent on contact. In *Chemical Signals*,
D. Müller-Schwarze and R. M. Silverstein (Eds.). New York: Plenum Press.

Morse, D. E., N. Hooker, L. Jensen, and H. Duncan. 1979. Induction of larval abalone settling
and metamorphosis by gamma-aminobutyric acid and its congeners from crustose red
algae: 2. Applications to cultivation, seed-production and bioassays; principal causes of
mortality and interference. *Proc. World Maricult. Soc.* 10(1–4):81–91.

Muller, W. A., and T. Leitz. 2002. Metamorphosis in the Cnidaria. *Can. J. Zool.*
80(10):1755–1771.

Negri, A. P., N. S. Webster, R. T. Hill, and A. J. Heyward. 2001. Metamorphosis of broadcast
spawning corals in response to bacteria isolated from crustose algae. *Mar. Ecol. Prog.
Ser.* 223:121–131.

Nielsen, S. J., T. Harder, and P. D. Steinberg. 2015. Sea urchin larvae decipher the epiphytic
bacterial community composition when selecting sites for attachment and metamorphosis. *FEMS Microbiol. Ecol.* 91:1–9.

Pagett, H. E., J. L. Abrahams, J. Bones, N. O'Donoghue, J. Marles-Wright, R. J. Lewis,
J. R. Harris, G. S. Caldwell, P. M. Rudd, and A. S. Clare. 2012. Structural charac-
terisation of the N-glycan moiety of the barnacle settlement-inducing protein complex
(SIPC). *J. Exp. Biol.* 215(7):1192–1198.

Pawlik, J. R. 1986. Chemical induction of larval settlement and metamorphosis in the reef-
building tube worm *Phragmatopoma californica* (Sabellariidae: Polychaeta). *Mar.
Biol.* 91(1):59–68.

Pearce, C. M., and R. E. Scheibling. 1990. Induction of metamorphosis of larvae of the
green sea-urchin, *Strongylocentrotus-droebachiensis*, by coralline red algae. *Biol. Bull.*
179(3):304–311.

Pennington, J. T., M. N. Tamburri, and J. P. Barry. 1999. Development, temperature tolerance,
and settlement preference of embryos and larvae of the articulate brachiopod *Laqueus
californianus*. *Biol. Bull.* 196(3):245–256.

Poloczanska, E. S., C. J. Brown, W. J. Sydeman, W. Kiessling, D. S. Schoeman, P. J. Moore,
K. Brander et al. 2013. Global imprint of climate change on marine life. *Nat. Clim.
Change* 3(10):919–925.

Puglisi, M. P., J. M. Sneed, K. H. Sharp, R. Ritson-Williams, and V. J. Paul. 2014. Marine
chemical ecology in benthic environments. *Nat. Prod. Rep.* 31(11):1510–1553.

Purahong, W., M. Schloter, M. J. Pecyna, D. Kapturska, V. Daumlich, S. Mital, F. Buscot,
M. Hofrichter, J. L. M. Gutknecht, and D. Kruger. 2014. Uncoupling of microbial com-
munity structure and function in decomposing litter across beech forest ecosystems in
Central Europe. *Sci. Rep.* 4:7014.

Qian, P. Y. 1999. Larval settlement of polychaetes. *Hydrobiologia* 402:239–253.

Qian, P. Y., V. Thiyagarajan, S. C. K. Lau, and S. C. K. Cheung. 2003. Relationship between bacterial community profile in biofilm and attachment of the acorn barnacle *Balanus amphitrite*. *Aquat. Microb. Ecol.* 33(3):225–237.

Rahim, S., J. Y. Li, and H. Kitamura. 2004. Larval metamorphosis of the sea urchins, *Pseudocentrotus depressus* and *Anthocidaris crassispina* in response to microbial films. *Mar. Biol.* 144(1):71–78.

Raimondi, P. T., and A. N. C. Morse. 2000. The consequences of complex larval behavior in a coral. *Ecology* 81(11):3193–3211.

Ritson-Williams, R., S. M. Shjegstad, and V. J. Paul. 2009. Larval metamorphosis of *Phestilla* spp. in response to waterborne cues from corals. *J. Exp. Mar. Biol. Ecol.* 375(1–2):84–88.

Rivera-Ingraham, G. A., F. Espinosa, and J. C. Garcia-Gomez. 2011. Effect of gamma-amino butyric acid on limpet populations: Towards the future management and conservation of endangered patellid species. *J. Chem. Ecol.* 37(1):1–9.

Rodriguez, S. R., F. P. Ojeda, and N. C. Inestrosa. 1993. Settlement of benthic marine invertebrates. *Mar. Ecol. Prog. Ser.* 97(2):193–207.

Santagata, S. 2004. A waterborne behavioral cue for the actinotroch larva of *Phoronis pallida* (Phoronida) produced by *Upogebia pugettensis* (Decapoda: Thalassinidea). *Biol. Bull.* 207(2):103–115.

Sebesvari, Z., R. Neumann, T. Brinkhoff, and T. Harder. 2013. Single-species bacteria in sediments induce larval settlement of the infaunal polychaetes *Polydora cornuta* and *Streblospio benedicti*. *Mar. Biol.* 160(5):1259–1270.

Sharp, K. H., J. M. Sneed, K. B. Ritchie, L. McDaniel, and V. J. Paul. 2015. Induction of larval settlement in the reef coral *Porites astreoides* by a cultivated marine *Roseobacter* strain. *Biol. Bull.* 228:98–107.

Shepherd, S. A., and J. A. Turner. 1985. Studies on southern Australian abalone (genus. *Haliotis*). VI. Habitat preference, abundance and predators of juveniles. *J. Exp. Mar. Biol. Ecol.* 93(3):285–298.

Shikuma, N. J., M. Pilhofer, G. L. Weiss, M. G. Hadfield, G. J. Jensen, and D. K. Newman. 2014. Marine tubeworm metamorphosis induced by arrays of bacterial phage tail-like structures. *Science* 343(6170):529–533.

Shimeta, J., J. Cutajar, M. G. Watson, and T. Vlamis. 2012. Influences of biofilm-associated ciliates on the settlement of marine invertebrate larvae. *Mar. Ecol. Prog. Ser.* 449:1–12.

Siboni, N., D. Abrego, C. A. Motti, J. Tebben, and T. Harder. 2014. Gene expression patterns during the early stages of chemically induced larval metamorphosis and settlement of the coral *Acropora millepora*. *PLoS One* 9(3):e91082.

Slattery, M. 1992. Larval settlement and juvenile survival in the red abalone (*Haliotis rufescens*), an examination of inductive cues and substrate selection. *Aquaculture* 102(1–2):143–153.

Sneed, J. M., K. H. Sharp, K. B. Ritchie, and V. J. Paul. 2014. The chemical cue tetrabro-mopyrrole from a biofilm bacterium induces settlement of multiple Caribbean corals. *Proc. R. Soc. B Biol. Sci.* 281(1786):20133086.

Soo, P., and P. A. Todd. 2014. The behaviour of giant clams (Bivalvia: Cardiidae: Tridacninae). *Mar. Biol.* 161(12):2699–2717.

Steinberg, P. D., R. De Nys, and S. Kjelleberg. 2001. Chemical mediation of surface colonization. In *Marine Chemical Ecology*, J. B. McClintock and B. J. Baker (Eds.). Boca Raton, FL: CRC Press.

Swanson, R. L., D. J. Marshall, and P. D. Steinberg. 2007. Larval desperation and histamine: How simple responses can lead to complex changes in larval behaviour. *J. Exp. Biol.* 210(18):3228–3235.

Swanson, R. L., J. E. Williamson, R. De Nys, N. Kumar, M. P. Bucknall, and P. D. Steinberg. 2004. Induction of settlement of larvae of the sea urchin *Holopneustes purpurascens* by histamine from a host alga. *Biol. Bull.* 206(3):161–172.

Swanson, R. L., M. Byrne, T. A. A. Prowse, B. Mos, S. A. Dworjanyn, and P. D. Steinberg. 2012. Dissolved histamine: A potential habitat marker promoting settlement and metamorphosis in sea urchin larvae. *Mar. Biol.* 159(4):915–925.

Swanson, R. L., R. de Nys, M. J. Huggett, J. K. Green, and P. D. Steinberg. 2006. In situ quantification of a natural settlement cue and recruitment of the Australian sea urchin *Holopneustes purpurascens*. *Mar. Ecol. Prog. Ser.* 314:1–14.

Szewzyk, U., C. Holmstrom, M. Wrangstadh, M. O. Samuelsson, J. S. Maki, and S. Kjelleberg. 1991. Relevance of the exopolysaccharide of marine *Pseudomonas* sp. strain S9 for attachment of *Ciona intestinalis* larvae. *Mar. Ecol. Prog. Ser.* 75(2–3):259–265.

Takahashi, T., and N. Takeda. 2015. Insight into the molecular and functional diversity of cnidarian neuropeptides. *Int. J. Mol. Sci.* 16(2):2610–2625.

Takahashi, Y., K. Itoh, M. Ishii, M. Suzuki, and Y. Itabashi. 2002. Induction of larval settlement and metamorphosis of the sea urchin *Strongylocentrotus intermedius* by glycoglycerolipids from the green alga *Ulvella lens*. *Mar. Biol.* 140(4):763–771.

Tamburri, M. N., R. K. Zimmerfaust, and M. L. Tamplin. 1992. Natural sources and properties of chemical inducers. Mediating settlement of oyster larvae: A re-examination. *Biol. Bull.* 183(2):327–338.

Taris, N., T. Comtet, R. Stolba, R. Lasbleiz, J. A. Pechenik, and F. Viard. 2010. Experimental induction of larval metamorphosis by a naturally-produced halogenated compound (dibromomethane) in the invasive mollusc *Crepidula fornicata* (L.). *J. Exp. Mar. Biol. Ecol.* 393(1–2):71–77.

Tebben, J., C. A. Motti, N. Siboni, D. M. Tapiolas, A. P. Negri, P. J. Schupp, M. Kitamura, M. Hatta, M. Steinberg, and T. Harder. 2015. Chemical mediation of coral larval settlement by crustose coralline algae. *Sci. Rep.* 5:10803.

Tebben, J., D. M. Tapiolas, C. A. Motti, D. Abrego, A. P. Negri, L. L. Blackall, P. D. Steinberg, and T. Harder. 2011. Induction of larval metamorphosis of the coral *Acropora millepora* by tetrabromopyrrole isolated from a *Pseudoalteromonas* bacterium. *PLoS One* 6(4):e19082.

Tegtmeyer, K., and D. Rittschof. 1988. Synthetic peptide analogs to barnacle settlement pheromone. *Peptides* 9(6):1403–1406.

Thiyagarajan, V. 2010. A review on the role of chemical cues in habitat selection by barnacles: New insights from larval proteomics. *J. Exp. Mar. Biol. Ecol.* 392(1–2):22–36.

Toupoint, N., V. Mohit, I. Linossier, N. Bourgougnon, B. Myrand, F. Olivier, C. Lovejoy, and R. Tremblay. 2012. Effect of biofilm age on settlement of *Mytilus edulis*. *Biofouling* 28(9):985–1001.

Tran, C., and M. G. Hadfield. 2011. Larvae of *Pocillopora damicornis* (Anthozoa) settle and metamorphose in response to surface-biofilm bacteria. *Mar. Ecol. Prog. Ser.* 433:85–96.

Tsukamoto, S., H. Kato, H. Hirota, and N. Fusetani. 1999. Lumichrome—A larval metamorphosis-inducing substance in the ascidian *Halocynthia roretzi*. *Eur. J. Biochem.* 264(3):785–789.

Tsukamoto, S., H. Hirota, H. Kato, and N. Fusetani. 1993. Urochordamines A and B: Larval settlement/metamorphosis-promoting, pteridine-containing physostigmine alkaloids from the tunicate *Ciona savignyi*. *Tetrahedron Lett.* 34(30):4819–4822.

Tsukamoto, S., H. Kato, H. Hirota, and N. Fusetani. 1995. Pipecolate derivatives, anthosamines A and B, inducers of larval metamorphosis in ascidians, from a marine sponge *Anthosigmella* aff. *raromicrosclera*. *Tetrahedron* 51(24):6687–6694.

Tung, C.-H., and A. C. Alfaro. 2011. Initial attachment, metamorphosis, settlement, and survival of black-footed abalone, *Haliotis iris*, on microalgal biofilms containing different amino acid compositions. *J. World Aquacult. Soc.* 42(2):167–183.

Unabia, C. R. C., and M. G. Hadfield. 1999. Role of bacteria in larval settlement and metamorphosis of the polychaete *Hydroides elegans*. *Mar. Biol.* 133(1):55–64.

Vasquez, H. E., K. Hashimoto, A. Yoshida, K. Hara, C. C. Imai, H. Kitamura, and C. G. Satuito. 2013. A glycoprotein in shells of conspecifics induces larval settlement of the Pacific oyster *Crassostrea gigas*. *PLoS One* 8(12):e82358.

Veitch, F. P., and H. Hidu. 1971. Gregarious setting in the American oyster *Crassostrea virginica* Gmelin: I. Properties of a partially purified "Setting factor". *Chesapeake Sci.* 12(3):173–178.

Wang, C., W.-Y. Bao, Z.-Q. Gu, Y.-F. Li, X. Liang, Y. Ling, S.-L. Cai, H.-D. Shen, and J.-L. Yang. 2012. Larval settlement and metamorphosis of the mussel *Mytilus coruscus* in response to natural biofilms. *Biofouling* 28(3):249–256.

Webster, N. S., L. D. Smith, A. J. Heyward, J. E. M. Watts, R. I. Webb, L. L. Blackall, and A. P. Negri. 2004. Metamorphosis of a scleractinian coral in response to microbial biofilms. *Appl. Environ. Microbiol.* 70(2):1213–1221.

Whalan, S., P. Ettinger-Epstein, C. Battershill, and R. de Nys. 2008. Larval vertical migration and hierarchical selectivity of settlement in a brooding marine sponge. *Mar. Ecol. Prog. Ser.* 368:145–154.

Whalan, S., and N. S. Webster. 2014. Sponge larval settlement cues: The role of microbial biofilms in a warming ocean. *Sci. Rep.* 4:4072.

Wieczorek, S. K., and C. D. Todd. 1997. Inhibition and facilitation of bryozoan and ascidian settlement by natural multi-species biofilms: Effects of film age and the roles of active and passive larval attachment. *Mar. Biol.* 128(3):463–473.

Williamson, J. E., R. De Nys, N. Kumar, D. G. Carson, and P. D. Steinberg. 2000. Induction of metamorphosis in the sea urchin *Holopneustes purpurascens* by a metabolite complex from the algal host *Delisea pulchra*. *Biol. Bull.* 198(3):332–345.

Yang, J.-L., X. Li, X. Liang, W.-Y. Bao, H.-D. Shen, and J.-L. Li. 2014. Effects of natural biofilms on settlement of plantigrades of the mussel *Mytilus coruscus*. *Aquaculture* 424:228–233.

Yang, J.-L., P.-J. Shen, X. Liang, Y.-F. Li, W.-Y. Bao, and J.-L. Li. 2013. Larval settlement and metamorphosis of the mussel *Mytilus coruscus* in response to monospecific bacterial biofilms. *Biofouling* 29(3):247–259.

Young, C. M., and L. F. Braithwaite. 1980. Orientation and current-induced flow in the stalked ascidian, *Styela montereyensis*. *Biol. Bull.* 159(2):428–440.

Young, C. M., and F. S. Chia. 1982. Factors controlling spatial distribution of the sea cucumber *Psolus chitonoides*: Settling and post-settling behavior. *Mar. Biol.* 69(2):195–205.

Yule, A. B., and G. Walker. 1985. Settlement of *Balanus balanoides*: The effect of cyprid antennular secretion. *J. Mar. Biol. Assoc. U.K.* 65(3):707–712.

Yvin, J. C., L. Chevolot, A. M. Chevolot-Magueur, and J. C. Cochard. 1985. First isolation of jacaranone from an alga, *Delesseria sanguinea*. A metamorphosis inducer of *Pecten* larvae. *J. Nat. Prod.* 48(5):814–816.

Zimmer, R. K., and C. A. Butman. 2000. Chemical signaling processes in the marine environment. *Biol. Bull.* 198(2):168–187.

Zimmerfaust, R. K., and M. N. Tamburri. 1994. Chemical identity and ecological implications of a waterborne, larval settlement cue. *Limnol. Oceanogr.* 39(5):1075–1087.

Zobell, C. E., and E. C. Allen. 1935. The significance of marine bacteria in the fouling of submerged surfaces. *J. Bacteriol.* 29(3):239–251.

11 Chemical Ecology of Marine Bryozoans

Grace Lim-Fong and Rian Kabir

CONTENTS

11.1 INTRODUCTION

Organisms belonging to the Phylum Bryozoa ("bryo" = moss; "zoa" = animals) are colonial invertebrates that can be found in freshwater and marine environments. The first bryozoan was noted in the fossil record in the early Ordovician (~480 million years ago), and all three classes of bryozoans observed in the fossil record contain extant representatives. One of the three classes, the Phylactolaemata, is exclusively freshwater and will not be considered in this chapter. The other two classes, the Cyclostomata and Gymnolaemata, are marine taxa and will be described extensively in this chapter.

Bryozoans are anatomically united, together with the Phyla Phoronida and Brachiopoda, by the presence of a crown of hollow tentacles, called the lophophore (Figure 11.1a); however, bryozoans can be distinguished from phoronids and brachiopods by their adult colonial lifestyle. While there is ongoing debate about the

(a)   (b)

FIGURE 11.1 (a) Light micrograph of a colony of *Bugula neritina*. Lophophores are found on feeding zooids, and ovicells (brood chambers) are derived from maternal and its adjacent zooid. (b) Scanning electron micrograph of a colony of *B. pacifica*. Spines (sp) and avicularia (av) are thought to act as a physical defense in some bryozoans.

phylogenetic placement of Bryozoa in the metazoan phylogeny, recent multi-locus molecular analysis supports the monophyly of the phylum (Fuchs et al. 2009).

Practically all bryozoans, with the exception of the genus *Monobryozoon* (Berge et al. 1985), are colonial, made up of clonal individuals, called zooids (Figure 11.1a). Each zooid is comprised of a calcified "shell" called the cystid that encases the soft body parts called the polypide. Bryozoan colonies can be found in diverse marine benthic environments, ranging from man-made hard substrates (like pier pilings) to natural soft sediments. Colonies can also adopt diverse growth forms, ranging from erect tree-like growth or encrusting growth (Figure 11.2). There are several different kinds of zooids in a colony, but most predominant is the autozooid (i.e., feeding zooid). Each autozooid contains a functional gut, and the mouth is situated at the

(a)   (b)

FIGURE 11.2 Different growth forms of bryozoans. (a) Encrusting growth of *Membranipora* spp. on a blade of the giant kelp (*Macrocystis* sp.) washed ashore in California, USA; (b) Tree-like growth of two *Bugula neritina* colonies, a brown morph and a purple-red morph, sampled in Virginia, USA. Scale bar represents approximately 1 cm.

base of the lophophore, whose cilia beat to generate a current to bring plankton to the mouth. The zooids in a colony are connected to each other by a "vascular" system, comprised of funicular cords that facilitate nutrient transport across the colony. It is presumed that the funicular cords can also translocate natural products.

Bryozoan colonies are hermaphroditic, and fertilization of sperm and egg can occur internally or externally. In some taxa, internally fertilized eggs are brooded in specialized zooids called ovicells (Figure 11.1a) until they develop into embryos and are released as larvae. Larvae that result from sexual reproduction swim in the water column before settling on a substrate and undergoing metamorphosis to form an ancestrula, the first autozooid. The colony grows through asexual budding of the ancestrula and subsequent zooids.

11.2   NATURAL PRODUCTS FROM MARINE BRYOZOANS

Many natural products have been isolated from marine bryozoans: from 1990 to the present, approximately 120 new metabolites from 17 different chemical families (including revisions of incorrect structures) derived from bryozoans were reported in the annual *Marine Natural Products* review published in the journal *Natural Product Reports*. This number, however, is dwarfed by the total number of marine natural products characterized: in 2012 alone, over 1,200 compounds were reported from marine sources (Blunt et al. 2014).

Much of the effort in isolating marine natural products stems from a drug discovery perspective (Mayer et al. 2013), and at least one bryozoan metabolite, bryostatin 1, has entered the "preclinical pipeline" as an anticancer drug (Mayer and Gustafson 2004); however, the ecological role of many of these 120 bryozoan metabolites is unknown. In this chapter, we will primarily describe metabolites whose ecological roles have been demonstrated and will discuss the kinds of future studies that would further illuminate our understanding of the chemical ecology of bryozoans. The majority of the known ecological roles of bryozoan-derived metabolites fall into three categories: settlement inhibition, anti-predation, and anti-fouling. We will consider these three roles and describe examples of bryozoan-derived metabolites that fall in these categories.

11.3   SETTLEMENT INHIBITION

All marine bryozoans can reproduce sexually and form ciliated planktonic larvae. Some of these larvae, for example those of *Electra pilosa*, are planktotrophic and have a fully functional gut. Consequently, they can remain pelagic for relatively long periods of time (a few weeks) before finding substrate on which to settle (Goldson et al. 2001). Other bryozoan taxa, such as *Bugula neritina*, have lecithotrophic larvae that derive all of their energy from the parental colony and are thus only able to swim for hours before running out of energy reserves. Extended time in the water column reduces postmetamorphic fitness of lecithotrophic larvae and prolongs exposure to pelagic predators (Wendt 1996). Once metamorphosis is initiated, subsequent life stages of the bryozoan are sessile, and the bryozoan is committed to that space. Therefore, a bryozoan larva encounters a trade-off between rapid settlement and finding an appropriate substrate on which to develop into an adult colony. Larvae have been described to become "desperate" after prolonged swimming and are more likely to settle on suboptimal substrates after a long time in the water column (Marshall and Keough 2003).

Bryozoan larvae show preferences for different kinds of substrate. For example, *B. neritina* larvae prefer roughened, biofilmed, and downward-facing substrates over pristine substrates facing upward (Marshall and Keough 2003; Burgess et al. 2012). The preferred substrates are thought to possess superior hydrodynamics and lower exposure to sunlight and predators (Burgess et al. 2012). Competition for optimal settlement substrate in the field is therefore not trivial, and some bryozoans, particularly those that produce non-feeding larvae, produce natural products to inhibit settlement of other benthic invertebrates.

The life stage in which an anti-settlement compound is produced deserves some consideration. All of the compounds described in the following paragraphs were isolated from whole adult colonies, so it is difficult to conclude if there was any spatial or ontological heterogeneity in the distribution of the compound. For an anti-settlement compound to be effective, it should either be coated on the larva or be actively secreted by the parental colony onto the surrounding substrate.

11.3.1 *AMATHIA* SPP.

Brominated alkaloids, volutamides A–E, were isolated from a North Carolina population of the erect bryozoan *Amathia convoluta* (Montanari et al. 1996), and seawater containing purified volumatide B or D (Figure 11.3a and b) inhibited settlement of larvae of a co-occurring hydroid, *Eudendrium carneum*. *Amathia* spp. larvae are lecithotrophic (Zimmer and Woollacott 1993) and therefore have limited dispersal potential, so it is likely that the production of anti-settlement compounds by adult colonies would preserve the surrounding substrate for conspecific recruits.

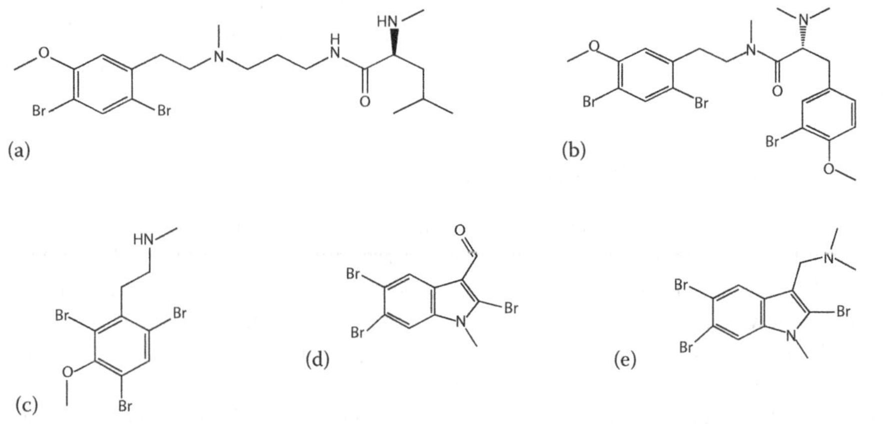

FIGURE 11.3 Natural products that inhibit settlement of other invertebrate larvae: (a) Volutamide B from *A. convoluta*; (b) Volutamide D from *A. convoluta*; (c) Convolutamine F from *A. convoluta*; (d) 2,5,6-tribromo-*N*-methylindole-3-carbaldehyde from *Zoobotryon verticullatum*; (e) 2,5,6-tribromo-1-methylgramine from *Z. pellucidum*.

It is unknown how much, if any, volutamides B and D are secreted into the seawater since they are relatively nonpolar. Interestingly, anti-settlement activity of other volutamides was not reported, which indicates that minor structural modifications (N-methyl in volutamide B vs. N-dimethyl in volutamide C) may have a significant effect on the ecological function of the compound. Floridian populations of *A. convoluta* do not contain volutamides (Montanari et al. 1996), suggesting that chemical defense is finely tuned to selective pressures from within a local environment. *Amathia convoluta* is also the source of the convolutamines A–H. Convolutamine F (Figure 11.3c) was demonstrated to inhibit cell division of sea urchin embryos (Kamano et al. 1999). Because most of the ecological activities were detected using taxon-specific bioassays that tend to vary across different investigators, it is virtually impossible to pinpoint the specificity of the mechanism of action of the different compounds. For example, it is not yet known if convolutamine F is active only against sea urchin embryos, or if it is also inhibitory against larvae of other taxa.

11.3.2 *ZOOBOTRYON* SPP.

The bryozoan *Zoobotryon verticillatum* from the Eastern Atlantic is the source of 2,5,6-tribromo-N-methylindole-3-carbaldehyde (Figure 11.3d), which delayed metamorphosis and impeded swimming in sea urchin larvae (Ortega et al. 1993). A related brominated indole, 2,5,6-tribromo-1-methylgramine (Figure 11.3e), was isolated from a congeneric species, *Z. pellicidum*. This compound inhibited settlement of larvae of the barnacle *Balanus amphitrite* and served as a starting point for structure-function investigations of related compounds in the search for novel antifouling compounds that can be added to marine coatings (Kawamata et al. 2006).

11.4   ANTIFOULING ACTIVITY

Seawater is teeming with plankton, some of which establish epibiotic relationships with bryozoans. Epibiosis typically follows a successional order: bacteria colonize a hard substrate within an hour of its introduction (e.g., growth of a new cystid), followed by eukaryotic epibionts, such as diatoms, after a few days (Wahl 1989). There are typically more than 10^6 bacterial cells in a milliliter of seawater (Zweifel and Hagstrom 1995); therefore, as soon as a larva is released from its parental colony or a new cystid begins to bud, an opportunity for epibiosis emerges. It is likely that most of these fouling bacteria are commensals, but the colonization of pathogenic bacteria is also possible. For example, chitin-degrading *Vibrio* sp. have been isolated from the Mediterranean bryozoan *Myriapora truncata* (Stabili et al. 2008). The covering of the mouth, the operculum, of the zooids in *M. truncata* are chitinous, and the *Vibrio* sp. have been localized to the operculum, suggesting a pathogenic association. Bryozoans therefore need to protect their surfaces from pathogenic or parasitic epibionts, while allowing mutualistic ones to thrive.

   Current understanding of antifouling properties of bryozoan natural products, like that of anti-settlement activities, is hampered by the lack of ecologically relevant bioassays. Most of the antifouling compounds that have been identified to date are antibacterial molecules because there are well-established protocols to test for antibacterial activity (Payne et al. 2007); however, the test organisms tend to be clinically relevant, such as *Escherichia coli* and *Staphylococcus aureus*, rather than co-occurring marine bacteria. Nevertheless, the compounds identified by these drug discovery-oriented bioassays are prime candidates for further testing of ecological roles of these bryozoan bioactive metabolites.

11.4.1 *Amathia* spp.

The genus *Amathia* has been a rich source of bryozoan natural products. Volutamides B and D from *A. convoluta* inhibit settlement of hydroid larvae (see previous Settlement Inhibition section), amathaspiramide A and E, isolated from a New Zealand population of *A. wilsoni* (Morris and Prinsep 1999) are active against the bacterium *Bacillus subtilis*, whereas amathaspiramide B and C were not.

Amathia alternata from the Western Atlantic is the source of a family of compounds called alternatamides, some of which (alternatamide A–C) inhibit the growth of Gram-positive bacteria such as *S. aureus* and *Enterococcus faecalis* but has no activity against Gram-negative bacteria (Lee et al. 1997). Thus, it is likely that antibacterial compounds produced by bryozoans can selectively target a suite of bacteria while leaving others unaffected. Tested strains were bacteria with clinical significance, but it is unknown if alternatamide A–C affect ecologically relevant bacteria, especially marine bacteria encountered by *A. alternata*.

11.4.2 *Sessibugula translucens*

Tambjamines A–D (Figure 11.4a) were first isolated from the nudibranch *Tambje abdere* (Carté and Faulkner 1983) but were later found to be also present in the congener *T. eliore*, its predatory nudibranch, *Roboaster tigris*, and its bryozoan prey, *Sessibugula translucens*. The presence of the tambjamines in these four organisms suggest that *S. translucens* is the source of the compounds since nudibranchs are known to accumulate and sequester natural products from their diet (Faulkner 1999). Dimerization of the bipyrroles results in a green pigment, which could account for the green coloration of *S. translucens* colonies. These compounds are inhibitory towards a marine bacterium *Vibrio anguillarum* (Carte and Faulkner 1983); however, it is unknown if the antimicrobial activity is the primary role of these compounds since these compounds are also feeding deterrents (see section on Anti-Predation).

11.4.3 *Flustra foliacea*

Flustra foliacea is the source of several brominated alkaloids and a terpene (Figure 11.4b–i), which inhibit the growth of bacteria isolated from the bryozoan (Peters et al. 2003). Moreover, some of these compounds also interfere with bacterial cell-cell communication processes, which might be vital to the ability of the bacteria to form a biofilm on the surface of *F. foliacea*. A correlation between antibacterial activity and anti-settlement activity was observed: colony segments of *F. foliacea* that exhibited antibacterial activity were also less likely to be settled on by larvae of the bryozoan *Scrupocellaria reptans* (Al-ogily and Knight-Jones 1977). It is unknown if the same metabolite is responsible for these two ecological functions.

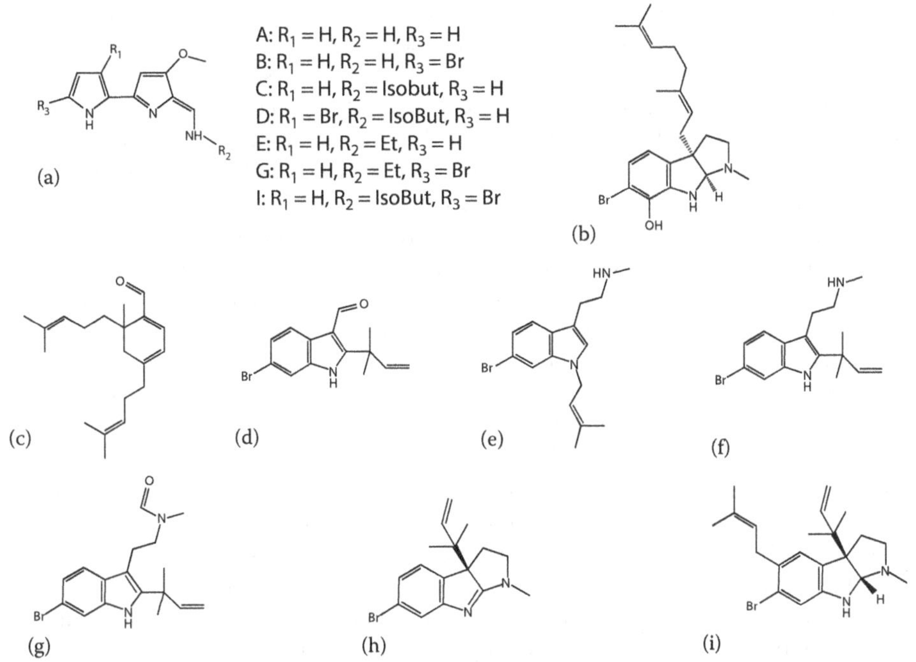

FIGURE 11.4 Natural products that have antimicrobial activity towards marine bacteria: (a) Tambjamines A-E, G, I from *Sessibugula transluscens*; (b) (3aR*,8aR*)-6-bromo-3a-[(2E)-3,7-dimethyl-2,6-octadienyl]-1,2,3,3a,8,8a-hexahydropyrrolo[2,3-b]-indol-7-ol from *Flustra foliacea*; (c) 4,6-bis(4-methylpent-3-en-1-yl)-6-methylcyclohexa-1,3-diene-carbaldehyde from *F. foliacea*; (d) 6-bromo-2-(1,1-dimethyl-2-propenyl)-1H-indole-3-carbaldehyde from *F. foliacea*; (e) Deformylflustrabromine B from *F. foliacea*; (f) Desformylflustrabromine from *F. foliacea*; (g) Flustrabromine from *F. foliacea*; (h) Flustramine C from *F. foliacea*; (i) flustramine D from *F. foliacea*.

11.5 ANTI-PREDATION

Bryozoans are prey to many marine taxa: fish, crustaceans, gastropods, and echinoids, to name a few (see Lidgard 2008 for review and meta-analysis). Some of these predators are generalist predators, whereas some predators, notably nudibranchs, feed exclusively on bryozoans. Predation on bryozoans deserves to be viewed in the context of the colonial lifestyle of adult bryozoans. Larger predators, such as fish, can consume an entire colony, whereas smaller predators, like several worm phyla and gastropods, typically only prey on one or a few zooids at a time; therefore, predation need not be lethal with respect to the colony.

The nutritive value of bryozoans has been proposed to influence how much predation they experience. Some bryozoans, both erect and encrusting, are heavily calcified, requiring that predators possess specialized feeding mechanisms for the extraction of the polypide (soft tissue). One way predators can circumvent the calcified armor is to wait until the polypide emerges from the cystid before attacking the bryozoan. Regardless of how a predator accesses the soft tissue, because of the exoskeleton, bryozoan adult colonies are less nutritious than most other soft-bodied sessile marine invertebrates (Lippert and Iken 2003). Some predators exhibit preference for different parts of the colony. For example, nudibranch predators prefer eating young and non-reproductive zooids on the colony perimeter of the encrusting bryozoan *Dendrobeania lichenoides* compared to the zooids in the middle of the colony (Harvell 1984). It is unknown why the zooids in the middle of the colony are avoided, but the author suggests that the embryos might contain a feeding deterrent.

Many bryozoan species have sharp spines and bird's head structures called avicularia that can potentially deter predators (Figure 11.1b). Though these structures are found in many bryozoans, their role in defense has only been reported in a few. Notably, the spines in *Membranipora membranacea* protect them against their specialized nudibranch predator (Harvell 1986). These spines are induced in the presence of a chemical cue exuded by the predator, and there is good reason why their anti-predation device is inducible: spined colonies grow more slowly than unspined ones. Thus, bryozoans possess both chemical and mechanical anti-predation devices.

These anti-predation strategies apply to adult colonies but rarely to larvae or juveniles. Unlike partial predation of adult colonies, the larval or young juvenile (ancestrula) stage of bryozoans is susceptible to complete predation, since each larva is in effect a solitary individual. Larvae also typically are less physically defended than adult colonies from predation as they lack spines and avicularia and are less calcified. Moreover, lecithotrophic bryozoan larvae and juveniles are nutrient-dense compared to their adult counterparts (Lopanik et al. 2007), and thus can be valuable prey items.

While several bryozoan natural products have been demonstrated to exhibit anti-feeding properties, the majority of these studies deserve follow-up to establish if the molecules act as anti-predatory compounds in the environment.

11.5.1 *Sessibugula translucens*

The tambjamines (Figure 11.4a), whose antimicrobial activity was described in a previous section and which were first isolated from nudibranchs, were found to have originated from the bryozoan *S. translucens* (Carté and Faulkner 1983). These compounds deter feeding by a generalist predator, the spotted kelpfish (*Gibbonsia elegans*) but are clearly not deterrent to their specialized nudibranch predators since these metabolites are found at higher concentrations (percent dry weight) in the nudibranchs *T. eliore* and *T. abdere* than in the bryozoan (Carté and Faulkner 1986). Therefore, the tambjamines are able to protect *S. translucens* from generalist predators, but some predators have evolved to specialize in feeding on the bryozoan almost exclusively and have even used the compounds as a way to chemodetect their bryozoan prey (Carté and Faulkner 1986). Amathamides share a similar story (see the following *Amathia* spp. section).

Tambjamines have also been isolated from another bryozoan, *Bugula dentata* (Blackman and Li 1994; Carbone et al. 2010), and another *Tambje* nudibranch

(Carbone et al. 2010), a colonial ascidian (Lindquist and Fenical 1991), and a marine bacterium (Franks et al. 2005). Several tambjamines are thought to play a role in defending the ascidian larva, since it is found in high enough quantities in the larvae as to be unpalatable to co-occurring generalist fish predators (Lindquist and Fenical 1991; Paul et al. 1990). The presence of identical compounds across phylogenetically diverse taxa such as ascidians, bryozoans, and a bacterium raises the possibility of an endosymbiotic origin of the tambjamines (further discussed in the section Biosynthetic Origins).

11.5.2 AMATHIA SPP.

Amathamide C (Figure 11.5a), isolated from *A. wilsoni*, deters feeding by a generalist fish predator, but is readily consumed by an epibiont of *A. wilsoni*, the sea spider *Stylopallene longicauda* (Sherwood et al. 1998), which is hypothesized to use amathamide C as a chemical defense for itself. High concentrations of amathamides have been shown to persist in the sea spider, even after starvation, indicating that the sea spider sequesters amathamides from its prey. Interestingly, amathamides are found in higher concentrations on the zooidal surface at the distal ends of *A. wilsoni* colonies. This localization pattern correlates with the presence of a rod-shaped epibiotic bacterium, strongly suggesting a bacterial role in the production of amathamides (Walls et al. 1995).

Some compounds may possibly play multiple ecological roles. Volutamide B, isolated from *A. convoluta*, not only inhibits settlement of hydroid larvae (Figure 11.5b; see previous section on Settlement Inhibition) but also deters feeding by pinfish (Montanari et al. 1996). A related compound, volutamide C (Figure 11.5b) deterred

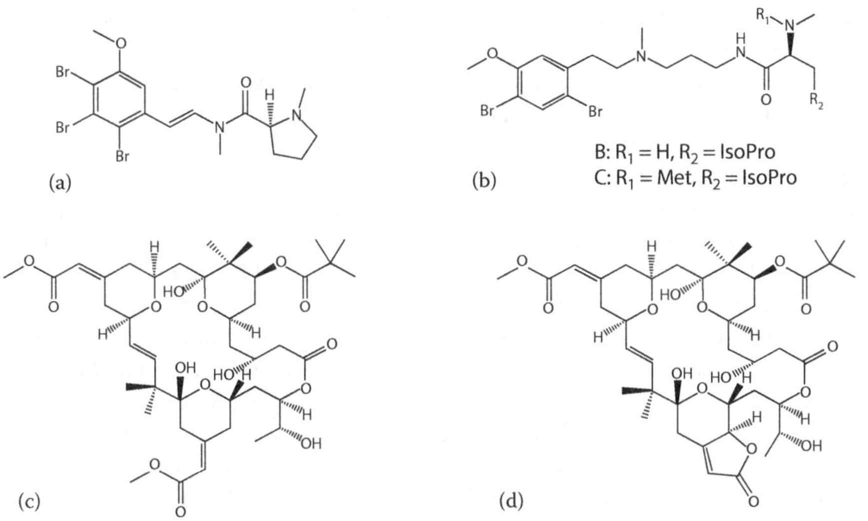

FIGURE 11.5 Natural products that function as feeding deterrents: (a) Amathamide C from *Amathia wilsoni*; (b) Volutamides B & C from *A. convoluta*; (c) Bryostatin 10 from *Bugula neritina*; (d) Bryostatin 20 from *B. neritina*.

feeding by the sea urchin *Arbacia punctulata* (Montanari et al. 1996). The multiple ecological functions of a natural product may indicate a similar molecular mechanism underpinning the ecological outcome. Moreover, the use of highly related or identical molecules for different functions may be an example of molecular exaptation, that is, the shift in the function of a molecule over evolutionary time.

11.5.3 *BUGULA NERITINA*

Bugula neritina is an erect bryozoan with a cosmopolitan distribution in subtropical and temperate waters and is the source of a family of polyketides called bryostatins (Trindade-Silva et al. 2010). Twenty bryostatins have been characterized (Lopanik et al. 2004a) and most of them possess bioactivity. Bryostatin 1 has anti-cancer (Pettit et al. 1982), anti-Alzheimer's (Etcheberrigaray et al. 2004), and anti-viral (Mehla et al. 2010) properties, and it is currently being tested in clinical trials in patients diagnosed with Alzheimer's disease or human immunodeficiency virus (HIV) infection (clinicaltrials.gov).

While the biomedical potential of the bryostatins has been explored for over four decades, an ecological role of bryostatins was only demonstrated in the 2000s (Lopanik et al. 2004b), even though *B. neritina* larvae were previously known to be distasteful to generalist predators (Lindquist and Hay 1996). The link between the unpalatability of larvae and bryostatins was established by Lopanik and colleagues (2004b), who found that bryostatins 10 (Figure 11.5c) and 20 (Figure 11.5d) deterred feeding by a generalist predator (the co-occurring pinfish, *Lagodon rhomboides*). The defensive bryostatins have been localized to the surface of *B. neritina* larvae (Sharp et al. 2007), consistent with the observation that larvae attacked by pinfish were promptly regurgitated (Lindquist and Hay 1996). These larvae were able to complete metamorphosis, indicating that bryostatins confer effective defense to the bryozoan.

Because most known polyketides have a microbial origin, it was hypothesized that a microbial symbiont of *B. neritina* was the true biosynthetic source of the bryostatins. Symbiont elimination experiments support this hypothesis as antibiotic-treated *B. neritina* contains no detectable bryostatins (Lopanik et al. 2004b). Genomic analysis of the unculturable symbiont of *B. neritina, Candidatus Endobugula sertula*, also revealed the presence of a polyketide synthase gene cluster that likely encodes a bryostatin biosynthetic complex (Sudek et al. 2007; unpublished data).

A more detailed examination of the distribution of bryostatins among *B. neritina* populations revealed that different populations of *B. neritina* contained different suites of bryostatins, and that the different chemotypes were correlated to different host-symbiont combinations (Davidson and Haygood 1999; McGovern and Hellberg 2003). Bryostatin 1 is found predominantly in Eastern Pacific populations at lower depths (Davidson and Haygood 1999), whereas bryostatin 10 appears to be ubiquitous (Pettit et al. 1987; Kamano et al. 1995; Lopanik et al. 2004b). The incidence of *Ca. E. sertula* is lowered at higher latitudes (Linneman et al. 2014), and therefore, these populations are devoid of bryostatins (McGovern and Hellberg 2003). It is likely that bryostatins are not needed for defense at higher latitudes because the predation pressure is less intense than in warmer waters.

11.6 CHARACTERIZING ECOLOGICALLY RELEVANT MOLECULES

In the preceding sections, we have described bryozoans whose ecologically relevant compounds have been characterized. Apart from these studies, there are several studies in which natural compounds have been implicated in mediating interactions between the bryozoan and its environment, but structures of the compounds have not yet been elucidated.

Dichloromethane extracts of *B. pacifica* and another bryozoan, *Tricellaria occidentalis*, both found in the Northeastern Pacific, were tested against co-occurring marine bacteria and clinical bacterial strains to establish if either bryozoan contained anti-fouling compounds (Shellenberger and Ross 1998). *Bugula pacifica* extracts were slightly inhibitory toward two of the marine bacteria tested, whereas *T. occidentalis* extracts contained virtually no activity. The authors postulated that the low degree of surface fouling by bacteria on *B. pacifica* compared to *T. occidentalis* might be due to the presence of antibacterial compounds in *B. pacifica*.

A similar observation was made in Tasmanian waters, where there was a correlation between antibacterial activity and the degree of fouling in four bryozoan species. *Amathia wilsoni* and *Orthoscuticella ventricosa* exhibited high antibacterial activity and low levels of fouling whereas the opposite was observed for *Cellaria pilosa* and *Bugularia dissimilis* (Walls et al. 1993).

A survey of feeding repellence towards generalist invertebrate predators in thirteen Antarctic bryozoans was implemented by Figuerola and colleagues (2013). They found that extracts from the majority of tested Antarctic bryozoans deterred feeding by the sea star *Odontaster validus* and the amphipod *Cheirimedon femoratus*. Even though trade-offs in resource allocation between chemical versus mechanical defense have been previously documented in some benthic invertebrates (Stachowicz and Lindquist 2000), there was no inverse correlation between chemical defense and mechanical defense, such as spines and avicularia, in these bryozoans.

11.7 DIRECTIONS: DELVING DEEPER INTO BRYOZOAN CHEMICAL ECOLOGY

11.7.1 LOCATION, LOCATION, LOCATION

Apart from the identity of the molecule involved in an ecological interaction, localizing the metabolite within the animal is useful because this piece of information can inform us about the functional significance of the metabolite. As mentioned previously, the high concentration of bryostatins on the outer coat of *B. neritina* larvae is consistent with their role as a feeding deterrent (Sharp et al. 2007).

Similarly, natural products are unlikely to be uniformly distributed throughout adult bryozoan colonies. Amathamides, isolated from *Amathia wilsoni* (see section on Anti-Predation) are found at higher concentrations on the distal ends (tips) of the colony compared to the proximal end (towards the base) of the colony (Walls et al. 1993). Because amathamide C is known to deter feeding by a general fish predator (Sherwood et al. 1998), it is quite likely that these fish predators avoid the most exposed part of the bryozoan and seek other prey.

Intraspecific variation in natural product chemistry has also been documented in several taxa (Davidson and Haygood 1999; Peters et al. 2004). The differences in chemistry may be due to genetic differences among morphologically identical populations due to different selective forces in different geographic regions, or due to the induction of a different set of compounds due to different environmental pressures (Blackman and Green 1987; Morris and Prinsep 1999). Whereas plasticity of a chemical defense has not been documented in bryozoans, one of the earliest examples of predator-induced defense was in the bryozoan *Membranipora membranacea*, which produces defensive spines in the presence of its nudibranch predator (Harvell 1986). Further ecological studies will certainly help us understand why the types and amounts of metabolites vary within and between apparently conspecific bryozoans.

11.7.2 BIOSYNTHETIC ORIGINS OF ECOLOGICALLY RELEVANT MOLECULES

After the discovery of a novel metabolite from marine invertebrates, it is not unusual for the researchers to ponder the biosynthetic origins of the compound, especially if it was isolated from an organism that is known to harbor microbial symbionts, or if the compound shares remarkable similarities with a known microbial metabolite (Proksch et al. 2002). However, direct evidence for the microbial involvement in the biosynthesis of a natural product is harder to come by. Generally, a metabolite is suspected to be of microbial origin if it (1) shared structural similarity with known microbial compounds; (2) is present in unrelated host taxa; (3) involves biosynthetic pathways that are normally found in microbes but rare in plants or animals (see Hildebrand et al. 2004 for an overview). It is worth considering that even if these three criteria are met, a eukaryotic origin cannot be discounted as convergent evolution can account for similar compounds in diverse taxa. Lateral gene transfer between a microbial donor and an invertebrate host recipient may also account for biosynthetic pathways that have microbial affiliations. Therefore, rigorous testing of a microbial origin hypothesis is essential.

The structural similarities between the tambjamines isolated from bryozoans (Carté and Faulkner 1983) and the tambjamine isolated from the marine bacterium *Pseudoalteromonas tunicata* (Franks et al. 2005) suggest a bacterial source of the tambjamines. Genomic analysis of *P. tunicata* uncovered a biosynthetic cluster that is consistent with the biosynthesis of tambjamines (Burke et al. 2007). While these studies support the symbiotic origins of tambjamines in bryozoans, no bacteria were observed microscopically in the tambjamine-producing ascidian *Atapazoa* sp. (Lindquist and Fenical 1991). Thus, the source of tambjamines in bryozoans is still unknown.

Recent advances in biotechnology have provided sophisticated tools to uncover the biosynthetic origins of natural products (Simmons et al. 2008). For example, MALDI-TOF imaging, a mass spectrometry method, can be used localize natural products in invertebrate tissue (Esquenazi et al. 2008). Shotgun metagenomic sequencing was used to characterize the biosynthetic capabilities of microbial symbionts (Donia et al. 2011), and heterologous expression of putative biosynthetic genes can confirm if the encoded enzyme performs the correct catalytic transformation (Lopanik et al. 2008). The increasing assessibility of metagenomic sequencing and assembly, coupled with gene fabrication (Mitchell 2011), might solve many questions regarding biosynthetic origins of bryozoan metabolites in the near future.

11.8 SUMMARY

While the rate at which novel bryozoan metabolites are being discovered is slowing down, the field of bryozoan chemical ecology is ripe for study. Most of these compounds were isolated from a drug discovery approach; therefore, the ecological functions of these known metabolites, for the most part, have not been comprehensively addressed. A multidisciplinary approach (field ecology studies combined with advanced spectroscopic and molecular methods and relevant bioassays) will likely be needed to understand the complex and subtle ecological roles of bryozoan metabolites.

REFERENCES

Al-ogily, S. M., and E. W. Knight-jones. 1977. Anti-fouling role of antibiotics produced by marine algae and bryozoans. *Nature* 265(5596):728–729.

Berge, J. A., H. P. Leinaas, and K. Sandøy. 1985. The solitary bryozoan, *Monobryozoon limicola* Franzén (Ctenostomata), a comparison of mesocosm and field samples from Oslofjorden, Norway. *Sarsia* 70(1):91–94. doi:10.1080/00364827.1985.10420621.

Blackman, A. J., and R. D. Green. 1987. Further amathamide alkaloids from the bryozoan *Amathiaw wilsoni. Aust. J. Chem.* 40(10):1655–1662.

Blackman, A. J., and C. P. Li. 1994. New tambjamine alkaloids from the marine bryozoan *Bugula dentata. Aust. J. Chem.* 47(8):1625–1629.

Blunt, J. W., B. R. Copp, R. A. Keyzers, M. H. G. Munro, and M. R. Prinsep. 2014. Marine natural products. *Nat. Prod. Rep.* 31(2):160–258.

Burgess, S. C., E. A. Treml, and D. J. Marshall. 2012. How do dispersal costs and habitat selection influence realized population connectivity? *Ecology* 93(6):1378–1387. doi:10.1890/11-1656.1.

Burke, C., T. Torsten, E. Suhelen, and S. Kjelleberg. 2007. The use of functional genomics for the identification of a gene cluster encoding for the biosynthesis of an antifungal tambjamine in the marine bacterium *Pseudoalteromonas tunicata. Environ. Microbiol.* 9(3):814–818.

Carbone, M., C. Irace, F. Costagliola, F. Castelluccio, G. Villani, G. Calado, V. Padula et al. 2010. A new cytotoxic tambjamine alkaloid from the Azorean nudibranch *Tambja ceutae. Bioorg. Med. Chem. Lett.* 20(8):2668–2670.

Carté, B., and D. J. Faulkner. 1983. Defensive metabolites from three nembrothid nudibranchs. *J. Org. Chem.* 48(14):2314–2318.

Carté, B., and D. J. Faulkner. 1986. Role of secondary metabolites in feeding associations between a predatory nudibranch, two grazing nudibranchs, and a bryozoan. *J. Chem. Ecol.* 12(3):795–804.

Davidson, S. K., and M. G. Haygood. 1999. Identification of sibling species of the bryozoan *Bugula neritina* that produce different anticancer bryostatins and harbor distinct strains of the bacterial symbiont 'Candidatus Endobugula sertula.' *Biol. Bull.* 196(3):273.

Donia, M. S., W. F. Fricke, F. Partensky, J. Cox, S. I. Elshahawi, J. R. White, A. M. Phillippy et al. 2011. Complex microbiome underlying secondary and primary metabolism in the tunicate-Prochloron symbiosis. *Proc. Natl. Acad. Sci. USA* 108(51):E1423–E1432. doi:10.1073/pnas.1111712108.

Esquenazi, E., C. Coates, L. Simmons, D. Gonzalez, W. H. Gerwick, and P. C. Dorrestein. 2008. Visualizing the spatial distribution of secondary metabolites produced by marine cyanobacteria and sponges via MALDI-TOF imaging. *Mol. BioSyst.* 4(6):562–570. doi:10.1039/B720018H.

Etcheberrigaray, R., M. Tan, I. Dewachter, C. Kuipéri, I. Van der Auwera, S. Wera, L. Qiao et al. 2004. Therapeutic effects of PKC activators in alzheimer's disease transgenic mice. *Proc. Natl. Acad. Sci. USA* 101(30):11141–11146. doi:10.1073/pnas.0403921101.

Faulkner, D. J. 1999. Marine natural products. *Nat. Prod. Rep.* 16:155–198. doi:10.1039/A804469D.

Figuerola, B., L. Núñez-Pons, J. Moles, and C. Avila. 2013. Feeding repellence in Antarctic bryozoans. *Naturwissenschaften* 100(11):1069–1081. doi:10.1007/s00114-013-1112-8.

Franks, A., P. Haywood, C. Holmström, S. Egan, S. Kjelleberg, and N. Kumar. 2005. Isolation and structure elucidation of a novel yellow pigment from the marine bacterium *Pseudoalteromonas tunicata. Molecules* 10(10):1286–1291. doi:10.3390/10101286.

Fuchs, J., M. Obst, and P. Sundberg. 2009. The first comprehensive molecular phylogeny of Bryozoa (Ectoprocta) based on combined analyses of nuclear and mitochondrial genes. *Mol. Phylogenetics Evol.* 52(1):225–233. doi:10.1016/j.ympev.2009.01.021.

Goldson, A. J., R. N. Hughes, and C. J. Gliddon. 2001. Population genetic consequences of larval dispersal mode and hydrography: A case study with bryozoans. *Mar. Biol.* 138(5):1037–1042.

Harvell, C. D. 1984. Why nudibranchs are partial predators: Intracolonial variation in bryozoan palatability. *Ecology* 63:716–724.

Harvell, C. D. 1986. The ecology and evolution of inducible defenses in a marine bryozoan: Cues, costs, and consequences. *Am. Nat.* 128:810–823.

Hildebrand, M., L. E. Waggoner, G. E. Lim, K. H. Sharp, C. P. Ridley, and M. G. Haygood. 2004. Approaches to identify, clone, and express symbiont bioactive metabolite genes. *Nat. Prod. Rep.* 21(1):122–142.

Kamano, Y., A. Kotake, H. Hashima, I. Hayakawa, H. Hiraide, H. Zhang, H. Kizu, K. Komiyama, M. Hayashi, and G. R. Pettit. 1999. Three new alkaloids, convolutamines F and G, and convolutamydine E, from the floridian marine bryozoan *Amathia convoluta. Collect. Czech. Chem. Commun.* 64(7):1147–1153.

Kamano, Y., H. Zhang, A. Hino, M. Yoshida, G. R. Pettit, C. L. Herald, and H. Itokawa. 1995. An improved source of bryostatin 10, *Bugula neritina* from the Gulf of Aomori, Japan. *J. Nat. Prod.* 58(12):1868–1875.

Kawamata, M., K. Kon-ya, and W. Miki. 2006. 5,6-dichloro-1-methylgramine, a nontoxic antifoulant derived from a marine natural product. *Prog. Mol. Subcell. Biol.* 42:125–139.

Lee, N.-K., W. Fenical, and N. Lindquist. 1997. Alternatamides AD: New bromotryptamine peptide antibiotics from the Atlantic marine bryozoan *Amathia alternata. J. Nat. Prod.* 60(7):697–699.

Lidgard, S. 2008. Predation on marine bryozoan colonies: Taxa, traits and trophic groups. *Mar. Ecol. Prog. Ser.* 359: 117–131. doi:10.3354/meps07322.

Lindquist, N., and W. Fenical. 1991. New tamjamine class alkaloids from the marine ascidian *Atapozoa* sp. and its nudibranch predators. Origin of the tambjamines in *Atapozoa. Experientia* 47(5):504–506. doi:10.1007/BF01959957.

Lindquist, N., and M. E. Hay. 1996. Palatability and chemical defense of marine invertebrate larvae. *Ecol. Monogr.* 66(4):431–450. doi:10.2307/2963489.

Linneman, J., D. Paulus, G. Lim-Fong, and N. B. Lopanik. 2014. Latitudinal variation of a defensive symbiosis in the *Bugula neritina* (Bryozoa) sibling species complex. *PLoS One* 9(10):e108783.

Lippert, H., and K. Iken. 2003. Palatability and nutritional quality of marine invertebrates in a sub-Arctic fjord. *J. Mar. Biol. Assoc. U.K.* 83(6):1215–1219. doi:10.1017/S0025315403008518.

Lopanik, N. B., N. M. Targett, and N. Lindquist. 2007. Ontogeny of a symbiont-produced chemical defense in *Bugula neritina* (Bryozoa). *Mar. Ecol. Prog. Ser.* 327:183–191.

Lopanik, N. B., J. A. Shields, T. J. Buchholz, C. M. Rath, J. Hothersall, M. G. Haygood, K. Håkansson, C. M. Thomas, and D. H. Sherman. 2008. In vivo and in vitro trans-acylation by BryP, the putative bryostatin pathway acyltransferase derived from an uncultured marine symbiont. *Chem. Biol.* 15(11):1175–1186. doi:10.1016/j.chembiol.2008.09.013.

Lopanik, N., K. R. Gustafson, and N. Lindquist. 2004a. Structure of bryostatin 20: A symbiont-produced chemical defense for larvae of the host bryozoan, *Bugula neritina. J. Nat. Prod.* 67(8):1412–1414.

Lopanik, N., N. Lindquist, and N. Targett. 2004b. Potent cytotoxins produced by a microbial symbiont protect host larvae from predation. *Oecologia* 139(1):131–139. doi:10.1007/s00442-004-1487-5.

Marshall, D. J., and M. J. Keough. 2003. Variation in the dispersal potential of non-feeding invertebrate larvae: The desperate larva hypothesis and larval size. *Mar. Ecol. Prog. Ser.* 255:145–153.

Mayer, A. M. S., and K. R. Gustafson. 2004. Marine pharmacology in 2001-2: Antitumour and cytotoxic compounds. *Eur. J. Cancer (Oxford, England: 1990)* 40(18):2676–2704. doi:10.1016/j.ejca.2004.09.005.

Mayer, A. M. S., A. D. Rodríguez, O. Taglialatela-Scafati, and N. Fusetani. 2013. Marine pharmacology in 2009–2011: Marine compounds with antibacterial, antidiabetic, antifungal, anti-inflammatory, antiprotozoal, antituberculosis, and antiviral activities; affecting the immune and nervous systems, and other miscellaneous mechanisms of action. *Mar. Drugs* 11(7):2510–2573. doi:10.3390/md11072510.

McGovern, T. M., and M. E. Hellberg. 2003. Cryptic species, cryptic endosymbionts, and geographical variation in chemical defences in the bryozoan *Bugula neritina. Mol. Ecol.* 12(5):1207–1215.

Mehla, R., S. Bivalkar-Mehla, R. Zhang, I. Handy, H. Albrecht, S. Giri, P. Nagarkatti, M. Nagarkatti, and A. Chauhan. 2010. Bryostatin modulates latent HIV-1 infection via PKC and AMPK signaling but inhibits acute infection in a receptor independent manner. *PLoS One* 5(6):e11160. doi:10.1371/journal.pone.0011160.

Mitchell, W. 2011. Natural products from synthetic biology. *Curr. Opin. Chem. Biol.* 15(4):505–515. doi:10.1016/j.cbpa.2011.05.017.

Montanari, A. M., W. Fenical, N. Lindquist, A. Y. Lee, and J. Clardy. 1996. Volutamides AE, halogenated alkaloids with antifeedant properties from the atlantic bryozoan *Amathia convoluta. Tetrahedron* 52(15):5371–5380.

Morris, B. D., and M. R. Prinsep. 1999. Amathaspiramides AF, novel brominated alkaloids from the marine bryozoan *Amathia wilsoni. J. Nat. Prod.* 62(5):688–693.

Ortega, M. J., E. Zubia, and J. Salvá. 1993. A new brominated indole-3-carbaldehyde from the marine bryozoan *Zoobotryon verticillatum. J. Nat. Prod.* 56(4):633–636.

Paul, V., N. Lindquist, and W. Fenical. 1990. Chemical defenses of the tropical ascidian *Atapozoa* sp. and its nudibranch predators *Nembrotha* spp. *Mar. Ecol. Prog. Ser.* 59:109–118. doi:10.3354/meps059109.

Payne, D. J., M. N. Gwynn, D. J. Holmes, and D. L. Pompliano. 2007. Drugs for bad bugs: Confronting the challenges of antibacterial discovery. *Nat. Rev. Drug Discov.* 6(1):29–40. doi:10.1038/nrd2201.

Peters, L., G. M. König, A. D. Wright, R. Pukall, E. Stackebrandt, L. Eberl, and K. Riedel. 2003. Secondary metabolites of *Flustra foliacea* and their influence on bacteria. *Appl. Environ. Microbiol.* 69(6):3469–3475.

Peters, L., A. Wright, A. Krick, and G. König. 2004. Variation of brominated indoles and terpenoids within single and different colonies of the marine bryozoan *Flustra foliacea. J. Chem. Ecol.* 30(6):1165–1181. doi:10.1023/B:JOEC.0000030270.65594.f4.

Pettit, G. R., C. L. Herald, D. L. Doubek, D. L. Herald, E. Arnold, and J. Clardy. 1982. Isolation and structure of bryostatin 1. *J. Am. Chem. Soc.* 104(24):6846–6848. doi:10.1021/ja00388a092.

Pettit, G. R., Y. Kamano, and C. L. Herald. 1987. Antineoplastic agents. 119. Isolation and structure of bryostatins 10 and 11. *J. Org. Chem.* 52(13):2848–2854. doi:10.1021/jo00389a036.

Proksch, P., R. Edrada, and R. Ebel. 2002. Drugs from the seas–current status and microbiological implications. *Appl. Microbiol. Biotechnol.* 59(2–3):125–134.

Sharp, K. H., S. K. Davidson, and M. G. Haygood. 2007. Localization of *Candidatus Endobugula sertula* and the bryostatins throughout the life cycle of the bryozoan *Bugula neritina*. *ISME J.* 1(8):693–702.

Shellenberger, J. S., and J. R. P. Ross. 1998. Antibacterial activity of two species of bryozoans from northern puget sound. https://research.wsulibs.wsu.edu:8443/jspui/handle/2376/1225.

Sherwood, J., J. T. Walls, and D. A. Ritz. 1998. Amathamide alkaloids in the pycnogonid, *Stylopallene longicauda*, epizoic on the chemically defended bryozoan, *Amathia wilsoni*. In *Papers Proc. R. Soc. Tas.* 132:65–70. http://eprints.utas.edu.au/13627/.

Simmons, T. L., R. C. Coates, B. R. Clark, N. Engene, D. Gonzalez, E. Esquenazi, P. C. Dorrestein, and W. H. Gerwick. 2008. Biosynthetic origin of natural products isolated from marine microorganism–invertebrate assemblages. *Proc. Natl. Acad. Sci.* 105(12):4587–4594. doi:10.1073/pnas.0709851105.

Stabili, L., C. Gravili, S. M. Tredici, S. Piraino, A. Talà, F. Boero, and P. Alifano. 2008. Epibiotic vibrio luminous bacteria isolated from some hydrozoa and bryozoa species. *Microb. Ecol.* 56(4):625–636.

Stachowicz, J. J., and N. Lindquist. 2000. Hydroid defenses against predators: The importance of secondary metabolites versus nematocysts. *Oecologia* 124(2):280–288. doi:10.1007/s004420000372.

Sudek, S., N. B. Lopanik, L. E. Waggoner, M. Hildebrand, C. Anderson, H. Liu, A. Patel, D. H. Sherman, and M. G. Haygood. 2007. Identification of the putative bryostatin polyketide synthase gene cluster from *Candidatus Endobugula sertula*, the uncultivated microbial symbiont of the marine bryozoan *Bugula neritina*. *J. Nat. Prod.* 70(1):67–74.

Trindade-Silva, A. E., G. E. Lim-Fong, K. H. Sharp, and M. G. Haygood. 2010. Bryostatins: Biological context and biotechnological prospects. *Curr. Opin. Biotechnol.* 21(6):834–842. doi:10.1016/j.copbio.2010.09.018.

Wahl, M. 1989. Marine epibiosis. I. Fouling and antifouling: Some basic aspects. *Mar. Ecol. Prog. Ser.* 58:175–189.

Walls, J. T., A. J. Blackman, and D. A. Ritz. 1995. Localisation of the amathamide alkaloids in surface bacteria of *Amathia wilsoni* Kirkpatrick, 1888 (Bryozoa: Ctenostomata). *Hydrobiologia* 297(2):163–172. doi:10.1007/BF00017482.

Walls, J. T., D. A. Ritz, and A. J. Blackman. 1993. Fouling, surface bacteria and antibacterial agents of four bryozoan species found in Tasmania, Australia. *J. Exp. Mar. Biol. Ecol.* 169(1):1–13.

Wendt, D. E. 1996. Effect of larval swimming duration on success of metamorphosis and size of the ancestrular lophophore in *Bugula neritina* (Bryozoa). *Biol. Bull.* 191(2):224–233.

Zimmer, R. L., and R. M. Woollacott. 1993. Anatomy of the larva of *Amathia vidovici* (Bryozoa: Ctenostomata) and phylogenetic significance of the vesiculariform larva. *J. Morphol.* 215(1):1–29. doi:10.1002/jmor.1052150102.

Zweifel, U. L., and A. Hagstrom. 1995. Total counts of marine bacteria include a large fraction of non-nucleoid-containing bacteria (ghosts). *Appl. Environ. Microbiol.* 61(6):2180–2185.

12 Spatial and Temporal Variability in Sponge Chemical Defense

Sven Rohde and Peter J. Schupp

CONTENTS

12.1 SPATIAL AND TEMPORAL VARIABILITY IN SPONGE CHEMICAL DEFENSE

The following chapter is meant to provide an overview on the variability of sponge chemical defenses and is not intended to provide a detailed review on the chemical ecology of sponges, as there have been several reviews on sponge chemical ecology (Wulff 2006, Pawlik 2011) in recent years. Furthermore, several more general reviews on invertebrate chemical ecology including sponges have also been published in recent years (Paul and Puglisi 2004, Paul et al. 2011). Most studies on sponge ecology have treated sponges as a static system in a very dynamic environment. Only in the last decade has it become obvious that many sponge species respond dynamically to environmental factors and allocate physiological processes specifically to varying stressors. This chapter presents a detailed review on the recent literature examining the spatial (geographic, intraspecific, and intraindividual) and temporal variability of secondary metabolites in sponges. Adaptive variability of defenses in response to biotic or abiotic factors or as result of physiologically or genetically variation can increase the effectiveness of defenses and save limited resources. However, it also can impede the detection or assessment of defenses and complicate the ecological evaluation of defensive effects. Recognizing the variability of sponge chemical defenses might help to explain why sponge chemical defenses observed in feeding assays or manipulative field studies do not necessarily correlate with observed deterrence in the field or abundances of sponges in the various habitats. A better understanding of the various factors affecting secondary metabolite production might also provide new insights to the success of sponges as one of the dominant sessile benthic invertebrates.

12.1.1 IMPORTANCE OF SPONGES IN STRUCTURING BENTHIC INVERTEBRATE COMMUNITIES

Sponges are an important component of the sessile benthic invertebrate community in the world oceans, occurring from the Arctic and Antarctic to temperate and tropical regions. They remain an important component of the sessile benthic invertebrate fauna due to both their biomass and their ability to influence the coupling between benthic and pelagic processes (Bell 2008). While sponges are an important component of shallow-water benthic communities, they are often one of the dominant benthic organisms in deep-water habitats where light-dependent benthic organisms are absent. The ability of sponges to filter large quantities of water (>100,000 L/kg sponge/day; Vogel 1977, Weisz et al. 2008) makes them also an important link between benthic and pelagic processes (Gili and Coma 1998, Bell 2008, Maldonado et al. 2011, 2012). The importance of sponges in benthic pelagic coupling has been recognized by recent publications on the "sponge loop," which describes the uptake of dissolved organic matter (DOC) by sponges from the surrounding water and the conversion of it to particulate organic matter (POM) via shedding of filter cells as detritus. The produced detritus can then be utilized by other benthic organisms (de Goeij et al. 2013, Rix et al. 2016).

The importance of sponges as a major component of marine benthic communities has been further emphasized in recent years, as coral reefs worldwide are being exposed to multiple natural and anthropogenic stressors (e.g., climate change, predator outbreaks, overfishing, disease), often resulting in a pronounced decline of live coral cover and an increase of algae and invertebrate competitors (Cote et al. 2005, Hughes et al. 2010, De'ath et al. 2012). This has led to so-called phase shifts where corals no longer represent the most abundant organisms on the reef, but rather algae and sponges have become the dominating sessile organisms (Norström et al. 2009, Hughes et al. 2010, Bell et al. 2013). Especially on coral reefs in the Caribbean, phase shifts from coral- to sponge-dominated reefs have been well documented (Bell and Smith 2004, Pawlik 2011, Loh and Pawlik 2014, Loh et al. 2015). There, benthic cover of algae and sponges on most reefs has increased to 29% and 16%, respectively (Loh et al. 2015). In the Pacific and Indian Oceans, phase shifts have been documented as well, for example from the Barrier Reef of Madagascar, Quirimba Archipelago in Mozambique, or the Indonesian Wakatobi region (Barnes 1999, Barnes and Bell 2002, Bell and Smith 2004), although reports of shifts from coral- to sponge-dominated reefs are less common. In addition, benthic cover of sponges on such reefs has been reported to be between 3% and 11% and therefore lower compared to most affected reefs in the Caribbean, where sponge cover reaches on average 16% (Loh et al. 2015, Pawlik et al. 2016). Given that these drastic changes in benthic community composition continue over the next decades, the ecological significance of sponges in structuring coral reef ecosystems will therefore likely gain even greater importance in the future. While stressors contributing to coral decline have been well documented (e.g., increasing sea surface temperatures, eutrophication, sedimentation; Cote et al. 2005, Hughes et al. 2010, De'ath et al. 2012), factors contributing to the increase of sponges in benthic cover on coral reefs are not yet well understood (e.g., role of top-down versus bottom-up factors; Lesser and Slattery 2013, Pawlik et al. 2013, 2015).

12.1.2 PRODUCTION OF SECONDARY METABOLITES

Production of bioactive secondary metabolites is certainly one factor contributing to sponges being successful spatial competitors in many habitats. The use of secondary metabolites in chemical defense against potential predators (Pawlik et al. 1995), microbial pathogenesis (Newbold et al. 1999, Kelly et al. 2003, Rohde et al. 2015), and space competitors (Pawlik et al. 2007) has been well documented for many sponge species from polar (McClintock et al. 2005, Peters et al. 2009a, Nunez-Pons and Avila 2014, Berne et al. 2015), temperate (Battershill and Bergquist 1990, Becerro et al. 2003), and especially tropical regions (Pawlik et al. 1995, Becerro et al. 2003, Burns et al. 2003, Thoms and Schupp 2007, Rohde et al. 2015). The majority of literature on chemical defenses today has examined sponge crude extracts or isolated secondary metabolites as constitutive defense, where it is assumed that production of bioactive secondary metabolites is continuous and concentrations of these compounds fairly stable.

12.1.3　Chemical Defense Against Sponge Predators

Most of the studies on sponge chemical defenses have focused on predator deterrence, mainly against fishes and mostly on tropical coral reef habitats. The majority of these studies examined fish deterrence of sponge crude extracts in laboratory feeding experiments using the blue head wrasse *Thalassoma bifasciatum* (Pawlik et al. 1995, Lindel et al. 2000) or the pufferfish *Canthigaster solandri* (Rohde et al. 2012), while only a few examined chemical deterrence both in laboratory feeding assays and in the field (Schupp et al. 1999, Thoms and Schupp 2008, Rohde and Schupp 2011). Feeding assays examining sponge secondary metabolites with more generalist sponge predators such as urchins or starfish are fairly uncommon (Wulff 1994, Rohde et al. 2015) and feeding experiments examining highly specialized spongivorous predators such as angelfishes or mollusks (e.g., couries and nudibranchs) are restricted to field observations (Hill 1998, Pawlik and Deignan 2015). Transplant experiments with whole sponges have provided insights on feeding rates of various reef fishes, especially the more abundant species such as parrotfishes (Wulff 1997), which are not by definition spongivorous, but ingest sponges incidentally during their intense feeding episodes. However, such experiments and observational data from field surveys do not unambiguously demonstrate that avoidance of certain sponge species is caused through production of feeding deterrent secondary metabolites, as other factors such as physical defense mechanisms (Hill et al. 2005, Jones et al. 2005) or nutritional value (Chanas and Pawlik 1995) can influence feeding behavior as well. To gain a better understanding of sponge chemical defense mechanisms in the future, experiments involving generalist and specialist sponge predators should be considered.

12.1.3.1　Variability of Chemical Defenses According to the Allocation Cost Model

Assessment of sponge chemical defenses is further complicated if an allocation cost model is applied (Cronin 2001). Accordingly, resources are divided between various life processes including nutrient acquisition, growth, maintenance, reproduction, and defense. Resource allocation will be shifted depending on the biotic and abiotic conditions an organism is exposed to. Following a storm more resources might be allocated to maintenance and growth to heal and regrow damaged parts, rather than allocating resources to chemical defense (for details see Cronin 2001). For example, chemically undefended sponges likely invest more resources in reproduction and growth to regrow lost biomass following a storm or predator attack. Chemically defended sponges on the other hand will invest continuously energy towards production of defense metabolites and therefore will have fewer resources for growth and reproduction. Pawlik and colleagues (2008) showed indirect evidence for such a resource tradeoff by investigating sponge recruitment patterns on an artificial reef (ship wreck) in the Caribbean. They found that initial colonization by sponges on the wreck was almost exclusively by chemically non-defended sponges (96%) while nearby coral reef habitats showed only very low abundances of non-defended sponges (15%). After 18 months, the first chemically defended sponges had recruited to the site, transforming the site to a similar sponge community as nearby reefs. The authors concluded that resource tradeoff between growth or reproduction and chemical defenses exists, as recruitment combines the resource allocation of reproduction and growth.

Besides abiotic and biotic factors influencing the production of deterrent secondary metabolites there is also intraspecimen variability in chemical defenses, as compounds are allocated differently even within an organism (see Section 12.2.3 for details). Another level of variability in chemical defenses can be due to reactions of the sponge following predator attack. Given sponges have the metabolic framework, they can also increase synthesis of defense metabolites (induced defenses, see Section 12.3.2.1) or convert less active precursors metabolites to bioactive secondary metabolites (activated defenses, see Section 12.3.2.2).

Provided that production of sponge secondary metabolites is costly and resources (e.g., food or nutrients) are limited, shifting of resources between important life processes seems to provide an evolutionary advantage and might aid sponges in being successful competitors with other sessile benthic invertebrates. The recent publications on the sponge loop (de Goeij et al. 2013, Rix et al. 2016) emphasize that uptake of DOM and conversion to POM (most likely via shedding of sponge cells) is an important mechanism in the biochemical recycling of nutrients, especially on oligotrophic coral reefs. The constant loss of carbon and nitrogen via shedding of POM by the sponge is likely limiting resources for the various life processes. Therefore, having the ability to modify chemical defenses spatially and temporally according to the present life history (e.g., growth or reproduction) and abiotic and biotic conditions (e.g., presence or absence of predators) should enable sponges to respond to environmental challenges in more successful ways.

12.2   SPATIAL VARIABILITY OF CHEMICAL DEFENSE

12.2.1   REGIONAL VERSUS LOCAL VARIABILITY

The variability of secondary metabolite production in sponges can be very pronounced both on an inter- and intraspecific level (Turon et al. 1996, Betancourt-Lozano et al. 1998, Turon et al. 2009).

Interspecific differences in chemical defense can affect the distribution of sponge species and the community composition. High densities of spongivores (e.g., angel- and parrotfishes, sea urchins, turtles) restrict undefended species and favor the abundance of well-defended species. Consequently, predator-rich coral reefs harbor a better-defended sponge community than protected mangrove habitats (Pawlik 1998). Similar patterns have been shown for reefs with high abundances of spongivorous fish compared to overfished areas (Loh and Pawlik 2014). Regional and local predation factors determine the sponge community patterns with regard to their defensive potential (Pawlik et al. 1995, 1998, Peters et al. 2009b). However, this review discusses rather how abiotic and biotic factors affect the intraspecific variability of sponge defenses.

Intraspecific production of secondary metabolites may vary between populations or even between individuals (Turon et al. 1996, Page et al. 2005). Biological or environmental factors including nutrient availability, light, UV radiation, and temperature can affect this plasticity (Uriz et al. 1995, Becerro and Paul 2004, Turon et al. 2009).

Investigations on the variability of secondary metabolites within a sponge species are scarce and controversial. Some studies observed a very low variability in sponge metabolite concentrations. The brominated pyrrole alkaloids hymenialdisine, debromohymenialdisine, dibromophakellin, and 3-bromohymenialdisin occur in very similar concentrations within specimens of the sponge *Axinella carteri*, collected along sites as far as 2000 km (Supriyono et al. 1995). In contrast, a recent study on bromotyrosine derivatives from *Aplysina fulva* demonstrated chemical variability among specimens collected in multiple locations in Brazil and the United States (Nunez et al. 2008). The concentration of brominated compounds in specimens of the sponge *Aplysina aerophoba* collected around the Canary Islands showed significant variation (Teeyapant and Proksch 1993). Certain brominated alkaloids of *A. aerophoba* revealed the greatest variation between sites less than 500 m apart, while other specimens had the highest compound variation between regions over 2500 km apart (Sacristan-Soriano et al. 2011). An extensive study on the intraspecific chemical diversity of the Mediterranean sponge *Spongia lamella* investigated the chemical profiles of seven populations spreading over 1200 km (Noyer and Becerro 2012). They calculated the chemical diversity by taking into account the number and concentrations of chemical compounds and concluded that diversity differed significantly between sponge populations, but not at a larger regional scale. However, the abundance of some metabolites varied significantly between sponge populations and the chemical dissimilarity increased with increasing geographic distance. A study on *Stylissa massa* followed a similar trend. Several alkaloids had the greatest variation in specimens collected from different sites around the island of Guam, while another compound, dibromohymenialdisine, varied more among locations across the Pacific region. However, the fact that small-scale spatial variation affects metabolite concentrations of *S. massa* more profoundly than large-scale variation is not a universally valid pattern. Other studies reported opposing results, producing a diverse pattern regarding metabolite concentrations as function of location (Puglisi et al. 2000, Slattery et al. 2001, Fahey and Garson 2002, Page et al. 2005, Jumaryatno et al. 2007). Studies on biogeographic variation face the constraint that many factors such as water temperature, food availability, and light exposure vary highly between locations and can affect the production of secondary metabolites in sponges or other organisms. Moreover, temporal variation like seasonal patterns can confound a spatial analysis (Turon et al. 1996, Page et al. 2005, Abdo et al. 2007). Within-species variation of secondary metabolites can be so high as to hinder the identification of extrinsic factors (Targett et al. 1992, Rohde et al. 2012) and whether such variability in secondary metabolites is genetically or environmentally controlled is still unknown. Additionally, so far most studies on spatial and temporal variation of metabolite concentration were restricted to the quantitative analysis of variation without studying the causing factors of spatial and temporal variation. While correlative studies can indicate potential relations of environmental or biotic factors with secondary metabolites, more experimental studies are necessary to prove vigorously what the causes for metabolite variation are.

12.2.2 HABITAT VARIABILITY

Depth gradients are accompanied by gradients of other abiotic and biotic factors. Light and food availability as well as abundance of predators and competitors can change with depth and consequently affect the secondary metabolite production. Studies on depth-related changes in metabolite concentrations demonstrated ambiguous results. Several studies found that metabolite concentrations did not vary significantly with depth in sponges (Becerro and Paul 2004, Abdo et al. 2007, Putz et al. 2009). Others showed complex variations of metabolite concentrations with regard to depth. In the sponge *Mycale hentscheli*, depth did not affect the cytotoxic compound mycalamide A, but there was a relationship with depth for pateamine and peloruside A, but this varied among populations at different times (Page et al. 2005). The highest peloruside A concentrations were detected in sponges from 8 to 10 m depth and coincided with the depth of the density boundary layer and highest chlorophyll concentration at this site. A transplantation experiment along a depth gradient with the sponge *Aplysina cavernicola* also revealed no changes in the metabolite profile in transplanted sponges (Thoms et al. 2003). However, the composition of diterpenes in *Rhopaloeides odorabile* varied with environmental factors such as light intensity and depth after a year (Thompson et al. 1987). Reduced light availability due to increasing depth or natural shading reduced diterpene concentrations significantly. In *Stylissa massa* (Figure 12.1a), the alkaloid hymenidin varied with depth, although concentrations did not adapt over a 3-month duration of a transplantation study, suggesting that such metabolic changes may occur more slowly (Rohde et al. 2012). The sponge *Plakortis angulospiculatus* revealed lower concentrations of the defensive compound plakortide F with increasing depth, which correlated with lower spongivory rates (Slattery et al. 2016). It is obvious that depth-correlating factors like light, competition, fouling, or predation pressure vary according to the site and habitat and, consequently, hinder general depth-related patterns in metabolite concentrations.

Shallow-water sponges can be exposed to high intensities of UV radiation. Some marine organisms produce metabolites that are suggested to provide protection against UV radiation (Pavia et al. 1997, Dunlap and Shick 1998) and their levels could be adjusted to changing intensities of UV radiation. It is known that the photosymbionts of corals produce mycosporine-like amino acids (MAAs) (Shick et al. 1999), thus providing UV-protection for the coral host. Symbiotic cyanobacteria are also known to produce MAAs (Gröniger et al. 2000), and MAAs have been found in marine sponges (Bandaranayake et al. 1996, 1997, McClintock and Karentz 1997). Additionally, Steindler et al. (2002) found a higher abundance of photosymbiotic sponge species in the intertidal compared to the subtidal and hypothesized that this may be related to a higher necessity for symbiont-derived substances that provide protection from UV radiation. The sponge *S. massa* contains the UV-absorbing metabolites hymenialdisine and debromohymenialdisine; however, sponges from very shallow water (<1 m) had similar concentrations of hymenialdisine and debromohymenialdisine compared to sponges from deeper (20 m), UV-protected habitats (Rohde et al. 2012). Furthermore, the experimental removal of UV radiation did not affect the concentration of these metabolites, indicating that an induced defense against UV radiation is not expressed in *S. massa*.

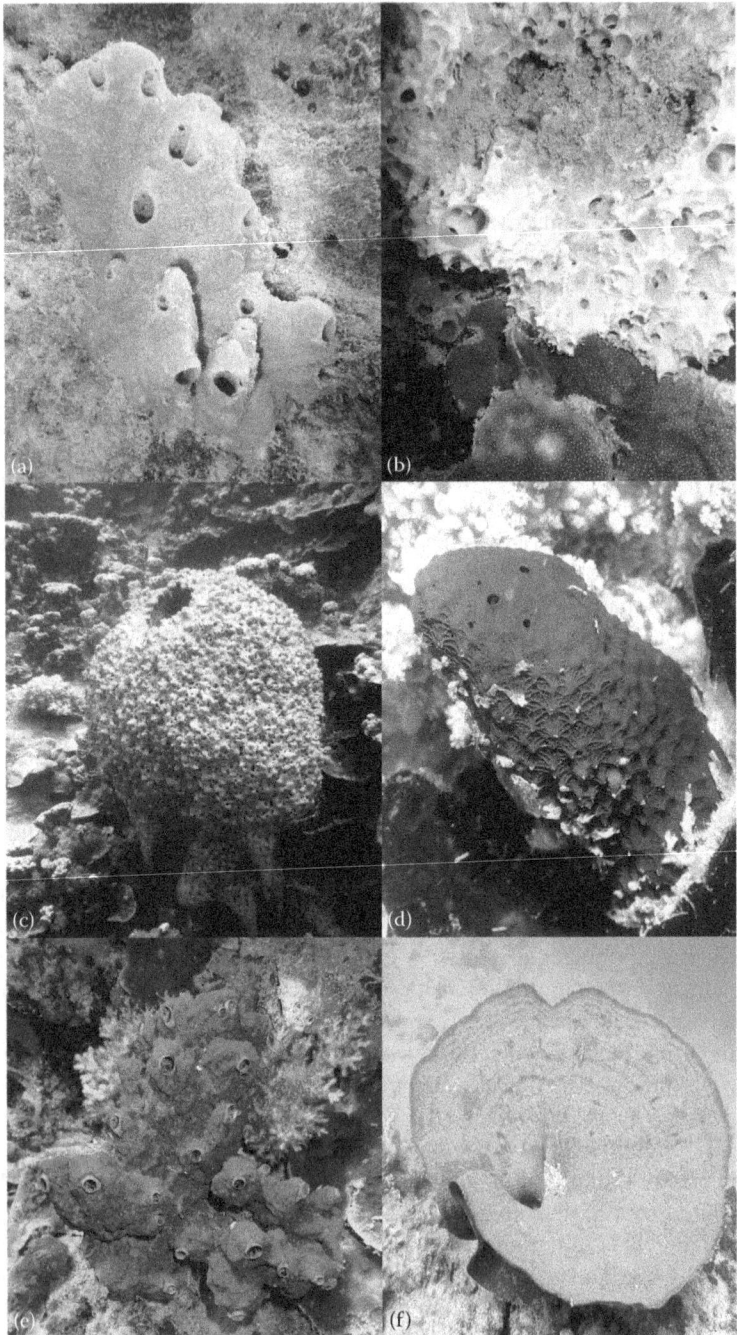

FIGURE 12.1   Intensively studied sponge species in the field of chemical ecology. (a) *Stylissa massa*, (b) *Dysidea avara*, (c) *Melophlus sarasinorum*, (d) *Hyrtios erecta*, (e) *Aplysinella* sp., (f) *Ianthella basta*.

12.2.3 INTRAINDIVIDUAL VARIABILITY, DIFFERENTIAL COMPOUND ALLOCATION, OPTIMAL DEFENSE THEORY

Spatial sponge metabolite variation has not only been reported on a regional, local, or habitat scale, but also on an intraindividual scale. The nonrandom allocation of secondary metabolites within the individual has been attributed to a defensive role against biotic stressors, for example fouling, predation, and competition. The theoretical framework to explain the evolution of intraindividual variation of bioactivity is basing on the Optimal Defense Theory (ODT) that states that defenses are expressed in a way that maximizes fitness and minimizes costs (Rhoades 1979). Since sponges in general lack distinct morphological tissues, differential expression of defensive compounds within a sponge has mostly been addressed to a preferential accumulation to the peripheral or apical parts, which leads to an optimized defense against foulers, predators, or competitors.

Even though interactions with foulers and competitors are almost exclusively driven by surface-associated chemistry, specific studies on surface-associated antifouling and allelopathic activities are scarce. It has been shown that the Antarctic sponge *Phorbas glaberrima* possesses the highest antifouling activity in the outer layer, but a sympatric species, *Cinachyra barbata*, showed antifouling activity in the internal instead of the external part (Angulo-Preckler et al. 2015). This has been attributed to similar encounter rates of microbes in the inner layer (choanosome) due to the sponge's pumping activity during feeding (exposing the choanosome to seawater bacteria), while the outer layer (ectosome) was protected by dense spicule masses. Similarly, higher concentrations of brominated alkaloids were found in the choanosome of *Aplysina aerophoba* compared to the ectosome (Sacristán-Soriano et al. 2016). Increased allelopathic activity in the ectosome has been shown in the sponge *Dysidea avara* (see Figure 12.1b, De Caralt et al. 2013) and toxicity in the sponge *Crambe crambe* was also higher in the ectosome compared to the choanosome (Becerro et al. 1997). More studies have been done on the differential intraindividual expression of antipredatory compounds. The basic idea of optimizing defenses is to concentrate them in tissues that have the highest chance of encountering predators (i.e., exposed body parts or the ectosome) or which have the highest fitness value for the organism. Studies on intraspecimen variation of defenses in sponges are few and focused mainly on the concentrations of secondary metabolites. The Antarctic sponge *Latrunculia apicalis* accumulates more defensive alkaloids in its outer than inner tissue layers, providing a defense against a predatory sea star (Furrow et al. 2003). Becerro et al. (1998) and Schupp et al. (1999) found higher concentrations of defensive metabolites in more apical parts compared to basal parts of the investigated sponges, while two other studies examining eight sponge species found no evidence that the ectosome was better defended than the choanosome (Swearingen and Pawlik 1998, Burns et al. 2003). The picture gets more complicated adding the results of Freeman and Gleason (2010) who found one sponge species with higher concentrations of fish deterrent metabolites in the outermost layer, while two other species had higher concentrations in the deeper tissue layer. A study of 27 Antarctic sponges revealed that 68% contained deterrent secondary metabolites against starfish predators in their outermost layer, while only 62% of these defended species were also defended in their inner tissue (Peters et al. 2009b). A study on the chemical and structural defense of the sponge *Melophlus sarasinorum* (Figure 12.1c) concluded that the equal distribution of chemical defenses throughout *M. sarasinorum* was in accordance with ODT in regards to fish predation, while structural defense supports ODT by being restricted to the surface layer, which experiences the highest predation risks from mesograzers (Rohde and Schupp 2011). All these studies demonstrated clearly that defensive compounds can be distributed within a sponge in a very specific pattern. However, the interpretation of these allocations is a complicated and elaborate matter. To assess the allocation of defensive compounds with regards to the ODT, the fitness value of different body parts needs to be estimated including the relative risk of predator attacks or the microbial encounter rates. Data for these characteristics are mostly lacking, making interpretations of demonstrated patterns to some extent hypothetical.

12.3 TEMPORAL VARIABILITY OF CHEMICAL DEFENSE

12.3.1 SEASONAL VARIATION

Not only can spatial patterns be attributed to compound variation, but temporal patterns can also play a major role. Seasonal variations of the relative composition of biologically active metabolites have been described for several sponges. The sponge *M. hentscheli* revealed generally higher concentrations of bioactive compounds during spring and summer (Page et al. 2005). A similar seasonal pattern in toxicity was described for the sponge *Latrunculi* sp. nov. (*wellingtonensis*), suggesting that higher degrees of biochemical activity within the sponge in spring prevented fouling of the surface (Duckworth and Battershill 2001). The sponge *Crambe crambe* demonstrated high toxicity in the peripheral ectosome during summer and autumn, which correlated with seasonal variation of competitors (Turon et al. 1996). However, not only competitors or foulers can induce a seasonal pattern; abiotic factors were also discussed as controlling factors for compound variation. High water temperature seems to increase the production of some sponge metabolites. The cytotoxic natural product salicylihalamide A from the sponge *Haliclona* sp. correlated with water temperature leading to highest concentrations in the summer (Abdo et al. 2007). Also, lysophospholipids from the sponge *Oscarella tuberculata* showed the highest expression rates in the summer (Ivanisevic et al. 2011a), which also correlated with bioactivity measurements (Ivanisevic et al. 2011b). The main compounds from the sponge *Dysidea avara*, avarol and 5'-monoacetylavarol, varied with time; however, no direct relationship between metabolite production and seawater temperature was found (De Caralt et al. 2013). A two-year lasting study on *Aplysina aerophoba* revealed an even more differentiated pattern. Brominated alkaloids were shown to increase in concentration in the ectosome during the summer months, while the compound concentrations in the choanosome followed no pattern (Sacristan-Soriano et al. 2012). Other species like *Polymastia croceus* showed no seasonal pattern of bioactivity (Duckworth and Battershill 2001). However, all these studies on seasonal variations used only correlative evidence to explain the observed patterns. As mentioned previously, this way it is hardly possible to identify a causing factor. Whether increased water temperature in the summer, or variation in predator, competitor, or fouler densities affect metabolite production can only be vigorously demonstrated by experimental designs.

12.3.2 INDUCIBLE AND ACTIVATED DEFENSES

12.3.2.1 Inducible Defenses

Inducible defense represents a mechanism by which a given genotype can respond to an environmental stressor through an increased production of defensive compounds (Karban and Baldwin 1997). Induction of defense includes that defensive compounds are only produced on demand, thereby saving resources and increasing the prey's chemical variability (Cronin 2001, Rohde et al. 2004).

Very few studies have investigated the chemical variation in sponges in response to fouling or competitive pressures. Luter and Duckworth (2010) tried to identify changes in bioactivity in the sponges *Coscinoderma matthewsi*, *Hyrtios erecta* (Figure 12.1d), and *Ianthella basta* (Figure 12.1f) caused by neighboring organisms. They found that the bioactivity of sponges was highly variable among individuals, regardless of levels of competition, suggesting that bioactivity is not influenced by surrounding competition.

Richelle-Maurer et al. (2003) investigated the sponge *Agelas conifer* and found a rapid increase in oroidin (Figure 12.2a) and sceptrin (Figure 12.2b) in response to simulated predation, but there was no overall change in these compounds in a longer-term predation-exclusion experiment. The sponge *Plakortis angulospiculatus*, a common sponge species that occurs along a depth gradient from shallow to mesophotic reefs in the Caribbean, developed chemical deterrence to predatory fishes following transplantation from predator-poor mesophotic depths to predator-exposed shallow reefs (Slattery et al. 2016). An accompanied survey of bioactive extracts indicated that concentration differences of the defensive metabolite, plakortide F, were responsible for altered deterrence between shallow and mesophotic reefs.

A comprehensive study on inducible defenses in sponges investigated eight Pacific sponge species with regard to antipredatory and antimicrobial defenses (Rohde et al. 2015). Of these eight species, one sponge species, *Stylissa massa*, showed induced defense against feeding by two omnivore species, the pufferfish *Canthigaster solandri* and the sea urchin *Diadema savignyi*. Additionally, induction of defense in *Melophlus sarasinorum* reduced palatability towards predation by sea urchins. The prevalence of inducible defense in the tested sponge species seems to be relatively low compared to the prevalence in marine macroalgae (Toth and Pavia 2007, Rohde and Wahl 2008).

Defense induction is a cost-effective defense adaptation, since metabolically costly defensive metabolites are produced on demand only. However, the evolution of inducible defenses against predation requires special ecological conditions (reviewed in Harvell and Tollrian 1999, Miner et al. 2005, Pigliucci 2005). Predation pressure must be variable since constant pressure would select for a constitutive defense. Reducing defense should enhance fitness in the absence of consumers. Finally, a reliable cue for triggering induction must exist. The tropics are characterized by a relative constant predation pressure compared to temperate regions,

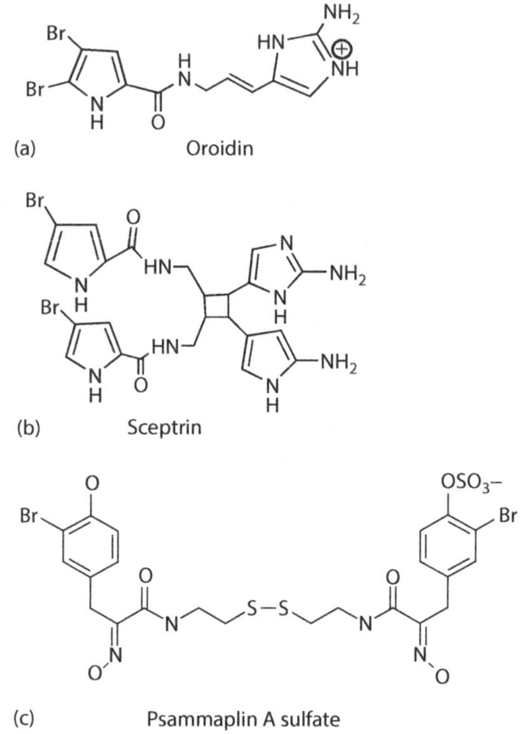

(a)   Oroidin

(b)   Sceptrin

(c)   Psammaplin A sulfate

FIGURE 12.2 Dominant compounds that have been isolated and studied in multiple sponge species.

where seasonality provides an additional source for variability in predation pressure (Harvell and Tollrian 1999). Consequently, it may be an ecological pattern rather than a research bias that only little prevalence of inducible defenses in the tropics has been demonstrated.

Interestingly, the screening for inducible defenses in sponges revealed that 50% of the tested sponge species demonstrated induced antimicrobial defense (Rohde et al. 2015). Simulated predation increased the antimicrobial activity of *Aplysinella* sp. (Figure 12.1e), *Cacospongia* sp., *M. sarasinorum*, and *S. massa*. With increased wounding intensity, the risk of pathogen infection via feeding scars rises, while wounding can also lead to increased antimicrobial defenses through induced defense mechanisms. This may also be an interesting fact for researchers investigating bioactive compounds for the development of new pharmacologically active compounds. They could increase amounts of antimicrobial compounds by inducing compound production prior to collection.

12.3.2.2 Activated Defenses

A special phenomenon with regard to compound allocation within individuals has been described as activated defense. Defense activation describes the wound-activated conversion of inactive or less active compounds to more active forms with defensive functions (Paul and Van Alstyne 1992). Activation requires that the conversion is enzyme-catalyzed and that the precursor compound is stored either in different cells or different cell compartments than the catalyzing enzyme so as to prevent continuous conversion of the precursors (i.e., organisms need to provide the cytological prerequisites in their tissue to allow activated defensive mechanisms). Activated defense mechanisms are well known from terrestrial plants (Karban and Baldwin 1997, Heil and Baldwin 2002) and in several marine algae (Paul and Van Alstyne 1992, Cetrulo and Hay 2000, Jung and Pohnert 2001). However, examples of activated defense against predation in sponges are limited to the Tetractinellid sponge *Melophlus sarasinorum* (Rohde et al. 2015) and two examples of the Verongid sponges *Aplysina* spp. (Teeyapant and Proksch 1993, Thoms et al. 2006), and *Aplysinella* sp. (Thoms and Schupp 2008). *Aplysinella* sp. contains the bromotyrosine compound psammaplin A sulfate (Figure 12.2c), which is converted via cleaving of the sulfate group to psammaplin A, a highly potent antifeeding compound. The advantages of activated defenses include the avoidance or reduction of autotoxicity, as only less active compounds have to be stored for bioconversion (Van Alstyne et al. 2001). However, the high number of constitutive defenses in sponges indicates that sponges deal well with potential autotoxic effects of defensive compounds. Further studies on the autotoxic effects of precursor and converted defensive compounds are needed to reveal whether constitutive defenses are more common due to ecological or physiological reasons (Rohde et al. 2015).

12.4 CONCEPTUAL MODEL AND NEW RESEARCH DIRECTIONS

It has to be recognized from the aforementioned literature that sponge chemical defenses are indeed variable in multiple ways and not as static as once was generally assumed.

12.4.1 CONCEPTUAL MODEL

Based on known abiotic and biotic interactions affecting the defense variation, we propose the following conceptual model that summarizes the conditions favoring the production of sponge secondary metabolites both in space and time (Figure 12.3). Predation pressure and/or consumption traits can increase defense variability by favoring the evolution of facultative defense mechanisms like induced or activated defenses (top-down control in Figure 12.3). Even though the bacterial community within a sponge holobiont appears to be specific and quite stable, abiotic factors can affect the bacterial community composition and, consequently, the production of secondary metabolites. Anthropogenic effects like climate change or eutrophication will increase the variability in environmental conditions and increase the variability of secondary metabolite production, either directly in the sponge or via changes in the associated bacterial composition or metabolism (Figure 12.3). Overfishing represents an anthropogenic effect that decreases the intensity and frequency of predators (mainly macro- and mesopredators such as spongivores, fishes, or urchins) on the sponge and, consequently, rather may decrease the variability of sponge defense. On the other hand, eutrophication or warming increase environmental gradients,

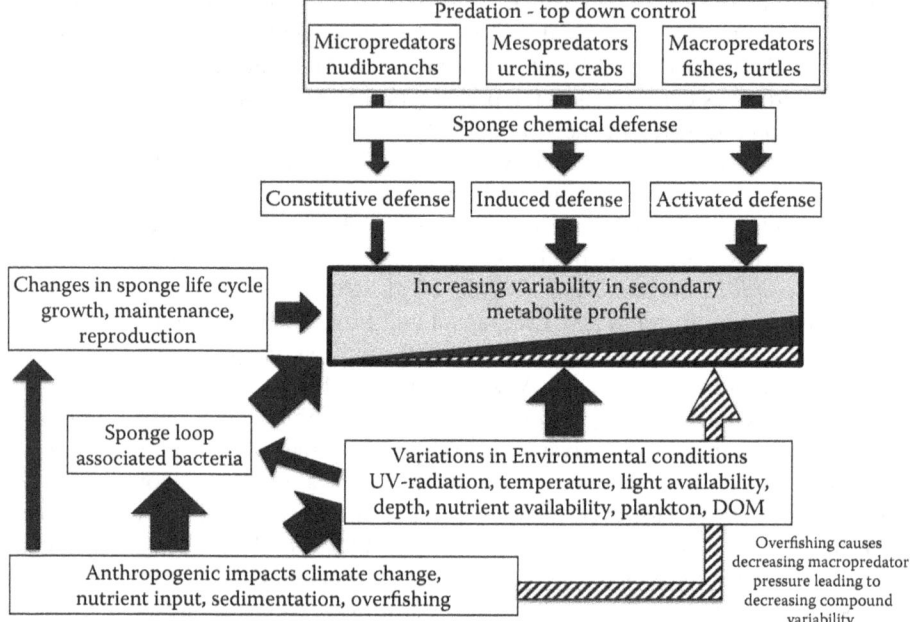

FIGURE 12.3 A conceptual model of the interacting effects of abiotic and biotic factors on the variability of sponge chemical defenses. The size of the arrows indicates the contribution of the various factors to the variability of sponge secondary metabolites. Hatched arrows indicate a decrease, while solid arrows resemble an increase in secondary metabolite variability.

therefore leading to a higher variation of defensive compounds. The way in which environmental factors affect metabolite production appears to be dependent on the sponge life history (as sponges allocate resources between reproduction, maintenance, and growth) and it is hard to predict this variability with the limited current knowledge.

To gain a better understanding of sponge chemical ecology and the chemical defense mechanisms that structure sponge populations and benthic sessile invertebrate communities as a whole, multifactorial studies are needed that include physiological, ecological, and evolutionary traits. Chemical defense and allelopathic mechanisms are likely contributing to phase-shifts from coral- to sponge-dominated reefs, thereby even restructuring one of the most diverse marine ecosystems. Based on the available literature today, some research foci emerge that are missing in the holistic concepts of sponge ecology and should be considered if we want to gain a better understanding of the variability of sponge chemical defenses.

12.4.2 PREDATOR–PREY EXPERIMENTS

Much of the data today on palatable and defended sponges comes from (a) transplant and caging experiments to exclude fish predators, (b) fish feeding observations, and (c) laboratory experiments using omnivorous fishes. There are hardly any studies testing extracts or pure compounds with other invertebrate predators, except urchins and starfish (McClintock et al. 2005, Rohde et al. 2015), although other mobile invertebrates such as crabs and nudibranchs are also known to prey on sponges (Becerro et al. 1998, Waddell and Pawlik 2000a, 2000b, Manzo et al. 2007).

Fish feeding assays testing the deterrent effects of sponge crude extracts or in a limited number of pure compounds have mainly been conducted with omnivorous species in aquaria. Feeding assays testing extracts of isolated secondary metabolites with true spongivorous fishes (e.g., angelfishes) have been lacking. The same seems to hold for parrotfishes, which are herbivorous, but consume sponges is most likely incidental. As they are often one of the most abundant consumers, they will likely have one of the greatest effects on sponge populations. Experiments testing feeding preference with such generalist species and specialists (angelfishes) should allow observers to determine how relevant defenses on a community level are. Assays with pure compounds and a range of generalist and specialist predators could reveal if one compound provides protection against multiple predators or if several different compounds are needed for an effective sponge defense.

Given the variability of secondary metabolites in time, geographic regions, and within specimens, researcher should consider using multiple specimens for each sponge species and also testing different parts within an individual sponge to properly determine extract and compound concentrations. New techniques such as high-resolution matrix-assisted laser desorption ionization imaging mass spectrometry (MALDI-IMS) could provide such information therefore allowing to test feeding deterrent compounds at more ecologically relevant concentrations (see Section 12.4.3 for details).

12.4.3 Investigations on Abiotic Factors Affecting Secondary Metabolite Production in Sponges

Sponges respond phenotypically plastic to environmental conditions like predation or to changes in light (depth) and nutrients. Not only abundance and growth patterns are subject to environmental factors, but also chemical metabolite concentrations. Therefore, the recent history of the sponge is affecting chemical activities and should be considered in all studies. Specific effects of light or nutrient availability are scarcely investigated but seem to be crucial in understanding metabolite variation. In particular, the effects of nutrient availability on metabolite production are barely known. Plant defense theories discuss and have observed positive and negative effects of nutrient limitation on chemical defenses. We lack this kind of information in sponges. There is no information on how food quality affects the production of chemical metabolites. Considering the recent publications on the sponge loop and the importance of DOM as nutrient source, it will be interesting to see if DOM molecules are used in secondary metabolite synthesis.

12.4.4 Use of New Analytical Technologies for Investigations on Metabolite Variability in Sponges

In the last decade there have been a multitude of technological advances in the hardware available to natural products chemists and ecologist, such as ultra-high-performance liquid chromatography (UHPLC) and various advances in mass spectrometry (MS) and nuclear magnetic resonance spectroscopy (NMR). These techniques allow an improved isolation of compounds that are present in the organisms at even minor concentrations, as well as the characterization of their chemical structures and detection and quantification of the compounds in the organisms. Especially, advances in MALDI-IMS offer new possibilities to detect and observe the distribution of secondary metabolites on and within organisms.

Provided that compound allocation can vary within a sponge (see Section 12.2.3 for details), current methodology of extraction sponge pieces will only provide average compound concentrations of the extracted parts. This could, however, highly under- or overestimate the real compound concentrations if these are allocated specifically to the outermost layer (pinacoderm) or inner sponge tissue (choanoderm). MALDI-IMS enables researchers to visualize where on or in the sponge compounds are located. When investigating sponge chemical defenses, it is certainly important to know where the compounds are located and at what concentrations. For example, (Yarnold et al. 2012) could demonstrate using MALDI-IMS experiments on various sponge sections of the tropical sponge *Stylissa flabelata* that compounds were differently allocated within the sponge and that bioactivities of certain compounds matched their distribution within the sponge. Dibromophakellin was localized within the external pinacoderm, sceptrin was highly abundant within the mesohyl (connective tissue between pinacoderm and choanoderm), while dibromopalauamine and konbuacidin B were only found in the choanoderm.

Sceptrin, being inactive against fouling organism (Proksch et al. 2010), was also not present in the outer sponge parts (picanoderm). Rather it could be a defense against potential pathogens following tissue damage from predation, as it has pronounced antimicrobial activity and concentrations increased following wounding (Richelle-Maurer et al. 2003). Besides these known compounds they also identified several unknown pyrrole-2-aminoimidazole alkaloids in the picanoderm, where these compounds potentially act as antimicrobial compounds against pathogens, biofouling, or predation. Another study could demonstrate using MALDI-IMS that the polyketide misakinolides from the sponge *Theonella swinhoei* WA were not produced by the sponge itself, but rather data suggested production by an uncultivated "*Entotheonella*" symbiont (Ueoka et al. 2015). Using catalyzed reporter deposition-fluorescence in situ hybridization (CARD-FISH) to visualize the bacterial symbionts in sections of *T. swinhoei* WA and MALDI-IMS to visualize the spatial distribution of misakinolides, they found both the bacterial symbiont and the polyketide compounds to be localized in a thin margin around the water channels in the sponge and the pinacoderm.

The first example demonstrates nicely how MALDI-IMS can be used to investigate the allocation of secondary metabolites within organisms and how the presence of compounds within certain parts (tissue) of an individual can corroborate observed bioactivities and ecological function of the compounds. Co-localization studies of bacterial symbionts and bioactive compounds like the one from Ueoka et al. (2015) can give first insights into the role of sponge-associated bacteria and who the actual producers of bioactive compounds are. Identifying the producer of bioactive compounds has become of increasing importance not just in ecology, but also for researchers investigating sponges for the production of novel bioactive compounds for drug development (Wilson et al. 2014). As sponges harbor large amounts of distinct sponge-specific bacteria (Taylor et al. 2007, Thomas et al. 2016), efforts to identify the actual producer of biomedically important secondary metabolites have increased in recent years.

IMS with its recent advances in laser-technology (in frequency and pixel size) and MS-technology (in resolution and mass accuracy) represents a powerful tool to investigate the chemical ecology of sponge secondary metabolites. The insights on compound location and concentrations within a sponge allow testing of isolated compounds at ecologically relevant concentrations, which is rather different to the current practice of extracting sponge pieces and using averaged tissue concentrations. Given that sponge chemical defenses are more variable than thought in the past, IMS should allow new and more detailed insights into sponge chemical defenses and interactions with competitors.

REFERENCES

Abdo, D. A., C. A. Motti, C. N. Battershill, and E. S. Harvey. 2007. Temperature and spatio-temporal variability of salicylihalamide A in the sponge *Haliclona* sp. *J. Chem. Ecol.* 33:1635–1645.

Angulo-Preckler, C., C. Cid, F. Oliva, and C. Avila. 2015. Antifouling activity in some benthic Antarctic invertebrates by "in situ" experiments at Deception Island, Antarctica. *Mar. Environ. Res.* 105:30–38.

Bandaranayake, W. M., J. E. Bemis, and D. J. Bourne. 1996. Ultraviolet absorbing pigments from the marine sponge *Dysidea herbacea*: Isolation and structure of a new mycosporine. *Comp. Biochem. Physiol. C Toxicol. Pharmacol.* 115:281–286.

Bandaranayake, W. M., D. J. Bourne, and R. G. Sim. 1997. Chemical composition during maturing and spawning of the sponge *Dysidea herbacea* (Porifera: Demospongiae). *Comp. Biochem. Physiol. B Biochem. Mol. Biol.* 118:851–859.

Barnes, D. K. A. 1999. High diversity of tropical intertidal zone sponges in temperature, salinity and current extremes. *Afr. J. Ecol.* 37:424–434.

Barnes, D. K. A., and J. J. Bell. 2002. Coastal sponge communities of the West Indian Ocean: Taxonomic affinities, richness and diversity. *Afr. J. Ecol.* 40:337–349.

Battershill, C. N., and P. R. Bergquist. 1990. The influence of storms on asexual reproduction, recruitment, and survivorship of sponges. In *New Perspectives in Sponge Biology*, K. Rützler (Ed.), pp. 3966–3403. Washington, DC: Smithsonian Institution Press.

Becerro, M. A., and V. J. Paul. 2004. Effects of depth and light on secondary metabolites and cyanobacterial symbionts of the sponge *Dysidea granulosa*. *Mar. Ecol. Prog. Ser.* 280:115–128.

Becerro, M. A., V. J. Paul, and J. Starmer. 1998. Intracolonial variation in chemical defenses of the sponge *Cacospongia* sp. and its consequences on generalist fish predators and the specialist nudibranch predator *Glossodoris pallida*. *Mar. Ecol. Prog. Ser.* 168:187–196.

Becerro, M. A., R. W. Thacker, X. Turon, M. J. Uriz, and V. J. Paul. 2003. Biogeography of sponge chemical ecology: Comparisons of tropical and temperate defenses. *Oecologia* 135:91–101.

Becerro, M. A., M. J. Uriz, and X. Turon. 1997. Chemically-mediated interactions in benthic organisms: The chemical ecology of *Crambe crambe* (Porifera, Poecilosclerida). *Hydrobiologia* 355:77–89.

Bell, J. J. 2008. The functional roles of marine sponges. *Estuar. Coast. Shelf Sci.* 79:341–353.

Bell, J. J., S. K. Davy, T. Jones, M. W. Taylor, and N. S. Webster. 2013. Could some coral reefs become sponge reefs as our climate changes? *Global Change Biol.* 19:2613–2624.

Bell, J. J., and D. Smith. 2004. Ecology of sponge assemblages (Porifera) in the Wakatobi region, south-east Sulawesi, Indonesia: Richness and abundance. *J. Mar. Biol. Assoc. U.K.* 84:581–591.

Berne, S., M. Kalauz, M. Lapat, L. Savin, D. Janussen, D. Kersken, J. Ambrožič Avguštin et al. 2015. Screening of the Antarctic marine sponges (Porifera) as a source of bioactive compounds. *Polar Biol.* 39:947–959.

Betancourt-Lozano, M., F. Gonzalez-Farias, B. Gonzalez-Acosta, A. Garcia-Gasca, and J. R. Bastida-Zavala. 1998. Variation of antimicrobial activity of the sponge *Aplysina fistularis* (Pallas, 1766) and its relation to associated fauna. *J. Exp. Mar. Biol. Ecol.* 223:1–18.

Burns, E., I. Ifrach, S. Carmeli, J. R. Pawlik, and M. Ilan. 2003. Comparison of anti-predatory defenses of Red Sea and Caribbean sponges. I. Chemical defense. *Mar. Ecol. Prog. Ser.* 252:105–114.

Cetrulo, G. L., and M. E. Hay. 2000. Activated chemical defenses in tropical versus temperate seaweeds. *Mar. Ecol. Prog. Ser.* 207:243–253.

Chanas, B., and J. R. Pawlik. 1995. Defenses of Caribbean sponges against predatory reef fish. 2. Spicules, tissue toughness, and nutritional quality. *Mar. Ecol. Prog. Ser.* 127:195–211.

Cote, I. M., J. A. Gill, T. A. Gardner, and A. R. Watkinson. 2005. Measuring coral reef decline through meta-analyses. *Philos. Trans. R. Soc. B Biol. Sci.* 360:385–395.

Cronin, G. 2001. Resource allocation in seaweeds and marine invertebrates: Chemical defense patterns in relation to defense theories. In *Marine Chemical Ecology*, J. McClintock and B. Baker (Eds.), pp. 325–353. Boca Raton, FL: CRC Press.

De'ath, G., K. E. Fabricius, H. Sweatman, and M. Puotinen. 2012. The 27-year decline of coral cover on the Great Barrier Reef and its causes. *Proc. Natl. Acad. Sci. USA* 109:17995–17999.

De Caralt, S., D. Bry, N. Bontemps, X. Turon, M. J. Uriz, and B. Banaigs. 2013. Sources of secondary metabolite variation in *Dysidea avara* (Porifera: Demospongiae): The importance of having good neighbors. *Mar. Drugs* 11:489–503.

de Goeij, J. M., D. van Oevelen, M. J. A. Vermeij, R. Osinga, J. J. Middelburg, A. F. P. M. de Goeij, and W. Admiraal. 2013. Surviving in a marine desert: The sponge loop retains resources within coral reefs. *Science* 342:108–110.

Duckworth, A. R., and C. N. Battershill. 2001. Population dynamics and chemical ecology of New Zealand demospongiae *Latrunculia* sp. nov. and *Polymastia croceus* (Poecilosclerida: Latrunculiidae: Polymastiidae). *New Zeal. J. Mar. Freshwater Res.* 35:935–949.

Dunlap, W. C., and J. M. Shick. 1998. Ultraviolet radiation-absorbing mycosporine-like amino acids in coral reef organisms: A biochemical and environmental perspective. *J. Phycol.* 34:418–430.

Fahey, S. J., and M. J. Garson. 2002. Geographic variation of natural products of tropical nudibranch *Asteronotus cespitosus*. *J. Chem. Ecol.* 28:1773–1785.

Freeman, C. J., and D. F. Gleason. 2010. Chemical defenses, nutritional quality, and structural components in three sponge species: *Ircinia felix, I. campana*, and *Aplysina fulva*. *Mar. Biol.* 157:1083–1093.

Furrow, F. B., C. D. Amsler, J. B. McClintock, and B. J. Baker. 2003. Surface sequestration of chemical feeding deterrents in the Antarctic sponge *Latrunculia apicalis* as an optimal defense against sea star spongivory. *Mar. Biol.* 143:443–449.

Gili, J. M., and R. Coma. 1998. Benthic suspension feeders: Their paramount role in littoral marine food webs. *Trends Ecol. Evol.* 13:316–321.

Gröniger, A., R. P. Sinha, M. Klisch, and D. P. Hader. 2000. Photoprotective compounds in cyanobacteria, phytoplankton and macroalgae—A database. *J. Photochem. Photobiol. B Biol.* 58:115–122.

Harvell, C., and R. Tollrian. 1999. Why inducible defenses? In *The Ecology and Evolution of Inducible Defenses*, R. Tollrian and C. Harvell (Eds.), pp. 3–9. Princeton, NJ: Princeton University Press.

Heil, M., and I. T. Baldwin. 2002. Fitness costs of induced resistance: Emerging experimental support for a slippery concept. *Trends Plant Sci.* 7:61–67.

Hill, M. S. 1998. Spongivory on Caribbean reefs releases corals from competition with sponges. *Oecologia* 117:143–150.

Hill, M. S., N. A. Lopez, and K. A. Young. 2005. Anti-predator defenses in western North Atlantic sponges with evidence of enhanced defense through interactions between spicules and chemicals. *Mar. Ecol. Prog. Ser.* 291:93–102.

Hughes, T. P., N. A. J. Graham, J. B. C. Jackson, P. J. Mumby, and R. S. Steneck. 2010. Rising to the challenge of sustaining coral reef resilience. *Trends Ecol. Evol.* 25:633–642.

Ivanisevic, J., T. Perez, A. V. Ereskovsky, G. Barnathan, and O. P. Thomas. 2011a. Lysophospholipids in the Mediterranean sponge *Oscarella tuberculata*: Seasonal variability and putative biological role. *J. Chem. Ecol.* 37:537–545.

Ivanisevic, J., O. P. Thomas, L. Pedel, N. Penez, A. V. Ereskovsky, G. Culioli, and T. Perez. 2011b. Biochemical trade-offs: Evidence for ecologically linked secondary metabolism of the sponge *Oscarella balibaloi*. *PLoS One* 6:e28059.

Jones, A. C., J. E. Blum, and J. R. Pawlik. 2005. Testing for defensive synergy in Caribbean sponges: Bad taste or glass spicules? *J. Exp. Mar. Biol. Ecol.* 322:67–81.

Jumaryatno, P., B. L. Stapleton, J. N. A. Hooper, D. J. Brecknell, J. T. Blanchfield, and M. J. Garson. 2007. A comparison of sesquiterpene scaffolds across different populations of the tropical marine sponge *Acanthella cavernosa*. *J. Nat. Prod.* 70:1725–1730.

Jung, V., and G. Pohnert. 2001. Rapid wound-activated transformation of the green algal defensive metabolite caulerpenyne. *Tetrahedron* 57:7169–7172.

Karban, R., and I. T. Baldwin. 1997. *Induced Response to Herbivory*. Chicago, IL: University of Chicago Press.

Kelly, S. R., P. R. Jensen, T. P. Henkel, W. Fenical, and J. R. Pawlik. 2003. Effects of Caribbean sponge extracts on bacterial attachment. *Aquat. Microb. Ecol.* 31:175–182.

Lesser, M. P., and M. Slattery. 2013. Ecology of Caribbean sponges: Are top-down or bottom-up processes more important? *PLoS One* 8:e79799.

Lindel, T., H. Hoffmann, M. Hochgurtel, and J. R. Pawlik. 2000. Structure-activity relationship of inhibition of fish feeding by sponge-derived and synthetic pyrrole-imidazole alkaloids. *J. Chem. Ecol.* 26:1477–1496.

Loh, T.-L., S. E. McMurray, T. P. Henkel, J. Vicente, and J. R. Pawlik. 2015. Indirect effects of overfishing on Caribbean reefs: Sponges overgrow reef-building corals. *PeerJ* 3:e901.

Loh, T.-L., and J. R. Pawlik. 2014. Chemical defenses and resource trade-offs structure sponge communities on Caribbean coral reefs. *Proc. Natl. Acad. Sci. USA* 111:4151–4156.

Luter, H. M., and A. R. Duckworth. 2010. Influence of size and spatial competition on the bioactivity of coral reef sponges. *Biochem. Syst. Ecol.* 38:146–153.

Maldonado, M., L. Navarro, A. Grasa, A. Gonzalez, and I. Vaquerizo. 2011. Silicon uptake by sponges: A twist to understanding nutrient cycling on continental margins. *Sci. Rep.* 1:30.

Maldonado, M., M. Ribes, and F. C. van Duyl. 2012. Nutrient fluxes through sponges: Biology, budgets, and ecological implications. *Adv. Mar. Biol.* 62:113–182.

Manzo, E., M. Gavagnin, M. J. Somerville, S. C. Mao, M. L. Ciavatta, E. Mollo, P. J. Schupp, M. J. Garson, Y. W. Guo, and G. Cimino. 2007. Chemistry of *Glossodoris* nudibranchs: Specific occurrence of 12-keto scalaranes. *J. Chem. Ecol.* 33:2325–2336.

McClintock, J. B., C. D. Amsler, B. J. Baker, and R. W. van Soest. 2005. Ecology of antarctic marine sponges: An overview. *Integr. Comp. Biol.* 45:359–368.

McClintock, J. B., and D. Karentz. 1997. Mycosporine-like amino acids in 38 species of subtidal marine organisms from McMurdo Sound, Antarctica. *Antarct. Sci.* 9:392–398.

Miner, B. G., S. E. Sultan, S. G. Morgan, D. K. Padilla, and R. A. Relyea. 2005. Ecological consequences of phenotypic plasticity. *Trends Ecol. Evol.* 20:685–692.

Newbold, R. W., P. R. Jensen, W. Fenical, and J. R. Pawlik. 1999. Antimicrobial activity of Caribbean sponge extracts. *Aquat. Microb. Ecol.* 19:279–284.

Norström, A. V., M. Nyström, J. Lokrantz, and C. Folke. 2009. Alternative states on coral reefs: Beyond coral–macroalgal phase shifts. *Mar. Ecol. Prog. Ser.* 376:295–306.

Noyer, C., and M. A. Becerro. 2012. Relationship between genetic, chemical, and bacterial diversity in the Atlanto-Mediterranean bath sponge *Spongia lamella. Hydrobiologia* 687:85–99.

Nunez-Pons, L., and C. Avila. 2014. Defensive metabolites from Antarctic invertebrates: Does energetic content interfere with feeding repellence? *Mar. Drugs* 12:3770–3791.

Nunez, C. V., E. V. R. de Almelda, A. C. Granato, S. O. Marques, K. O. Santos, F. R. Pereira, M. L. Macedo et al. 2008. Chemical variability within the marine sponge *Aplysina fulva. Biochem. Syst. Ecol.* 36:283–296.

Page, M., L. West, P. Northcote, C. Battershill, and M. Kelly. 2005. Spatial and temporal variability of cytotoxic metabolites in populations of the New Zealand sponge *Mycale hentscheli. J. Chem. Ecol.* 31:1161–1174.

Paul, V. J., and M. P. Puglisi. 2004. Chemical mediation of interactions among marine organisms. *Nat. Prod. Rep.* 21:189–209.

Paul, V. J., R. Ritson-Williams, and K. Sharp. 2011. Marine chemical ecology in benthic environments. *Nat. Prod. Rep.* 28:345–387.

Paul, V. J., and K. L. Van Alstyne. 1992. Activation of chemical defense in the tropical green algae *Halimeda* spp. *J. Exp. Mar. Biol. Ecol.* 160:191–203.

Pavia, H., G. Cervin, A. Lindgren, and P. Aberg. 1997. Effects of UV-B radiation and simulated herbivory on phlorotannins in the brown alga *Ascophyllum nodosum. Mar. Ecol. Prog. Ser.* 157:139–146.

Pawlik, J. R. 1998. Coral reef sponges: Do predatory fishes affect their distribution? *Limnol. Oceanogr.* 43:1396–1399.

Pawlik, J. R. 2011. The chemical ecology of sponges on Caribbean Reefs: Natural products shape natural systems. *Bioscience* 61:888–898.

Pawlik, J. R., D. E. Burkepile, and R. V. Thurber. 2016. A vicious circle? Altered carbon and nutrient cycling may explain the low resilience of Caribbean coral reefs. *Bioscience* 66:470–476.

Pawlik, J. R., B. Chanas, R. J. Toonen, and W. Fenical. 1995. Defenses of Caribbean sponges against predatory reef fish. 1. Chemical deterrency. *Mar. Ecol. Prog. Ser.* 127:183–194.

Pawlik, J. R., and L. K. Deignan. 2015. Cowries graze verongid sponges on Caribbean reefs. *Coral Reefs* 34:663–663.

Pawlik, J. R., T. P. Henkel, S. E. McMurray, S. Lopez-Legentil, T. L. Loh, and S. Rohde. 2008. Patterns of sponge recruitment and growth on a shipwreck corroborate chemical defense resource trade-off. *Mar. Ecol. Prog. Ser.* 368:137–143.

Pawlik, J. R., T.-L. Loh, S. E. McMurray, and C. M. Finelli. 2013. Sponge communities on Caribbean coral reefs are structured by factors that are top-down, not bottom-up. *PLoS One* 8:e62573.

Pawlik, J. R., S. E. McMurray, P. Erwin, and S. Zea. 2015. Trophic ecology of sponges from shallow to mesophotic depths (3 to 150 m): Comment on Pawlik et al. (2015). *Mar. Ecol. Prog. Ser.* 527:275–279.

Pawlik, J. R., L. Steindler, T. P. Henkel, S. Beer, and M. Ilan. 2007. Chemical warfare on coral reefs: Sponge metabolites differentially affect coral symbiosis in situ. *Limnol. Oceanogr.* 52:907–911.

Peters, K. J., C. D. Amsler, J. B. McClintock, and B. J. Baker. 2009a. Potential chemical defenses of Antarctic sponges against sympatric microorganisms. *Polar Biol.*
33:649–658.

Peters, K. J., C. D. Amsler, J. B. McClintock, R. W. M. van Soest, and B. J. Baker. 2009b.
Palatability and chemical defenses of sponges from the western Antarctic Peninsula.
Mar. Ecol. Prog. Ser. 385:77–85.

Pigliucci, M. 2005. Evolution of phenotypic plasticity: Where are we going now? *Trends
Ecol. Evol.* 20:481–486.

Proksch, P., A. Putz, S. Ortlepp, J. Kjer, and M. Bayer. 2010. Bioactive natural products from
marine sponges and fungal endophytes. *Phytochem. Rev.* 9:475–489.

Puglisi, M. P., V. J. Paul, and M. Slattery. 2000. Biogeographic comparisons of chemical and
structural defenses of the Pacific gorgonians *Annella mollis* and *A. reticulata. Mar.
Ecol. Prog. Ser.* 207:263–272.

Putz, A., A. Kloeppel, M. Pfannkuchen, F. Brummer, and P. Proksch. 2009. Depth-related
alkaloid variation in Mediterranean *Aplysina* sponges. *Z. Naturforsch. C* 64:279–287.

Rhoades, D. F. 1979. Evolution of plant chemical defenses against herbivores. In *Herbivores:
Their Interaction with Secondary Plant Metabolites,* G. A. Rosenthal and D. H. Janzen
(Eds.), pp. 3–54. Chicago, IL: University of Chicago Press.

Richelle-Maurer E., M. J. De Kluijver, S. Feio, S. Gaudencio, H. Gaspar, R. Gomez,
R. Tavares, G. Van de Vyver, R. W. M. Van Soest. 2003. Localization and ecological
significance of oroidin and sceptrin in the Caribbean sponge *Agelas conifera. Biochem.
Syst. Ecol.* 31:1073–1109.

Rix, L., J. M. de Goeij, C. E. Mueller, U. Struck, J. J. Middelburg, F. C. van Duyl, F. A.
Al-Horani, C. Wild, M. S. Naumann, and D. van Oevelen. 2016. Coral mucus fuels the
sponge loop in warm- and cold-water coral reef ecosystems. *Sci. Rep.* 6:18715.

Rohde, S., D. J. Gochfeld, S. Ankisetty, B. Avula, P. J. Schupp, and M. Slattery. 2012. Spatial
variability in secondary metabolites of the Indo-Pacific Sponge *Stylissa massa.
J. Chem. Ecol.* 38:463–475.

Rohde, S., M. Molis, and M. Wahl. 2004. Regulation of anti-herbivore defence by *Fucus
vesiculosus* in response to various cues. *J. Ecol.* 92:1011–1018.

Rohde, S., S. Nietzer, and P. J. Schupp. 2015. Prevalence and mechanisms of dynamic chemical defenses in tropical sponges. *PLoS One* 10:e0132236.

Rohde, S., and P. J. Schupp. 2011. Allocation of chemical and structural defenses in the
sponge *Melophlus sarasinorum. J. Exp. Mar. Biol. Ecol.* 399:76–83.

Rohde, S., and M. Wahl. 2008. Antifeeding defense in Baltic macroalgae: Induction by direct
grazing vs. waterborne cues. *J. Phycol.* 44:85–90.

Sacristan-Soriano, O., B. Banaigs, and M. A. Becerro. 2011. Relevant spatial scales of chemical variation in *Aplysina aerophoba. Mar. Drugs* 9:2499–2513.

Sacristan-Soriano, O., B. Banaigs, and M. A. Becerro. 2012. Temporal trends in the secondary metabolite production of the sponge *Aplysina aerophoba. Mar. Drugs* 10:677–693.

Sacristán-Soriano, O., B. Banaigs, and M. A. Becerro. 2016. Can light intensity cause shifts
in natural product and bacterial profiles of the sponge *Aplysina aerophoba? Mar. Ecol.*
37:88–105.

Schupp, P., C. Eder, V. Paul, and P. Proksch. 1999. Distribution of secondary metabolites
in the sponge *Oceanapia* sp. and its ecological implications. *Mar. Biol.* 135:573–580.

Shick, J. M., S. Romaine-Lioud, C. Ferrier-Pages, and J. P. Gattuso. 1999. Ultraviolet-B radiation stimulates shikimate pathway-dependent accumulation of mycosporine-like amino
acids in the coral *Stylophora pistillata* despite decreases in its population of symbiotic
dinoflagellates. *Limnol. Oceanogr.* 44:1667–1682.

Slattery, M., D. Gochfeld, M. C. Diaz, R. Thacker, and M. Lesser. 2016. Variability in chemical defense across a shallow to mesophotic depth gradient in the Caribbean sponge
Plakortis angulospiculatus. Coral Reefs 35:11–22.

Slattery, M., J. Starmer, and V. J. Paul. 2001. Temporal and spatial variation in defensive metabolites of the tropical Pacific soft corals *Sinularia maxima* and *S. polydactyla*. *Mar. Biol.* 138:1183–1193.

Steindler, L., S. Beer, and M. Ilan. 2002. Photosymbiosis in intertidal and subtidal tropical sponges. *Symbiosis* 33:263–273.

Supriyono, A., B. Schwarz, V. Wray, L. Witte, W. E. G. Muller, R. Vansoest, W. Sumaryono, and P. Proksch. 1995. Bioactive alkaloids from the tropical marine sponge *Axinella carteri*. *Z. Naturforsch. C* 50:669–674.

Swearingen, D. C., and J. R. Pawlik. 1998. Variability in the chemical defense of the sponge *Chondrilla nucula* against predatory reef fishes. *Mar. Biol.* 131:619–627.

Targett, N. M., L. D. Coen, A. A. Boettcher, and C. E. Tanner. 1992. Biogeographic comparisons of marine algal polyphenolics—Evidence against a latitudinal trend. *Oecologia* 89:464–470.

Taylor, M. W., R. Radax, D. Steger, and M. Wagner. 2007. Sponge-associated microorganisms: Evolution, ecology, and biotechnological potential. *Microbiol. Mol. Biol. Rev.* 71:295–347.

Teeyapant, R., and P. Proksch. 1993. Biotransformation of brominated compounds in the marine sponge *Verongia aerophoba*—Evidence for an induced chemical defense. *Naturwissenschaften* 80:369–370.

Thomas, T., L. Moitinho-Silva, M. Lurgi, J. R. Bjork, C. Easson, C. Astudillo-Garcia, J. B. Olson et al. 2016. Diversity, structure and convergent evolution of the global sponge microbiome. *Nat. Commun.* 7:11870.

Thompson, J. E., P. T. Murphy, P. R. Bergquist, and E. A. Evans. 1987. Environmentally induced variation in diterpene composition of the marine sponge *Rhopaloeides odorabile*. *Biochem. Syst. Ecol.* 15:595–606.

Thoms, C., R. Ebel, and P. Proksch. 2006. Activated chemical defense in *Aplysina* sponges revisited. *J. Chem. Ecol.* 32:97–123.

Thoms, C., M. Horn, M. Wagner, U. Hentschel, and P. Proksh. 2003. Monitoring microbial diversity and natural product profiles of the sponge *Aplysina cavernicola* following transplantation. *Mar. Biol.* 142:685–692.

Thoms, C., and P. Schupp. 2007. Chemical defense strategies in sponges: A review. In *Porifera Research: Biodiversity, Innovation and Sustainability*, M. R. Custodio, E. Lobo-Hajdu, and G. Muricy (Eds.), pp. 627–637. Serie Livros 28. Rio de Janeiro, Brazil: Museu Nacional.

Thoms, C., and P. J. Schupp. 2008. Activated chemical defense in marine sponges—A case study on *Aplysinella rhax*. *J. Chem. Ecol.* 34:1242–1252.

Toth, G. B., and H. Pavia. 2007. Induced herbivore resistance in seaweeds: A meta-analysis. *J. Ecol.* 95:425–434.

Turon, X., M. A. Becerro, and M. J. Uriz. 1996. Seasonal patterns of toxicity in benthic invertebrates: The encrusting sponge *Crambe crambe* (Poecilosclerida). *Oikos* 75:33–40.

Turon, X., R. Martí, and M. J. Uriz. 2009. Chemical bioactivity of sponges along an environmental gradient in a Mediterranean cave. *Sci. Mar.* 73:387–397.

Ueoka, R., A. R. Uria, S. Reiter, T. Mori, P. Karbaum, E. E. Peters, E. J. Helfrich et al. 2015. Metabolic and evolutionary origin of actin-binding polyketides from diverse organisms. *Nat. Chem. Biol.* 11:705–712.

Uriz, M. J., X. Turon, M. A. Becerro, J. Galera, and J. Lozano. 1995. Patterns of resource allocation to somatic, defensive and reproductive functions in the mediterranean encrusting sponge *Crambe crambe* (Demospongiae, Poecilosclerida). *Mar. Ecol. Prog. Ser.* 124:159–170.

Van Alstyne, K. L., G. V. Wolfe, T. L. Freidenburg, A. Neill, and C. Hicken. 2001. Activated defense systems in marine macroalgae: Evidence for an ecological role for DMSP cleavage. *Mar. Ecol. Prog. Ser.* 213:53–65.

Vogel, S. 1977. Current-induced flow through living sponges in nature. *Proc. Natl. Acad. Sci. USA* 74:2069–2071.

Waddell, B., and J. R. Pawlik. 2000a. Defenses of Caribbean sponges against invertebrate predators. I. Assays with hermit crabs. *Mar. Ecol. Prog. Ser.* 195:125–132.

Waddell, B., and J. R. Pawlik. 2000b. Defenses of Caribbean sponges against invertebrate predators. II. Assays with sea stars. *Mar. Ecol. Prog. Ser.* 195:133–144.

Weisz, J. B., N. Lindquist, and C. S. Martens. 2008. Do associated microbial abundances impact marine demosponge pumping rates and tissue densities? *Oecologia* 155:367–376.

Wilson, M. C., T. Mori, C. Ruckert, A. R. Uria, M. J. Helf, K. Takada, C. Gernert et al. 2014. An environmental bacterial taxon with a large and distinct metabolic repertoire. *Nature* 506:58–62.

Wulff, J. L. 1994. Sponge feeding by Caribbean angelfishes, trunkfishes, and filefishes. In *Sponges in Time and Space*, R. W. M. Van Soest, T. Van Kempen, and J. Braekman (Eds.), pp. 265–271. Rotterdam, the Netherlands: Balkema.

Wulff, J. L. 1997. Parrotfish predation on cryptic sponges of Caribbean coral reefs. *Mar. Biol.* 129:41–52.

Wulff, J. L. 2006. Ecological interactions of marine sponges. *Can. J. Zool.* 84:146–166.

Yarnold, J. E., B. R. Hamilton, D. T. Welsh, G. F. Pool, D. J. Venter, and A. R. Carroll. 2012. High resolution spatial mapping of brominated pyrrole-2-aminoimidazole alkaloids distributions in the marine sponge *Stylissa flabellata* via MALDI-mass spectrometry imaging. *Mol. Biosyst.* 8:2249–2259.

Index

Note: Page numbers in italic and bold refer to figures and tables respectively.